Fiber-Optic
Communication Systems

WILEY SERIES IN MICROWAVE AND OPTICAL ENGINEERING

KAI CHANG, Editor
Texas A&M University

Fiber-Optic Communication Systems

Second Edition

GOVIND P. AGRAWAL
The Institute of Optics
University of Rochester
Rochester, NY

A WILEY-INTERSCIENCE PUBLICATION
JOHN WILEY & SONS, INC.
NEW YORK / CHICHESTER / WEINHEIM / BRISBANE / SINGAPORE / TORONTO

Copyright © 1997 by John Wiley & Sons, Inc.

Library of Congress Cataloging in Publication Data:
Agrawal, G. P. (Govind P.), 1951–
    Fiber-optic communication systems / Govind P. Agrawal. — 2nd ed.
      p. cm. — (Wiley series in microwave and optical engineering)
    "A Wiley-Interscience publication."
    Includes index.
    ISBN 0-471-17540-4 (cloth : alk. paper)
    1. Optical communications. 2. Fiber optics. I. Title.
  II. Series.
  TK5103.59.A37 1997
  621.382′75—dc21   97-4040

Printed in the United States of America

10 9 8 7 6 5 4 3 2

For My Parents

Preface

Since the publication of the first edition of this book in 1992, the state of the art of fiber-optic communication systems has advanced dramatically despite the relatively short period of only five years between the two editions. As an example, the highest bit rate of commercial point-to-point links in 1992 was 2.5 Gb/s. By 1996, wavelength-multiplexed systems with a total capacity of 40 Gb/s were available commercially, with the prospect of systems operating at 100 Gb/s or more in sight. In fact, the next transpacific cable (TPC–6), which will transport data at a bit rate of 100 Gb/s, is scheduled to be deployed in 1998 and will be operational by the year 2000. Moreover, three postdeadline papers at the Optical Fiber Communication Conference (OFC'96), held in February 1996 at San Jose, California, demonstrated that lightwave systems operating at a bit rate of 1 Tb/s are within reach by using wavelength-division and time-division multiplexing techniques. Just a few years ago it was unimaginable that lightwave systems would approach a bit rate 1 Tb/s before the end of the twentieth century.

Because of the rapid advances that have occurred in fiber-optic communication technology over the last five years, the publisher and I deemed it necessary to bring out this second edition in order to continue to provide a comprehensive and up-to-date account of fiber-optic communication systems, as stated in the preface of the first edition. The result is in your hands. The primary objective of the book remains the same. Specifically, it should be able to serve both as a textbook and as a reference monograph. For this reason the emphasis is on the physical understanding, but the engineering aspects are also discussed throughout the text.

Because of the large amount of material that needed to be added to provide a comprehensive coverage, the book size has increased considerably. Although all chapters have been updated, the major changes have occurred in the last four chapters. Almost half of the material in Chapter 7 is new because of recent advances in the use of wavelength- and time-division multiplexing techniques for optical networks. Chapter 9 is completely new. It covers a multitude of dispersion-compensation techniques that have been discovered and implemented over the last five years in an attempt to make use of the worldwide installed base of more than 50 million kilometers of standard telecommunication fiber. More than half of the material in Chapter 10 is also new in order to describe the rapid development of soliton-based communication technology. The contents of the book reflect the state of the art of lightwave transmission systems in 1996.

I am acutely aware of the problem caused by an enlarged revised edition: How can a teacher fit all this material in a one-semester course on *optical communications*? I teach such a course to the graduate students of the Institute of Optics at University of Rochester and have to struggle with the same question. In fact, it is impossible to cover the entire book in one semester. The best solution is to offer a two-semester course covering Chapters 1 through 5 during the first semester, leaving the remainder for the second semester. However, not many universities may have the luxury of offering a two-semester course on optical communications. The book can be used for a one-semester course provided that the instructor makes a selection of topics. I can offer my selection as an example. Chapter 3 can be largely skipped, especially if students have taken a laser course previously. Chapter 6 can also be skipped without affecting the continuity. If only parts of Chapters 7 through 10 are covered to provide students a glimpse of recent advances, the material can easily fit in a single one-semester course offered either at the senior level for undergraduates or to graduate students.

Several of my colleagues have helped me in preparing the second edition. I thank R. J. Essiambre, G. R. Gray, and G. H. M. van Tartwijk for reading several chapters and making helpful suggestions. R. J. Essiambre also helped in writing parts of Chapter 10. I am grateful to teachers who adopted this book for their courses and provided occasional feedback. Last, but not least, I thank my wife, Anne, and my daughters, Sipra, Caroline, and Claire, for understanding why I needed to spend many weekends on the book instead of spending time with them.

GOVIND P. AGRAWAL

Rochester, NY
March 1997

Preface to the First Edition

The use of optical fibers for information transmission has become widespread during the decade of the 1980s, as is evident from the installation of fiber-optic telecommunication networks throughout the world. It is further exemplified by the deployment of undersea fiber cables cross both the Atlantic and Pacific oceans. The pace of technological advances in the design of fiber-optic communication systems has been very rapid throughout the 1980s. The trend is continuing during the 1990s, as is apparent from the current emphasis on the research and development of multichannel lightwave systems, erbium-doped fiber amplifiers, and soliton communication systems. An example of how lightwave technology is influencing our society is provided by the recent use of optical fibers by the caʋle-television industry for analog video distribution through a technique known as subcarrier multiplexing. This change from coaxial cables to optical fibers can increase the transmission capacity by an order of magnitude or more, making it possible to transmit hundreds of video channels to each subscriber. It also enables us to make the transition from analog to digital video, and eventually to high-definition television. Another example is provided by broadband integrated-services digital networks, whose advent is expected to affect the telecommunication industry considerably. Indeed, fiber-optic communication systems can be thought of as an integral part of the information age.

Despite the enormous progress realized in the field of optical fiber communications, it is difficult to convey the sense and importance of this progress to a student or to a scientist who is not an expert in the field. The reason simply is that most of the material is available only in the form of research papers. The objective of this book is to provide a comprehensive, up-to-date account of fiber-optic communication systems in such a way that it can serve as both a textbook and a reference monograph. The emphasis is on physical understanding, but the engineering aspects are also discussed throughout the text.

Many universities in the United States and elsewhere offer a course on optical communications as part of their curriculum in electrical engineering, physics, or optics. I have taught such a course for several years to graduate students at the Institute of Optics. Unfortunately, it is very difficult to find a suitable textbook for the course. Most textbooks on optical fiber communications have become outdated as a result of rapid progress in this growing field. This book is intended to fulfill the acute need for a graduate-level textbook in the field of optical communications. An attempt is made to include as much recent material as possible so that students are exposed to the recent advances in this exciting field. The book can also serve as a reference text for researchers already engaged in or wishing to enter the field of optical fiber communications. The reference list at the end of each chapter is more elaborate than what is common for a typical textbook. The listing of recent research papers should be useful for researchers using this book as a reference. At the same time, students can benefit from it if they are assigned problems requiring reading of original research papers. A set of problems is included at the end of each chapter to help both teacher and student. Although written primarily for graduate students, the book can also be used for an undergraduate course at the senior level with an appropriate selection of topics. Parts of the book can be used for related courses. For example, Chapter 2 can be used for a course on optical waveguides, and Chapter 3 can be useful for a course on optoelectronics.

A large number of people have contributed to this book either directly or indirectly. It is impossible to mention all of them by name. I thank the students who took my course on optical communications and helped improve my class notes through interesting discussions. Thanks are due to T. G. Brown and G. R. Gray for reading parts of the manuscript. I appreciate the help of Karen Rolfe, who typed the manuscript and made numerous revisions with a smile. Last, but not least, I thank my wife, Anne, and my daughters, Sipra, Caroline, and Claire, for putting up with my preoccupation with the book, which certainly took away time we could have spent together.

GOVIND P. AGRAWAL

Rochester, NY
March 1992

Contents

Fiber-Optic
Communication Systems

Chapter 1

INTRODUCTION

A communication system transmits information from one place to another, whether separated by a few kilometers or by transoceanic distances. Information is often carried by an electromagnetic carrier wave whose frequency can vary from a few megahertz to several hundred terahertz. Optical communication systems use high carrier frequencies (~ 100 THz) in the visible or near-infrared region of the electromagnetic spectrum. They are sometimes called lightwave systems to distinguish them from microwave systems, whose carrier frequency is typically smaller by five orders of magnitude (~ 1 GHz). Fiber-optic communication systems are lightwave systems that employ optical fibers for information transmission. Such systems have been deployed worldwide since 1980 and have indeed revolutionized the technology behind telecommunications. Indeed, the lightwave technology, together with microelectronics, is believed to be a major factor in the advent of the "information age." The objective of this book is to describe fiber-optic communication systems in a comprehensive manner. The emphasis is on the fundamental aspects, but the engineering issues are also discussed. The purpose of this introductory chapter is to present the basic concepts and to provide the background material. Section 1.1 gives a historical perspective on the development of optical communication systems. In Section 1.2 we cover concepts such as analog and digital signals, channel multiplexing, and modulation formats. Relative merits of guided and unguided optical communication systems are discussed in Section 1.3. In Section 1.4 we describe the components of a fiber-optic communication system. The concept of channel capacity is introduced in Section 1.5.

1.1 HISTORICAL PERSPECTIVE

The use of light for communication purposes dates back to antiquity if we interpret optical communications in a broad sense. Most civilizations have used fire and smoke signals to convey a single piece of information (such as victory in a war). Essentially the same idea was used up to the end of the eighteenth cen-

1

tury through signaling lamps, flags, and other semaphore devices. The idea was extended further, following a suggestion of Claude Chappe in 1792, to transmit mechanically coded messages over long distances (~ 100 km) by the use of intermediate relay stations (acting as *regenerators* or *repeaters*). The role of light was simply to make the coded signals visible so that they could be intercepted by relay stations. Such optical communication systems were inherently slow. Indeed, in modern-day terminology, the effective bit rate was less than 1 bit per second ($B < 1$ b/s).

1.1.1   Need for Fiber-Optic Communications

The advent of telegraphy [1] in the 1830s replaced the use of light by electricity and began the era of electrical communications. The bit rate B could be increased to ~ 10 b/s by the use of new coding techniques, such as *Morse code* [1]. The use of intermediate relay stations allowed communication over long distances (~ 1000 km). Indeed, the first successful transatlantic telegraph cable went into operation in 1866. Interestingly enough, telegraphy used essentially a digital scheme through two electrical pulses of different durations (dots and dashes of the Morse code). The invention of the telephone [2] in 1876 brought a major change inasmuch as electric signals were transmitted in analog form through a continuously varying electric current. Analog electrical techniques were to dominate communication systems for a century or so.

The development of worldwide telephone networks during the twentieth century led to many advances in the design of electrical communication systems. The use of coaxial cables in place of wire pairs increased system capacity considerably. The first coaxial-cable system, put into service in 1940, was a 3-MHz system capable of transmitting 300 voice channels or a single television channel. The bandwidth of such systems is limited by the frequency-dependent cable losses, which increase rapidly for frequencies beyond 10 MHz. This limitation led to the development of microwave communication systems in which an electromagnetic carrier wave with frequencies of ~ 1 to 10 GHz is used to transmit the signal by using suitable modulation techniques. The first microwave system operating at the carrier frequency of 4 GHz was put into service in 1948. Since then, both coaxial and microwave systems have evolved considerably and are able to operate at bit rates of ~ 100 Mb/s. The most advanced coaxial system, put into service in 1975, operates at a bit rate of 274 Mb/s. A severe drawback of such a high-speed coaxial system is the small *repeater spacing* (~ 1 km), which makes the system relatively expensive to operate. Microwave communication systems generally allow for a larger repeater spacing, but their bit rate is also limited by the carrier frequency of such waves. A commonly used figure of merit for communication systems is the *bit rate–distance product*, BL, where B is the bit rate and L is the repeater spacing. Figure 1.1 shows how the BL product has increased through technological advances during the last century and a half. Communication systems with $BL \sim 100$ (Mb/s)-km were available by 1970 and were limited to such values because of fundamental limitations.

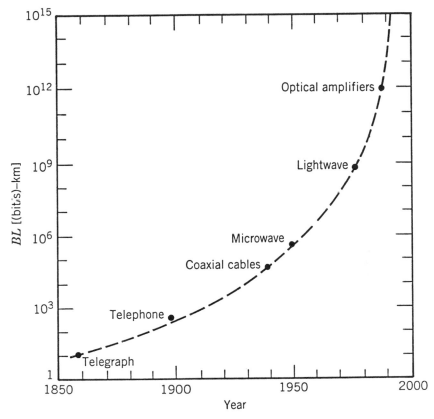

Figure 1.1 Increase in bit rate–distance product BL during the period 1850–2000. The emergence of a new technology is marked by a solid circle.

It was realized during the second half of the twentieth century that an increase of several orders of magnitude in the BL product would be possible if optical waves were used as the carrier. However, neither a coherent optical source nor a suitable transmission medium was available during the 1950s. The invention of the laser and its demonstration [3] in 1960 solved the first problem. Attention was then focused on finding ways for using laser light for optical communications. Many ideas were advanced during the 1960s [4], the most noteworthy being the idea of light confinement using a sequence of gas lenses [5]. It was suggested [6] in 1966 that optical fibers might be the best choice, as they are capable of guiding the light in a manner similar to the guiding of electrons in copper wires. The main problem was the high loss of optical fibers—fibers available during the 1960s had losses in excess of 1000 dB/km. A breakthrough occurred in 1970 when the fiber loss could be reduced to about 20 dB/km in the wavelength region near 1 μm [7]. At about the same time, GaAs semiconductor lasers, operating continuously at room temperature, were demonstrated [8]. The simultaneous availability of a *compact optical source* and a *low-loss optical*

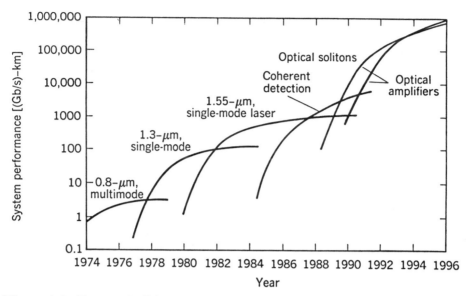

Figure 1.2 Progress in lightwave communication technology over the period 1974–1996. Different curves show increase in the bit rate–distance product BL for five generations of fiber-optic communication systems.

fiber led to a worldwide effort for developing fiber-optic communication systems. Figure 1.2 shows the progress in the performance of lightwave systems realized after 1974 through several generations of development. The progress has indeed been rapid, as is evident by a many-order-of-magnitude increase in the bit rate–distance product over a period of less than 25 years.

1.1.2 Five Generations of Lightwave Systems

The commercial deployment of lightwave systems followed the research and development closely. After many field trials, the first-generation lightwave systems operating near 0.8 μm became available commercially in 1980 [9]. They operated at a bit rate of 45 Mb/s and allowed a repeater spacing of about 10 km. The larger repeater spacing compared with that of a coaxial system was an important motivation for system designers, as it decreased the installation and maintenance costs associated with each repeater.

It was clear during the 1970s that the repeater spacing could be increased considerably by operating the lightwave system in the wavelength region near 1.3 μm, where fiber loss is below 1 dB/km. Furthermore, optical fibers exhibit minimum dispersion in this wavelength region. This realization led to a worldwide effort for the development of InGaAsP semiconductor lasers and detectors operating near 1.3 μm. The second generation of fiber-optic communication systems became available in the early 1980s, but the bit rate of early systems was limited to below 100 Mb/s because of dispersion in multimode fibers [10]. This limitation was overcome by the use of *single-mode fibers*. A laboratory experiment in 1981 demonstrated 2-Gb/s transmission over 44 km of single-mode fiber [11]. The introduction of commercial systems soon followed. By 1987, second-generation 1.3-μm lightwave systems, operating at bit rates of up to 1.7 Gb/s with a repeater spacing of about 50 km, were commercially available.

The repeater spacing of the second-generation lightwave systems was limited by the fiber loss at the operating wavelength of 1.3 μm (typically, 0.5 dB/km). The loss of silica fibers is minimum near 1.55 μm. Indeed, a loss of 0.2 dB/km in this spectral region was realized in 1979 [12]. However, the introduction of third-generation lightwave systems operating at 1.55 μm was considerably delayed by large fiber dispersion near 1.55 μm. Conventional InGaAsP semiconductor lasers could not be used because of pulse spreading occurring as a result of simultaneous oscillation of several longitudinal modes. The dispersion problem can be overcome either by using dispersion-shifted fibers designed to have minimum dispersion near 1.55 μm or by limiting the laser spectrum to a single longitudinal mode. Both approaches were followed during the 1980s. By 1985, laboratory experiments [13], [14] indicated the possibility of transmitting information at bit rates of up to 4 Gb/s over distances in excess of 100 km. Third-generation 1.55-μm systems operating at 2.5 Gb/s became available commercially in 1990. Such systems are capable of operating at a bit rate of up to 10 Gb/s [15]. The best performance is achieved using dispersion-shifted fibers together with single-longitudinal-mode lasers.

A drawback of third-generation 1.55-μm systems is that the signal is regenerated periodically by using electronic repeaters spaced apart typically by 60–70 km. The repeater spacing can be increased by making use of a homodyne or heterodyne detection scheme since its use improves receiver sensitivity. Such systems are referred to as coherent lightwave systems. Coherent systems were under development worldwide during the 1980s, and their potential benefits were demonstrated in many system experiments [16]. However, commercial introduction of such systems had been delayed by the advent of fiber amplifiers in 1989.

The fourth generation of lightwave systems makes use of *optical amplification* for increasing the repeater spacing and of *wavelength-division multiplexing* (WDM) for increasing the bit rate. In such systems, fiber loss is compensated periodically by using erbium-doped fiber amplifiers spaced 60–100 km apart. Such amplifiers were developed during the 1980s and became available commercially by 1990. In 1991, an experiment showed the possibility of data transmission over 21,000 km at 2.5 Gb/s, and over 14,300 km at 5 Gb/s, by using a recirculating-loop configuration [17]. This performance indicated that an amplifier-based, all-optical, submarine transmission system was feasible for intercontinental communication. By 1996, not only transmission over 11,300 km at a bit rate of 5 Gb/s had been demonstrated by using actual submarine cables [18], but a commercial transpacific cable (TPC-5) also became operational. Figure 1.3 shows the international network of undersea lightwave systems [19] operational in 1996. Many other transoceanic lightwave systems have been planned. The 27,300-km fiber-optic link around the globe (known as FLAG) will begin operation in 1997, linking many Asian and European countries at

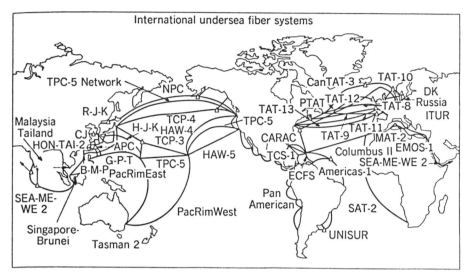

Figure 1.3 International undersea network of fiber-optic communication systems as installed by the end of 1996. (After Ref. [19]. ©1995 AT&T. Reprinted with permission.)

5 Gb/s, with several sections operating at 10 Gb/s [20]. Another fiber-optic network, known as *Africa One*, will circle the African continent and cover a total transmission distance of about 35,000 km [21]. Clearly, the fourth-generation systems have revolutionized the state of the art of lightwave systems.

The current emphasis of fourth-generation lightwave systems is on increasing the system capacity by transmitting multiple channels through the WDM technique. Optical amplifiers are ideal for multichannel lightwave systems since all channels can be amplified simultaneously without requiring demultiplexing of individual channels. In a 1996 demonstration, twenty 5-Gb/s channels were transmitted over 9100 km in a recirculating-loop configuration, resulting in a total bit rate of 100 Gb/s and the BL product of 910 (Tb/s)-km [22]. In another record experiment, a total bit rate of 1.1 Tb/s was achieved by multiplexing 55 channels, each operating at 20 Gb/s [23]. Despite the use of dispersion-compensation schemes, dispersive effects limited the total transmission distance to 150 km. Commercial WDM systems operating at a bit rate of up to 40 Gb/s were available by the end of 1996. A transpacific system (TPC-6) operating at 100 Gb/s is scheduled to begin operation by the year 2000. The bit rate–distance product for such a system exceeds 900 (Tb/s)-km, indicating the progress realized over a 20-year period.

The fifth generation of fiber-optic communication systems is concerned with finding a solution to the fiber-dispersion problem. Optical amplifiers solve the loss problem but, at the same time, make the dispersion problem worse since the dispersive effects accumulate over multiple amplification stages. Several dispersion-compensation techniques have been developed, as discussed in Chapter 9. An ultimate solution is based on the novel concept of *optical solitons*, optical pulses that preserve their shape during propagation in a lossless fiber by counteracting the effect of dispersion through the fiber nonlinearity. Although the basic idea was proposed [24] as early as 1973, it was only in 1988 that a laboratory experiment [25] demonstrated the feasibility of data transmission over 4000 km by compensating the fiber loss through stimulated Raman scattering. Erbium-doped fiber amplifiers were used for soliton amplification starting in 1989. Since then, many system experiments have demonstrated the eventual potential of soliton communication systems. By 1994, solitons were transmitted over 35,000 km at 10 Gb/s and over 24,000 km at 15 Gb/s [26]. In a 1996 recirculating-loop experiment [27], soliton transmission over 9400 km was demonstrated at a bit rate of 70 Gb/s by multiplexing seven 10-Gb/s channels.

Even though the fiber-optic communication technology is barely two decades old, it has progressed rapidly and has reached a certain stage of maturity. This is also apparent from the publication of a large number of books on optical communications since 1991 [28]–[52]. This text, first published in 1992, is intended to present an up-to-date account of fiber-optic communications systems with an emphasis on recent developments.

1.2   BASIC CONCEPTS

In this section we introduce a few basic concepts common to all communication systems. We begin with a description of analog and digital signals and describe how an analog signal can be converted into digital form. We then consider time- and frequency-division multiplexing of input signals, and conclude with a discussion of various modulation formats.

1.2.1   Analog and Digital Signals

In any communication system, information to be transmitted is generally available as an electrical signal that may take *analog* or *digital* form [53]. In the analog case, the signal (e. g., electric current) varies continuously with time, as shown schematically in Fig. 1.4(a). Familiar examples include audio and video signals resulting when a microphone converts voice or a video camera converts an image into an electrical signal. By contrast, the digital signal takes only a few discrete values. In the *binary representation* of a digital signal only two values are possible. The simplest case of a binary digital signal is one in which the electric current is either on or off, as shown in Fig. 1.4(b). These two possibilities are called "bit 1" and "bit 0" (*bit* is a contracted form of binary dig*it*). Each bit lasts for a certain period of time T_B, known as the bit period or *bit slot*. Since one bit of information is conveyed in a time interval T_B, the bit rate B, defined as the number of bits per second, is simply $B = T_B^{-1}$. A well-known example of digital signals is provided by computer data. Each letter of the alphabet together with other common symbols (decimal numerals, punctuation marks, etc.) is assigned a code number (ASCII code) in the range 0–127 whose

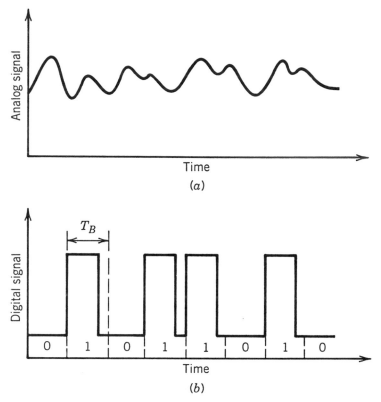

Figure 1.4 Representation of (a) an analog signal and (b) a digital signal.

binary representation corresponds to a 7-bit digital signal. The original ASCII code has been extended to represent 256 characters transmitted through 8-bit bytes. Both analog and digital signals are characterized by their bandwidth, which is a measure of the spectral contents of the signal. The *signal bandwidth* represents the range of frequencies contained within the signal and is determined mathematically through its Fourier transform.

An analog signal can be converted into digital form by sampling it at regular intervals of time [53]. Figure 1.5 shows the conversion method schematically. The sampling rate is determined by the bandwidth Δf of the analog signal. According to the *sampling theorem* [54]–[56], a bandwidth-limited signal can be fully represented by discrete samples, without any loss of information, provided that the sampling frequency f_s satisfies the *Nyquist criterion* [57], $f_s \geq 2\Delta f$. The first step consists of sampling the analog signal at the right frequency. The sampled values can take any value in the range $0 \leq A \leq A_{\max}$, where $A_{\max}$ is the maximum amplitude of the given analog signal. Let us assume that $A_{\max}$ is divided into M discrete (not necessarily equally spaced) intervals. Each sampled value is quantized to correspond to one of these discrete values. Clearly, this procedure leads to additional noise, known as *quantization noise*, which adds to

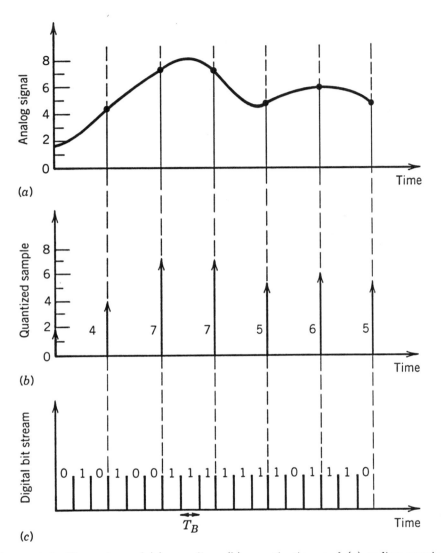

Figure 1.5 Three steps of (a) sampling, (b) quantization, and (c) coding required for converting an analog signal into a binary digital signal.

the noise already present in the analog signal.

The effect of quantization noise can be minimized by choosing the number of discrete levels such that $M > A_{max}/A_N$, where A_N is the root-mean-square noise amplitude of the analog signal. The ratio A_{max}/A_N is called the dynamic range and is related to the *signal-to-noise ratio* (SNR) by the relation

$$\text{SNR} = 20 \log_{10}(A_{max}/A_N), \qquad (1.2.1)$$

where SNR is expressed in decibel (dB) units. Any ratio R can be converted into decibels by using the general definition $10 \log_{10} R$ (see Appendix B). Equation

(1.2.1) contains a factor of 20 in place of 10 simply because the SNR for electrical signals is defined with respect to the electrical power, whereas A is related to the electric current (or voltage).

The quantized sampled values can be converted into digital format by using a suitable conversion technique. In one scheme, known as *pulse-position modulation*, pulse position within the bit slot is a measure of the sampled value. In another, known as *pulse-duration modulation*, the pulse width is varied from bit to bit in accordance with the sampled value. These techniques are rarely used in practical optical communication systems, since it is difficult to maintain the pulse position or pulse width to high accuracy during propagation inside the fiber. The technique used almost universally, known as *pulse-code modulation* (PCM), is based on a binary scheme in which information is conveyed by the absence or the presence of pulses that are otherwise identical. A binary code is used to convert each sampled value into a string of "1" and "0" bits. The number of bits m needed to code each sample is related to the number of quantized signal levels M by the relation

$$M = 2^m \quad \text{or} \quad m = \log_2 M. \tag{1.2.2}$$

The bit rate associated with the PCM digital signal is thus given by

$$B = mf_s \geq (2\Delta f)\log_2 M, \tag{1.2.3}$$

where the Nyquist criterion, $f_s \geq 2\Delta f$, was used. By noting that $M > A_{\max}/A_N$ and using Eq. (1.2.1) together with $\log_2 10 = 3.33$,

$$B > (\Delta f/3)\,\text{SNR}, \tag{1.2.4}$$

where the SNR is expressed in decibel (dB) units.

Equation (1.2.4) provides the minimum bit rate required for digital representation of an analog signal of bandwidth Δf and a specific SNR. For SNR $>$ 30 dB the required bit rate exceeds $10\Delta f$, indicating a considerable increase in the bandwidth requirements of digital signals. Despite this increase, the digital format is almost always used for optical communication systems. This choice is made because of the superior performance of digital transmission systems. Lightwave systems offer such an enormous increase in the system capacity (by a factor $\sim 10^5$) compared with microwave systems that some bandwidth can be traded for improved performance.

As an illustration of Eq. (1.2.4), consider the digital conversion of an audio signal generated in a telephone. The analog audio signal contains frequencies in the range 0.3-3.4 kHz with a bandwidth $\Delta f = 3.1$ kHz and has a SNR of about 30 dB. Equation (1.2.4) indicates that $B > 31$ kb/s. In practice, a digital audio channel operates at 64 kb/s. The analog signal is sampled at intervals of 125 μs (sampling rate $f_s = 8$ kHz), and each sample is represented by 8 bits. The required bit rate for a digital video signal is higher by more than a factor of 1000. The analog television signal has a bandwidth ~ 4 MHz with a SNR of about 50 dB. The minimum bit rate from Eq. (1.2.4) is 66 Mb/s. In practice, a digital video signal requires a bit rate of 100 Mb/s or more unless it is compressed by using a standard format (such as MPEG-2).

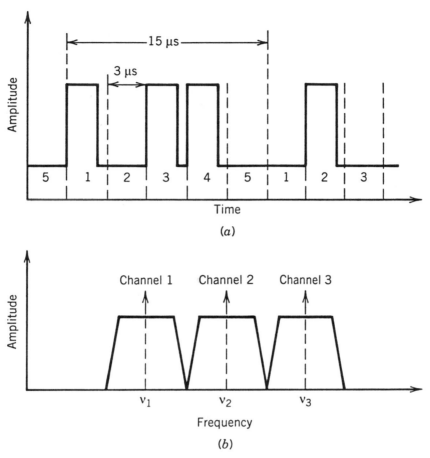

Figure 1.6 (a) Time-division multiplexing of five digital voice channels operating at 64 kb/s; (b) frequency-division multiplexing of three analog signals.

1.2.2 Channel Multiplexing

As seen in the preceding discussion, a digital voice channel operates at 64 kb/s. Most fiber-optic communication systems are capable of transmitting at a rate of more than 100 Mb/s. To utilize the system capacity fully, it is necessary to transmit many channels simultaneously through multiplexing. There are mainly two ways to accomplish this: *time-division multiplexing* (TDM) and *frequency-division multiplexing* (FDM). In the case of TDM, bits associated with different channels are interleaved in the time domain to form a composite bit stream. For example, the bit slot is about 15 μs for a single voice channel operating at 64 kb/s. Five such channels can be multiplexed through TDM if the bit streams of successive channels are delayed by 3 μs. Figure 1.6(a) shows the resulting bit stream schematically at a composite bit rate of 320 kb/s. In the case of FDM, the channels are spaced apart in the frequency domain. Each channel

Table 1.1 SONET/SDH bit rates

SONET	SDH	B (Mb/s)	Channels
OC-1		51.84	672
OC-3	STM-1	155.52	2,016
OC-12	STM-4	622.08	8,064
OC-48	STM-16	2,488.32	32,256
OC-192	STM-64	9,953.28	129,024

is carried by its own carrier wave. The carrier frequencies are spaced more than the channel bandwidth so that the channel spectra do not overlap, as seen Fig. 1.6(b). FDM is suitable for both analog and digital signals and is used in broadcasting of radio and television channels. TDM is readily implemented for digital signals and is commonly used for telecommunication networks.

It is important to realize that TDM and FDM can be implemented in both the electrical and optical domains; optical FDM is often referred to as WDM. Chapter 7 is devoted to optical-domain multiplexing techniques. This section covers electrical TDM since it is employed universally to multiplex a large number of voice channels into a single electrical bit stream.

The concept of TDM has been used to form *commercial digital hierarchies*. In North America and Japan, the first level corresponds to multiplexing of 24 voice channels with a composite bit rate of 1.544 Mb/s (hierarchy DS-1), whereas in Europe 30 voice channels are multiplexed, resulting in a composite bit rate of 2.048 Mb/s. The bit rate of the multiplexed signal is slightly larger than the simple product of 64 kb/s with the number of channels because of extra control bits that are added for separating (demultiplexing) the channels at the receiver end. The second-level hierarchy is obtained by multiplexing 4 DS-1 TDM channels. This results in a bit rate of 6.312 Mb/s (hierarchy DS-2) for North America or Japan and 8.448 Mb/s for Europe. This procedure is continued to obtain higher-level hierarchies. For example, at the fifth level of hierarchy, the bit rate becomes 565 Mb/s for Europe and 396 Mb/s for Japan.

The lack of an international standard in the telecommunication industry during 1980s led to the advent of a new standard, first called the *synchronous optical network* (SONET) and later termed the *synchronous digital hierarchy* (SDH) [58]–[60]. It defines a synchronous frame structure for transmitting TDM digital signals. The basic building block of the SONET has a bit rate of 51.84 Mb/s. The corresponding optical signal is referred to as OC-1, where OC stands for "optical carrier." The basic building block of the SDH has a bit rate of 155.52 Mb/s and is referred to as STM-1, where STM stands for a *synchronous transport module*. A useful feature of the SONET and SDH is that higher levels have a bit rate that is an exact multiple of the basic bit rate. Table 1.1 lists the correspondence between SONET and SDH bit rates for several levels. The SDH provides an international standard that appears to be well adopted. Indeed, lightwave systems operating at the STM-16 level ($B = 2.488$ Gb/s) have been

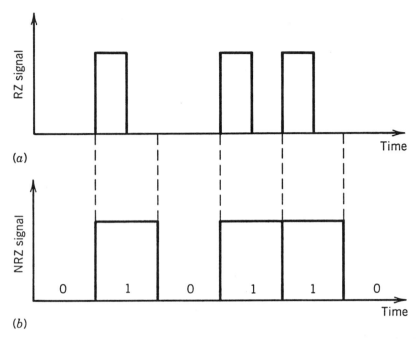

Figure 1.7 Digital bit stream 010110... coded by using (a) return-to-zero (RZ) and (b) nonreturn-to-zero (NRZ) formats.

available commercially since 1990. STM-64 systems operating near 10 Gb/s have also been developed [15]. Progress in the *asynchronous-transfer-mode* (ATM) networks is also likely to affect lightwave technology considerably [61].

1.2.3 Modulation Formats

The first step in the design of an optical communication system is to decide how the electrical signal would be converted into an optical bit stream. Normally, the output of an optical source such as a semiconductor laser is modulated by applying the electrical signal either directly to the optical source or to an external modulator. There are two choices for the modulation format of the resulting optical bit stream. These are shown in Fig. 1.7 and are known as the *return-to-zero* (RZ) and *nonreturn-to-zero* (NRZ) formats. In the RZ format, each optical pulse representing bit 1 is shorter than the bit slot, and its amplitude returns to zero before the bit duration is over. In the NRZ format, the optical pulse remains on throughout the bit slot and its amplitude does not drop to zero between two or more successive 1 bits. As a result, pulse width varies depending on the bit pattern, whereas it remains the same in the case of RZ format. An advantage of the NRZ format is that the bandwidth associated with the bit stream is smaller than that of the RZ format by about a factor of 2 simply because on–off transitions occur fewer times. However, its use requires tighter control of the pulse width and may lead to bit-pattern-dependent effects if the optical pulse spreads during transmission. The NRZ format is often used in practice because of a smaller signal bandwidth associated with it. The RZ format is required for soliton communication systems (discussed in Chapter 10).

An important issue is related to the choice of the physical variable that is modulated to encode the data on the optical carrier. The optical carrier wave before modulation is of the form

$$\mathbf{E}(t) = \hat{e}A\cos(\omega_0 t + \phi), \qquad (1.2.5)$$

where $\mathbf{E}$ is the electric field vector, $\hat{e}$ is the polarization unit vector, A is the amplitude, ω_0 is the carrier frequency, and ϕ is the phase. The spatial dependence of $\mathbf{E}$ is suppressed for simplicity of notation. One may choose to modulate the amplitude A, the frequency ω_0, or the phase ϕ. In the case of analog modulation, the three modulation choices are known as *amplitude modulation* (AM), *frequency modulation* (FM), and *phase modulation* (PM). The same modulation techniques can be applied in the digital case and are called *amplitude-shift keying* (ASK), *frequency-shift keying* (FSK) and *phase-shift keying* (PSK), depending on whether the amplitude, frequency, or phase of the carrier wave is shifted between the two levels of a binary digital signal. The simplest technique consists of simply changing the signal intensity between two levels, one of which is set to zero, and is often called *on–off keying* (OOK) to reflect the on–off nature of the resulting optical signal. Most digital lightwave systems employ OOK in combination with PCM.

1.3   OPTICAL COMMUNICATION SYSTEMS

As mentioned earlier, optical communication systems differ in principle from microwave systems only in the frequency range of the carrier wave used to carry the information. The optical carrier frequency is typically ~ 100 THz, in contrast with the microwave carrier frequencies, ~ 1–10 GHz. An increase in the information capacity of optical communication systems by a factor of $\sim 10{,}000$ is expected simply because of such high carrier frequencies used for lightwave systems. This increase can be understood by noting that the bandwidth of the modulated carrier can be up to a few percent of the carrier frequency. Taking, for illustration, 1% as the limiting value, optical communication systems have the potential of carrying information at bit rates ~ 1 Tb/s. It is this enormous potential bandwidth of optical communication systems that is the driving force behind the worldwide development and deployment of lightwave systems. Current state-of-the-art systems operate at bit rates ~ 10 Gb/s, indicating that there is considerable room for improvement.

Figure 1.8 shows a generic block diagram of an optical communication system. It consists of a transmitter, a communication channel, and a receiver, the three elements common to all communication systems. Optical communication systems can be classified into two broad categories: *guided* and *unguided*. As the name implies, in the case of guided lightwave systems, the optical beam emitted by the transmitter remains spatially confined. This is achieved by using optical

Figure 1.8 Generic optical communication system.

fibers, as discussed in Chapter 2. Since all guided optical communication systems currently use optical fibers, the commonly used term for them is fiber-optic communication systems. The term *lightwave system* is also sometimes used for fiber-optic communication systems, although it should generally include both guided and unguided systems.

In the case of unguided optical communication systems, the optical beam emitted by the transmitter spreads in space, similar to the spreading of microwaves. However, unguided optical systems are less suitable for broadcasting applications than microwave systems because optical beams spread mainly in the forward direction (as a result of their short wavelength). Their use generally requires accurate pointing between the transmitter and the receiver. In the case of terrestrial propagation, the signal in unguided systems can deteriorate considerably by scattering within the atmosphere. This problem, of course, disappears in *free-space communications* above the earth atmosphere (e.g., intersatellite communications). Although free-space optical communication systems are needed for certain applications and have been studied extensively [62], most terrestrial applications make use of *fiber-optic communication systems*. This book does not consider unguided optical communication systems.

The application of optical fiber communications is in general possible in any area that requires transfer of information from one place to another. However, fiber-optic communication systems have been developed mostly for telecommunications applications. This is understandable in view of the existing worldwide telephone networks which are used to transmit not only voice signals but also computer data and fax messages. The telecommunication applications can be broadly classified into two categories, *long-haul* and *short-haul*, depending on whether the optical signal is transmitted over relatively long or short distances compared with typical intercity distances ($\sim$ 100 km). Long-haul telecommunication systems require high-capacity trunk lines and benefit most by the use of fiber-optic lightwave systems. Indeed, the technology behind optical fiber communication is often driven by long-haul applications. Each successive generation of lightwave systems is capable of operating at higher bit rates and over longer distances. Periodic regeneration of the optical signal by using repeaters is still required for most long-haul systems. However, more than an order-of-magnitude increase in both the repeater spacing and the bit rate compared with those of coaxial systems has made the use of lightwave systems very attractive for long-haul applications. Furthermore, transmission distances of thousands of kilometers can be realized by using optical amplifiers. As shown in Fig. 1.3, a large number of transoceanic lightwave systems have already been installed to create an international fiber-optic network.

Short-haul telecommunication applications cover intracity and local-loop traffic. Such systems typically operate at low bit rates over distances of less than 10 km. The use of single-channel lightwave systems for such applications is not very cost-effective, and multichannel networks with multiple services should be considered. The concept of a *broadband integrated-services digital network* requires a high-capacity communication system capable of carrying multiple services. The ATM technology also demands high bandwidths [61]. Only fiber-optic communication systems are likely to meet such wideband distribution requirements. Multichannel lightwave systems and their applications in local-area networks are discussed in Chapter 7.

1.4 LIGHTWAVE SYSTEM COMPONENTS

The generic block diagram of Fig. 1.8 applies to a fiber-optic communication system, the only difference being that the communication channel is an optical fiber cable. The other two components, the optical transmitter and the optical receiver, are designed to meet the needs of such a specific communication channel. In this section we discuss the general issues related to the role of optical fiber as a communication channel and to the design of transmitters and receivers. The objective is to provide an introductory overview, as the three components are discussed in detail in Chapters 2–4.

1.4.1 Optical Fibers as a Communication Channel

The role of communication channel is to transport the optical signal from transmitter to receiver without distorting it. Most lightwave systems use optical fibers as the communication channel because fibers can transmit light with a relatively small amount of power loss. In Chapter 2 we discuss the properties of optical fibers in detail. Fiber loss is, of course, an important design issue, as it determines directly the repeater spacing of a long-haul lightwave system. Another important design issue is *fiber dispersion*, which leads to broadening of individual optical pulses inside the fiber. If optical pulses spread significantly outside their allocated bit slot, the transmitted signal is severely degraded. Eventually, it becomes impossible to recover the original signal with high accuracy. The problem is most severe in the case of multimode fibers, since pulses spread rapidly (typically at a rate of ~ 10 ns/km) because of different speeds associated with different fiber modes. It is for this reason that most optical communication systems use single-mode fibers. *Material dispersion* (related to the frequency dependence of the refractive index) still leads to pulse broadening (typically < 0.1 ns/km), but it is small enough to be acceptable for most applications and can be reduced further by controlling the spectral width of the optical source. Nevertheless, as discussed in Chapter 2, material dispersion sets the ultimate limit on the bit rate and the transmission distance of fiber-optic communication systems.

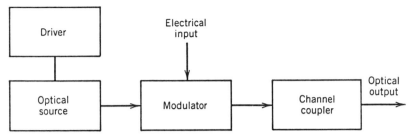

Figure 1.9 Components of an optical transmitter.

1.4.2 Optical Transmitters

The role of an *optical transmitter* is to convert the electrical signal into optical form and to launch the resulting optical signal into the optical fiber. Figure 1.9 shows the block diagram of an optical transmitter. It consists of an optical source, a modulator, and a channel coupler. Semiconductor lasers or light-emitting diodes are used as optical sources because of their compatibility with the optical-fiber communication channel; both are discussed in detail in Chapter 3. The optical signal is generated by modulating the optical carrier wave. Although an external modulator is sometimes used, it can be dispensed with in most cases, since the output of a semiconductor optical source can be modulated directly by varying the injection current. Such a scheme simplifies the transmitter design and is generally cost-effective. The coupler is typically a microlens that focuses the optical signal onto the entrance plane of an optical fiber with the maximum possible efficiency.

The *launched power* is an important design parameter, as it indicates how much fiber loss can be tolerated. It is often expressed in units of dBm with 1 mW as the reference level. The general definition is (see Appendix B)

$$\text{power (dBm)} = 10 \log_{10}\left(\frac{\text{power}}{1 \text{ mW}}\right). \tag{1.4.1}$$

Thus, 1 mW is 0 dBm, but 1 μW corresponds to -30 dBm. The launched power is rather low (< -10 dBm) in the case of light-emitting diodes, whereas semiconductor lasers can launch powers ~ 10 dBm. Since light-emitting diodes are also limited in their modulation capabilities, most high-performance lightwave systems use semiconductor lasers as optical sources. The bit rate of optical transmitters is often limited by electronics rather than by the semiconductor laser itself. With proper design, optical transmitters can be made to operate at a bit rate of up to 20 Gb/s.

1.4.3 Optical Receivers

An *optical receiver* converts the optical signal received at the output end of the optical fiber back into the original electrical signal. Figure 1.10 shows the block diagram of an optical receiver. It consists of a coupler, a photodetector,

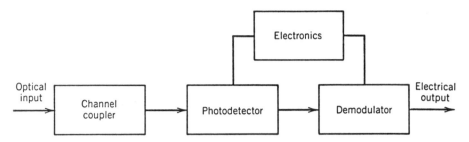

Figure 1.10 Components of an optical receiver.

and a demodulator. The coupler focuses the received optical signal onto the photodetector. Semiconductor photodiodes are used as photodetectors because of their compatibility with the whole system; they are discussed in Chapter 4. The design of the demodulator depends on the modulation format used by the lightwave system. The use of FSK and PSK formats, generally appropriate for coherent communication systems, requires heterodyne or homodyne demodulation techniques; such coherent receivers have many components and are relatively expensive. Often, the received signal is in the form of optical pulses representing 1 and 0 bits and is converted directly into an electric current. Such a scheme is referred to as *intensity modulation with direct detection* (IM/DD), in contrast with coherent detection. Demodulation is done by a decision circuit that identifies bits as 1 or 0, depending on the amplitude of the electric current. The accuracy of the decision circuit depends on the SNR of the electrical signal generated at the photodetector.

The performance of a digital lightwave system is characterized through the *bit-error rate* (BER). Although the BER can be defined as the number of errors made per second, such a definition makes the BER bit-rate dependent. It is customary to define the BER as the average probability of incorrect bit identification. Therefore, a BER of 10^{-6} corresponds to on average one error per million bits. Most lightwave systems specify a BER of 10^{-9} as the operating requirement; some even require a BER as small as 10^{-14}.

An important parameter that is indicative of receiver performance is called *receiver sensitivity*. It is usually defined as the minimum average received optical power for which the BER is 10^{-9}. Receiver sensitivity depends on the SNR, which in turn depends on various noise sources that corrupt the signal received. Even for a perfect receiver, some noise is introduced by the process of photodetection itself. This is referred to as the *quantum noise* or the *shot noise*, as it has its origin in the particle nature of electrons. Optical receivers operating at the shot-noise limit are called *quantum-noise-limited receivers*. No practical receiver operates at the quantum-noise limit, since many other noise sources decrease the SNR considerably below the shot-noise limit. Some of the noise sources such as *thermal noise* and *amplifier noise* are internal to the receiver. Others originate at the transmitter or during propagation inside the fiber. For instance, the optical signal launched by the transmitter has inherent intensity and phase fluctuations that have their origin in the fundamental process of spontaneous emission. Chromatic dispersion in optical fibers can add additional noise through phenomena such as intersymbol interference and mode-partition noise. The receiver sensitivity is determined by a cumulative effect of all possible noise mechanisms that degrade the SNR at the decision circuit. In general, it also depends on the bit rate, since the contribution of some noise sources (e.g. shot noise) increases in proportion to the signal bandwidth. In Chapters 4 and 6 we discuss noise and sensitivity of IM/DD and coherent receivers, respectively, by considering the SNR and the BER in digital lightwave systems.

1.5 CHANNEL CAPACITY

The performance of any communication system is ultimately limited by the SNR of the received signal. This limitation can be stated more formally by using the concept of *channel capacity* introduced within the framework of *information theory* [54]–[56]. It turns out that a maximum possible bit rate exists for error-free transmission of a binary digital signal in the presence of Gaussian noise. This rate is called the channel capacity. More specifically, the channel capacity of a noisy communication channel is given by

$$C = \Delta f_{ch} \log_2(1 + S/N), \qquad (1.5.1)$$

where Δf_{ch} is the channel bandwidth, S is the average signal power, and N is the average noise power.

An important implication of Eq. (1.5.1) is that the channel capacity cannot be increased indefinitely simply by increasing the bandwidth Δf_{ch}. This statement becomes clear if we note that the shot noise, present even in a perfect system, increases linearly with Δf_{ch}. Indeed, by using $N = N_0 \Delta f_{ch}$, where N_0 is the spectral density of shot noise, and taking the limit of infinite Δf_{ch}, the channel capacity is found to be limited by

$$C \leq C_{\max} = (S/N_0) \log_2 e. \qquad (1.5.2)$$

The channel bandwidth Δf_{ch} of optical communication systems is larger by a factor of nearly 10,000 than that of microwave systems. However, the channel capacity is not necessarily increased by the same factor because of the fundamental limitation imposed by Eq. (1.5.1). It is nonetheless considerably higher for optical communication systems. Current lightwave systems operate well below the channel capacity, as typical bit rates are below 10 Gb/s. Considerable increase in the transmission capacity of lightwave systems is expected to occur through multiple channels transmitted over the same fiber by the use of WDM techniques. Multichannel communication systems are discussed in Chapter 7. The importance of Eq. (1.5.1) is that it provides the maximum value of the system capacity possible for a given communication channel under the best operating conditions.

PROBLEMS

1.1 Calculate the carrier frequency for optical communication systems operating at 0.88, 1.3, and 1.55 μm. What is the photon energy (in eV) in each case?

1.2 Calculate the transmission distance over which the optical power will attenuate by a factor of 10 for three fibers with losses of 0.2, 20, and 2000 dB/km. Assuming that the optical power decreases as $\exp(-\alpha L)$, calculate α (in cm^{-1}) for the three fibers.

1.3 Assume that a digital communication system can be operated at a bit rate of up to 1% of the carrier frequency. How many audio channels at 64 kb/s can be transmitted over a microwave carrier at 5 GHz and an optical carrier at 1.55 μm?

1.4 A 1-hour lecture script is stored on the computer hard disk in the ASCII format. Estimate the total number of bits assuming a delivery rate of 200 words per minute and on average 5 letters per word. How long will it take to transmit the script at a bit rate of 1 Gb/s?

1.5 A 1.55-μm digital communication system operating at 1 Gb/s receives an average power of -40 dBm at the detector. Assuming that 1 and 0 bits are equally likely to occur, calculate the number of photons received within each 1 bit.

1.6 An analog voice signal that can vary over the range 0–50 mA is digitized by sampling it at 8 kHz. The first four sample values are 10, 21, 36, and 16 mA. Write the corresponding digital signal (a string of 1 and 0 bits) by using a 4-bit representation for each sample.

1.7 Sketch the variation of optical power with time for a digital NRZ bit stream 010111101110 by assuming a bit rate of 2.5 Gb/s. What is the duration of the shortest and widest optical pulse?

1.8 A 1.55-μm fiber-optic communication system is transmitting digital signals over 100 km at 2 Gb/s. The transmitter launches 2 mW of average power into the fiber cable, having a net loss of 0.3 dB/km. How many photons are incident on the receiver during a single 1 bit? Assume that 0 bits carry no power, while 1 bits are in the form of a rectangular pulse occupying the entire bit slot (NRZ format).

1.9 A 0.8-μm optical receiver needs at least 1000 photons to detect the 1 bits accurately. What is the maximum possible length of the fiber link for a 100-Mb/s optical communication system designed to transmit -10 dBm of average power? The fiber loss is 2 dB/km at 0.8 μm. Assume the NRZ format and a rectangular pulse shape.

1.10 A 1.3-μm optical transmitter is used to obtain a digital bit stream at a bit rate of 2 Gb/s. Calculate the number of photons contained in a single 1 bit when the average power emitted by the transmitter is 4 mW. Assume that the 0 bits carry no energy.

REFERENCES

[1] A. Jones, *Historical Sketch of the Electrical Telegraph*, Putnam, New York, 1852.

[2] A. G. Bell, U.S. Patent No. 174,465 (1876).

[3] T. H. Maiman, *Nature* **187**, 493 (1960).

[4] W. K. Pratt, *Laser Communication Systems*, Wiley, New York, 1969.

[5] S. E. Miller, *Sci. Am.* **214** (1), 19 (1966).

[6] K. C. Kao and G. A. Hockham, *Proc. IEE* **113**, 1151 (1966); A. Werts, *Onde Electr.* **45**, 967 (1966).

[7] F. P. Kapron, D. B. Keck, and R. D. Maurer, *Appl. Phys. Lett.* **17**, 423 (1970).

[8] I. Hayashi, M. B. Panish, P. W. Foy, and S. Sumski, *Appl. Phys. Lett.* **17**, 109 (1970).

[9] R. J. Sanferrare, *AT&T Tech. J.* **66**, 95 (1987).

[10] D. Gloge, A. Albanese, C. A. Burrus, E. L. Chinnock, J. A. Copeland, A. G. Dentai, T. P. Lee, T. Li, and K. Ogawa, *Bell Syst. Tech. J.* **59**, 1365 (1980).

[11] J. I. Yamada, S. Machida, and T. Kimura, *Electron. Lett.* **17**, 479 (1981).

[12] T. Miya, Y. Terunuma, T. Hosaka, and T. Miyoshita, *Electron. Lett.* **15**, 106 (1979).

[13] A. H. Gnauck, B. L. Kasper, R. A. Linke, R. W. Dawson, T. L. Koch, T. J. Bridges, E. G. Burkhardt, R. T. Yen, D. P. Wilt, J. C. Campbell, K. C. Nelson, and L. G. Cohen, *J. Lightwave Technol.* **3**, 1032 (1985).

[14] K. L. Monham, R. Plastow, A. C. Carter, and R. C. Goodfellow, *Electron. Lett.* **21**, 619 (1985).

[15] K. Nakagawa, *Trans. IECE Jpn. Pt. J* **78B**, 713 (1995).

[16] R. A. Linke and A. H. Gnauck, *J. Lightwave Technol.* **6**, 1750 (1988); P. S. Henry, *Coherent Lightwave Communications*, IEEE Press, New York, 1990.

[17] N. S. Bergano, J. Aspell, C. R. Davidson, P. R. Trischitta, B. M. Nyman, and F. W. Kerfoot, *Electron. Lett.* **27**, 1889 (1991).

[18] T. Otani, K. Goto, H. Abe, M. Tanaka, H. Yamamoto, and H. Wakabayashi, *Electron. Lett.* **31**, 380 (1995).

[19] J. M. Sipress, *AT&T Tech. J.* **73** (1), 4 (1995).

[20] T. Welsh, R. Smith, H. Azami, and R. Chrisner. *IEEE Commun. Mag.* **34** (2), 30 (1996).

[21] W. C. Marra and J. Schesser, *IEEE Commun. Mag.* **34** (2), 50 (1996).

[22] N. S. Bergano and C. R. Davidson, *J. Lightwave Technol.* **14**, 1299 (1996).

[23] H. Onaka, S. Kinoshita, and T. Chikama, *Fujitsu Sci. Tech. J.* **32**, 36 (1996).

[24] A. Hasegawa and F. Tappert, *Appl. Phys. Lett.* **23**, 142 (1973).

[25] L. F. Mollenauer and K. Smith, *Opt. Lett.* **13**, 675 (1988).

[26] L. F. Mollenauer, *Opt. Photon. News* **11** (4), 15 (1994).

[27] L. F. Mollenauer, P. V. Mamyshev, and M. J. Neubelt, *Electron. Lett.* **32**, 471 (1996).

[28] H. B. Killen, *Fiber-Optic Communications*, Prentice Hall, Upper Saddle River, NJ, 1991.

[29] T. Li, Ed., *Topics in Lightwave Transmission Systems*, Academic Press, San Diego, CA, 1991.

[30] G. E. Keiser, *Optical Fiber Communications*, 2nd ed., McGraw-Hill, New York, 1991.

[31] G. P. Agrawal, *Fiber-Optic Communication Systems*, Wiley, New York, 1992.

[32] J. C. Palais, *Fiber-Optic Communications*, 3rd ed., Prentice Hall, Upper Saddle River, NJ, 1992.

[33] J. M. Senior, *Optical Fiber Communications*, 2nd ed., Prentice Hall, Upper Saddle River, NJ, 1992.

[34] J. Gowar, *Optical Communication Systems*, 2nd ed., Prentice Hall, Upper Saddle River, NJ, 1993.

[35] P. E. Green, Jr., *Fiber-Optic Networks*, Prentice Hall, Upper Saddle River, NJ, 1993.

[36] N. Kashima, *Optical Transmission for the Subscriber Loop*, Artec House, Boston, 1993.

[37] L. D. Green, *Fiber Optic Communications*, CRC Press, Boca Raton, FL, 1993.

[38] P. W. Hooijmans, *Coherent Optical System Design*, Wiley, New York, 1994.

[39] G. Jacobsen, *Noise in Digital Optical Transmission Systems*, Artec House, Boston, 1994.

[40] S. Betti, G. de Marchis, and E. Iannone, *Coherent Optical Communication Systems*, Wiley, New York, 1995.

[41] R. M. Gagliardi and S. Karp, *Optical Communications*, Wiley, New York, 1995.

[42] G. Einarrson, *Principles of Lightwave Communication Systems*, Wiley, New York, 1995.

[43] S. Ryu, *Coherent Lightwave Communication Systems*, Artec House, Boston, 1995.

[44] N. Kashima, *Passive Optical Components for Optical Fiber Transmission*, Artec House, Boston, 1995.

[45] D. W. Smith, *Optical Network Technology*, Chapman & Hall, New York, 1995.

[46] D. J. G. Mestdagh, *Fundamentals of Multiaccess Optical Fiber Networks*, Artec House, Boston, 1995.

[47] J. Franz, *Optical Communication Systems*, Narosa Publishing House, New Delhi, 1996.

[48] M. M.-K. Liu, *Principles and Applications of Optical Communications*, Irwin, Chicago, 1996.

[49] M. Cvijetic, *Coherent and Nonlinear Lightwave Communications*, Artec House, Boston, 1996.

[50] L. Kazovsky, S. Bendetto, and A. E. Willner, *Optical Fiber Communication Systems*, Artec House, Boston, 1996.

[51] R. Papannareddy *Introduction to Lightwave Communication Systems*, Artec House, Boston, 1997.

[52] K. Nosu *Optical FDM Network Technologies*, Artec House, Boston, 1997.

[53] M. Schwartz, *Information Transmission, Modulation, and Noise*, 4th ed., McGraw-Hill, New York, 1990.

[54] C. E. Shannon, *Proc. IRE* **37**, 10 (1949).

[55] A. J. Jerri, *Proc. IEEE* **65**, 1565 (1977).

[56] L. W. Couch II, *Modern Communication Systems: Principles and Applications*, 4th ed., Prentice Hall, Upper Saddle River, NJ, 1995.

[57] H. Nyquist, *Trans. AIEE* **47**, 617 (1928).

[58] R. Ballart and Y.-C. Ching, *IEEE Commun. Mag.* **27** (3), 8 (1989).

[59] T. Miki, Jr. and C. A. Siller, Eds., *IEEE Commun. Mag.* **28** (8), 1 (1990).

[60] T. Miki, Jr. and M. Shafi, *SONET/SDH: A Sourcebook of Synchronous Networking*, IEEE Press, Piscataway, NJ, 1996.

[61] R. O. Onvural, *Asynchronous Transfer Mode Networks*, Artec House, Boston, 1995.

[62] S. G. Lambert and W. L. Casey, *Laser Communications in Space*, Artec House, Boston, 1995.

Chapter 2

OPTICAL FIBERS

The phenomenon of *total internal reflection*, responsible for guiding of light in optical fibers, has been known since 1854 [1]. Although glass fibers were made [2]–[4] in the 1920s, their use became practical only in the 1950s, when the use of a cladding layer led to considerable improvement in their guiding characteristics [5]–[7]. Before 1970, optical fibers were used mainly for medical imaging over short distances [8]. Their use for communication purposes was considered impractical because of high losses (~ 1000 dB/km). However, the situation changed drastically in 1970 when, following an earlier suggestion [9], the loss of optical fibers was reduced to about 20 dB/km [10]. Further progress resulted by 1979 in a loss of only 0.2 dB/km near the 1.55-μm spectral region [11]. The availability of low-loss fibers led to a revolution in the field of lightwave technology and started the era of optical fiber communications. Several books devoted entirely to optical fibers [12]–[21] cover numerous advances made in their design and understanding. This chapter focuses on the role of optical fibers as a communication channel in lightwave systems. In Section 2.1 we use geometrical-optics description to explain the guiding mechanism and introduce the related basic concepts. Maxwell's equations are used in Section 2.2 to describe wave propagation in optical fibers. The origin of fiber dispersion is discussed in Section 2.3, and Section 2.4 considers limitations on the bit rate and the transmission distance imposed by fiber dispersion. The loss mechanisms in optical fibers are discussed in Section 2.5, and Section 2.6 is devoted to a discussion of the nonlinear effects. The last section covers manufacturing details and includes a discussion of the design of fiber cables.

2.1 GEOMETRICAL-OPTICS DESCRIPTION

In its simplest form an optical fiber consists of a cylindrical core of silica glass surrounded by a cladding whose refractive index is lower than that of the core. Because of an abrupt index change at the core–cladding interface, such fibers are called *step-index fibers*. In a different type of fiber, known as *graded-index*

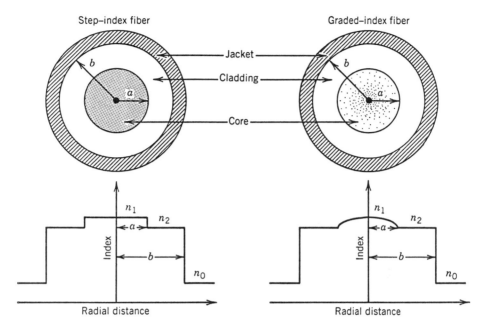

Figure 2.1 Cross section and refractive-index profile for step-index and graded-index fibers.

fiber, the refractive index decreases gradually inside the core. Figure 2.1 shows schematically the index profile and the cross section for the two kinds of fibers. Considerable insight in the guiding properties of optical fibers can be gained by using a ray picture based on geometrical optics [22]. The geometrical-optics description, although approximate, is valid when the core radius a is much larger than the light wavelength λ. When the two become comparable, it is necessary to use the wave-propagation theory of Section 2.2.

2.1.1   Step-Index Fibers

Consider the geometry of Fig. 2.2, where a ray making an angle θ_i with the fiber axis is incident at the core center. Because of refraction at the fiber-air interface, the ray bends toward the normal. The angle θ_r of the refracted ray is given by [22]

$$n_0 \sin \theta_i = n_1 \sin \theta_r, \qquad (2.1.1)$$

where n_1 and n_0 are the refractive indices of the fiber core and air, respectively. The refracted ray hits the core–cladding interface and is refracted again. However, refraction is possible only for an angle of incidence ϕ such that $\sin \phi < n_2/n_1$. For angles larger than a *critical angle* ϕ_c, defined by [22]

$$\sin \phi_c = n_2/n_1, \qquad (2.1.2)$$

where n_2 is the cladding index, the ray experiences total internal reflection at the core-cladding interface. Since such reflections occur throughout the fiber

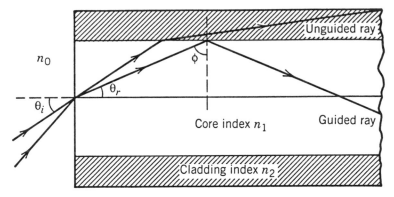

Figure 2.2 Light confinement through total internal reflection in step-index fibers. Rays for which $\phi < \phi_c$ are refracted out of the core.

length, all rays with $\phi > \phi_c$ remain confined to the fiber core. This is the basic mechanism behind light confinement in optical fibers.

One can use Eqs. (2.1.1) and (2.1.2) to find the maximum angle that the incident ray should make with the fiber axis to remain confined inside the core. Noting that $\theta_r = \pi/2 - \phi_c$ for such a ray and substituting it in Eq. (2.1.1), we obtain

$$n_0 \sin \theta_i = n_1 \cos \phi_c = (n_1^2 - n_2^2)^{1/2}. \tag{2.1.3}$$

In analogy with lenses, $n_0 \sin \theta_i$ is known as the *numerical aperture* (NA) of the fiber. It represents the light-gathering capacity of an optical fiber. For $n_1 \simeq n_2$ the NA can be approximated by

$$\text{NA} = n_1 (2\Delta)^{1/2}, \qquad \Delta = (n_1 - n_2)/n_1, \tag{2.1.4}$$

where Δ is the fractional index change at the core–cladding interface. Clearly, Δ should be made as large as possible in order to couple maximum light into the fiber. However, such fibers are not useful for the purpose of optical communications because of a phenomenon known as multipath dispersion, also called *intermodal dispersion* (the concept of fiber modes is introduced in Section 2.2).

Multipath dispersion can be understood by referring to Fig. 2.2, where different rays travel along paths of different lengths. As a result, these rays disperse in time at the output end of the fiber even if they were coincident at the input end and traveled at the same speed inside the fiber. A short pulse (called an *impulse*) would broaden considerably as a result of different path lengths. One can estimate the extent of pulse broadening simply by considering the shortest and longest ray paths. The shortest path occurs for $\theta_i = 0$ and is just equal to the fiber length L. The longest path occurs for θ_i given by Eq. (2.1.3) and has a length $L/\sin \phi_c$. By taking the velocity of propagation $v = c/n_1$, the time delay is given by

$$\Delta T = \frac{n_1}{c} \left(\frac{L}{\sin \phi_c} - L \right) = \frac{L\, n_1^2}{c\, n_2} \Delta. \tag{2.1.5}$$

The time delay between the two rays taking the shortest and longest paths is a measure of broadening experienced by an impulse launched at the fiber input.

We can relate ΔT to the information-carrying capacity of the fiber measured through the bit rate B. Although a precise relation between B and ΔT depends on many details, such as the pulse shape, it is clear intuitively that ΔT should be less than the allocated bit slot ($T_B = 1/B$). Thus, an order-of-magnitude estimate of the bit rate is obtained from the condition $B\Delta T < 1$. By using Eq. (2.1.5) we obtain

$$BL < \frac{n_2}{n_1^2} \frac{c}{\Delta}.\tag{2.1.6}$$

This condition provides a rough estimate of a fundamental limitation of step-index fibers. As an illustration, consider an unclad glass fiber with $n_1 = 1.5$ and $n_2 = 1$. The bit rate–distance product of such a fiber is limited to quite small values since $BL < 0.4$ (Mb/s)-km. Considerable improvement occurs for cladded fibers with a small index step. Most fibers for communication applications are designed with $\Delta < 0.01$. As an example, $BL < 100$ (Mb/s)-km for $\Delta = 2 \times 10^{-3}$. Such fibers can communicate data at a bit rate of 10 Mb/s over distances up to 10 km and may be suitable for some local-area networks.

Two remarks are in order concerning the validity of Eq. (2.1.6). First, it is obtained by considering only rays that pass through the fiber axis after each total internal reflection. Such rays are called *meridional rays*. In general, the fiber also supports *skew rays*, which travel at angles oblique to the fiber axis. Skew rays scatter out of the core at bends and irregularities and are not expected to contribute significantly to Eq. (2.1.6). Second, even the oblique meridional rays suffer higher losses than paraxial meridional rays because of scattering. Equation (2.1.6) provides a conservative estimate since all rays are treated equally. The effect of intermodal dispersion can be considerably reduced by using graded-index fibers, which are discussed in the next subsection. It can be eliminated entirely by using the single-mode fibers discussed in Section 2.2.

2.1.2   Graded-Index Fibers

The refractive index of the core in graded-index fibers is not constant but decreases gradually from its maximum value n_1 at the core center to its minimum value n_2 at the core–cladding interface. Most graded-index fibers are designed to have a nearly quadratic decrease and are analyzed by using α-profile, given by

$$n(\rho) = \begin{cases} n_1[1 - \Delta(\rho/a)^\alpha] & : \quad \rho < a, \\ n_1(1 - \Delta) = n_2 & : \quad \rho \geq a, \end{cases}\tag{2.1.7}$$

where a is the core radius. The parameter α determines the index profile. A step-index profile is approached in the limit of large α. A *parabolic-index fiber* corresponds to $\alpha = 2$.

It is easy to understand qualitatively why intermodal or multipath dispersion is reduced for graded-index fibers. Figure 2.3 shows schematically paths for three different rays. Similar to the case of step-index fibers, the path is longer for

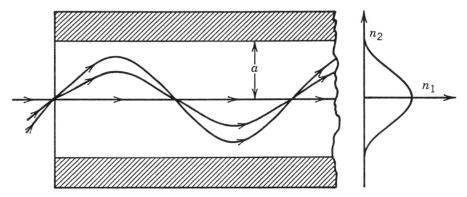

Figure 2.3 Ray trajectories in a graded-index fiber.

more oblique rays. However, the ray velocity changes along the path because of variations in the refractive index. More specifically, the ray propagating along the fiber axis takes the shortest path but travels most slowly as the index is largest along this path. Oblique rays have a large part of their path in a medium of lower refractive index, where they travel faster. It is therefore possible for all rays to arrive together at the fiber output by a suitable choice of the refractive-index profile.

Geometrical optics can be used to show that a parabolic-index profile leads to nondispersive pulse propagation within the *paraxial approximation*. The trajectory of a paraxial ray is obtained by solving [22]

$$\frac{d^2\rho}{dz^2} = \frac{1}{n}\frac{dn}{d\rho}, \tag{2.1.8}$$

where ρ is the radial distance of the ray from the axis. By using Eq. (2.1.7) for $\rho < a$ with $\alpha = 2$, Eq. (2.1.8) reduces to an equation of harmonic oscillator and has the general solution

$$\rho = \rho_0 \cos(pz) + (\rho'_0/p) \sin(pz), \tag{2.1.9}$$

where $p = (2\Delta/a^2)^{1/2}$ and ρ_0 and ρ'_0 are the position and the direction of the input ray, respectively. Equation (2.1.9) shows that all rays recover their initial positions and directions at distances $z = 2m\pi/p$, where m is an integer (see Fig. 2.3). Such a complete restoration of the input implies that a parabolic-index fiber does not exhibit intermodal dispersion.

The conclusion above holds only within the paraxial and the geometrical-optics approximations, both of which must be relaxed for practical fibers. Intermodal dispersion in graded-index fibers has been studied extensively by using wave-propagation techniques [13], [15]. The quantity $\Delta T/L$, where ΔT is the maximum multipath delay in a fiber of length L, is found to vary considerably with α. Figure 2.4 shows this variation for $n_1 = 1.5$ and $\Delta = 0.01$. The minimum dispersion occurs for $\alpha = 2(1 - \Delta)$ and depends on Δ as [23]

$$\Delta T/L = n_1 \Delta^2/8c. \tag{2.1.10}$$

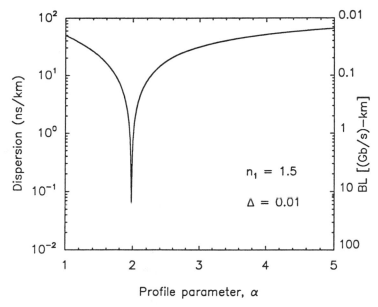

Figure 2.4 Variation of intermodal dispersion $\Delta T/L$ with the profile parameter α for a graded-index fiber with $n_1 = 1.5$ and $\Delta = 0.01$. The right scale shows the corresponding bit rate–distance product.

The limiting bit-rate–distance product is obtained by using the criterion $\Delta T < 1/B$ and is given by

$$BL < 8c/n_1\Delta^2. \qquad (2.1.11)$$

The right scale in Fig. 2.4 shows the BL product as a function of α. Graded-index fibers with a suitably optimized index profile can communicate data at a bit rate of 100 Mb/s over distances up to 100 km. The BL product of such fibers is improved by nearly three orders of magnitude over that of step-index fibers. Indeed, the first generation of lightwave systems used graded-index fibers. Further improvement is possible only by using single-mode fibers whose core radius is comparable to the light wavelength. Geometrical optics cannot be used to for such fibers.

Although graded-index fibers are rarely used for long-haul links, the use of graded-index *plastic* optical fibers for data-link applications has attracted considerable attention during the 1990s [24]–[26]. Such fibers have a relatively large core (up to 1-mm diameter), resulting in a high numerical aperture and high coupling efficiency. They also exhibit a high loss, which typically exceeds 50 dB/km. The BL product of plastic fibers, however, exceeds 2 (Gb/s)-km because of a graded-index profile [24]. As a result, they can be used to transmit data at bit rates > 1 GB/s over short distances of 1 km or less. In a 1996 demonstration, a 10-Gb/s signal was transmitted over 0.5 km with a bit-error rate of less than 10^{-11} [26]. Graded-index plastic optical fibers provide an ideal solution for transferring data among computers in a local-area network.

2.2 WAVE PROPAGATION

In this section we consider propagation of light in step-index fibers by using Maxwell's equations for electromagnetic waves. These equations are introduced in Section 2.2.1. The concept of fiber modes is discussed in Section 2.2.2, where the fiber is shown to support a finite number of guided modes. Section 2.2.3 focuses on how a step-index fiber can be designed to support only a single mode and discusses the properties of single-mode fibers.

2.2.1 Maxwell's Equations

Like all electromagnetic phenomena, propagation of optical fields in fibers is governed by *Maxwell's equations*. For a nonconducting medium without free charges, these equations take the form [27] (in SI units; see Appendix A)

$$\nabla \times \mathbf{E} = -\partial \mathbf{B}/\partial t, \tag{2.2.1}$$

$$\nabla \times \mathbf{H} = \partial \mathbf{D}/\partial t, \tag{2.2.2}$$

$$\nabla \cdot \mathbf{D} = 0, \tag{2.2.3}$$

$$\nabla \cdot \mathbf{B} = 0, \tag{2.2.4}$$

where $\mathbf{E}$ and $\mathbf{H}$ are the electric and magnetic field vectors, respectively, and $\mathbf{D}$ and $\mathbf{B}$ are the corresponding flux densities. The flux densities are related to the field vectors by the constitutive relations [27]

$$\mathbf{D} = \epsilon_0 \mathbf{E} + \mathbf{P}, \tag{2.2.5}$$

$$\mathbf{B} = \mu_0 \mathbf{H} + \mathbf{M}, \tag{2.2.6}$$

where ϵ_0 is the vacuum permittivity, μ_0 is the vacuum permeability, and $\mathbf{P}$ and $\mathbf{M}$ are the induced electric and magnetic polarizations, respectively. For optical fibers $\mathbf{M} = 0$ because of the nonmagnetic nature of silica glass.

Evaluation of the electric polarization $\mathbf{P}$ requires a microscopic quantum-mechanical approach. Although such an approach is essential when the optical frequency is near a medium resonance, a phenomenological relation between $\mathbf{P}$ and $\mathbf{E}$ can be used far from medium resonances. This is the case for optical fibers in the wavelength region 0.5–2 μm, a range that covers the low-loss region of optical fibers that is of interest for fiber-optic communication systems. In general, the relation between $\mathbf{P}$ and $\mathbf{E}$ can be nonlinear. Although the nonlinear effects in optical fibers are of considerable interest [28] and are covered in Section 2.6, they can be ignored in a discussion of fiber modes. $\mathbf{P}$ is then related to $\mathbf{E}$ by the relation

$$\mathbf{P}(\mathbf{r}, t) = \epsilon_0 \int_{-\infty}^{\infty} \chi(\mathbf{r}, t - t') \mathbf{E}(\mathbf{r}, t') \, dt'. \tag{2.2.7}$$

Linear susceptibility χ is, in general, a second-rank tensor but reduces to a scalar for an isotropic medium such as silica glass. Optical fibers become slightly birefringent because of unintentional variations in the core shape or in local strain; such birefringent effects are considered in Section 2.2.3. Equation (2.2.7) assumes a spatially local response. However, it includes the delayed nature of the temporal response, a feature that has important implications for optical fiber communications through chromatic dispersion.

Equations (2.2.1)–(2.2.7) provide a general formalism for studying wave propagation in optical fibers. In practice, it is convenient to use a single field variable $\mathbf{E}$. By taking the curl of Eq. (2.2.1) and using Eqs. (2.2.2), (2.2.5), and (2.2.6), we obtain the wave equation

$$\nabla \times \nabla \times \mathbf{E} = -\frac{1}{c^2}\frac{\partial^2 \mathbf{E}}{\partial t^2} - \mu_0 \frac{\partial^2 \mathbf{P}}{\partial t^2}, \tag{2.2.8}$$

where the speed of light in vacuum is defined as usual by $c = (\mu_0\epsilon_0)^{-1/2}$. By introducing the Fourier transform of $\mathbf{E}(\mathbf{r},t)$ through the relation

$$\tilde{\mathbf{E}}(\mathbf{r},\omega) = \int_{-\infty}^{\infty} \mathbf{E}(\mathbf{r},t)\exp(i\omega t)\,dt, \tag{2.2.9}$$

as well as a similar relation for $\mathbf{P}(\mathbf{r},t)$, and by using Eq. (2.2.7), Eq. (2.2.8) can be written in the frequency domain as

$$\nabla \times \nabla \times \tilde{\mathbf{E}} = -\epsilon(\mathbf{r},\omega)(\omega^2/c^2)\tilde{\mathbf{E}}, \tag{2.2.10}$$

where the frequency-dependent *dielectric constant* is defined as

$$\epsilon(\mathbf{r},\omega) = 1 + \tilde{\chi}(\mathbf{r},\omega), \tag{2.2.11}$$

and $\tilde{\chi}(\mathbf{r},\omega)$ is the Fourier transform of $\chi(\mathbf{r},t)$. In general, $\epsilon(\mathbf{r},\omega)$ is complex. Its real and imaginary parts are related to the *refractive index n* and the *absorption coefficient* α by the definition

$$\epsilon = (n + i\alpha c/2\omega)^2. \tag{2.2.12}$$

By using Eqs. (2.2.11) and (2.2.12), n and α are related to $\tilde{\chi}$ as

$$\begin{aligned} n &= (1 + \mathrm{Re}\,\tilde{\chi})^{1/2}, &\tag{2.2.13}\\ \alpha &= (\omega/nc)\,\mathrm{Im}\,\tilde{\chi}, &\tag{2.2.14} \end{aligned}$$

where Re and Im stand for the real and imaginary parts, respectively. Both n and α are frequency dependent. The frequency dependence of n is referred to as *chromatic dispersion* or simply as *material dispersion*. In Section 2.3, fiber dispersion is shown to limit the performance of fiber-optic communication systems in a fundamental way.

Two further simplifications can be made before solving Eq. (2.2.10). First, ϵ can be taken to be real and replaced by n^2 because of low optical losses in silica fibers. Second, since $n(\mathbf{r},\omega)$ is independent of the spatial coordinate $\mathbf{r}$ in both the core and the cladding of a step-index fiber, one can use the identity

$$\nabla \times \nabla \times \tilde{\mathbf{E}} = \nabla(\nabla \cdot \tilde{\mathbf{E}}) - \nabla^2\tilde{\mathbf{E}} = -\nabla^2\tilde{\mathbf{E}}, \tag{2.2.15}$$

where we used Eq. (2.2.3) and the relation $\tilde{\mathbf{D}} = \epsilon\tilde{\mathbf{E}}$ to set $\nabla \cdot \tilde{\mathbf{E}} = 0$. This simplification is made even for graded-index fibers. Equation (2.2.15) then holds approximately as long as the index changes occur over a length scale much longer than the wavelength. By using Eq. (2.2.15) in Eq. (2.2.10), we obtain

$$\nabla^2\tilde{\mathbf{E}} + n^2(\omega)k_0^2\tilde{\mathbf{E}} = 0, \qquad (2.2.16)$$

where the free-space wave number k_0 is defined as

$$k_0 = \omega/c = 2\pi/\lambda, \qquad (2.2.17)$$

and λ is the vacuum wavelength of the optical field oscillating at the frequency ω. Equation (2.2.16) is solved next to obtain the optical modes of step-index fibers.

2.2.2 Fiber Modes

The concept of the mode is a general concept in optics occurring also, for example, in the theory of lasers. An *optical mode* refers to a specific solution of the wave equation (2.2.16) that satisfies the appropriate boundary conditions and has the property that its spatial distribution does not change with propagation. The fiber modes can be classified as guided modes, leaky modes, and radiation modes [14], [19]. As one might expect, signal transmission in fiber-optic communication systems takes place through the guided modes only. The following discussion focuses exclusively on the guided modes of a step-index fiber.

To take advantage of the cylindrical symmetry, Eq. (2.2.16) is written in the cylindrical coordinates ρ, ϕ, and z as

$$\frac{\partial^2 E_z}{\partial\rho^2} + \frac{1}{\rho}\frac{\partial E_z}{\partial\rho} + \frac{1}{\rho^2}\frac{\partial^2 E_z}{\partial\phi^2} + \frac{\partial^2 E_z}{\partial z^2} + n^2k_0^2 E_z = 0, \qquad (2.2.18)$$

where for a step-index fiber of core radius a, the refractive index n is of the form

$$n = \begin{cases} n_1 & : \quad \rho \le a, \\ n_2 & : \quad \rho > a. \end{cases} \qquad (2.2.19)$$

For simplicity of notation, the tilde over $\tilde{\mathbf{E}}$ has been dropped and the frequency dependence of all variables is implicitly understood. Equation (2.2.18) is written for the axial component E_z of the electric field vector. Similar equations can be written for the other five components of $\mathbf{E}$ and $\mathbf{H}$. However, it is not necessary to solve all six equations since only two components out of six are independent. It is customary to choose E_z and H_z as the independent components and obtain E_ρ, E_ϕ, H_ρ, and H_ϕ in terms of them. Equation (2.2.18) is easily solved by using the method of separation of variables and writing E_z as

$$E_z(\rho,\phi,z) = F(\rho)\Phi(\phi)Z(z). \qquad (2.2.20)$$

By using Eq. (2.2.20) in Eq. (2.2.18), we obtain the three ordinary differential equations:

$$d^2 Z/dz^2 + \beta^2 Z = 0, \qquad (2.2.21)$$

$$d^2 \Phi/d\phi^2 + m^2 \Phi = 0, \qquad (2.2.22)$$

$$\frac{d^2 F}{d\rho^2} + \frac{1}{\rho}\frac{dF}{d\rho} + \left(n^2 k_0^2 - \beta^2 - \frac{m^2}{\rho^2}\right) F = 0. \qquad (2.2.23)$$

Equation (2.2.21) has a solution of the form $Z = \exp(i\beta z)$, where β has the physical significance of the propagation constant. Similarly, Eq. (2.2.22) has a solution $\Phi = \exp(im\phi)$, but the constant m is restricted to take only integer values since the field must be periodic in ϕ with a period of 2π.

Equation (2.2.23) is the well-known differential equation for the Bessel functions [29]. Its general solution in the core and cladding regions is given by

$$F(\rho) = \begin{cases} AJ_m(\kappa\rho) + A'Y_m(\kappa\rho) & : \quad \rho \leq a, \\ CK_m(\gamma\rho) + C'I_m(\gamma\rho) & : \quad \rho > a, \end{cases} \qquad (2.2.24)$$

where A, A', C, and C' are constants and J_m, Y_m, K_m, and I_m are different kinds of Bessel functions [29]. The parameters κ and γ are defined by

$$\kappa^2 = n_1^2 k_0^2 - \beta^2, \qquad (2.2.25)$$

$$\gamma^2 = \beta^2 - n_2^2 k_0^2. \qquad (2.2.26)$$

Considerable simplification occurs when we use the boundary condition that the optical field for a guided mode should be finite at $\rho = 0$ and decay to zero at $\rho = \infty$. Since $Y_m(\kappa\rho)$ has a singularity at $\rho = 0$, $F(0)$ can remain finite only if $A' = 0$. Similarly $F(\rho)$ vanishes at infinity only if $C' = 0$. The general solution of Eq. (2.2.18) is thus of the form

$$E_z = \begin{cases} AJ_m(\kappa\rho)\exp(im\phi)\exp(i\beta z) & : \quad \rho \leq a, \\ CK_m(\gamma\rho)\exp(im\phi)\exp(i\beta z) & : \quad \rho > a. \end{cases} \qquad (2.2.27)$$

The same method can be used to obtain H_z which also satisfies Eq. (2.2.18). Indeed, the solution is the same but with different constants B and D, that is,

$$H_z = \begin{cases} BJ_m(\kappa\rho)\exp(im\phi)\exp(i\beta z) & : \quad \rho \leq a, \\ DK_m(\gamma\rho)\exp(im\phi)\exp(i\beta z) & : \quad \rho > a. \end{cases} \qquad (2.2.28)$$

The other four components E_ρ, E_ϕ, H_ρ, and H_ϕ can be expressed in terms of E_z and H_z by using Maxwell's equations. In the core region, we obtain

$$E_\rho = \frac{i}{\kappa^2}\left(\beta\frac{\partial E_z}{\partial\rho} + \mu_0\frac{\omega}{\rho}\frac{\partial H_z}{\partial\phi}\right), \qquad (2.2.29)$$

$$E_\phi = \frac{i}{\kappa^2}\left(\frac{\beta}{\rho}\frac{\partial E_z}{\partial\phi} - \mu_0\omega\frac{\partial H_z}{\partial\rho}\right), \qquad (2.2.30)$$

$$H_\rho = \frac{i}{\kappa^2}\left(\beta\frac{\partial H_z}{\partial\rho} - \epsilon_0 n^2\frac{\omega}{\rho}\frac{\partial E_z}{\partial\phi}\right), \qquad (2.2.31)$$

$$H_\phi = \frac{i}{\kappa^2}\left(\frac{\beta}{\rho}\frac{\partial H_z}{\partial\phi} + \epsilon_0 n^2\omega\frac{\partial E_z}{\partial\rho}\right). \qquad (2.2.32)$$

These equations can be used in the cladding region after replacing κ^2 by $-\gamma^2$.

Equations (2.2.27)–(2.2.32) express the electromagnetic field in the core and cladding regions of an optical fiber in terms of four constants A, B, C, and D. These constants are determined by applying the boundary condition that the tangential components of $\mathbf{E}$ and $\mathbf{H}$ be continuous across the core–cladding interface. By requiring the continuity of E_z, H_z, E_ϕ, and H_ϕ at $\rho = a$, we obtain a set of four homogeneous equations [19] satisfied by A, B, C, and D. These equations have a nontrivial solution only if the determinant of the coefficient matrix vanishes. After considerable algebraic details [19]–[21], this condition leads us to the following eigenvalue equation:

$$\left[\frac{J'_m(\kappa a)}{\kappa J_m(\kappa a)} + \frac{K'_m(\gamma a)}{\gamma K_m(\gamma a)}\right]\left[\frac{J'_m(\kappa a)}{\kappa J_m(\kappa a)} + \frac{n_2^2}{n_1^2}\frac{K'_m(\gamma a)}{\gamma K_m(\gamma a)}\right] = \left[\frac{2m\beta(n_1 - n_2)}{a\kappa^2\gamma^2}\right]^2,$$
(2.2.33)

where a prime indicates differentiation with respect to the argument.

For a given set of the parameters k_0, a, n_1, and n_2, the eigenvalue equation (2.2.33) can be solved numerically to determine the propagation constant β. In general, it may have multiple solutions for each integer value of m. It is customary to enumerate these solutions in descending numerical order and denote them by β_{mn} for a given m ($n = 1, 2, \ldots$). Each value β_{mn} corresponds to one possible mode of propagation of the optical field whose spatial distribution is obtained from Eqs. (2.2.27)–(2.2.32). Since the field distribution does not change with propagation except for a phase factor and satisfies all boundary conditions, it is an optical mode of the fiber. In general, both E_z and H_z are nonzero (except for $m = 0$), in contrast with the planar waveguides, for which one of them can be taken to be zero. Fiber modes are therefore referred to as *hybrid modes* and are denoted by HE_{mn} or EH_{mn}, depending on whether H_z or E_z dominates. In the special case $m = 0$, HE_{0n} and EH_{0n} are also denoted by TE_{0n} and TM_{0n}, respectively, since they correspond to transverse-electric ($E_z = 0$) and transverse-magnetic ($H_z = 0$) modes of propagation. A different notation LP_{mn} is also used for weakly guiding fibers [30] for which both E_z and H_z are nearly zero (LP stands for linearly polarized modes).

A mode is uniquely determined by its propagation constant β. It is useful to introduce a quantity $\bar{n} = \beta/k_0$, called the *mode index* or *effective index* and having the physical significance that each fiber mode propagates with an effective refractive index $\bar{n}$ whose value lies in the range $n_1 > \bar{n} > n_2$. A mode ceases to be guided when $\bar{n} \le n_2$. This can be understood by noting that the optical field of guided modes decays exponentially inside the cladding layer since [29]

$$K_m(\gamma\rho) = (\pi/2\gamma\rho)^{1/2}\exp(-\gamma\rho) \quad \text{for} \quad \gamma\rho \gg 1. \tag{2.2.34}$$

When $\bar{n} \le n_2$, $\gamma^2 \le 0$ from Eq. (2.2.26) and the exponential decay does not occur. The mode is said to reach cutoff when γ becomes zero or when $\bar{n} = n_2$. From Eq. (2.2.25), $\kappa = k_0(n_1^2 - n_2^2)^{1/2}$ when $\gamma = 0$. A parameter that plays an important role in determining the *cutoff condition* is defined as

$$V = k_0 a(n_1^2 - n_2^2)^{1/2} \approx (2\pi/\lambda)an_1\sqrt{2\Delta}. \tag{2.2.35}$$

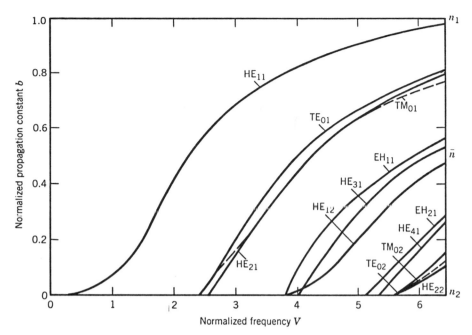

Figure 2.5 Normalized propagation constant b as a function of normalized frequency V for a few low-order fiber modes. The right scale shows the mode index $\bar{n}$. (After Ref. [31]. ©1981 Academic Press. Reprinted with permission.)

It is called the *normalized frequency* ($V \propto \omega$) or simply the V parameter. It is also useful to introduce a *normalized propagation constant b* defined by

$$b = \frac{\beta/k_0 - n_2}{n_1 - n_2} = \frac{\bar{n} - n_2}{n_1 - n_2}. \qquad (2.2.36)$$

Figure 2.5 shows a plot of b as a function of V for a few low-order fiber modes obtained by solving the eigenvalue equation (2.2.33). A fiber with a large value of V supports many modes. A rough estimate of the number of modes for such a multimode fiber is given by $V^2/2$ [23]. For example, a typical multimode fiber with $a = 25$ μm and $\Delta = 5 \times 10^{-3}$ has $V \simeq 18$ at $\lambda = 1.3$ μm and would support about 162 modes. However, the number of modes decreases rapidly as V is reduced. As seen in Fig. 2.5, a fiber with $V = 5$ supports seven modes. Below a certain value of V all modes except the HE_{11} mode reach cutoff. Such fibers support a single mode and are called single-mode fibers. The properties of single-mode fibers are described next.

2.2.3 Single-Mode Fibers

Single-mode fibers support only the HE_{11} mode, also known as the fundamental mode of the fiber. The fiber is designed such that all higher-order modes are cut off at the operating wavelength. As seen in Fig. 2.5, the V parameter determines the number of modes supported by a fiber. The cutoff condition of various modes is also determined by V. The fundamental mode has no cutoff and is always supported by a fiber.

Single-Mode Condition

The *single-mode condition* is determined by the value of V at which the TE_{01} and TM_{01} modes reach cutoff (see Fig. 2.5). The eigenvalue equations for these two modes can be obtained by setting $m = 0$ in Eq. (2.2.36) and are given by

$$\kappa J_0(\kappa a) K_0'(\gamma a) + \gamma J_0'(\gamma a) K_0(\gamma a) = 0, \qquad (2.2.37)$$
$$\kappa n_2^2 J_0(\kappa a) K_0'(\gamma a) + \gamma n_1^2 J_0'(\gamma a) K_0(\gamma a) = 0. \qquad (2.2.38)$$

A mode reaches cutoff when $\gamma = 0$. Since $\kappa a = V$ when $\gamma = 0$, the cutoff condition for both modes is simply given by

$$J_0(V) = 0. \qquad (2.2.39)$$

The smallest value of V for which $J_0(V) = 0$ is 2.405. A fiber designed such that $V < 2.405$ supports only the fundamental HE_{11} mode. This is the single-mode condition.

   We can use Eq. (2.2.35) to estimate the core radius of single-mode fibers used in lightwave systems. For the operating wavelength range 1.3–1.6 μm, the fiber is generally designed to become single mode for $\lambda > 1.2$ μm. By taking $\lambda = 1.2$ μm, $n_1 = 1.45$, and $\Delta = 5 \times 10^{-3}$, Eq. (2.2.35) shows that $V < 2.405$ for a core radius $a < 3.2$ μm. The required core radius can be increased to about 4 μm by decreasing Δ to 3×10^{-3}. Indeed, most telecommunication fibers are designed with $a \approx 4$ μm.

Mode Index

The mode index $\bar{n}$ at the operating wavelength can be obtained by using Eq. (2.2.36), according to which

$$\bar{n} = n_2 + b(n_1 - n_2) \approx n_2(1 + b\Delta) \qquad (2.2.40)$$

and by using Fig. 2.5, which provides b as a function of V for the HE_{11} mode. An analytic approximation for b is [15]

$$b(V) \approx (1.1428 - 0.9960/V)^2 \qquad (2.2.41)$$

and is accurate to within 0.2% for V in the range 1.5–2.5.

Field Distribution

The field distribution of the fundamental mode is obtained by using Eqs. (2.2.27)–(2.2.32). The axial components E_z and H_z are quite small for $\Delta \ll 1$. Hence, the HE_{11} mode is approximately linearly polarized for weakly guiding fibers. It is also denoted as LP_{01}, following an alternative terminology in which all fibers

modes are assumed to be linearly polarized [30]. One of the transverse components can be taken as zero for a linearly polarized mode. If we set $E_y = 0$, the
E_x component of the electric field for the HE_{11} mode is given by [15]

$$E_x = E_0 \begin{cases} [J_0(\kappa\rho)/J_0(\kappa a)]\exp(i\beta z) & : \quad \rho \leq a, \\ [K_0(\gamma\rho)/K_0(\gamma a)]\exp(i\beta z) & : \quad \rho > a, \end{cases} \qquad (2.2.42)$$

where E_0 is a constant related to the power carried by the mode. The dominant
component of the corresponding magnetic field is given by $H_y = n_2(\epsilon_0/\mu_0)^{1/2}E_x$.
This mode is linearly polarized along the x axis. The fiber also supports another
mode linearly polarized along the y axis. In this sense a single-mode fiber actually supports two orthogonally polarized modes that are degenerate, as they
have the same mode index $\bar{n}$.

Birefringence

The degenerate nature of the orthogonally polarized modes holds only for an
ideal single-mode fiber with a perfectly cylindrical core of uniform diameter.
Real fibers exhibit considerable variation in the shape of their core along the
fiber length. They may also experience nonuniform stress such that the cylindrical symmetry of the fiber is broken. Degeneracy between the orthogonally
polarized fiber modes is removed because of these factors, and the fiber acquires
birefringence. The *degree of birefringence* is defined by

$$B = |\bar{n}_x - \bar{n}_y|, \qquad (2.2.43)$$

where $\bar{n}_x$ and $\bar{n}_y$ are the mode indices for the orthogonally polarized fiber modes.
Birefringence leads to a periodic power exchange between the two polarization
components. The period, referred to as the *beat length*, is given by

$$L_B = \lambda/B. \qquad (2.2.44)$$

Typically, $B \sim 10^{-7}$, and $L_B \sim 10$ m for $\lambda \sim 1$ μm. From a physical viewpoint,
linearly polarized light remains linearly polarized only when it is polarized along
one of the principal axes. Otherwise, its state of polarization changes along the
fiber length from linear to elliptical, and then back to linear, in a periodic
manner over the length L_B. Figure 2.6 shows schematically such a periodic
change in the state of polarization for a fiber of constant birefringence B. The
fast axis in this figure corresponds to the axis along which the mode index is
smaller. The other axis is called the *slow axis*.

In conventional single-mode fibers, B is not constant along the fiber but
changes randomly because of fluctuations in the core shape and the nonuniform
stress acting on the core. As a result, light launched into the fiber with linear
polarization quickly reaches a state of arbitrary polarization. The unknown polarization state is generally not a problem for optical communication systems
whose receivers detect the total intensity directly. However, it becomes of concern for coherent communication systems (see Section 6.5), for which it becomes

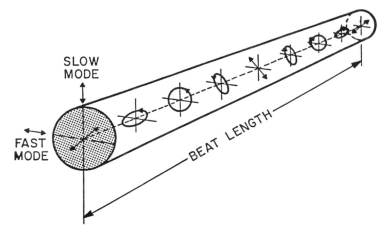

Figure 2.6 State of polarization in a birefringent fiber over one beat length. Input beam is linearly polarized at 45° with respect to the slow and fast axes.

necessary to employ polarization-insensitive coherent receivers. It is possible to make fibers for which random fluctuations in the core shape and size are not the governing factor in determining the state of polarization. Such fibers are called polarization-preserving fibers. A large amount of birefringence is introduced intentionally in these fibers through design modifications so that small random birefringence fluctuations do not affect the light polarization significantly. Typically, $B \sim 10^{-4}$ for such fibers.

Spot Size

Since the field distribution given by Eq. (2.2.42) is cumbersome to use in practice, it is often approximated by a *Gaussian distribution* of the form

$$E_x = A \exp(-\rho^2/w^2) \exp(i\beta z), \qquad (2.2.45)$$

where w is the *field radius* and is referred to as the *spot size*. It is determined by fitting the exact distribution to the Gaussian function or by following a variational procedure [32]. Figure 2.7 shows the dependence of w/a on the V parameter. A comparison of the actual field distribution with the fitted Gaussian is also shown for $V = 2.4$. The quality of fit is generally quite good for values of V in the neighborhood of 2. The spot size w can be determined from Fig. 2.7. It can also be determined from an analytic approximation accurate to within 1% for $1.2 < V < 2.4$ and given by [32]

$$w/a \approx 0.65 + 1.619V^{-3/2} + 2.879V^{-6}. \qquad (2.2.46)$$

The fraction of the power contained in the core can be obtained by using Eq. (2.2.45) and is given by the *confinement factor*

$$\Gamma = \frac{P_{\text{core}}}{P_{\text{total}}} = \frac{\int_0^a |E_x|^2 \rho \, d\rho}{\int_0^\infty |E_x|^2 \rho \, d\rho} = 1 - \exp\left(-\frac{2a^2}{w^2}\right). \qquad (2.2.47)$$

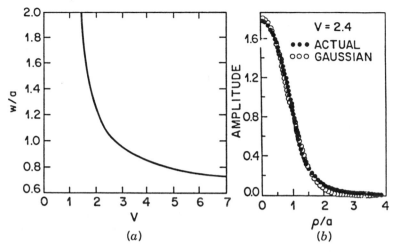

Figure 2.7 (a) Normalized spot size w/a as a function of the V parameter obtained by fitting the fundamental fiber mode to a Gaussian distribution; (b) quality of fit for $V = 2.4$. (After Ref. [32]. ©1978 OSA. Reprinted with permission.)

Equations (2.2.46) and (2.2.47) determine the fraction of the mode power contained inside the core for a given value of V. Although nearly 75% of the mode power resides in the core for $V = 2$, this percentage drops down to 20% for $V = 1$. For this reason most telecommunication single-mode fibers are designed to operate in the range $2 < V < 2.4$.

2.3 DISPERSION IN SINGLE-MODE FIBERS

It was seen in Section 2.1 that intermodal dispersion in multimode fibers leads to considerable broadening of short optical pulses (~ 10 ns/km). In the geometrical-optics description, such broadening was attributed to different paths followed by different rays. In the modal description it is related to the different mode indices (or group velocities) associated with different modes. The main advantage of single-mode fibers is that intermodal dispersion is absent simply because the energy of the injected pulse is transported by a single mode. However, pulse broadening does not disappear altogether. The group velocity associated with the fundamental mode is frequency dependent because of chromatic dispersion. As a result, different spectral components of the pulse travel at slightly different group velocities, a phenomenon referred to as *group-velocity dispersion* (GVD), *intramodal dispersion*, or simply *fiber dispersion*. Intramodal dispersion has two contributions, material dispersion and waveguide dispersion. We consider both of them and discuss how GVD limits the performance of lightwave systems employing single-mode fibers.

2.3.1 Group-Velocity Dispersion

Consider a single-mode fiber of length L. A specific spectral component at the frequency ω would arrive at the output end of the fiber after a time delay $T = L/v_g$, where v_g is the *group velocity*, defined as [22]

$$v_g = (d\beta/d\omega)^{-1}. \tag{2.3.1}$$

By using $\beta = \bar{n}k_0 = \bar{n}\omega/c$ in Eq. (2.3.1), one can show that $v_g = c/\bar{n}_g$, where $\bar{n}_g$ is the *group index* given by

$$\bar{n}_g = \bar{n} + \omega(d\bar{n}/d\omega). \tag{2.3.2}$$

The frequency dependence of the group velocity leads to pulse broadening simply because different spectral components of the pulse disperse during propagation and do not arrive simultaneously at the fiber output. If $\Delta\omega$ is the spectral width of the pulse, the extent of pulse broadening for a fiber of length L is governed by

$$\Delta T = \frac{dT}{d\omega}\Delta\omega = \frac{d}{d\omega}\left(\frac{L}{v_g}\right)\Delta\omega = L\frac{d^2\beta}{d\omega^2}\Delta\omega = L\beta_2\Delta\omega, \tag{2.3.3}$$

where Eq. (2.3.1) was used. The parameter $\beta_2 = d^2\beta/d\omega^2$ is known as the GVD parameter. It determines how much an optical pulse would broaden on propagation inside the fiber.

In some optical communication systems, the frequency spread $\Delta\omega$ is determined by the range of wavelengths $\Delta\lambda$ emitted by the optical source. It is customary to use $\Delta\lambda$ in place of $\Delta\omega$. By using $\omega = 2\pi c/\lambda$ and $\Delta\omega = (-2\pi c/\lambda^2)\Delta\lambda$, Eq. (2.3.3) can be written as

$$\Delta T = \frac{d}{d\lambda}\left(\frac{L}{v_g}\right)\Delta\omega = DL\Delta\lambda, \tag{2.3.4}$$

where

$$D = \frac{d}{d\lambda}\left(\frac{1}{v_g}\right) = -\frac{2\pi c}{\lambda^2}\beta_2. \tag{2.3.5}$$

D is called the *dispersion parameter* and is expressed in units of ps/(km-nm).

The effect of dispersion on the bit rate B can be estimated by using the criterion $B\Delta T < 1$ in a manner similar to that used in Section 2.1. By using ΔT from Eq. (2.3.4) this condition becomes

$$BL|D|\Delta\lambda < 1. \tag{2.3.6}$$

Equation (2.3.6) provides an order-of-magnitude estimate of the BL product offered by single-mode fibers. The wavelength dependence of D is studied in the next two subsections. For standard silica fibers, D is relatively small in the wavelength region near 1.3 μm [$D \sim 1$ ps/(km-nm)]. For a semiconductor laser, the spectral width $\Delta\lambda$ is 2–4 nm even when the laser operates in several longitudinal modes. The BL product of such lightwave systems can exceed

100 (Gb/s)-km. Indeed, 1.3-μm telecommunication systems typically operate at a bit rate of 2 Gb/s with a repeater spacing of 40–50 km. The BL product of single-mode fibers can exceed 1 (Tb/s)-km when single-mode semiconductor lasers (see Section 3.3) are used to reduce $\Delta\lambda$ below 1 nm.

The dispersion parameter D can vary considerably when the operating wavelength is shifted from 1.3 μm. The wavelength dependence of D is governed by the frequency dependence of the mode index $\bar{n}$. From Eq. (2.3.5), D can be written as

$$D = -\frac{2\pi c}{\lambda^2}\frac{d}{d\omega}\left(\frac{1}{v_g}\right) = -\frac{2\pi}{\lambda^2}\left(2\frac{d\bar{n}}{d\omega} + \omega\frac{d^2\bar{n}}{d\omega^2}\right), \tag{2.3.7}$$

where Eq. (2.3.2) was used. If we substitute $\bar{n}$ from Eq. (2.2.43) and use Eq. (2.2.35), D can be written as the sum of two terms,

$$D = D_M + D_W, \tag{2.3.8}$$

where the *material dispersion* D_M and the *waveguide dispersion* D_W are given by

$$D_M = -\frac{2\pi}{\lambda^2}\frac{dn_{2g}}{d\omega} = \frac{1}{c}\frac{dn_{2g}}{d\lambda}, \tag{2.3.9}$$

$$D_W = -\frac{2\pi\Delta}{\lambda^2}\left[\frac{n_{2g}^2}{n_2\omega}\frac{Vd^2(Vb)}{dV^2} + \frac{dn_{2g}}{d\omega}\frac{d(Vb)}{dV}\right]. \tag{2.3.10}$$

Here n_{2g} is the group index of the cladding material and the parameters V and b are given by Eqs. (2.2.35) and (2.2.36), respectively. In obtaining Eqs. (2.3.8)–(2.3.10) the parameter Δ was assumed to be frequency independent. A third term known as differential material dispersion should be added to Eq. (2.3.8) when $d\Delta/d\omega \neq 0$. Its contribution is, however, negligible in practice.

2.3.2 Material Dispersion

Material dispersion occurs because the refractive index of silica, the material used for fiber fabrication, changes with the optical frequency ω. On a fundamental level, the origin of material dispersion is related to the characteristic resonance frequencies at which the material absorbs the electromagnetic radiation. Far from the medium resonances, the refractive index $n(\omega)$ is well approximated by the *Sellmeier equation* [33]

$$n^2(\omega) = 1 + \sum_{j=1}^{M}\frac{B_j\omega_j^2}{\omega_j^2 - \omega^2}, \tag{2.3.11}$$

where ω_j is the resonance frequency and B_j is the oscillator strength. Here n stands for n_1 or n_2, depending on whether the dispersive properties of the core or the cladding are considered. The sum in Eq. (2.3.11) extends over all material resonances that contribute in the frequency range of interest. In the case of optical fibers, the parameters B_j and ω_j are obtained empirically

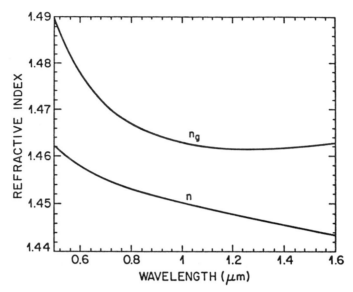

Figure 2.8 Variation of refractive index n and group index n_g with wavelength for fused silica.

by fitting the measured dispersion curves to Eq. (2.3.11) with $M = 3$. They depend on the amount of dopants and have been tabulated for several kinds of fibers [12]. For pure silica these parameters are found to be $B_1 = 0.6961663$, $B_2 = 0.4079426$, $B_3 = 0.8974794$, $\lambda_1 = 0.0684043$ μm, $\lambda_2 = 0.1162414$ μm, and $\lambda_3 = 9.896161$ μm, where $\lambda_j = 2\pi c/\omega_j$ with $j = 1$–3 [33]. The group index $n_g = n + \omega(dn/d\omega)$ can be obtained by using these parameter values.

Figure 2.8 shows the wavelength dependence of n and n_g in the range 0.5–1.6 μm for fused silica. Material dispersion D_M is related to the slope of n_g by the relation $D_M = c^{-1}(dn_g/d\lambda)$ [Eq. (2.3.9)]. It turns out that $dn_g/d\lambda = 0$ at $\lambda = 1.276$ μm. This wavelength is referred to as the *zero-dispersion wavelength* λ_{ZD}, since $D_M = 0$ at $\lambda = \lambda_{ZD}$. The dispersion parameter D_M is negative below λ_{ZD} and becomes positive above that. In the wavelength range 1.25–1.66 μm it can be approximated by an empirical relation

$$D_M \approx 122(1 - \lambda_{ZD}/\lambda). \tag{2.3.12}$$

It should be stressed that $\lambda_{ZD} = 1.276$ μm only for pure silica. It can vary in the range 1.27–1.29 μm for optical fibers whose core and cladding are doped to vary the refractive index. The zero-dispersion wavelength of optical fibers also depends on the core radius a and the index step Δ through the waveguide contribution to the total dispersion.

2.3.3 Waveguide Dispersion

The contribution of waveguide dispersion D_W to the dispersion parameter D is given by Eq. (2.3.10) and depends on the V parameter of the fiber. Figure 2.9

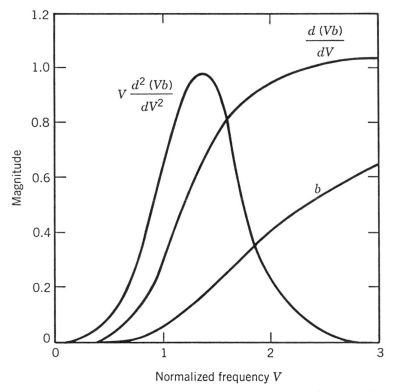

Figure 2.9 Variation of b and its derivatives $d(Vb)/dV$ and $V[d^2(Vb)/dV^2]$ with the V parameter. (After Ref. [30]. ©1971 OSA. Reprinted with permission.)

shows how $d(Vb)/dV$ and $V d^2(Vb)/dV^2$ change with V. Since both derivatives are positive, D_W is negative in the entire wavelength range 0–1.6 μm. On the other hand, D_M is negative for wavelengths below λ_{ZD} and becomes positive above that. Figure 2.10 shows D_M, D_W, and their sum $D = D_M + D_W$, for a typical single-mode fiber. The main effect of waveguide dispersion is to shift λ_{ZD} by an amount 30–40 nm so that the total dispersion is zero near 1.31 μm. It also reduces D from its material value D_M in the wavelength range 1.3–1.6 μm that is of interest for optical communication systems. Typical values of D are in the range 15–18 ps/(km-nm) near 1.55 μm. This wavelength region is of considerable interest for lightwave systems, since, as discussed in Section 2.5, the fiber loss is minimum near 1.55 μm. High values of D limit the performance of 1.55-μm lightwave systems.

Since the waveguide contribution D_W depends on fiber parameters such as the core radius a and the index difference Δ, it is possible to design the fiber such that λ_{ZD} is shifted into the vicinity of 1.55 μm [34], [35]. Such fibers are called *dispersion-shifted fibers*. It is also possible to tailor the waveguide contribution such that the total dispersion D is relatively small over a wide wavelength range extending from 1.3 to 1.6 μm [36]–[38]. Such fibers are called *dispersion-flattened*

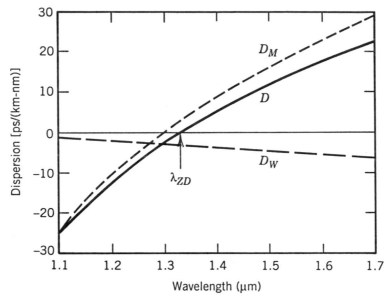

Figure 2.10 Total dispersion D and relative contributions of material dispersion D_M and waveguide dispersion D_W for a conventional single-mode fiber. The zero-dispersion wavelength shifts to a higher value because of the waveguide contribution.

fibers. Figure 2.11 shows typical examples of the wavelength dependence of D for standard (conventional), dispersion-shifted, and dispersion-flattened fibers. The design of dispersion-modified fibers involves the use of multiple cladding layers and a tailoring of the refractive-index profile [34]–[40]. Design issues are discussed in Section 2.7. Table 2.1 lists the operating characteristics of several commercially available fibers having minimum dispersion near 1.3- and 1.55-μm wavelengths. Since 1991, waveguide dispersion has been used to produce fibers whose GVD decreases along the fiber length through axial variations in the core radius. Such fibers, called *dispersion-decreasing fibers,* have found applications in the context of solitons (see Chapter 10).

Table 2.1 Characteristics of several commercial fibers

Fiber Type	NA	Δ (%)	$2w$ (μm)	λ_{ZD} (μm)	GVD Slope S [ps/(km-nm^2)]
Corning SMF-28	0.13	0.36	9.3	1.312	0.090
AT&T Matched-Clad	0.12	0.33	9.3	1.312	0.088
LITESPEC GSM-13	0.12	0.33	9.3	1.312	0.087
Corning SMF-DS	0.17	0.90	8.1	1.550	0.075
AT&T TrueWave	0.16	0.75	8.4	1.530	0.095
LITESPEC DSM-15	0.17	0.90	8.0	1.555	0.072

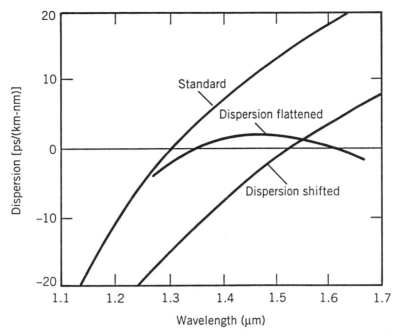

Figure 2.11 Typical wavelength dependence of the dispersion parameter D for standard, dispersion-shifted, and dispersion-flattened fibers.

2.3.4   Higher-Order Dispersion

It appears from Eq. (2.3.6) that the BL product of a single-mode fiber can be increased indefinitely by operating at the zero-dispersion wavelength λ_{ZD} where $D = 0$. The dispersive effects, however, do not disappear completely at $\lambda = \lambda_{ZD}$. Optical pulses still experience broadening because of higher-order dispersive effects. This feature can be understood by noting that D cannot be made zero at all wavelengths contained within the pulse spectrum centered at λ_{ZD}. Clearly, the wavelength dependence of D will play a role in pulse broadening. Higher-order dispersive effects are governed by the *dispersion slope* $S = dD/d\lambda$. The parameter S is also called a *differential-dispersion parameter* or *second-order dispersion parameter*. By using Eq. (2.3.5) it can be written as

$$S = (2\pi c/\lambda^2)^2\beta_3 + (4\pi c/\lambda^3)\beta_2\,, \tag{2.3.13}$$

where $\beta_3 = d\beta_2/d\omega = d^3\beta/d\omega^3$. At $\lambda = \lambda_{ZD}$, $\beta_2 = 0$, and S is proportional to β_3. Typical values of S at $\lambda = \lambda_{ZD}$ are listed in Table 2.1 for both standard and dispersion-shifted fibers. For a source of spectral width $\Delta\lambda$, the effective value of dispersion parameter becomes $D = S\Delta\lambda$. The limiting bit rate–distance product can be estimated by using Eq. (2.3.6) with this value of D, or by using

$$BL|S|(\Delta\lambda)^2 < 1\,. \tag{2.3.14}$$

For a multimode semiconductor laser with $\Delta\lambda = 2$ nm and a dispersion-shifted fiber with $S = 0.05$ ps/(km-nm^2) at $\lambda = 1.55$ μm, the BL product can approach 5 (Tb/s)-km. Further improvement is possible by using single-mode semiconductor lasers.

2.3.5 Polarization-Mode Dispersion

A potential source of pulse broadening is related to fiber birefringence. As discussed in Section 2.2.3, small departures from perfect cylindrical symmetry lead to birefringence because of different mode indices associated with the orthogonally polarized components of the fundamental fiber mode. If the input pulse excites both polarization components, it becomes broader at the fiber output since the two components disperse along the fiber because of their different group velocities. This phenomenon, referred to as *polarization-mode dispersion* (PMD), has been studied extensively during the 1990s because of its importance for periodically amplified lightwave systems [41]–[47].

Similar to the case of GVD, pulse broadening can be estimated from the time delay ΔT between the two polarization components during propagation of the pulse. For a fiber of length L, ΔT is given by

$$\Delta T = \left| \frac{L}{v_{gx}} - \frac{L}{v_{gy}} \right| = L|\beta_{1x} - \beta_{1y}| = L\Delta\beta_1, \qquad (2.3.15)$$

where the subscripts x and y identify the two orthogonally polarized modes and $\Delta\beta_1$ is related to the fiber birefringence. Equation (2.3.1) was used to relate the group velocity v_g to the propagation constant β. Similar to the case of intermodal dispersion discussed in Section 2.1.1, the quantity $\Delta T/L$ is a measure of PMD. For polarization-preserving fibers, $\Delta T/L$ is quite large (~ 1 ns/km) when the two components are equally excited at the fiber input but can be reduced to zero by launching light along one of the principal axes.

Equation (2.3.15) cannot be used directly to estimate PMD for standard telecommunication fibers because of random coupling between the two modes, induced by random perturbations of birefringence occurring along the fiber. The coupling tends to equalize the propagation times for the two polarization components. In fact, PMD is characterized by the root-mean-square (RMS) value of ΔT obtained after averaging over random perturbations. The result is found to be [42], [46]

$$\sigma_T^2 = \langle (\Delta T)^2 \rangle = \frac{1}{2}\Delta\beta_1^2 h^2 \left[\frac{2L}{h} - 1 + \exp\left(-\frac{2L}{h} \right) \right]. \qquad (2.3.16)$$

where h is the decorrelation length, with typical values in the range 1–10 m [47]. For polarization-preserving fibers, the decorrelation length is infinitely large, and the PMD σ_T increases linearly with the fiber length, as expected. In contrast, for $h \ll L$,

$$\sigma_T \approx \Delta\beta_1 \sqrt{hL} = D_p\sqrt{L}, \qquad (2.3.17)$$

where D_p is the PMD parameter with typical values in the range $D_p = 0.1$–1 ps/$\sqrt{\text{km}}$. Because of its $\sqrt{L}$ dependence, PMD-induced pulse broadening is relatively small compared with the GVD effects. However, PMD can become a limiting factor for fiber-optic communication systems designed to operate over long distances near the zero-dispersion wavelength of the fiber [44], [45].

2.4 DISPERSION-INDUCED LIMITATIONS

Pulse broadening, discussed in Section 2.3.1 [see Eq. (2.3.4)], is based on an intuitive phenomenological approach. It provides a first-order estimate for pulses whose spectral width is dominated by the spectrum of the optical source rather than by the Fourier spectrum of the pulse. In general, the extent of pulse broadening depends on the width and the shape of the input pulse. In this section we discuss pulse broadening by using the wave equation (2.2.16).

2.4.1 Basic Propagation Equation

The discussion of fiber modes in Section 2.2.2 showed that each frequency component of the optical field propagates in a single-mode fiber as

$$\tilde{\mathbf{E}}(\mathbf{r},\omega) = \hat{\mathbf{x}} F(x,y)\tilde{B}(0,\omega)\exp(i\beta z), \tag{2.4.1}$$

where $\hat{\mathbf{x}}$ is the polarization unit vector, $\tilde{B}(0,\omega)$ is the initial amplitude, and β is the propagation constant. $F(x,y)$ is the field distribution of the fundamental fiber mode that can often be approximated by a Gaussian distribution [see Eq. (2.2.45)]. In general, $F(x,y)$ also depends on ω, but this dependence can be ignored for pulses whose spectral width $\Delta\omega \ll \omega_0$, a condition generally satisfied in practice. Here ω_0 is the frequency at which the pulse spectrum is centered; it is referred to as the center frequency or carrier frequency. Different spectral components propagate inside the fiber according to the simple relation

$$\tilde{B}(z,\omega) = \tilde{B}(0,\omega)\exp(i\beta z). \tag{2.4.2}$$

The amplitude in the time domain can be obtained by taking the inverse Fourier transform and is given by

$$B(z,t) = \frac{1}{2\pi}\int_{-\infty}^{\infty}\tilde{B}(z,\omega)\exp(-i\omega t)\,d\omega. \tag{2.4.3}$$

The initial spectral amplitude $\tilde{B}(0,\omega)$ is just the Fourier transform of the input amplitude $B(0,t)$.

Pulse broadening results from the frequency dependence of β. For quasi-monochromatic pulses with $\Delta\omega \ll \omega_0$, it is useful to expand $\beta(\omega)$ in a Taylor series around the carrier frequency ω_0 and retain terms up to third order, that is,

$$\beta(\omega) = \bar{n}(\omega)\frac{\omega}{c} \approx \beta_0 + \beta_1(\Delta\omega) + \frac{1}{2}\beta_2(\Delta\omega)^2 + \frac{1}{6}\beta_3(\Delta\omega)^3, \tag{2.4.4}$$

where $\Delta\omega = \omega - \omega_0$ and $\beta_m = (d^m\beta/d\omega^m)_{\omega=\omega_0}$. From Eq. (2.3.1) $\beta_1 = 1/v_g$, where v_g is the group velocity. The GVD coefficient β_2 is related to the dispersion parameter D by Eq. (2.3.5), whereas β_3 is related to the dispersion slope S by Eq. (2.3.13). We substitute Eqs. (2.4.2) and (2.4.4) in Eq. (2.4.3) and introduce a *slowly varying amplitude* $A(z,t)$ of the pulse envelope by the relation

$$B(z,t) = A(z,t) \exp[i(\beta_0 z - \omega_0 t)]. \qquad (2.4.5)$$

The amplitude $A(z,t)$ is found to be given by

$$A(z,t) = \frac{1}{2\pi} \int_{-\infty}^{\infty} d(\Delta\omega)\tilde{A}(0,\Delta\omega) \times$$

$$\exp\left[i\beta_1 z\Delta\omega + \frac{i}{2}\beta_2 z(\Delta\omega)^2 + \frac{i}{6}\beta_3 z(\Delta\omega)^3 - i\Delta\omega\, t\right], \quad (2.4.6)$$

where $\tilde{A}(0,\Delta\omega) = G(0,\omega - \omega_0)$ is the Fourier transform of $A(0,t)$.

By calculating $\partial A/\partial z$ and noting that $\Delta\omega$ is replaced by $i(\partial A/\partial t)$ in the time domain, Eq. (2.4.6) can be written as [28]

$$\frac{\partial A}{\partial z} + \beta_1 \frac{\partial A}{\partial t} + \frac{i}{2}\beta_2 \frac{\partial^2 A}{\partial t^2} - \frac{1}{6}\beta_3 \frac{\partial^3 A}{\partial t^3} = 0. \qquad (2.4.7)$$

This is the basic propagation equation that governs pulse evolution inside a single-mode fiber. In the absence of dispersion ($\beta_2 = \beta_3 = 0$), the optical pulse propagates without change in its shape such that $A(z,t) = A(0, t - \beta_1 z)$. By making the transformation to a reference frame moving with the pulse and introducing the new coordinates

$$t' = t - \beta_1 z \qquad \text{and} \qquad z' = z, \qquad (2.4.8)$$

Eq. (2.4.7) can be written as

$$\frac{\partial A}{\partial z'} + \frac{i}{2}\beta_2 \frac{\partial^2 A}{\partial t'^2} - \frac{1}{6}\beta_3 \frac{\partial^3 A}{\partial t'^3} = 0. \qquad (2.4.9)$$

For simplicity of notation, the prime over z' and t' is dropped in the following discussion.

2.4.2   Chirped Gaussian Pulses

As an application of Eq. (2.4.9), consider propagation of Gaussian input pulses in optical fibers by taking the initial amplitude as [28]

$$A(0,t) = A_0 \exp\left[-\frac{1+iC}{2}\left(\frac{t}{T_0}\right)^2\right], \qquad (2.4.10)$$

where A_0 is the peak amplitude. The parameter T_0 represents the half-width at $1/e$-intensity point. It is related to the full width at half maximum (FWHM) of the pulse by the relation

$$T_{\text{FWHM}} = 2(\ln 2)^{1/2}T_0 \approx 1.665 T_0. \qquad (2.4.11)$$

The parameter C governs the *linear frequency chirp* imposed on the pulse. A pulse is said to be chirped if its carrier frequency changes with time. The frequency change is related to the phase derivative and is given by

$$\delta\omega(t) = -\frac{\partial\phi}{\partial t} = \frac{C}{T_0^2}\,t, \tag{2.4.12}$$

where ϕ is the phase of $A(0,t)$. The time-dependent frequency shift $\delta\omega$ is called the *chirp*. Its inclusion is important since semiconductor lasers generally emit pulses that are considerably chirped (see Section 3.3.7). The Fourier spectrum of a chirped pulse is broader than that of the unchirped pulse. This can be seen by taking the Fourier transform of Eq. (2.4.10), so that

$$\tilde{A}(0,\omega) = A_0 \left(\frac{2\pi T_0^2}{1 + iC}\right)^{1/2} \exp\left[-\frac{\omega^2 T_0^2}{2(1 + iC)}\right]. \tag{2.4.13}$$

The spectral half-width (at $1/e$-intensity point) is given by

$$\Delta\omega_0 = (1 + C^2)^{1/2} T_0^{-1}. \tag{2.4.14}$$

In the absence of frequency chirp ($C = 0$), the spectral width satisfies the relation $\Delta\omega_0 T_0 = 1$. Such a pulse has the narrowest spectrum and is called *transform-limited*. The spectral width is enhanced by a factor of $(1 + C^2)^{1/2}$ in the presence of linear chirp, as seen in Eq. (2.4.14).

The pulse-propagation equation (2.4.9) can be easily solved in the Fourier domain. Its solution is [see Eq. 2.4.6)]

$$A(z,t) = \frac{1}{2\pi} \int_{-\infty}^{\infty} \tilde{A}(0,\omega) \exp\left(\frac{i}{2}\beta_2 z\omega^2 + \frac{i}{6}\beta_3 z\omega^3 - i\omega t\right) d\omega, \tag{2.4.15}$$

where $\tilde{A}(0,\omega)$ is given by Eq. (2.4.13) for the Gaussian input pulse. Let us first consider the case in which the carrier wavelength is far away from the zero-dispersion wavelength so that the contribution of the β_3 term is negligible. The integration in Eq. (2.4.15) can be performed analytically, and the result is

$$A(z,t) = \left(\frac{A_0 T_0}{[T_0^2 - i\beta_2 z(1 + iC)]^{1/2}}\right) \exp\left(-\frac{(1 + iC)t^2}{2[T_0^2 - i\beta_2 z(1 + iC)]}\right). \tag{2.4.16}$$

Equation (2.4.16) shows that a Gaussian pulse remains Gaussian on propagation. The pulse width changes with z as

$$\frac{T_1}{T_0} = \left[\left(1 + \frac{C\beta_2 z}{T_0^2}\right)^2 + \left(\frac{\beta_2 z}{T_0^2}\right)^2\right]^{1/2}, \tag{2.4.17}$$

where T_1 is the half-width defined similar to T_0. Figure 2.12 shows the broadening factor T_1/T_0 as a function of the propagation distance z/L_D, where $L_D = T_0^2/|\beta_2|$ is called the *dispersion length*. An unchirped pulse ($C = 0$) broadens as $[1 + (z/L_D)^2]^{1/2}$ and its width increases by a factor of $\sqrt{2}$ at $z = L_D$.

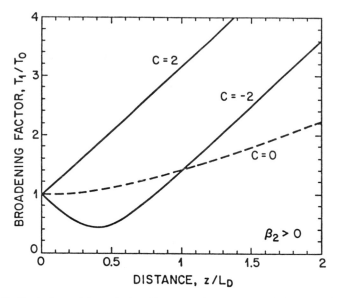

Figure 2.12 Variation of broadening factor with propagated distance for a chirped Gaussian input pulse. Dashed curve corresponds to the case of an unchirped Gaussian pulse. For $\beta_2 < 0$ the same curves are obtained if the sign of the chirp parameter C is reversed.

The chirped pulse, on the other hand, may broaden or compress depending on whether β_2 and C have the same or opposite signs. For $\beta_2 C > 0$ the chirped Gaussian pulse broadens monotonically at a rate faster than the unchirped pulse. For $\beta_2 C < 0$, the pulse width initially decreases and becomes minimum at a distance

$$z_{\min} = \left[|C|/(1 + C^2) \right] L_D. \tag{2.4.18}$$

The minimum value depends on the chirp parameter as

$$T_1^{\min} = T_0/(1 + C^2)^{1/2}. \tag{2.4.19}$$

As we shall see later, initial narrowing of suitably chirped pulses can be used to advantage in the design of optical communication systems.

Equation (2.4.17) can be generalized to include higher-order dispersion governed by β_3 in Eq. (2.4.15). The integral can still be performed in closed form in terms of an Airy function [49]. However, a Gaussian input pulse does not remain Gaussian on propagation and develops a long tail with an oscillatory structure. Such pulses cannot be properly characterized by their FWHM. A proper measure of the pulse width is the RMS width of the pulse defined by

$$\sigma = \left[\langle t^2 \rangle - \langle t \rangle^2 \right]^{1/2}, \tag{2.4.20}$$

where angle brackets denote averaging with respect to the intensity profile, i.e.,

$$\langle t^m \rangle = \frac{\int_{-\infty}^{\infty} t^m |A(z,t)|^2 \, dt}{\int_{-\infty}^{\infty} |A(z,t)|^2 \, dt}. \tag{2.4.21}$$

The *broadening factor* defined as σ/σ_0, where σ_0 is the RMS width of the input Gaussian pulse ($\sigma_0 = T_0/\sqrt{2}$), is given by [48]

$$\frac{\sigma}{\sigma_0} = \left[\left(1 + \frac{C\beta_2 L}{2\sigma_0^2}\right)^2 + \left(\frac{\beta_2 L}{2\sigma_0^2}\right)^2 + (1 + C^2)\frac{1}{2}\left(\frac{\beta_3 L}{4\sigma_0^3}\right)^2\right]^{1/2}, \qquad (2.4.22)$$

where L is the fiber length.

The foregoing discussion assumes that the optical source used to produce the input pulses is nearly monochromatic such that its spectral width satisfies $\Delta\omega_L \ll \Delta\omega_0$ (under continuous-wave or CW operation), where $\Delta\omega_0$ is given by Eq. (2.4.14). This condition is often not satisfied in practice. To account for the source spectral width, one must treat the optical field as a stochastic process and consider the coherence properties of the source through the mutual coherence function [22]. The average in Eq. (2.4.20) then includes an average over the statistical nature of the source. For a Gaussian source spectrum with a RMS spectral width σ_ω, it is possible to obtain the broadening factor in an analytic form with the result [48]

$$\frac{\sigma}{\sigma_0} = \left[\left(1 + \frac{C\beta_2 L}{2\sigma_0^2}\right)^2 + (1 + V_\omega^2)\left(\frac{\beta_2 L}{2\sigma_0^2}\right)^2 + (1 + C^2 + V_\omega^2)\frac{1}{2}\left(\frac{\beta_3 L}{4\sigma_0^3}\right)^2\right]^{1/2},$$
$$(2.4.23)$$

where V_ω is defined by $V_\omega = 2\sigma_\omega\sigma_0$. Equation (2.4.23) provides an expression for dispersion-induced broadening of Gaussian input pulses under quite general conditions. The next section uses it to obtain the limiting bit rate of optical communication systems employing single-mode fibers.

2.4.3 Limitations on the Bit Rate

The limitation imposed on the bit rate by fiber dispersion can be quite different depending on the source spectral width. It is instructive to consider the following two cases separately.

Optical Sources with a Large Spectral Width

This case corresponds to $V_\omega \gg 1$ in Eq. (2.4.23). Consider first the case of a lightwave system operating away from the zero-dispersion wavelength so that the β_3 term can be neglected. The effects of frequency chirp are negligible for sources with a large spectral width. By setting $C = 0$ in Eq. (2.4.23), we obtain

$$\frac{\sigma}{\sigma_0} = \left[1 + \left(\frac{\beta_2 L\sigma_\omega}{\sigma_0}\right)^2\right]^{1/2} = \left[1 + \left(\frac{DL\sigma_\lambda}{\sigma_0}\right)^2\right]^{1/2}, \qquad (2.4.24)$$

where σ_λ is the RMS source spectral width in wavelength units. The output pulse width is thus given by

$$\sigma = (\sigma_0^2 + \sigma_D^2)^{1/2}, \qquad (2.4.25)$$

where $\sigma_D \equiv |D|L\sigma_\lambda$ is a measure of dispersion-induced broadening.

We can relate σ to the bit rate by using the criterion that the broadened pulse should remain inside the allocated bit slot, $T_B = 1/B$, where B is the bit rate. A commonly used criterion is $\sigma \leq T_B/4$; for Gaussian pulses at least 95% of the pulse energy then remains within the bit slot. The limiting bit rate is given by $4B\sigma \leq 1$. In the limit $\sigma_D \gg \sigma_0$, $\sigma \approx \sigma_D = |D|L\sigma_\lambda$, and the condition becomes

$$BL|D|\sigma_\lambda \leq 1/4. \qquad (2.4.26)$$

This condition should be compared with Eq. (2.3.6) obtained heuristically; the two become identical if we interpret $\Delta\lambda$ as $4\sigma_\lambda$ in Eq. (2.3.6).

For a lightwave system operating exactly at the zero-dispersion wavelength $\beta_2 = 0$ in Eq. (2.4.23). By setting $C = 0$ as before and taking $V_\omega \gg 1$, Eq. (2.4.23) can be approximated by

$$\frac{\sigma}{\sigma_0} = \left[1 + \frac{1}{2}\left(\frac{\beta_3 L\sigma_\omega^2}{\sigma_0}\right)^2\right]^{1/2} = \left[1 + \frac{1}{2}\left(\frac{SL\sigma_\lambda^2}{\sigma_0}\right)^2\right]^{1/2}, \qquad (2.4.27)$$

where Eq. (2.3.13) was used to relate β_3 to the dispersion slope S. The output pulse width is thus given by

$$\sigma = [\sigma_0^2 + (SL\sigma_\lambda^2)^2/2]^{1/2} = (\sigma_0^2 + \sigma_D^2)^{1/2}, \qquad (2.4.28)$$

where now $\sigma_D \equiv |S|L\sigma_\lambda^2/\sqrt{2}$. As before, we can relate σ to the limiting bit rate by the condition $4B\sigma \leq 1$. When $\sigma_D \gg \sigma_0$, the limitation on the bit rate is governed by

$$BL|S|\sigma_\lambda^2 \leq 1/\sqrt{8}. \qquad (2.4.29)$$

This condition should be compared with Eq. (2.3.14) obtained heuristically by using simple physical arguments.

As an example, consider the case of a light-emitting diode (see Section 3.2) for which $\sigma_\lambda \approx 15$ nm. By using $D = 17$ ps/(km-nm) at 1.55 μm, Eq. (2.4.26) yields $BL < 1$ (Gb/s)-km. However, if the system is designed to operate at the zero-dispersion wavelength, BL can be increased to 20 (Gb/s)-km for a typical value $S = 0.08$ ps/(km-nm^2).

Optical Sources with a Small Spectral Width

This case corresponds to $V_\omega \ll 1$ in Eq. (2.4.23). As before, if we neglect the β_3 term and set $C = 0$, Eq. (2.4.23) can be approximated by

$$\sigma = [\sigma_0^2 + (\beta_2 L/2\sigma_0)^2]^{1/2} = (\sigma_0^2 + \sigma_D^2)^{1/2}. \qquad (2.4.30)$$

A comparison with Eq. (2.4.25) reveals a major difference between the two cases. For a narrow source spectrum, dispersion-induced broadening depends on the initial width σ_0, whereas it is independent of σ_0 when the spectral width of the optical source dominates. In fact, σ can be minimized by choosing an optimum value of σ_0. The minimum value of σ is found to occur for $\sigma_0 =$

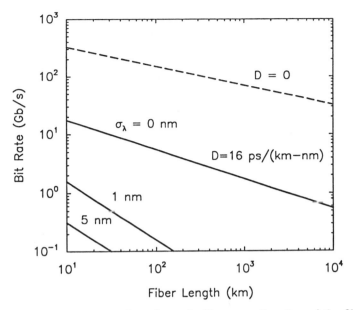

Figure 2.13 Limiting bit rate of single-mode fibers as a function of the fiber length for $\sigma_\lambda = 0$, 1, and 5 nm. The case $\sigma_\lambda = 0$ corresponds to the case of an optical source whose spectral width is much smaller than the bit rate.

$\sigma_D = (|\beta_2|L/2)^{1/2}$ and is given by $\sigma = (|\beta_2|L)^{1/2}$. The limiting bit rate can be obtained by using $4B\sigma \le 1$ and leads to the condition

$$B\sqrt{|\beta_2|L} \le 1/4. \tag{2.4.31}$$

The main difference from Eq. (2.4.26) is that B scales as $L^{-1/2}$ rather than L^{-1}. Figure 2.13 compares the decrease in the bit rate with increasing L by choosing $D = 16$ ps/(km-nm) for $\sigma_\lambda = 0$, 1, and 5 nm. Equation (2.4.31) was used for the case $\sigma_\lambda = 0$.

For a lightwave system operating close to the zero-dispersion wavelength, $\beta_2 \approx 0$ in Eq. (2.4.23). By using $V_\omega \ll 1$ and $C = 0$, the pulse width is then given by

$$\sigma = [\sigma_0^2 + (\beta_3 L/4\sigma_0^2)^2/2]^{1/2} = (\sigma_0^2 + \sigma_D^2)^{1/2}. \tag{2.4.32}$$

Similar to the case of Eq. (2.4.30), σ can be minimized by optimizing the input pulse width σ_0. The minimum value of σ_0 is found to occur for $\sigma_0 = (|\beta_3|L/4)^{1/3}$ and is given by

$$\sigma = (3/2)^{1/2}(|\beta_3|L/4)^{1/3}. \tag{2.4.33}$$

The limiting bit rate is obtained by using the condition $4B\sigma \le 1$, or

$$B(|\beta_3|L)^{1/3} \le 0.324. \tag{2.4.34}$$

The dispersive effects are most forgiving in this case. For a typical value $\beta_3 = 0.1$ ps^3/km, the bit rate can be as large as 150 Gb/s for $L = 100$ km. It decreases

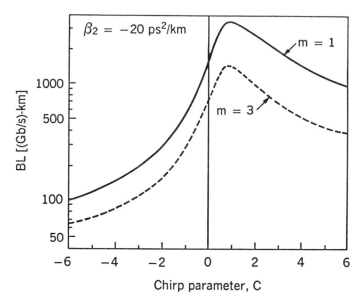

Figure 2.14 Dispersion-limited BL product as a function of the chirp parameter for Gaussian (solid curve) and super-Gaussian (dashed curve) input pulses. (After Ref. [50]. ©1986 OSA. Reprinted with permission.)

to only about 70 Gb/s even when L increases by a factor of 10 because of the $L^{-1/3}$ dependence of the bit rate on the fiber length. The dashed line in Fig. 2.13 shows this dependence by using Eq. (2.4.34) with $\beta_3 = 0.1$ ps^3/km. Clearly, the performance of lightwave systems can be improved considerably by operating near the zero-dispersion wavelength of the fiber and by using optical sources with a relatively narrow spectral width.

Effect of Frequency Chirp

The input pulse in all preceding cases has been assumed to be an unchirped Gaussian pulse. In practice, optical pulses emitted from semiconductor lasers are non-Gaussian and exhibit considerable chirp. A super-Gaussian model has been used to study the bit-rate limitation imposed by fiber dispersion for such input pulses [50]. Equation (2.4.10) is then replaced by

$$A(0,T) = A_0 \exp\left[-\frac{1+iC}{2}\left(\frac{t}{T_0}\right)^{2m}\right], \qquad (2.4.35)$$

where the parameter m controls the pulse shape. Chirped Gaussian pulses correspond to $m = 1$. For large value of m the pulse becomes nearly rectangular, with sharp leading and trailing edges. The output pulse shape can be obtained by solving Eq. (2.4.9) numerically. The limiting bit rate–distance product BL is found by requiring that the RMS pulse width does not increase above a tolerable broadening factor. Figure 2.14 shows the BL product as a function of the chirp parameter C for Gaussian ($m = 1$) and super-Gaussian ($m = 3$) input pulses. In both cases the fiber length L at which the pulse broadens by 20% was obtained for $T_0 = 125$ ps and $\beta_2 = -20$ ps^2/km. As expected, the BL product is smaller for super-Gaussian pulses because such pulses broaden more rapidly than Gaussian pulses. The BL product is reduced dramatically for negative values of the chirp parameter C. This is due to enhanced broadening occurring when $\beta_2 C$ is positive (see Fig. 2.12). Unfortunately, C is generally negative for directly modulated semiconductor lasers with a typical value of -6 at 1.55 μm. Since $BL < 100$ (Gb/s)-km under such conditions, fiber dispersion limits the bit rate to about 2 Gb/s for $L = 50$ km. This problem can be overcome by using dispersion-shifted fibers or by using a dispersion-compensation scheme (see Chapter 9).

2.4.4 Fiber Bandwidth

The concept of fiber bandwidth originates from the general theory of time-invariant linear systems [51]. If the optical fiber can be treated as a *linear system*, its input and output powers should be related by a general relation

$$P_{\text{out}}(t) = \int_{-\infty}^{\infty} h(t - t') P_{\text{in}}(t') \, dt'. \tag{2.4.36}$$

For an impulse $P_{\text{in}}(t) = \delta(t)$, where $\delta(t)$ is the delta function, and $P_{\text{out}}(t) = h(t)$. For this reason, $h(t)$ is called the *impulse response* of the linear system. Its Fourier transform,

$$H(f) = \int_{-\infty}^{\infty} h(t) \exp(i2\pi f t) \, dt, \tag{2.4.37}$$

provides the frequency response and is called the *transfer function*. In general, $|H(f)|$ falls off with increasing f, indicating that the high-frequency components of the input signal are attenuated by the fiber. In effect, the optical fiber acts as a *bandpass filter*. The *fiber bandwidth* $f_{3\,\text{dB}}$ corresponds to the frequency $f = f_{3\,\text{dB}}$ at which $|H(f)|$ is reduced by a factor of 2 or by 3 dB:

$$|H(f_{3\,\text{dB}})/H(0)| = 1/2. \tag{2.4.38}$$

Note that $f_{3\,\text{dB}}$ is the optical bandwidth of the fiber as the optical power drops by 3 dB at this frequency compared with the zero-frequency response. In the theory of electrical communications, the bandwidth of a linear system is generally defined as the frequency at which electrical power drops by 3 dB. Since the optical power converts to electric current, the 3-dB electrical bandwidth is obtained by replacing the factor of 1/2 in Eq. (2.4.38) by $1/\sqrt{2}$.

Optical fibers cannot generally be treated as linear systems, and Eq. (2.4.36) does not hold for them [52]. However, they can be approximated by a linear system if the source spectral width $\Delta\omega_L$ is much larger than the signal spectral width $\Delta\omega_0$ ($V_\omega \gg 1$). One can then consider propagation of different spectral components independently and add the power carried by them linearly to obtain the output power. For a Gaussian spectrum, the transfer function $H(f)$ is found to be given by [53]

$$H(f) = \frac{1}{(1 + i\,f/f_2)^{1/2}} \exp\left(-\frac{1}{2}\frac{(f/f_1)^2}{1 + i\,f/f_2}\right), \qquad (2.4.39)$$

where

$$f_1 = (2\pi\beta_2 L\sigma_\omega)^{-1} = (2\pi|D|L\sigma_\lambda)^{-1}, \qquad (2.4.40)$$

$$f_2 = (2\pi\beta_3 L\sigma_\omega^2)^{-1} = [2\pi(S + 2|D|/\lambda)L\sigma_\lambda^2]^{-1}, \qquad (2.4.41)$$

and we used Eqs. (2.3.5) and (2.3.13) to introduce the dispersion parameters D and S.

For lightwave systems operating far away from the zero-dispersion wavelength ($f_1 \ll f_2$), the transfer function is approximately Gaussian. By using Eqs. (2.4.38) and (2.4.39) with $f \ll f_2$, the fiber bandwidth is given by

$$f_{3\,\text{dB}} = (2\ln 2)^{1/2} f_1 \approx 0.188(|D|L\sigma_\lambda)^{-1}. \qquad (2.4.42)$$

If we use $\sigma_D = |D|L\sigma_\lambda$ from Eq. (2.4.25), we obtain the relation $f_{3\,\text{dB}}\sigma_D \approx 0.188$ between the fiber bandwidth and dispersion-induced pulse broadening. We can also get a relation between the bandwidth and the bit rate B by using Eqs. (2.4.26) and (2.4.42). The relation is $B \leq 1.33 f_{3\,\text{dB}}$ and shows that the fiber bandwidth is an approximate measure of the maximum possible bit rate of dispersion-limited lightwave systems. In fact, Fig. 2.13 can be used to estimate $f_{3\,\text{dB}}$ and its variation with the fiber length under different operating conditions.

For lightwave systems operating at the zero-dispersion wavelength, the transfer function is obtained from Eq. (2.4.39) by setting $D = 0$. The use of Eq. (2.4.38) then provides the following expression for the fiber bandwidth

$$f_{3\,\text{dB}} = \sqrt{15}f_2 \approx 0.616(SL\sigma_\lambda^2)^{-1}. \qquad (2.4.43)$$

The limiting bit rate can be related to $f_{3\,\text{dB}}$ by using Eq. (2.4.29) and is given by $B \leq 0.574 f_{3\,\text{dB}}$. Again, the fiber bandwidth provides a measure of the dispersion-limited bit rate. As a numerical estimate, consider a 1.55-μm lightwave system employing dispersion-shifted fibers and multimode semiconductor lasers. By using $S = 0.05$ ps/(km-nm^2) and $\sigma_\lambda = 1$ nm as typical values, $f_{3\,\text{dB}}L \approx 32$ THz-km. By contrast, the bandwidth–distance product is reduced to 0.1 THz-km for standard fibers with $D = 18$ ps/(km-nm).

2.5   FIBER LOSS

Section 2.4 discussed how fiber dispersion limits the performance of fiber-optic communication systems by broadening optical pulses as they propagate inside the fiber. Fiber loss is another fundamental limiting factor, as it reduces the average power reaching the receiver. Since optical receivers need a certain minimum amount of power for recovering the signal accurately, the transmission distance is inherently limited by fiber loss. In fact, the use of silica fibers for optical communications became practical only when the loss was reduced to an acceptable level to achieve a transmission distance of 10 km or more. This section is devoted to a discussion of various loss mechanisms in optical fibers.

2.5.1 Attenuation Coefficient

Under quite general conditions, power attenuation inside an optical fiber is governed by

$$dP/dz = -\alpha P, \tag{2.5.1}$$

where α is the attenuation coefficient and P is the optical power. Although denoted by the same symbol as the absorption coefficient in Eq. (2.2.12), α in Eq. (2.5.1) includes not only material absorption but also other sources of power attenuation. If P is the power launched at the input of a fiber of length L, the output power P_{out} from Eq. (2.5.1) is given by

$$P_{\text{out}}(t) = P_{\text{in}} \exp(-\alpha L). \tag{2.5.2}$$

It is customary to express α in units of dB/km by using the relation

$$\alpha\,(\text{dB/km}) = -\frac{10}{L}\log_{10}\left(\frac{P_{\text{out}}}{P_{\text{in}}}\right) = 4.343\alpha, \tag{2.5.3}$$

and refer to it as the fiber loss.

Fiber loss depends on the wavelength of transmitted light. Figure 2.15 shows the loss spectrum $\alpha(\lambda)$ of a single-mode fiber with 9.4-μm core diameter, $\Delta = 1.9 \times 10^{-3}$, and 1.1-$\mu$m cutoff wavelength [11]. This fiber exhibits a loss of only about 0.2 dB/km in the wavelength region near 1.55 μm, the lowest value realized in 1979. This value is close to the fundamental limit of about 0.15 dB/km for silica fibers. The loss spectrum also exhibits a strong peak near 1.39 μm and several other smaller peaks. A secondary minimum is found to occur near 1.3 μm, where the fiber loss is below 0.5 dB/km. Since fiber dispersion is also minimum near 1.3 μm, this low-loss window is often used for optical communication systems. The loss is considerably higher for shorter wavelengths and exceeds 5 dB/km in the visible region of the optical spectrum. Several factors contribute to the loss; their relative contributions are also shown in Fig. 2.15. The two most important among them are material absorption and Rayleigh scattering; these are discussed in the following subsections.

2.5.2 Material Absorption

Optical fibers are made of *fused silica*. Material absorption can be divided into two categories. Intrinsic material absorption corresponds to the loss caused by pure silica whereas extrinsic absorption is related to the loss caused by impurities. Any material absorbs at certain wavelengths corresponding to the electronic and vibrational resonances associated with specific molecules. For

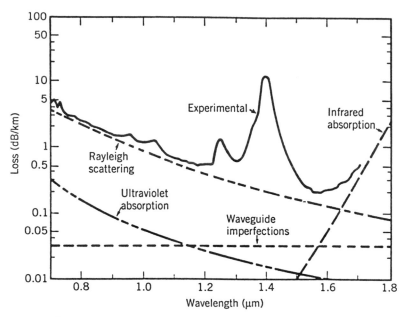

Figure 2.15 Spectral loss profile of a single-mode fiber. Wavelength dependence of fiber loss for several fundamental loss mechanisms is also shown. (After Ref. [11]. ©1979 IEE. Reprinted with permission.)

silica (SiO_2) molecules, electronic resonances occur in the ultraviolet region ($\lambda < 0.4$ μm), whereas vibrational resonances occur in the infrared region (($\lambda > 7$ μm). Because of the amorphous nature of fused silica, these resonances are in the form of absorption bands whose tails extend into the visible region. Figure 2.15 shows that intrinsic material absorption for silica in the wavelength range 0.8–1.6 μm is below 0.1 dB/km. In fact, it is less than 0.03 dB/km in the 1.3- to 1.6-μm wavelength window commonly used for lightwave systems.

Extrinsic absorption results from the presence of impurities. Transition-metal impurities such as Fe, Cu, Co, Ni, Mn, and Cr absorb strongly in the wavelength range 0.6–1.6 μm. Their amount should be reduced to below 1 part per billion to obtain a loss level below 1 dB/km. Such high-purity silica can be obtained by using modern techniques. The main source of extrinsic absorption in state-of-the-art silica fibers is the presence of water vapors. A vibrational resonance of the OH ion occurs at 2.73 μm. Its harmonic and combination tones with silica produce strong absorption at the 1.39-, 1.24-, and 0.95-μm wavelengths. The three spectral peaks seen in Fig. 2.15 occur near these wavelengths and are due to the presence of residual water vapor in silica. Even a concentration of 1 part per million can cause a loss of about 50 dB/km at 1.39 μm. Typically, the OH ion concentration should be reduced to below 10^{-8} to realize a low-loss fiber with a loss spectrum of Fig. 2.15. Dopants such as GeO_2, P_2O_5, and B_2O_3, used during fiber fabrication to produce the required index step, can also lead to additional losses.

2.5.3   Rayleigh Scattering

Rayleigh scattering is a fundamental loss mechanism arising from local micro-scopic fluctuations in density. Silica molecules move randomly in the molten state and freeze in place during fiber fabrication. Density fluctuations lead to random fluctuations of the refractive index on a scale smaller than the optical wavelength λ. Light scattering in such a medium is known as *Rayleigh scattering* [22]. The scattering cross section varies as λ^{-4}. As a result, the intrinsic loss of silica fibers from Rayleigh scattering can be written as

$$\alpha_R = C/\lambda^4, \tag{2.5.4}$$

where the constant C is in the range 0.7–0.9 (dB/km)-μm^4, depending on the constituents of the fiber core. These values of C correspond to $\alpha_R = 0.12$–0.16 dB/km at $\lambda = 1.55\ \mu m$, indicating that fiber loss in Fig. 2.15 is dominated by Rayleigh scattering near this wavelength.

The contribution of Rayleigh scattering can be reduced to below 0.01 dB/km for wavelengths longer than 3 μm. Silica fibers cannot be used in this wave-length region, since infrared absorption begins to dominate the fiber loss beyond 1.6 μm. Considerable effort has been directed [54]–[57] toward finding other suit-able materials with low absorption beyond 2 μm. Fluorozirconate (ZrF_4) fibers have an intrinsic material absorption of about 0.01 dB/km near 2.55 μm and have the potential for exhibiting loss much smaller than that of silica fibers. State-of-the-art fluoride fibers, however, exhibit a loss of about 1 dB/km be-cause of extrinsic losses. Chalcogenide and polycrystalline fibers exhibit min-imum loss in the far-infrared region near 10 μm. The theoretically predicted minimum value of fiber loss for such fibers is below 10^{-3} dB/km because of reduced Rayleigh scattering. However, practical loss levels remain higher than those of silica fibers [57].

2.5.4   Waveguide Imperfections

An ideal single-mode fiber with a perfect cylindrical geometry guides the optical mode without energy leakage into the cladding layer. In practice, imperfections at the core–cladding interface (e.g., random core-radius variations) can lead to additional losses which contribute to the net fiber loss. The physical process behind such losses is *Mie scattering* [22], occurring because of index inhomo-geneities on a scale longer than the optical wavelength. Care is generally taken to ensure that the core radius does not vary significantly along the fiber length during manufacture. Such variations can be kept below 1%, and the resulting scattering loss is typically below 0.03 dB/km.

Bends in the fiber constitute another source of scattering loss [58]. The reason can be understood by using the ray picture. Normally, a guided ray hits the core–cladding interface at an angle greater than the critical angle to experience total internal reflection. However, the angle decreases near a bend and may become smaller than the critical angle for tight bends. The ray would then escape out of the fiber. In the mode description, a part of the mode energy is scattered into the cladding layer. The bending loss is proportional to $\exp(-R/R_c)$, where R is the radius of curvature of the fiber bend and $R_c = a/(n_1^2 - n_2^2)$. For single-mode fibers, $R_c = 0.2$–0.4 μm typically, and the bending loss is negligible (< 0.01 dB/km) for bend radius $R > 5$ mm. Since most macroscopic bends exceed $R = 5$ mm, *macrobending losses* are negligible in practice.

A major source of fiber loss, particularly in cable form, is related to the random axial distortions that invariably occur during cabling when the fiber is pressed against a surface that is not perfectly smooth. Such losses are referred to as *microbending losses* and have been studied extensively [59]–[63]. Microbends cause an increase in the fiber loss for both multimode and single-mode fibers and can result in an excessively large loss (~ 100 dB/km) if precautions are not taken to minimize them. For single-mode fibers, microbending losses can be minimized by choosing the V parameter as close to the cutoff value of 2.405 as possible so that mode energy is confined primarily to the core. In practice, the fiber is designed to have V in the range 2.0–2.4 at the operating wavelength. Many other sources of optical loss exist in a fiber cable. These are related to splices and connectors used in forming the fiber link and are often treated as a part of the cable loss; microbending losses can also be included in the total cable loss.

The loss mechanisms discussed in this section are power independent. Optical fibers also exhibit nonlinear losses which become important at high power levels. These are discussed in the following section.

2.6 NONLINEAR OPTICAL EFFECTS

The response of any dielectric to light becomes nonlinear for intense electromagnetic fields, and optical fibers are no exception. Even though silica is intrinsically not a highly nonlinear material, the waveguide geometry that confines light to a small cross section over long fiber lengths makes nonlinear effects quite important in the design of modern lightwave systems [28]. We discuss next the nonlinear phenomena that are most relevant for fiber-optic communications.

2.6.1 Stimulated Light Scattering

Rayleigh scattering, discussed in Section 2.5.3, is an example of elastic scattering for which the frequency (or the photon energy) of scattered light remains unchanged. By contrast, the frequency of scattered light is shifted downward during inelastic scattering. Two examples of inelastic scattering are *Raman scattering* and *Brillouin scattering* [64]–[67]. Both of them can be understood as scattering of a photon to a lower energy photon such that the energy difference appears in the form of a phonon. The main difference between the two is that optical phonons participate in Raman scattering, whereas acoustic phonons participate in Brillouin scattering. Both scattering processes result in a loss of power at the incident frequency and constitute a loss mechanism for optical fibers. However, the scattering cross sections are sufficiently small that loss is negligible at low power levels.

At high power levels the nonlinear phenomena of stimulated Raman scattering (SRS) and stimulated Brillouin scattering (SBS) can lead to considerable fiber loss. The intensity of the scattered light in both cases grows exponentially once the incident power exceeds a threshold value [64]. SRS and SBS have been observed in optical fibers [65]–[67]. Even though SRS and SBS are quite similar in their origin, different dispersion relations for acoustic and optical phonons lead to some basic differences between the two [28]. A fundamental difference is that SBS in single-mode fibers occurs only in the backward direction, whereas SRS dominates in the forward direction.

The threshold power level for both SRS and SBS can be estimated by considering how the scattered-light intensity grows from noise [64]. In the case of SRS, the threshold power P_{th}, defined as the incident power at which half of the power is lost to SRS at the output end of a fiber of length L, is estimated from [28]

$$g_R P_{th} L_{\text{eff}} / A_{\text{eff}} \approx 16, \qquad (2.6.1)$$

where g_R is the peak value of the Raman gain, A_{eff} is the effective mode cross section, often referred to as the effective core area, and L_{eff} is the effective interaction length, defined by

$$L_{\text{eff}} = [1 - \exp(-\alpha L)]/\alpha, \qquad (2.6.2)$$

where α represents fiber loss. For optical communication systems, the fiber is sufficiently long that L_{eff} can be approximated by $1/\alpha$. If we replace A_{eff} by πw^2, where w is the spot size [see Eq. (2.2.45)], P_{th} is given by

$$P_{th} \approx 16\alpha(\pi w^2)/g_R. \qquad (2.6.3)$$

It is important to emphasize that Eq. (2.6.3) provides an order-of-magnitude estimate only as many approximations are made in its derivation [28]. The Raman gain coefficient $g_R \approx 1 \times 10^{-13}$ m/W for silica fibers [65] near 1 μm and scales inversely with the wavelength. If we use $\pi w^2 = 50$ μm^2 and $\alpha = 0.2$ dB/km as the representative values, P_{th} is about 570 mW near 1.55 μm. Since the launched power in optical communication systems is typically below 10 mW, SRS generally does not contribute to the fiber loss.

The situation is quite different for SBS [28]. The threshold power can be estimated by using a procedure similar to the case of SRS and is given by [64]

$$g_B P_{th} L_{\text{eff}} / A_{\text{eff}} \approx 21, \qquad (2.6.4)$$

where g_B is the Brillouin gain coefficient. As before, we can replace L_{eff} by $1/\alpha$ and A_{eff} by πw^2 so that $P_{th} \approx 21\alpha(\pi w^2)/g_B$. However, $g_B \approx 5 \times 10^{-11}$ m/W for silica fibers [66], a value larger than g_R by more than two orders of magnitude. As a result, P_{th} can be as low as ~ 1 mW [67], especially near 1.55 μm, where the fiber loss is minimum. Clearly, SBS can limit the launched power considerably because of its low threshold. The estimate of P_{th} above neglects the effect of spectral width associated with the incident light. Since the Brillouin-gain spectrum for silica fibers is quite narrow (< 100 MHz), the threshold power can be increased to 10 mW or more by intentionally increasing the gain bandwidth to 200–400 MHz through phase modulation. Nonetheless, SBS limits the launched power to below 100 mW in most lightwave communication systems.

Both SRS and SBS can be used to advantage in the design of optical communication systems, since they can amplify an optical field by transferring energy to it from a pump field whose wavelength is suitably chosen. SRS is especially useful because of an extremely large bandwidth (~ 10 THz) associated with the Raman-gain profile of silica [65]. It can be used to make fiber Raman amplifiers (see Section 8.3). SBS can also be used to make Brillouin amplifiers (see Section 8.4).

2.6.2 Nonlinear Refraction

The refractive index of silica was assumed to be power independent in the discussion of fiber modes in Section 2.2. Although this is a good approximation at low power levels, it becomes necessary to include the nonlinear contribution at high powers by taking [28]

$$n_j' = n_j + \bar{n}_2(P/A_{\text{eff}}), \qquad j = 1,\ 2, \tag{2.6.5}$$

where n_1' and n_2' are the core and cladding indices and $\bar{n}_2$ is the *nonlinear-index coefficient* ($\bar{n}_2 \approx 3 \times 10^{-20}$ m^2/W for silica fibers). Typically, the nonlinear contribution to the refractive index is quite small ($< 10^{-7}$). If we use first-order perturbation theory to obtain the fiber modes by using Eq. (2.6.5), we find that the propagation constant becomes power dependent and can be written as [28]

$$\beta' = \beta + \bar{\gamma}P, \tag{2.6.6}$$

where $\bar{\gamma} = k_0 \bar{n}_2 / A_{\text{eff}}$. By noting that the optical phase associated with the fiber mode increases linearly with z, the effect of nonlinear refraction is to produce a nonlinear phase shift given by

$$\phi_{NL} = \int_0^L (\beta' - \beta)\, dz = \int_0^L \bar{\gamma} P(z)\, dz = \bar{\gamma} P_{in} L_{\text{eff}}, \tag{2.6.7}$$

where $P(z) = P_{in} \exp(-\alpha z)$ accounts for the fiber loss and L_{eff} is defined in Eq. (2.6.2). In obtaining Eq. (2.6.7), P_{in} is assumed to be constant. In practice, time dependence of P_{in} makes ϕ_{NL} to vary with time, resulting in frequency chirping, which in turn affects the pulse shape through GVD. To reduce the impact of nonlinear refraction, it is necessary that $\phi_{NL} \ll 1$. By replacing L_{eff} by $1/\alpha$ for long fibers, this condition becomes

$$P_{in} \ll \alpha/\bar{\gamma}. \tag{2.6.8}$$

Typically, $\bar{\gamma} \approx 2$ W^{-1}km^{-1}. By using $\alpha = 0.2$ dB/km, the input power is limited to below 22 mW. Clearly, the power dependence of the refractive index can be a limiting factor for optical communication systems. The nonlinear phenomenon responsible for this limitation is referred to as *self-phase modulation* (SPM) since the phase shift ϕ_{NL} is induced by the optical field itself. SPM leads to considerable spectral broadening of pulses propagating inside the optical fiber [68].

The intensity dependence of the refractive index in Eq. (2.6.5) can also lead to another nonlinear phenomenon, known as *cross-phase modulation* (XPM). It occurs when two or more channels are transmitted simultaneously inside the fiber by using different carrier frequencies. The nonlinear phase shift for a specific channel depends not only on the power of that channel but also on the power of other channels [69]. The phase shift for the jth channel becomes [28]

$$\phi_j^{NL} = \bar{\gamma} L_{\text{eff}} \left(P_j + 2 \sum_{m \neq j}^{M} P_m \right), \qquad (2.6.9)$$

where M is the total number of channels and P_j is the channel power ($j = 1$ to M). The factor of 2 in Eq. (2.6.9) indicates that XPM is twice as effective as SPM for the same amount of power. The total phase shift now depends on the power in all channels and would vary from bit to bit depending on the bit pattern of the neighboring channels. If we assume equal channel powers, the phase shift in the worst case in which all channels simultaneously carry 1 bits is given by

$$\phi_j^{NL} = (\bar{\gamma}/\alpha)(2M - 1)P_j. \qquad (2.6.10)$$

To keep $\phi_j^{NL} \ll 1$, the channel power is limited to below 1 mW even for $M = 10$ if we use typical values of $\bar{\gamma}$ and α at 1.55 μm. Clearly, XPM can be a major power-limiting factor and is considered in Section 7.3 in the context of multichannel lightwave systems.

The preceding discussion has implicitly assumed that SPM and XPM act without significant dispersive effects and is valid for relatively wide optical pulses (> 100 ps). For shorter optical pulses the dispersive and nonlinear effects act together on the pulse and lead to new features. In particular, dispersion-induced broadening of optical pulses is considerably reduced in the presence of SPM and anomalous GVD [70]. In fact, an optical pulse can propagate without distortion if the peak power of the pulse is chosen to correspond to the fundamental soliton. Solitons and soliton-based communication systems are discussed in Chapter 10.

2.6.3 Four-Wave Mixing

The intensity dependence of the refractive index discussed in Section 2.6.2 has its origin in the third-order nonlinear susceptibility denoted by $\chi^{(3)}$. Another nonlinear phenomenon, known as *four-wave mixing* (FWM), also originates from a finite value of $\chi^{(3)}$ in silica fibers [28]. If three optical fields with carrier frequencies ω_1, ω_2, and ω_3 copropagate inside the fiber simultaneously, $\chi^{(3)}$ generates a fourth field whose frequency ω_4 is related to other frequencies by a relation $\omega_4 = \omega_1 \pm \omega_2 \pm \omega_3$. Several frequencies corresponding to different plus and minus sign combinations are possible in principle. In practice, most of these combinations do not build up because of a phase-matching requirement [28]. Frequency combinations of the form $\omega_4 = \omega_1 + \omega_2 - \omega_3$ are most troublesome for multichannel communication systems since they can become nearly phase-matched when channels wavelengths lie close to the zero-dispersion wavelength. In fact, commercial dispersion-shifted fibers are often designed to have some residual dispersion at the operating wavelength to suppress FWM.

On a fundamental level, a FWM process can be viewed as a scattering process in which two photons of energies $\hbar\omega_1$ and $\hbar\omega_2$ create two new photons of energies $\hbar\omega_3$ and $\hbar\omega_4$. The *phase-matching* condition then stems from the requirement of momentum conservation. The FWM process can also occur when the two initial photons are degenerate ($\hbar\omega_1 = \hbar\omega_2$), so that $\omega_4 = 2\omega_1 - \omega_3$.

FWM is not of concern for single-channel lightwave systems, but it becomes a limiting factor for multichannel systems making use of wavelength-division multiplexing. Significant amount of channel power may be transferred to neighboring channels through FWM. Such an energy transfer not only results in the power loss for a specific channel but also leads to *interchannel crosstalk* that degrades system performance. This issue is discussed in Section 7.3.2. FWM can also be useful for lightwave systems. It has been used for channel demultiplexing when time-division multiplexing is used in the optical domain (see Section 7.4). Since 1993, FWM in optical fibers has been used for generating a spectrally inverted signal through the process of *optical phase conjugation*. As discussed in Chapter 9, this technique is useful for dispersion compensation and can improve the performance of dispersion-limited lightwave systems.

2.7   FIBER MANUFACTURING

The final section is devoted to the engineering aspects of optical fibers. Manufacturing of fiber cables, suitable for installation in an actual lightwave system, involves sophisticated technology with attention to many practical details. Since such details are available in several texts [12]–[17], the discussion here is intentionally brief.

2.7.1   Design Issues

In its simplest form a step-index fiber consists of a cylindrical core surrounded by a cladding layer whose index is slightly lower than the core. Both core and cladding use silica as the base material; the difference in the refractive indices is realized by doping the core, or the cladding, or both. Dopants such as GeO_2 and P_2O_5 increase the refractive index of silica and are suitable for the core. On the other hand, dopants such as B_2O_3 and fluorine decrease the refractive index of silica and are suitable for the cladding. The major design issues are related to the refractive-index profile, the amount of dopants, and the core and cladding dimensions [71]–[75]. The diameter of the outermost cladding layer has the standard value of 125 μm for all communication-grade fibers.

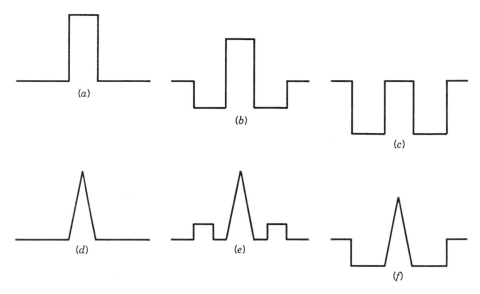

Figure 2.16 Several index profiles used in the design of single-mode fibers. Upper and lower rows correspond to standard and dispersion-shifted fibers, respectively.

Figure 2.16 shows typical index profiles that have been used for different kinds of fibers. The top row corresponds to standard fibers which are designed to have minimum dispersion near 1.3 μm with a cutoff wavelength in the range 1.1–1.2 μm. The simplest design [Fig. 2.16(a)] consists of a pure-silica cladding and a core doped with GeO_2 to obtain $\Delta \approx 3 \times 10^{-3}$. A commonly used variation [Fig. 2.16(b)] lowers the cladding index over a region adjacent to the core by doping it with fluorine. It is also possible to have an undoped core by using a design shown in Fig 2.16(c). The fibers of this kind are referred to as doubly clad or *depressed-cladding fibers* [71]. They are also called W fibers, reflecting the shape of the index profile. The bottom row in Fig. 2.16 shows three index profiles used for dispersion-shifted fibers with minimum dispersion occurring near 1.55 μm, where the fiber loss is also minimum. A triangular index profile with a depressed or raised cladding is often used for this purpose [72]–[74]. The refractive indices and the thickness of different layers are optimized to design a fiber with desirable dispersion characteristics [75]. Sometimes as many as four cladding layers are used for dispersion-flattened fibers (see Fig. 2.11).

2.7.2 Fabrication Methods

Fabrication of telecommunication-grade silica fibers involves two stages. In the first stage a vapor-deposition method is used to make a *cylindrical preform* with the desired refractive-index profile. It is typically 1 m long and 2 cm in diameter and contains core and cladding layers with correct relative dimensions. In the second stage, the preform is drawn into a fiber by using a precision-feed mechanism that feeds the preform into a furnace at the proper speed.

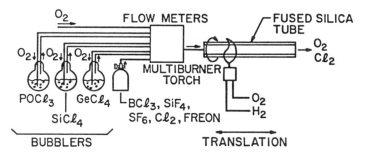

Figure 2.17 MCVD process commonly used for fiber fabrication. (After Ref. [76]. ©1985 Academic Press. Reprinted with permission.)

Several methods can be used to make the preform. The three commonly used methods [76]–[78] are modified chemical-vapor deposition (MCVD), outside-vapor deposition (OVD), and vapor-axial deposition (VAD). Figure 2.17 shows a schematic diagram of the MCVD process. In this process, successive layers of SiO_2 are deposited on the inside of a fused silica tube by mixing the vapors of $SiCl_4$ and O_2 at a temperature of about 1800°C. To ensure uniformity, a multi-burner torch is moved back and forth across the tube length using an automatic translation stage. The refractive index of the cladding layers is controlled by adding fluorine to the tube. When a sufficient cladding thickness has been deposited, the core is formed by adding the vapors of $GeCl_4$ or $POCl_3$. These vapors react with oxygen to form the dopants GeO_2 and $P2O_5$:

$$GeCl_4 + O_2 \quad \rightarrow \quad GeO_2 + 2Cl_2,$$
$$4POCl_3 + 3O_2 \quad \rightarrow \quad 2P_2O_5 + 6Cl_2.$$

The flow rate of $GeCl_4$ or $POCl_3$ determines the amount of dopant and the corresponding increase in the refractive index of the core. A graded-index fiber can be fabricated simply by varying the flow rate from layer to layer. When all layers forming the core have been deposited, the torch temperature is raised to collapse the tube into a solid rod of preform.

The MCVD process is also known as the *inner-vapor-deposition method*, as the core and cladding layers are deposited inside a silica tube. In a related process, known as the *plasma-activated chemical vapor deposition* process [79], the chemical reaction is initiated by a microwave plasma. By contrast, in the OVD and VAD processes the core and cladding layers are deposited on the outside of a rotating mandrel by using the technique of *flame hydrolysis*. The mandrel is removed prior to sintering. The porous soot boule is then placed in a sintering furnace to form a glass boule. The central hole allows an efficient way of reducing water vapors through dehydration in a controlled atmosphere of Cl_2–He mixture, although it results in a central dip in the index profile. The dip can be minimized by closing the hole during sintering.

The fiber drawing step [80] is essentially the same irrespective of the process used to make the preform. Figure 2.18 shows the drawing apparatus schematically. The preform is fed into a furnace in a controlled manner where it is

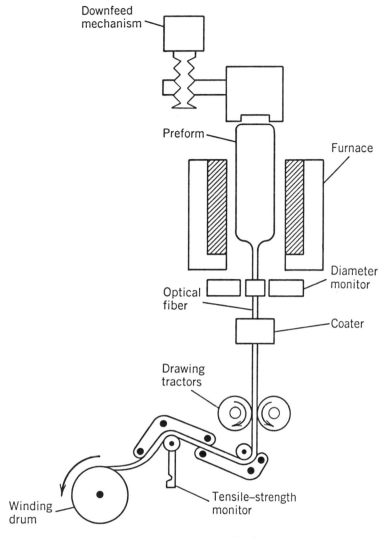

Figure 2.18 Apparatus used for fiber drawing.

heated to a temperature of about 2000°C. The melted preform is drawn into
a fiber by using a precision-feed mechanism. The fiber diameter is monitored
and controlled optically by using the diffraction pattern, produced by a laser. A
change in the diameter changes the diffraction pattern, which in turn changes
the photodiode current. This current change acts as a signal for a servocontrol
mechanism that adjusts the winding rate of the fiber. The fiber diameter can be
kept constant to within 0.1% by this technique. A polymer coating is applied
to the fiber during the drawing step. It serves a dual purpose, as it provides
mechanical protection and preserves the transmission properties of the fiber.
The typical diameter of the coated fiber is about 250 μm, although it can be

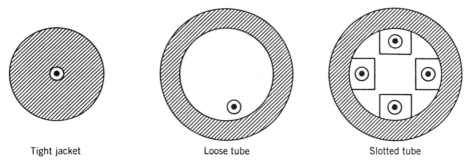

Figure 2.19 Typical designs for light-duty fiber cables.

as large as 900 μm when multiple coatings are used. The tensile strength of the fiber is monitored during its winding on the drum. The winding rate is typically 0.2–0.5 m/s. Several hours are required to draw fiber from a single preform. This brief discussion is intended to give a general idea. The fabrication of optical fiber generally requires attention to a large number of engineering details discussed in several texts [16], [17].

2.7.3   Cables and Connectors

Cabling of fibers is necessary to protect them from deterioration during transportation and installation [81]. Cable design depends on the type of application. For some applications it may be enough to buffer the fiber by placing it inside a plastic jacket. For others the cable must be made mechanically strong by using strengthening elements such as steel rods.

A light-duty cable is made by surrounding the fiber by a buffer jacket of hard plastic. Figure 2.19 shows three simple cable designs. A tight jacket can be provided by applying a buffer plastic coating of 0.5–1 mm thickness on top of the primary coating applied during the drawing process. In an alternative approach the fiber lies loosely inside a plastic tube. Microbending losses are nearly eliminated in this loose-tube construction, since the fiber can adjust itself within the tube. This construction can also be used to make multifiber cables by using a slotted tube with a different slot for each fiber.

Heavy-duty cables use steel or a strong polymer such as Kevlar to provide the mechanical strength. Figure 2.20 shows schematically three kinds of cables. In the loose-tube construction, fiberglass rods embedded in polyurethane and a Kevlar jacket provide the necessary mechanical strength [Fig. 2.20(a)]. The same design can be extended to multifiber cables by placing several loose-tube fibers around a central steel core [Fig. 2.20(b)]. When a large number of fibers need to be placed inside a single cable, a ribbon cable is used [Fig. 2.20(c)]. The ribbon is manufactured by packaging typically 12 fibers between two polyester tapes. Several ribbons are then stacked into a rectangular array which is placed inside a polyethylene tube. The mechanical strength is provided by using steel

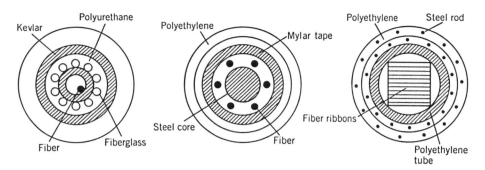

Figure 2.20 Typical designs for heavy-duty fiber cables.

rods in the two outermost polyethylene jackets. The outer diameter of such fiber cables is about 1–1.5 cm.

Connectors are needed to use optical fibers in an actual communication system. They can be divided into two categories. A permanent joint between two fibers is known as a fiber splice, and a detachable connection between them is realized by using a fiber connector. Connectors are used to link fiber cable with the transmitter (or the receiver), while splices are used to join fiber segments (usually 5–10 km long). The main issue in the use of splices and connectors is related to the loss. Some power is always lost, as the two fiber ends are never perfectly aligned in practice. Splice losses below 0.1 dB are routinely realized by using the technique of fusion splicing [82]. Connector losses are generally larger. State-of-the-art connectors provide an average loss of about 0.3 dB [83]. The technology behind the design of splices and connectors is quite sophisticated. For details, the reader is referred to Ref. [84], a book entirely devoted to this issue.

PROBLEMS

2.1 A multimode fiber with a 50-μm core diameter is designed to limit the intermodal dispersion to 10 ns/km. What is the numerical aperture of this fiber? What is the limiting bit rate for transmission over 10 km at 0.88 μm? Use 1.45 for the refractive index of the cladding.

2.2 Use the ray equation in the paraxial approximation [Eq. (2.1.8)] to prove that intermodal dispersion is zero for a graded-index fiber with a quadratic index profile.

2.3 Use Maxwell's equations to express the field components E_ρ, E_ϕ, H_ρ, and H_ϕ in terms of E_z and H_z and obtain Eqs. (2.2.32)–(2.2.35).

2.4 Derive the eigenvalue equation (2.2.36) by matching the boundary conditions at the core–cladding interface of a step-index fiber.

2.5 A single-mode fiber has an index step $n_1 - n_2 = 0.005$. Calculate the core radius if the fiber has a cutoff wavelength of 1 μm. Estimate the spot size

(FWHM) of the fiber mode and the fraction of the mode power inside the core when this fiber is used at 1.3 μm.

2.6 A 1.55-μm unchirped Gaussian pulse of 100-ps width (FWHM) is launched into a single-mode fiber. Calculate its FWHM after 50 km if the fiber has a dispersion of 16 ps/(km-nm). Neglect the source spectral width.

2.7 Derive an expression for the confinement factor Γ of single-mode fibers defined as the fraction of the total mode power contained inside the core. Use the Gaussian approximation for the fundamental fiber mode. Estimate Γ for $V = 2$.

2.8 A single-mode fiber is measured to have $\lambda^2(d^2n/d\lambda^2) = 0.02$ at 0.8 μm. Calculate the dispersion parameters β_2 and D.

2.9 Show that a chirped Gaussian pulse is compressed initially inside a single-mode fiber when $\beta_2 C < 0$. Derive expressions for the minimum width and the fiber length at which the minimum occurs.

2.10 Estimate the limiting bit rate for a 60-km single-mode fiber link at 1.3- and 1.55-μm wavelengths assuming transform-limited, 50-ps (FWHM) input pulses. Assume that $\beta_2 = 0$ and -20 ps^2/km and $\beta_3 = 0.1$ ps^3/km and 0 at 1.3- and 1.55-μm wavelengths, respectively. Also assume that $V_\omega \ll 1$.

2.11 A 0.88-μm communication system transmits data over a 10-km single-mode fiber by using 10-ns (FWHM) pulses. Determine the maximum bit rate if the LED has a spectral FWHM of 30 nm. Use $D = -80$ ps/(km-nm).

2.12 Use Eq. (2.4.23) to prove that the bit rate of an optical communication system operating at the zero-dispersion wavelength is limited by $BLS\sigma_\lambda^2 < 1$, where $S = dD/d\lambda$ and σ_λ is the RMS spectral width of the Gaussian source spectrum. Assume that $C = 0$ and $V_\omega \ll 1$ in the general expression of the output pulse width.

2.13 Repeat Problem 2.12 for the case of a single-mode semiconductor laser for which $V \ll 1$ and show that the bit rate is limited by $B(|\beta_3 L|)^{1/3} < 0.324$. What is the limiting bit rate for $L = 100$ km if $\beta_3 = 0.1$ ps^3/km?

2.14 An optical communication system is operating with chirped Gaussian input pulses. Assume that $\beta_3 = 0$ and $V_\omega \ll 1$ in Eq. (2.4.23) and obtain a condition on the bit rate in terms of the parameters C, β_2, σ_0, and L.

2.15 A 1.55-μm optical communication system operating at 5 Gb/s uses 100-ps (FWHM) Gaussian pulses chirped such that the chirp parameter $C = -6$. What is the dispersion-limited maximum fiber length, and how does it change if the pulses were unchirped? Neglect the laser linewidth and assume that $\beta_2 = -20$ ps^2/km.

2.16 A 1.3-μm lightwave system uses a 50-km fiber link and requires at least 0.3 mW at the receiver. The fiber loss is 0.5 dB/km. Fiber is spliced every 5 km and has two connectors of 1-dB loss at both ends. Splice loss is only 0.2 dB. Determine the minimum power that must be launched into the fiber.

2.17 Discuss various mechanisms, both intrinsic and extrinsic, that contribute to fiber loss. Which mechanism imposes a fundamental limit on fiber loss?

2.18 Calculate the threshold power for stimulated Brillouin scattering for a 50-km fiber link operating at 1.3 μm and having a loss of 0.5 dB/km. How much does the threshold power change if the operating wavelength is changed to 1.55 μm, where the fiber loss is only 0.2 dB/km? Assume that $A_{\text{eff}} = 20 \ \mu\text{m}^2$ and $g_B = 5 \times 10^{-11}$ m/W at both wavelengths.

2.19 Calculate the power launched into a 40-km-long single-mode fiber for which the SPM-induced nonlinear phase shift becomes 180°. Assume that $\lambda = 1.55 \ \mu$m, $A_{\text{eff}} = 40 \ \mu\text{m}^2$, $\alpha = 0.2$ dB/km, and $\bar{n}_2 = 3.2 \times 10^{-20}$ m^2/W.

2.20 Consider a 1.55-μm lightwave system designed to transmit five channels over a 20-km single-mode fiber. Each channel carries 1 mW of average power in the form of a PCM bit stream. Calculate the maximum and minimum nonlinear phase shift occurring because of the intensity dependence of the refractive index.

REFERENCES

[1] J. Tyndall, *Proc. Roy. Inst.* **1**, 446 (1854).

[2] J. L. Baird, British Patent 285,738 (1927).

[3] C. W. Hansell, U.S. Patent 1,751,584 (1930).

[4] H. Lamm, *Z. Instrumentenk.* **50**, 579 (1930).

[5] A. C. S. van Heel, *Nature* **173**, 39 (1954).

[6] B. I. Hirschowitz, L. E. Curtiss, C. W. Peters, and H. M. Pollard, *Gastroenterology* **35**, 50 (1958).

[7] N. S. Kapany, *J. Opt. Soc. Am.* **49**, 779 (1959).

[8] N. S. Kapany, *Fiber Optics: Principles and Applications*, Academic Press, San Diego, CA, 1967.

[9] K. C. Kao and G. A. Hockham, *Proc. IEE* **113**, 1151 (1966); A. Werts, *Onde Electr.* **45**, 967 (1966).

[10] F. P. Kapron, D. B. Keck, and R. D. Maurer, *Appl. Phys. Lett.* **17**, 423 (1970).

[11] T. Miya, Y. Terunuma, T. Hosaka, and T. Miyoshita, *Electron. Lett.* **15**, 106 (1979).

[12] M. J. Adams, *An Introduction to Optical Waveguides*, Wiley, New York, 1981.

[13] T. Okoshi, *Optical Fibers*, Academic Press, San Diego, CA, 1982.

[14] A. W. Snyder and J. D. Love, *Optical Waveguide Theory*, Chapman & Hall, London, 1983.

[15] L. B. Jeunhomme, *Single-Mode Fiber Optics*, Marcel Dekker, New York, 1983; 2nd ed., 1990.

[16] T. Li, Ed., *Optical Fiber Communications*, Vol. 1, Academic Press, San Diego, CA, 1985.

[17] T. Izawa and S. Sudo, *Optical Fibers: Materials and Fabrication*, Kluwer Academic, Boston, 1987.

[18] E. G. Neumann, *Single-Mode Fibers*, Springer-Verlag, New York, 1988.

[19] D. Marcuse, *Theory of Dielectric Optical Waveguides*, 2nd ed., Academic Press, San Diego, CA, 1991.

[20] G. Cancellieri, *Single-Mode Optical Fibers*, Pergamon Press, Elmsford, NY, 1991.

[21] J. A. Buck, *Fundamentals of Optical Fibers*, Wiley, New York, 1995.

[22] M. Born and E. Wolf, *Principles of Optics*, 6th ed., Pergamon Press, Elmsford, NY, 1980.

[23] J. Gower, *Optical Communication Systems*, 2nd ed., Prentice Hall, London, 1993.

[24] Y. Koike, T. Ishigure, and E. Nihei, *J. Lightwave Technol.* **13**, 1475 (1995).

[25] T. Ishigure, A. Horibe, E. Nihei, and Y. Koike, *J. Lightwave Technol.* **13**, 1686 (1995).

[26] U. Fiedler, G. Reiner, P. Schnitzer, and K. J. Ebeling, *IEEE Photon. Technol. Lett.* **8**, 746 (1996).

[27] P. Diament, *Wave Transmission and Fiber Optics*, Macmillan, New York, 1990, Chap. 3.

[28] G. P. Agrawal, *Nonlinear Fiber Optics*, 2nd ed., Academic Press, San Diego, CA, 1995.

[29] M. Abramowitz and I. A. Stegun, Eds., *Handbook of Mathematical Functions*, Dover, New York, 1970, Chap. 9.

[30] D. Gloge, *Appl. Opt.* **10**, 2252 (1971); **10**, 2442 (1971).

[31] D. B. Keck, in *Fundamentals of Optical Fiber Communications*, M. K. Barnoski, Ed., Academic Press, San Diego, CA, 1981.

[32] D. Marcuse, *J. Opt. Soc. Am.* **68**, 103 (1978).

[33] I. H. Malitson, *J. Opt. Soc. Am.* **55**, 1205 (1965).

[34] L. G. Cohen, C. Lin, and W. G. French, *Electron. Lett.* **15**, 334 (1979).

[35] C. T. Chang, *Electron. Lett.* **15**, 765 (1979); *Appl. Opt.* **18**, 2516 (1979).

[36] L. G. Cohen, W. L. Mammel, and S. Lumish, *Opt. Lett.* **7**, 183 (1982).

[37] S. J. Jang, L. G. Cohen, W. L. Mammel, and M. A. Shaifi, *Bell Syst. Tech. J.* **61**, 385 (1982).

[38] V. A. Bhagavatula, M. S. Spotz, W. F. Love, and D. B. Keck, *Electron. Lett.* **19**, 317 (1983).

[39] P. Bachamann, D. Leers, H. Wehr, D. V. Wiechert, J. A. van Steenwijk, D. L. A. Tjaden, and E. R. Wehrhahn, *J. Lightwave Technol.* **4**, 858 (1986).

[40] B. J. Ainslie and C. R. Day, *J. Lightwave Technol.* **4**, 967 (1986).

[41] C. D. Poole, J. H. Winters, and J. A. Nagel, *Opt. Lett.* **16**, 372 (1991).

[42] N. Gisin and J. P. Von der Weid, *J. Lightwave Technol.* **9**, 821 (1991).

[43] G. J. Foschini and C. D. Poole, *J. Lightwave Technol.* **9**, 1439, (1991).

[44] J. Zhou and M. J. O'Mahony, *IEEE Photon. Technol. Lett.* **6**, 1265 (1994).

[45] E. Lichtman, *J. Lightwave Technol.* **13**, 898 (1995).

[46] Y. Suetsugu, T. Kato, and M. Nishimura, *IEEE Photon. Technol. Lett.* **7**, 887 (1995).

[47] P. K. A. Wai and C. R. Menyuk, *J. Lightwave Technol.* **14**, 148 (1996).

[48] D. Marcuse, *Appl. Opt.* **19**, 1653 (1980); **20**, 3573 (1981).

[49] M. Miyagi and S. Nishida, *Appl. Opt.* **18**, 678 (1979); **18**, 2237 (1979).

[50] G. P. Agrawal and M. J. Potasek, *Opt. Lett.* **11**, 318 (1986).

[51] M. Schwartz, *Information, Transmission, Modulation, and Noise*, 4th ed., McGraw-Hill, New York, 1990, Chap. 2.

[52] M. J. Bennett, *IEE Proc.* **130**, Pt. H, 309 (1983).

[53] D. Gloge, K. Ogawa, and L. G. Cohen, *Electron. Lett.* **16**, 366 (1980).

[54] P. Klocek and G. H. Sigel, Jr., *Infrared Fiber Optics*, Vol. TT2, SPIE, Bellingham, WA, 1989.

[55] T. Katsuyama and H. Matsumura, *Infrared Optical Fibers*, Bristol, Philadelphia, 1989.

[56] J. A. Harrington, Ed., *Infrared Fiber Optics*, SPIE, Bellingham, WA, 1990.

[57] M. F. Churbanov, *J. Non-Cryst. Solids* **184**, 25 (1995).

[58] E. A. J. Marcatili, *Bell Syst. Tech. J.* **48**, 2103 (1969).

[59] W. B. Gardner, *Bell Syst. Tech. J.* **54**, 457 (1975).

[60] D. Marcuse, *Bell Syst. Tech. J.* **55**, 937 (1976).

[61] K. Petermann, *Electron. Lett.* **12**, 107 (1976); *Opt. Quantum Electron.* **9**, 167 (1977).

[62] K. Tanaka, S. Yamada, M. Sumi, and K. Mikoshiba, *Appl. Opt.* **16**, 2391 (1977).

[63] W. A. Gambling, H. Matsumura, and C. M. Rodgal, *Opt. Quantum Electron.* **11**, 43 (1979).

[64] R. G. Smith, *Appl. Opt.* **11**, 2489 (1972).

[65] R. H. Stolen, E. P. Ippen, and A. R. Tynes, *Appl. Phys. Lett.* **20**, 62 (1972).

[66] E. P. Ippen and R. H. Stolen, *Appl. Phys. Lett.* **21**, 539 (1972).

[67] D. Cotter, *Electron. Lett.* **18**, 495 (1982).

[68] R. H. Stolen and C. Lin, *Phys. Rev. A* **17**, 1448 (1978).

[69] A. R. Chraplyvy, D. Marcuse, and P. S. Henry, *J. Lightwave Technol.* **2**, 6 (1984).

[70] M. J. Potasek and G. P. Agrawal, *Electron. Lett.* **22**, 759 (1986).

[71] M. Monerie, *IEEE J. Quantum Electron.* **18**, 535 (1982); *Electron. Lett.* **18**, 642 (1982).

[72] M. A. Saifi, S. J. Jang, L. G. Cohen, and J. Stone, *Opt. Lett.* **7**, 43 (1982).

[73] Y. W. Li, C. D. Hussey, and T. A. Birks, *J. Lightwave Technol.* **11**, 1812 (1993).

[74] R. Lundin, *Appl. Opt.* **32**, 3241 (1993); *Appl. Opt.* **33**, 1011 (1994).

[75] S. P. Survaiya and R. K. Shevgaonkar, *IEEE Photon. Technol. Lett.* **8**, 803 (1996).

[76] S. R. Nagel, J. B. MacChesney, and K. L. Walker, in *Optical Fiber Communications*, Vol. 1, T. Li, Ed., Academic Press, San Diego, CA, 1985, Chap. 1.

[77] A. J. Morrow, A. Sarkar, and P. C. Schultz, in *Optical Fiber Communications*, Vol. 1, T. Li, Ed., Academic Press, San Diego, CA, 1985, Chap. 2.

[78] N. Niizeki, N. Ingaki, and T. Edahiro, in *Optical Fiber Communications*, Vol. 1, T. Li, Ed., Academic Press, San Diego, CA, 1985, Chap. 3.

[79] P. Geittner, H. J. Hagemann, J. Warnier, and H. Wilson, *J. Lightwave Technol.* **4**, 818 (1986).

[80] F. V. DiMarcello, C. R. Kurkjian, and J. C. Williams, in *Optical Fiber Communications*, Vol. 1, T. Li, Ed., Academic Press, San Diego, CA, 1985, Chap. 4.

[81] H. Murata, *Handbook of Optical Fibers and Cables*, Marcel Dekker, New York, 1996.

[82] S. C. Mettler and C. M. Miller, in *Optical Fiber Telecommunications II*, S. E. Miller and I. P. Kaminow, Eds., Academic Press, San Diego, CA, 1988, Chap. 6.

[83] W. C. Young and D. R. Frey, in *Optical Fiber Telecommunications II*, S. E. Miller and I. P. Kaminow, Eds., Academic Press, San Diego, CA, 1988, Chap. 7.

[84] C. M. Miller, S. C. Mettler, and I. A. White, *Optical Fiber Splices and Connectors*, Marcel Dekker, New York, 1986.

Chapter 3

OPTICAL
TRANSMITTERS

The role of the optical transmitter is to convert an electrical input signal into the corresponding optical signal and then launch it into the optical fiber serving as a communication channel. The major component of optical transmitters is an optical source. Fiber-optic communication systems often use semiconductor optical sources such as light-emitting diodes (LEDs) and semiconductor lasers because of several inherent advantages offered by them. Some of these advantages are compact size, high efficiency, good reliability, right wavelength range, small emissive area compatible with fiber-core dimensions, and possibility of direct modulation at relatively high frequencies. The last property is of considerable practical value, as it eliminates the need for an external modulator inside the optical transmitter. Although the operation of semiconductor lasers was demonstrated as early as 1962 [1]–[4], their use became practical only after 1970, when semiconductor lasers operating continuously at room temperature became available [5], [6]. Since then, semiconductor lasers have been developed extensively because of their importance for optical communications. Semiconductor lasers are also known as laser diodes or injection lasers. Their properties have been discussed in several books [7]–[17].

This chapter is devoted to optical transmitters, with emphasis on the operating characteristics of LEDs and semiconductor lasers. The basic concepts needed for an understanding of semiconductor optical sources are introduced in Section 3.1. LEDs are covered in Section 3.2, while Section 3.3 is devoted to semiconductor lasers. The emphasis is placed not only on the operating characteristics but also on the design aspects from the standpoint of fiber-optic communication systems. The discussion of semiconductor lasers includes a subsection on the design of lasers operating predominantly in a single longitudinal mode, required to reduce the impact of fiber dispersion. The design issues related to optical transmitters are considered in Section 3.4 with emphasis on source-fiber coupling and optoelectronic integration.

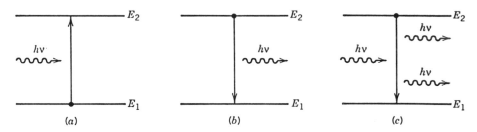

Figure 3.1 Three fundamental processes occurring between the two energy states of an atom: (a) spontaneous emission; (b) stimulated emission; and (c) absorption

3.1   BASIC CONCEPTS

Under normal conditions, all materials absorb light rather than emit it. The absorption process can be understood by referring to Fig. 3.1, where the energy levels E_1 and E_2 correspond to the ground state and the excited state of atoms of the absorbing medium. If the photon energy $h\nu$ of the incident light of frequency ν is about the same as the energy difference $E_g = E_2 - E_1$, the photon is absorbed by the atom, which ends up in the excited state. Incident light is attenuated as a result of many such absorption events occurring inside the medium.

The excited atoms eventually return to their normal "ground" state and emit light in the process. Light emission can occur through two fundamental processes known as *spontaneous emission* and *stimulated emission*. Both are shown schematically in Fig. 3.1. In the case of spontaneous emission, photons are emitted in random directions with no phase relationship among them. Stimulated emission, by contrast, is initiated by an existing photon. The remarkable feature of stimulated emission is that the emitted photon matches the original photon not only in energy (or in frequency), but also in its other characteristics, such as the direction of propagation. All lasers, including semiconductor lasers, emit light through the process of stimulated emission and are said to emit coherent light. In contrast, LEDs emit light through the incoherent process of spontaneous emission.

3.1.1   Emission and Absorption Rates

Before discussing the emission and absorption rates in semiconductors, it is instructive to consider a *two-level atomic system* interacting with a radiation field through transitions shown in Fig. 3.1. If N_1 and N_2 are the atomic densities in the ground and the excited states, respectively, and $\rho_{ph}(\nu)$ is the spectral density of the radiation energy, the rates of spontaneous emission, stimulated emission, and absorption can be written as [18]

$$R_{\text{spon}} = AN_2, \qquad R_{\text{stim}} = BN_2\rho_{ph}, \qquad R_{\text{abs}} = B'N_1\rho_{ph}, \qquad (3.1.1)$$

where A, B, and B' are constants. In *thermal equilibrium*, the atomic densities are distributed according to the *Boltzmann statistics* [17], i.e.,

$$N_2/N_1 = \exp(-E_g/k_BT) = \exp(-h\nu/k_BT), \qquad (3.1.2)$$

where k_B is the Boltzmann constant and T is the absolute temperature. Since N_1 and N_2 do not change with time in thermal equilibrium, the upward and downward transition rates should be equal, or

$$AN_2 + BN_2\rho_{ph} = B'N_1\rho_{ph}. \qquad (3.1.3)$$

By using Eq. (3.1.2) in Eq. (3.1.3), the spectral density ρ_{ph} becomes

$$\rho_{ph} = \frac{A/B}{(B'/B)\exp(h\nu/k_BT) - 1}. \qquad (3.1.4)$$

In thermal equilibrium, ρ_{ph} should be identical with the spectral density of blackbody radiation given by *Planck's formula* [17]

$$\rho_{ph} = \frac{8\pi h\nu^3/c^3}{\exp(h\nu/k_BT) - 1}. \qquad (3.1.5)$$

A comparison of Eqs. (3.1.4) and (3.1.5) provides the relations

$$A = (8\pi h\nu^3/c^3)B \qquad \text{and} \qquad B' = B. \qquad (3.1.6)$$

These relations were first obtained by Einstein [19]. For this reason, A and B are called *Einstein's coefficients*.

Two important conclusions can be drawn from Eqs. (3.1.1)–(3.1.6). First, R_{spon} can exceed both R_{stim} and R_{abs} considerably if $k_BT > h\nu$. Thermal sources operate in this regime. Second, for radiation in the visible or near-infrared region ($h\nu \sim 1$ eV), spontaneous emission always dominates over stimulated emission in thermal equilibrium at room temperature ($k_BT \approx 25$ meV) since

$$R_{stim}/R_{spon} = [\exp(h\nu/k_BT) - 1]^{-1} \ll 1. \qquad (3.1.7)$$

Thus, all lasers must operate away from thermal equilibrium. This is achieved by pumping lasers with an external energy source.

Even for an atomic system pumped externally, stimulated emission may not be the dominant process since it has to compete with the absorption process. R_{stim} can exceed R_{abs} only when $N_2 > N_1$. This condition is referred to as *population inversion* and is never realized for systems in thermal equilibrium [see Eq. (3.1.2)]. Population inversion is a prerequisite for laser operation. In atomic systems, it is achieved by using three- and four-level pumping schemes [17] such that an external energy source raises the atomic population from the ground state to an excited state lying above the energy state E_2 in Fig. 3.1.

The emission and absorption rates in semiconductors should take into account the energy bands associated with a semiconductor [20]. Figure 3.2 shows

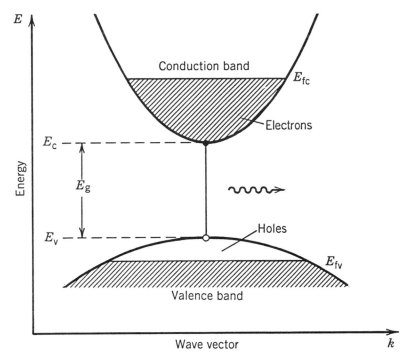

Figure 3.2 Conduction and valence bands of a semiconductor. Electrons in the conduction band and holes in the valence band can recombine and emit a photon through spontaneous emission as well as through stimulated emission.

the emission process schematically using the simplest band structure, consisting of parabolic conduction and valence bands in the energy–wavevector space (E–$\mathbf{k}$ diagram). Spontaneous emission can occur only if the energy state E_2 is occupied by an electron and the energy state E_1 is empty (i.e., occupied by a hole). The occupation probability for electrons in the conduction and valence bands is given by the *Fermi–Dirac distributions* [20]

$$f_c(E_2) = \{1 + \exp[(E_2 - E_{fc})/k_B T]\}^{-1}, \qquad (3.1.8)$$
$$f_v(E_1) = \{1 + \exp[(E_1 - E_{fv})/k_B T]\}^{-1}, \qquad (3.1.9)$$

where E_{fc} and E_{fv} are the Fermi levels. The total spontaneous emission rate at a frequency ω is obtained by summing over all possible transitions between the two bands such that $E_2 - E_1 = E_{ph} = \hbar\omega$, where $\omega = 2\pi\nu$, $\hbar = h/2\pi$, and E_{ph} is the energy of the emitted photon. The result is

$$R_{\text{spon}}(\omega) = \int_{E_c}^{\infty} A(E_1, E_2) f_c(E_2)[1 - f_v(E_1)]\rho_{cv}\, dE_2, \qquad (3.1.10)$$

where ρ_{cv} is the *joint density of states*, defined as the number of states per unit volume per unit energy range, and is given by [17]

$$\rho_{cv} = \frac{(2m_r)^{3/2}}{2\pi^2 \hbar^3} (\hbar\omega - E_g)^{1/2}. \tag{3.1.11}$$

In this equation, E_g is the bandgap and m_r is the reduced mass, defined as $m_r = m_c m_v/(m_c + m_v)$, where m_c and m_v are the effective masses of electrons and holes in the conduction and valence bands, respectively. Since ρ_{cv} is independent of E_2 in Eq. (3.1.10), it can be taken outside the integral. By contrast, $A(E_1, E_2)$ generally depends on E_2 and is related to the momentum matrix element in a semiclassical perturbation approach commonly used to calculate it [10].

The stimulated emission and absorption rates can be obtained in a similar manner and are given by

$$R_{\text{stim}}(\omega) = \int_{E_c}^{\infty} B(E_1, E_2) f_c(E_2)[1 - f_v(E_1)]\rho_{cv}\rho_{ph} \, dE_2, \tag{3.1.12}$$

$$R_{\text{abs}}(\omega) = \int_{E_c}^{\infty} B(E_1, E_2) f_v(E_1)[1 - f_c(E_2)]\rho_{cv}\rho_{ph} \, dE_2, \tag{3.1.13}$$

where $\rho_{ph}(\omega)$ is the spectral density of photons introduced in a manner similar to Eq. (3.1.1). The *population-inversion condition* $R_{\text{stim}} > R_{\text{abs}}$ is obtained by comparing Eqs. (3.1.12) and (3.1.13), resulting in $f_c(E_2) > f_v(E_1)$. If we use Eqs. (3.1.8) and (3.1.9), this condition is satisfied when

$$E_{fc} - E_{fv} > E_2 - E_1 > E_g. \tag{3.1.14}$$

Since the minimum value of $E_2 - E_1$ equals E_g, the separation between the Fermi levels must exceed the bandgap for population inversion to occur [21]. In thermal equilibrium, the two Fermi levels coincide ($E_{fc} = E_{fv}$). They can be separated by pumping energy into the semiconductor from an external energy source. The most convenient way to pump a semiconductor is by using a forward-biased p–n junction, discussed in the next subsection.

3.1.2 p–n Junctions

At the heart of a semiconductor optical source is the p–n junction, formed by bringing a p-type and an n-type semiconductor into contact. Recall that a semiconductor is made n-type or p-type by doping it with impurities whose atoms have an excess valence electron or one less electron compared with the semiconductor atoms. In the case of n-type semiconductor, the excess electrons occupy the conduction-band states, normally empty in undoped (intrinsic) semiconductors. The occupation probability is given by Eq. (3.1.8); the Fermi level moves toward the conduction band as the dopant concentration increases. (The Fermi level lies in the middle of the bandgap for intrinsic semiconductors.) For a heavily doped n-type semiconductor, the Fermi level E_{fc} lies inside the conduction band; such semiconductors are said to be degenerate. Similarly, the Fermi level E_{fv} moves toward the valence band for p-type semiconductors and

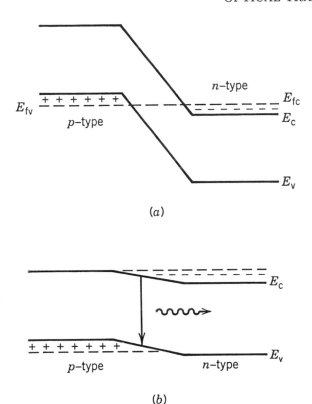

(a)

(b)

Figure 3.3 Energy-band diagram of a p–n junction (a) in thermal equilibrium and (b) under forward bias.

lies inside it under heavy doping. In thermal equilibrium, the Fermi level must be continuous across the p–n junction. This is achieved through diffusion of electrons and holes across the junction. The charged impurities left behind set up an electric field strong enough to prevent further diffusion of electrons and holds under equilibrium conditions. This field is referred to as the built-in electric field. Figure 3.3(a) shows the energy-band diagram of a p–n junction in thermal equilibrium.

When a p–n junction is forward biased by applying an external voltage, the built-in electric field is reduced. This reduction results in diffusion of electrons and holes across the junction. An electric current begins to flow as a result of carrier diffusion. The current I increases exponentially with the applied voltage V according to the well-known relation [20]

$$I = I_s[\exp(qV/k_BT) - 1], \qquad (3.1.15)$$

where I_s is the saturation current and depends on the diffusion coefficients associated with electrons and holes. As seen in Fig. 3.3(b), in a region surrounding the junction (known as the depletion region), electrons and holes are present simultaneously when the p–n junction is forward biased. These electrons and

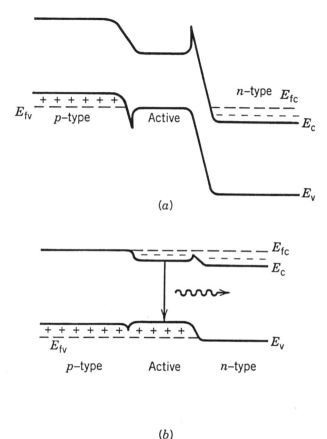

Figure 3.4 Energy-band diagram of a double-heterostructure p–n junction (a) in thermal equilibrium and (b) under forward bias.

holes can recombine through spontaneous or stimulated emission and generate light in a semiconductor optical source.

The p–n junction shown in Fig. 3.3 is called the *homojunction*, since the same semiconductor material is used on both sides of the junction. A problem with the homojunction is that electron–hole recombination occurs over a relatively wide region (~ 1–$10\ \mu$m) determined by the diffusion length of electrons and holes. Since the carriers are not confined to the immediate vicinity of the junction, it is difficult to realize high carrier densities. This carrier-confinement problem can be solved by sandwiching a thin layer between the p-type and n-type layers such that the bandgap of the sandwiched layer is smaller than the layers surrounding it. The middle layer may or may not be doped, depending on the device design; its role is to confine the carriers injected inside it under forward bias. The carrier confinement occurs as a result of bandgap discontinuity at the junction between two semiconductors which have the same crystalline structure (the same lattice constant) but different bandgaps. Such junctions are called *heterojunctions*,

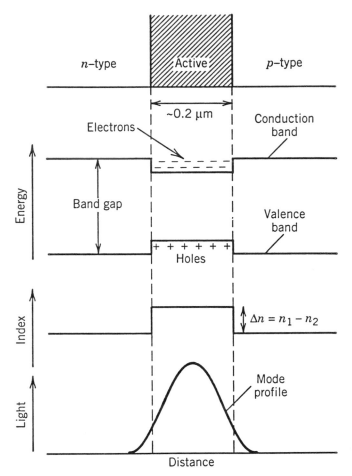

Figure 3.5 Simultaneous confinement of charged carriers and generated light by the use of a double-heterostructure design. The central layer has a lower bandgap and a higher refractive index than those of p-type and n-type cladding layers.

and such devices are called *double heterostructures*. Since the thickness of the sandwiched layer can be controlled externally (typically, $\sim 0.1~\mu$m), high carrier densities can be realized at a given injection current. Figure 3.4 shows the energy-band diagram of a double heterostructure with and without forward bias and should be compared with Fig. 3.3.

The use of a heterostructure geometry for semiconductor optical sources is doubly beneficial. As already mentioned, the bandgap difference between the two semiconductors helps to confine electrons and holes to the middle layer, also called the active layer since light is generated inside it as a result of electron–hole recombination. However, the active layer also has a slightly larger refractive index than the surrounding p-type and n-type cladding layers simply because its bandgap is smaller. As a result of the refractive-index difference, the active layer acts as a dielectric waveguide and supports optical modes whose number can be controlled by controlling the active-layer thickness (see Section 3.3). The main point is that a heterostructure confines the generated light to the active layer because of dielectric waveguiding. Figure 3.5 illustrates schematically simultaneous confinement of charged carriers and the light generated to the active region by the use of a heterostructure design. It is this feature that has made semiconductor lasers practical for their use in optical communications.

3.1.3 Nonradiative Recombination

When a p–n junction is forward-biased, electrons and holes are injected into the active region, where they recombine to produce light. In any semiconductor, electrons and holes can also recombine nonradiatively. Nonradiative recombination mechanisms [20] include recombination at traps or defects, surface recombination, and the Auger recombination. The last mechanism is particularly important for semiconductor optical sources [10] emitting light in the wavelength range 1.3–1.6 μm because of a relatively small bandgap of the active layer. In the Auger recombination process, the energy released during electron–hole recombination is given to another electron or hole as kinetic energy rather than producing light.

From the standpoint of device operation, all nonradiative processes are harmful, as they reduce the number of electron–hole pairs that emit light. Their effect can be incorporated through the *internal quantum efficiency*, defined as

$$\eta_{\text{int}} = \frac{R_{rr}}{R_{\text{tot}}} = \frac{R_{rr}}{R_{rr} + R_{nr}}, \tag{3.1.16}$$

where R_{rr} is the radiative recombination rate, R_{nr} is the nonradiative recombination rate, and $R_{\text{tot}} = R_{rr} + R_{nr}$ is the total recombination rate. It is customary to introduce the recombination times τ_{rr} and τ_{nr} by using the definition $R_{rr} = N/\tau_{rr}$ and $R_{nr} = N/\tau_{nr}$, where N is the carrier density. The internal quantum efficiency is then given by

$$\eta_{\text{int}} = \frac{\tau_{nr}}{\tau_{rr} + \tau_{nr}}. \tag{3.1.17}$$

The radiative and nonradiative recombination times vary from semiconductor to semiconductor. In general, τ_{rr} and τ_{nr} are comparable for direct-bandgap semiconductors, whereas τ_{nr} is a small fraction ($\sim 10^{-5}$) of τ_{rr} for semiconductors with an indirect bandgap. A semiconductor is said to have a direct bandgap if the conduction-band minimum and the valence-band maximum occur for the same value of the electron wave vector (see Fig. 3.2). The probability of radiative recombination is large in such semiconductors, since it is easy to conserve both energy and momentum during electron–hole recombination. By contrast, indirect-bandgap semiconductors require the assistance of a phonon for conserving momentum during electron–hole recombination. This feature reduces the probability of radiative recombination and increases τ_{rr} considerably compared with τ_{nr} in such semiconductors. As evident from Eq. (3.1.17), $\eta_{\text{int}} \ll 1$ under such conditions. Typically, $\eta_{\text{int}} \sim 10^{-5}$ for Si and Ge, the two semiconductors commonly used for electronic devices. Both are not suitable for optical sources because of their indirect bandgap. For direct-bandgap semiconductors such as GaAs and InP, $\eta_{\text{int}} \approx 0.5$ and approaches 1 when stimulated emission dominates.

The radiative recombination rate can be written as $R_{rr} = R_{\text{spon}} + R_{\text{stim}}$ in the general case in which radiative recombination occurs through spontaneous as well as stimulated emission. For LEDs, R_{stim} is negligible compared with R_{spon}, and R_{rr} in Eq. (3.1.16) can be replaced with R_{spon}. Typically, R_{spon} and R_{nr} are comparable in magnitude, resulting in an internal quantum efficiency of about 50%. However, η_{int} approaches 100% for semiconductor lasers as stimulated emission begins to dominate with an increase in the output power.

It is useful to define a quantity known as the *carrier lifetime* τ_c such that it represents the total recombination time of charged carriers in the absence of stimulated recombination. It is defined by the relation

$$R_{\text{spon}} + R_{nr} = N/\tau_c, \tag{3.1.18}$$

where N is the carrier density. If R_{spon} and R_{nr} vary linearly with N, τ_c becomes a constant. In practice, both of them increase nonlinearly with N such that $R_{\text{spon}} + R_{nr} = A_{nr}N + BN^2 + CN^3$, where A_{nr} is the nonradiative coefficient due to recombination at defects or traps, B is the spontaneous radiative recombination coefficient, and C is the Auger coefficient. The carrier lifetime then becomes N dependent and is obtained by using $\tau_c^{-1} = A_{nr} + BN + CN^2$. In spite of its N dependence, the concept of carrier lifetime τ_c is quite useful, as will become clear in Sections 3.2 and 3.3.

3.1.4 Semiconductor Materials

Almost any semiconductor with a direct bandgap can be used to make a p–n homojunction capable of emitting light through spontaneous emission. The choice is, however, considerably limited in the case of heterostructure devices because their performance depends on the quality of the heterojunction interface between two semiconductors of different bandgaps. To reduce the formation of lattice defects, the lattice constant of the two materials should match to better than 0.1%. Nature does not provide semiconductors whose lattice constants match to such precision. However, they can be fabricated artificially by forming ternary and quaternary compounds in which a fraction of the lattice sites in a naturally occurring binary semiconductor (e.g., GaAs) is replaced by other elements. In the case of GaAs, a ternary compound $\text{Al}_x\text{Ga}_{1-x}\text{As}$ can be made by replacing a fraction x of Ga atoms by Al atoms. The resulting semiconductor has nearly the same lattice constant, but its bandgap increases. The bandgap depends on the fraction x and can be approximated by a simple linear relation [8]

$$E_g(x) = 1.424 + 1.247x \qquad (0 < x < 0.45), \tag{3.1.19}$$

where E_g is expressed in electron-volt (eV) units.

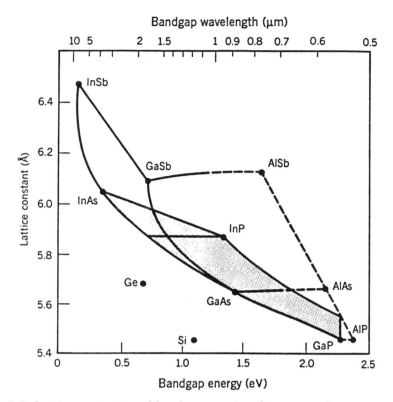

Figure 3.6 Lattice constants and bandgap energies of ternary and quaternary compounds formed by using nine binary semiconductors (group III–V). The shaded area corresponds to the quaternary compound $In_{1-x}Ga_xAs_yP_{1-y}$, where x and y are mixing fractions. The horizontal line passing through InP shows the quaternary compounds lattice matched to InP. The horizontal line connecting GaAs and AlAs corresponds to the ternary compound $Al_{1-x}Ga_xAs$. (After Ref. [17]. ©1991 Wiley. Reprinted with permission.

Figure 3.6 shows the interrelationship between the bandgap E_g and the lattice constant a for several ternary and quaternary compounds. Solid dots represent the binary semiconductors, and lines connecting them correspond to ternary compounds. The dashed portion of the line indicates that the resulting ternary compound has an indirect bandgap. The area of a closed polygon correspond to quaternary compounds. The bandgap is not necessarily direct for such semiconductors. The shaded area in Fig. 3.6 represents the ternary and quaternary compounds with a direct bandgap formed by using the elements indium (In), gallium (Ga), arsenic (As), and phosphorus (P).

The horizontal line connecting GaAs and AlAs corresponds to the ternary compound $Al_xGa_{1-x}As$, whose bandgap is direct for values of x up to about 0.45 and is given by Eq. (3.1.19). The active and cladding layers are formed such that x is larger for the cladding layers compared with the value of x for the active layer. The wavelength of the emitted light is determined by the bandgap since the photon energy is approximately equal to the bandgap. By using $E_g \approx h\nu = hc/\lambda$, one finds that $\lambda \approx 0.87$ μm for an active layer made of GaAs ($E_g = 1.424$ eV). The wavelength can be reduced to about 0.81 μm by using an active layer with $x = 0.1$. Optical sources based on GaAs typically operate in the range 0.81–0.87 μm and were used in the first generation of fiber-optic communication systems.

As discussed in Chapter 2, it is beneficial to operate lightwave systems in the wavelength range 1.3–1.6 μm, where both dispersion and loss of optical fibers are considerably reduced compared with the 0.85-μm region. InP is the base material for semiconductor optical sources emitting light in this wavelength region. As seen in Fig. 3.6 by the horizontal line passing through InP, the bandgap of InP can be reduced considerably by making the quaternary compound $In_{1-x}Ga_xAs_yP_{1-y}$ while the lattice constant remains matched to InP. The fractions x and y cannot be chosen arbitrarily but are related by $x/y = 0.45$ to ensure matching of the lattice constant. The bandgap of the quaternary compound can be expressed in terms of y only and is well approximated by [10]

$$E_g(y) = 1.35 - 0.72y + 0.12y^2, \qquad (3.1.20)$$

where $0 \leq y \leq 1$. The smallest bandgap occurs for $y = 1$. The corresponding ternary compound $In_{0.55}Ga_{0.45}As$ emits light near 1.65 μm ($E_g = 0.75$ eV). By a suitable choice of the mixing fractions x and y, $In_{1-x}Ga_xAs_yP_{1-y}$ sources can be designed to operate in the wide wavelength range 1.0–1.65 μm that includes the region 1.3–1.6 μm important for optical communication systems.

The fabrication of semiconductor optical sources requires epitaxial growth of multiple layers on a base substrate (GaAs or InP). The thickness and composition of each layer need to be controlled precisely. Several epitaxial growth techniques can be used for this purpose. The three primary techniques are known as liquid-phase epitaxy (LPE), vapor-phase epitaxy (VPE), and molecular-beam epitaxy (MBE) depending on whether the constituents of various layers are in the liquid form, vapor form, or in the form of a molecular beam. The VPE technique is also called chemical-vapor deposition. A variant of this technique is metal-organic chemical-vapor deposition (MOCVD), in which metal alkalis are used as the mixing compounds. Details of these techniques are available in the literature [10]. Both LPE and MOCVD are used to make commercial semiconductor lasers.

Both the MOCVD and MBE techniques provide an ability to control layer thickness to within 1 nm. It is quite common to use active layers of thickness ~ 10 nm. In such lasers, the thickness of the active layer is small enough that electrons and holes act as if they are confined in a quantum well. Such confinement leads to quantization of the energy bands into subbands. The main consequence is that the joint density of states ρ_{cv} acquires a staircase-like structure [20]. Such a modification of the density of states affects the gain characteristics considerably and improves the laser performance. Such *quantum-well lasers* have been studied extensively [12]. Often, multiple active layers of thickness 5–10 nm, separated by transparent barrier layers of about 10 nm thickness, are used to improve the device performance. Such lasers are called *multiquantum-well* (MQW) lasers. Another feature that has improved the performance of MQW lasers is the introduction of intentional, but controlled strain within active layers. The use of thin active layers permits a slight mismatch between lattice constants without introducing defects. The resulting strain changes the band structure and improves the laser performance [20]. Such semiconductor lasers are called *strained* MQW lasers.

3.2 LIGHT-EMITTING DIODES

A forward-biased p–n junction emits light through spontaneous emission, a phenomenon referred to as electroluminescence. In its simplest form, an LED is a forward-biased p–n homojunction. Radiative recombination of electron–hole pairs in the depletion region generates light; some of it escapes from the device and can be coupled into an optical fiber. The emitted light is incoherent with a relatively wide spectral width (30–60 nm) and a relatively large angular spread. In this section we discuss the characteristics and the design of LEDs from the standpoint of their application in optical communication systems [7], [18].

3.2.1 Light-Current Characteristics

It is easy to estimate the internal power generated by spontaneous emission. At a given current I the carrier-injection rate is I/q. In the steady state, the rate of electron–hole pairs recombining through radiative and nonradiative processes is equal to the carrier-injection rate I/q. Since the internal quantum efficiency η_{int} determines the fraction of electron–hole pairs that recombine through spontaneous emission, the rate of photon generation is simply $\eta_{\text{int}}I/q$. The internal optical power is thus given by

$$P_{\text{int}} = \eta_{\text{int}}(\hbar\omega/q)I, \tag{3.2.1}$$

where $\hbar\omega$ is the photon energy, assumed to be nearly the same for all photons. If η_{ext} is the fraction of photons escaping from the device, the emitted power is given by

$$P_e = \eta_{\text{ext}}P_{\text{int}} = \eta_{\text{ext}}\eta_{\text{int}}(\hbar\omega/q)I. \tag{3.2.2}$$

The quantity η_{ext} is called the *external quantum efficiency*. It can be calculated by taking into account internal absorption and reflection at the air-semiconductor interface. The calculation should also take into account the total internal reflection at the interface. As seen in Fig. 3.7, only light emitted within a cone of angle θ_c, where $\theta_c = \sin^{-1}(1/n)$ is the critical angle and n is the refractive index of the semiconductor material, escapes from the LED surface. Internal absorption can be avoided by using heterostructure LEDs in which the cladding layers surrounding the active layer are transparent to the radiation generated. The external quantum efficiency can then be written as

$$\eta_{\text{ext}} = \frac{1}{4\pi}\int_0^{\theta_c} T_f(\theta)(2\pi\sin\theta)\,d\theta, \tag{3.2.3}$$

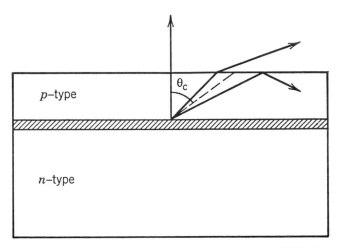

Figure 3.7 Total internal reflection at the output facet of an LED. Only light emitted within a cone of angle θ_c is transmitted, where θ_c is the critical angle for the semiconductor–air interface.

where we have assumed that the radiation is emitted uniformly in all directions over a solid angle of 4π. The Fresnel transmissivity T_f depends on the incidence angle θ. For normal incidence ($\theta = 0$), $T_f(0) = 4n/(n+1)^2$. For simplicity, if we replace $T_f(\theta)$ by $T_f(0)$ in Eq. (3.2.3), η_{ext} is given approximately by

$$\eta_{\text{ext}} = n^{-1}(n+1)^{-2}. \tag{3.2.4}$$

By using Eq. (3.2.4) in Eq. (3.2.2) we obtain the power emitted from one facet (see Fig. 3.7). If we use $n = 3.5$ as a typical value, $\eta_{\text{ext}} = 1.4\%$, indicating that only a small fraction of the internal power becomes the useful output power. A further loss in useful power occurs when the emitted light is coupled into an optical fiber. Because of the incoherent nature of the emitted light, an LED acts as a *Lambertian source* with an angular distribution $S(\theta) = S_0 \cos\theta$, where S_0 is the intensity in the direction $\theta = 0$. The coupling efficiency for such a source [18] is $\eta_c = (\text{NA})^2$. Since the numerical aperture NA for optical fibers is typically in the range 0.1–0.3, only a few percent of the emitted power is coupled into the fiber. Normally, the launched power for LEDs is 100 μW or less, even though the internal power can easily exceed 10 mW.

A measure of the LED performance is the total quantum efficiency η_{tot}, defined as the ratio of the emitted optical power P_e to the applied electrical power, $P_{\text{elec}} = V_0 I$, where V_0 is the voltage drop across the device. By using Eq. (3.2.2), η_{tot} is given by

$$\eta_{\text{tot}} = \eta_{\text{ext}}\eta_{\text{int}}(\hbar\omega/qV_0). \tag{3.2.5}$$

Typically, $\hbar\omega \approx qV_0$, and $\eta_{\text{tot}} \approx \eta_{\text{ext}}\eta_{\text{int}}$. The total quantum efficiency η_{tot}, also called the *power-conversion efficiency* or the *wall-plug efficiency*, is a measure of the overall performance of the device.

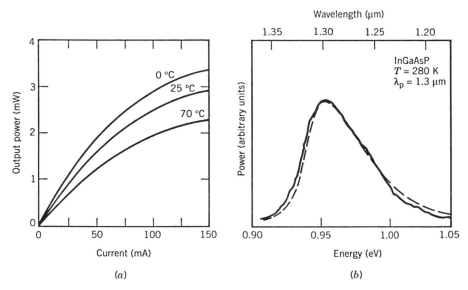

Figure 3.8 (a) Light–current curves at several temperatures; (b) spectrum of the emitted light for a typical 1.3-μm LED. Dashed curves shows the theoretical spectrum. (After Ref. [22]. ©1983 AT&T. Reprinted with permission.)

Another quantity sometimes used to characterize the LED performance is the *responsivity* R, defined as the ratio $R = P_e/I$. By using Eq. (3.2.2), R is given by

$$R = \eta_{\text{ext}}\eta_{\text{int}}(\hbar\omega/q). \qquad (3.2.6)$$

A comparison of Eqs. (3.2.5) and (3.2.6) shows that $R = \eta_{\text{tot}}V_0$. Typical values of R are ~ 0.01 W/A. The responsivity R remains constant as long as the linear relation between P_e and I holds. In practice, this linear relationship holds only over a limited current range [22]. Figure 3.8(a) shows the light–current (L–I) curves at several temperatures for a typical 1.3-μm LED. The responsivity of the device decreases at high currents above 80 mA because of bending of the L–I curve. One reason for this decrease is related to the increase in the active-region temperature. The internal quantum efficiency η_{int} is generally temperature dependent because of an increase in the nonradiative recombination rates at high temperatures.

3.2.2   Spectral Distribution

As discussed in Section 2.3, the spectral distribution of light source affects the performance of optical communication systems through fiber dispersion. The spectral distribution is governed by the spectrum of spontaneous emission $R_{\text{spon}}(\omega)$ given by Eq. (3.1.10). In general, $R_{\text{spon}}(\omega)$ is calculated numerically and depends on many material parameters. However, an approximate expression can be obtained if $A(E_1, E_2)$ is assumed to be nonzero only over a narrow energy range in the vicinity of the photon energy, and the Fermi functions are approximated by their exponential tails under the assumption of weak injection [20]. The result is

$$R_{\text{spon}}(\omega) = A_0 (\hbar\omega - E_g)^{1/2} \exp[-(\hbar\omega - E_g)/k_B T], \qquad (3.2.7)$$

where A_0 is a constant and E_g is the bandgap. It is easy to deduce that $R_{\text{spon}}(\omega)$ peaks when $\hbar\omega = E_g + k_B T/2$ and has a full width at half-maximum (FWHM) $\Delta\nu \approx 1.8 k_B T/h$. At room temperature ($T = 300$ K) the FWHM is about 11 THz. In practice, the spectral width is expressed in nanometers by using $\Delta\nu = (c/\lambda^2)\Delta\lambda$ and increases as λ^2 with an increase in the emission wavelength λ. As a result, $\Delta\lambda$ is larger for InGaAsP LEDs emitting at 1.3 μm by about a factor of 1.7 compared with GaAs LEDs. Figure 3.8(b) shows the output spectrum of a typical 1.3-μm LED and compares it with the theoretical curve obtained by using Eq. (3.2.7). Because of a large spectral width ($\Delta\lambda = 50$–60 nm), the bit rate–distance product is limited considerably by fiber dispersion when LEDs are used in optical communication systems. They are suitable primarily for local-area-network applications with bit rates of 10–100 Mb/s and transmission distances of a few kilometers.

3.2.3   Modulation Response

The modulation response of LEDs depends on carrier dynamics and is limited by the carrier lifetime τ_c defined by Eq. (3.1.18). It can be determined by using a *rate equation* for the carrier density N. Since electrons and holes are injected in pairs and recombine in pairs, it is enough to consider the rate equation for only one type of charge carrier. The rate equation should include all mechanisms through which electrons appear and disappear inside the active region. For LEDs it takes the simple form (since stimulated emission is negligible)

$$\frac{dN}{dt} = \frac{I}{qV} - \frac{N}{\tau_c}, \qquad (3.2.8)$$

where the last term includes both radiative and nonradiative recombination processes through the carrier lifetime τ_c, defined by Eq. (3.1.18). Consider sinusoidal modulation of the injected current by taking

$$I(t) = I_b + I_m \exp(i\omega_m t), \qquad (3.2.9)$$

where I_b is the bias current, I_m is the modulation current, and ω_m is the modulation frequency. Since Eq. (3.2.8) is linear, its general solution can be written as

$$N(t) = N_b + N_m \exp(i\omega_m t), \qquad (3.2.10)$$

where $N_b = \tau_c I_b/qV$, V is the volume of active region and N_m is given by

$$N_m(\omega_m) = \frac{\tau_c I_m/qV}{1 + i\omega_m \tau_c}. \qquad (3.2.11)$$

The modulated power P_m is related to $|N_m|$ linearly. One can define the LED transfer function $H(\omega_m)$ as

$$H(\omega_m) = \frac{N_m(\omega_m)}{N_m(0)} = \frac{1}{1 + i\omega_m\tau_c}. \tag{3.2.12}$$

In analogy with the case of optical fibers (see Section 2.4.4), the *3-dB modulation bandwidth* $f_{3\,\mathrm{dB}}$ is defined as the modulation frequency at which $|H(\omega_m)|$ is reduced by 3 dB or by a factor of 2. The result is

$$f_{3\,\mathrm{dB}} = \sqrt{3}\,(2\pi\tau_c)^{-1}. \tag{3.2.13}$$

Typically, τ_c is in the range 2–5 ns for InGaAsP LEDs. The corresponding LED modulation bandwidth is in the range 50–140 MHz. Note that Eq. (3.2.13) provides the optical bandwidth because $f_{3\,\mathrm{dB}}$ is defined as the frequency at which optical power is reduced by 3 dB. The corresponding electrical bandwidth can be obtained from Eq. (3.2.12) by considering the frequency at which $|H(\omega_m)|^2$ is reduced by 3 dB and is given by $(2\pi\tau_c)^{-1}$.

3.2.4 LED Structures

The LED structures can be classified as surface-emitting or edge-emitting, depending on whether the LED emits light from a surface that is parallel to the junction plane or from the edge of the junction region. Figure 3.9 shows schematically the design of two types of LEDs. Both types can be made using either a *p–n* homojunction or a heterostructure design in which the active region is surrounded by *p*- and *n*-type cladding layers. The heterostructure design leads to superior performance, as it provides a control over the emissive area and eliminates internal absorption because of the transparent cladding layers.

The *surface-emitting* design shown in Fig. 3.9 is referred to as a *Burrus-type* LED [23]. The emissive area of the device is limited to a small region whose lateral dimension is comparable to the fiber-core diameter. The use of a gold stud avoids power loss from the back surface. The coupling efficiency is improved by etching a well and bringing the fiber close to the emissive area. The power coupled into the fiber depends on many parameters, such as the numerical aperture of the fiber and the distance between fiber and LED. The addition of epoxy in the etched well tends to increase the external quantum efficiency as it reduces the refractive-index mismatch. Several variations of the basic design exist in the literature. In one variation, a truncated spherical microlens fabricated inside the etched well is used to couple light into the fiber [24]. In another variation, the fiber end is itself formed in the form of a spherical lens [25]. With a proper design, surface-emitting LEDs can couple up to 1% of the internally generated power into an optical fiber.

The *edge-emitting* LEDs employ a design commonly used for stripe-geometry semiconductor lasers (see Section 3.3.3). In fact, a semiconductor laser is converted into an LED by depositing an antireflection coating on its output facet to suppress lasing action. A high-reflection coating is often used on the other

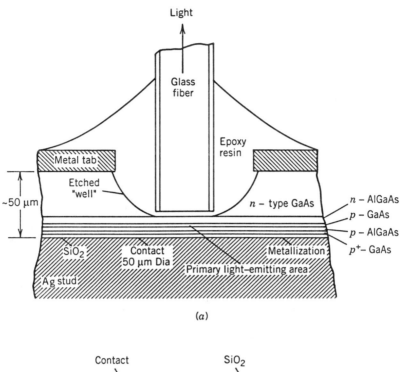

(a)

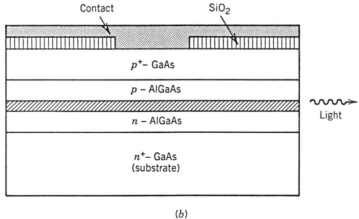

(b)

Figure 3.9 (a) Surface-emitting and (b) edge-emitting designs for LEDs with a double-heterostructure geometry.

facet to decrease the power emission from that facet. Beam divergence of edge-emitting LEDs differs from surface-emitting LEDs because of waveguiding in the plane perpendicular to the junction. Surface-emitting LEDs operate as a Lambertian source with angular distribution $S_e(\theta) = S_0 \cos\theta$ in both directions. The resulting beam divergence has a FWHM of 120° in each direction. In contrast, edge-emitting LEDs have a divergence of only about 30° in the direction perpendicular to the junction plane. Considerable light can be coupled into a fiber of even low numerical aperture ($<$ 0.3) because of reduced divergence and high radiance at the emitting facet [26], [27]. The modulation bandwidth of edge-emitting LEDs is generally larger ($\sim$ 200 MHz) than that of surface-emitting LEDs because of a reduced carrier lifetime at the same applied current [27]. The choice between the two designs is dictated, in practice, by a compromise between cost and performance.

In spite of a relatively low output power and a low bandwidth of LEDs compared with those of lasers, LEDs are useful for low-cost applications requiring data transmission at a bit rate of 100 Mb/s or less over a few kilometers. For this reason, several new LED structures have been developed during the 1990s. In one design, known as *resonant-cavity-enhanced* LED [28], two metal mirrors are fabricated around the epitaxially grown layers, and the device is bonded to a silicon substrate. In a variant of this idea, the bottom mirror is fabricated epitaxially by using a stack of alternating layers of two different semiconductors, while the top mirror consists of a deformable membrane suspended by an air gap [29]. The operating wavelength of such an LED can be tuned over 40 nm by changing the air-gap thickness. In another scheme, several quantum wells with different compositions and bandgaps are grown to form a MQW structure [30]. Since each quantum well emits light at a different wavelength, such LEDs can have an extremely broad spectrum (extending over a 500-nm wavelength range) and are useful for local-area WDM networks, discussed in Chapter 7.

3.3   SEMICONDUCTOR LASERS

Semiconductor lasers emit light through stimulated emission. As a result of the fundamental differences between spontaneous and stimulated emission, they are not only capable of emitting high powers ($\sim$ 100 mW), but also have other advantages related to the coherent nature of emitted light. A relatively narrow angular spread of the output beam compared with LEDs permits high coupling efficiency ($\sim$ 50%) into single-mode fibers. A relatively narrow spectral width of emitted light allows operation at high bit rates ($\sim$ 10 Gb/s), since fiber dispersion becomes less critical for such an optical source. Furthermore, semiconductor lasers can be modulated directly at high frequencies (up to 25 GHz) because of a short recombination time associated with stimulated emission. Most fiber-optic communication systems use semiconductor lasers as an optical source because of their superior performance compared with LEDs. In this section we describe the output characteristics of semiconductor lasers from the standpoint of their applications in lightwave systems. Further details can be found in Refs. [8]–[16], books devoted entirely to semiconductor lasers.

3.3.1   Optical Gain

As discussed in Section 3.1.1, stimulated emission can dominate only if the condition of population inversion is satisfied. For semiconductor lasers this

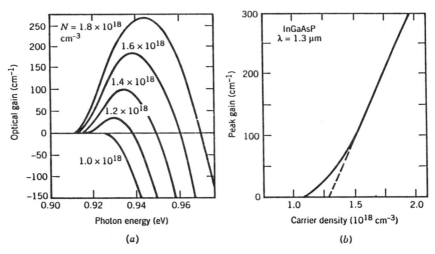

Figure 3.10 Calculated gain and its frequency dependence for a 1.3-μm InGaAsP active layer for several values of the injected carrier density N: (a) gain g as a function of the photon energy; (b) variation of peak gain g_p with N. The dashed line shows that g_p can be assumed to vary linearly with N in the high-gain region. (After Ref. [10]. ©1993 Van Nostrand Reinhold. Reprinted with permission.)

condition is realized by doping the p-type and n-type cladding layers so heavily that the Fermi-level separation exceeds the bandgap [see Eq. (3.1.14)] under forward biasing of the p–n junction. When the injected carrier density in the active layer exceeds a certain value, known as the transparency value, population inversion is realized and the active region exhibits optical gain. An input signal propagating inside the active layer would then amplify as $\exp(gz)$, where g is the *gain coefficient*. One can calculate g by noting that it is proportional to $R_{\text{stim}} - R_{\text{abs}}$, where R_{stim} and R_{abs} are given by Eqs. (3.1.12) and (3.1.13), respectively. In general, g is calculated numerically. Figure 3.10(a) shows the gain calculated for a 1.3-μm InGaAsP active layer at different values of the injected carrier density N. For $N = 1 \times 10^{18}$ cm^{-3}, $g < 0$, as population inversion has not yet occurred. As N increases, g becomes positive over a spectral range that increases with N. The peak value of the gain, g_p, also increases with N, together with a shift of the peak toward higher photon energies. The variation of g_p with N is shown in Fig. 3.10(b). For $N > 1.5 \times 10^{18}$ cm^{-3}, g_p varies almost linearly with N. Figure 3.10 shows that the optical gain in semiconductors increases rapidly once population inversion is realized. It is because of such a high gain that semiconductor lasers can be made with physical dimensions of less than 1 mm.

The nearly linear dependence of g_p on N suggests an empirical approach in which the peak gain is approximated by

$$g_p(N) = \sigma_g(N - N_T),   \tag{3.3.1}$$

where N_T is the transparency value of the carrier density and σ_g is the gain cross section; σ_g is also called the *differential gain*. Typical values of N_T and σ_g for

InGaAsP lasers [10] are in the range $1.0\text{–}1.5 \times 10^{18}$ cm^{-3} and $2\text{–}3 \times 10^{-16}$ cm^2, respectively. As seen in Fig. 3.10(b), the approximation (3.3.1) is reasonable in the high-gain region where g_p exceeds 100 cm^{-1}; most semiconductor lasers operate in this region. The use of Eq. (3.3.1) simplifies the analysis considerably, as band-structure details do not appear directly. The parameters σ_g and N_T can be estimated from numerical calculations such as those shown in Fig. 3.10(b) or can be measured experimentally.

Semiconductor lasers with a larger value of σ_g generally perform better, since the same amount of gain can be realized at a lower carrier density or, equivalently, at a lower injected current. Quantum-well semiconductor lasers usually have σ_g larger than that of the standard design by about a factor of 2 [12]. The linear approximation in Eq. (3.3.1) for the gain can still be used in a limited range. A better approximation to the peak gain in quantum-well lasers [20] replaces Eq. (3.3.1) with $g_p(N) = g_0[1 + \ln(N/N_0)]$, where $g_p = g_0$ at $N = N_0$ and $N_0 = eN_T \approx 2.718 N_T$ by using the definition $g_p = 0$ at $N = N_T$.

3.3.2 Feedback and Laser Threshold

The optical gain alone is not enough for laser operation. The other necessary ingredient is *optical feedback*—it converts an amplifier into an oscillator. In most lasers the feedback is provided by placing the gain medium inside an optical cavity formed by two mirrors, often referred to as the *Fabry–Perot* (FP) cavity. In the case of semiconductor lasers, external mirrors are not required since the two cleaved laser facets act as mirrors of reflectivity:

$$R_m = \left(\frac{n-1}{n+1}\right)^2, \qquad (3.3.2)$$

where n is the refractive index of the gain medium. Typically, $n = 3.5$, resulting in 30% facet reflectivity. Even though the FP cavity formed by two cleaved facets is relatively lossy, the gain is large enough that high losses can be tolerated. Figure 3.11 shows the basic structure of a semiconductor laser and the FP cavity associated with it.

The concept of *laser threshold* can be understood by noting that a certain fraction of photons generated by stimulated emission is lost because of cavity losses and needs to be replenished on a continuous basis. If the optical gain is not large enough to compensate for the cavity losses, the photon population cannot build up. Thus, a minimum amount of gain is necessary for the operation of a laser. This amount can be realized only when the laser is pumped above a threshold level. The current needed to reach the threshold is called the *threshold current*.

A simple way to obtain the threshold condition is to study how the amplitude of a plane wave changes during one round trip. Consider a plane wave of amplitude E_0, frequency ω, and wave number $k = n\omega/c$. During one round trip, its amplitude increases by $\exp[(g/2)(2L)]$ because of gain (g is the power gain) and its phase changes by $2kL$, where L is the length of the laser cavity. At the

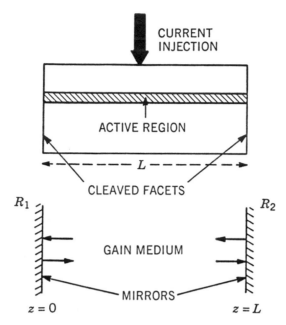

Figure 3.11 Structure of a semiconductor laser and the Fabry–Perot cavity associated with it. The cleaved facets act as partially reflecting mirrors.

same time, its amplitude changes by $\sqrt{R_1 R_2}\exp(-\alpha_{\text{int}}L)$ because of reflection at the laser facets and because of an internal loss α_{int} that includes free-carrier absorption, scattering, and other possible mechanisms. Here R_1 and R_2 are the reflectivities of the laser facets. Even though $R_1 = R_2$ in most cases, the two reflectivities can be different if laser facets are coated to change their natural reflectivity. In the steady state, the plane wave should remain unchanged after one round trip, i.e.,

$$E_0 \exp(gL)\sqrt{R_1 R_2}\exp(-\alpha_{\text{int}}L)\exp(2ikL) = E_0. \qquad (3.3.3)$$

By equating the amplitude and the phase on two sides, we obtain

$$g = \alpha_{\text{int}} + \frac{1}{2L}\ln\left(\frac{1}{R_1 R_2}\right) = \alpha_{\text{int}} + \alpha_{\text{mir}} = \alpha_{\text{cav}}, \qquad (3.3.4)$$

$$2kL = 2m\pi \quad \text{or} \quad \nu = \nu_m = mc/2nL, \qquad (3.3.5)$$

where $k = 2\pi n\nu/c$ and m is an integer. Equation (3.3.4) shows that gain equals total cavity loss α_{cav} at threshold and beyond. Equation (3.3.5) shows that the laser frequency ν must match one of the frequencies in the set ν_m, where m is an integer. These frequencies correspond to the *longitudinal modes* and are determined by the optical length nL. The spacing $\Delta\nu_L$ between the longitudinal modes is constant ($\Delta\nu_L = c/2nL$) if the frequency dependence of n is ignored. It is given by $\Delta\nu_L = c/2n_g L$ when material dispersion is included [10]. Here the

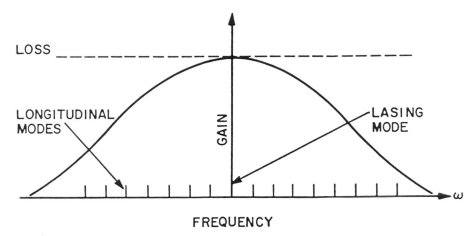

FREQUENCY

Figure 3.12 Gain and loss profiles in semiconductor lasers. Vertical bars show the location of longitudinal modes. The laser threshold is reached when the gain of the longitudinal mode closest to the gain peak equals loss.

group index n_g is defined as $n_g = n + \omega(dn/d\omega)$. Typically, $\Delta\nu_L = 100$–200 GHz for $L = 200$–$400\ \mu$m.

A FP semiconductor laser generally emits light in several longitudinal modes of the cavity. As seen in Fig. 3.12, the gain spectrum $g(\omega)$ of semiconductor lasers is wide enough (bandwidth ~ 10 THz) that many longitudinal modes of the FP cavity experience gain simultaneously. The mode closest to the gain peak becomes the dominant mode. Under ideal conditions, the other modes should not reach threshold since their gain always remains less than that of the main mode. In practice, the difference is extremely small (~ 0.1 cm^{-1}) and one or two neighboring modes on each side of the main mode carry a significant portion of the laser power together with the main mode. Such lasers are called multimode semiconductor lasers. Since each mode propagates inside the fiber at a slightly different speed because of group-velocity dispersion (see Section 2.3.1), the multimode nature of semiconductor lasers limits the bit-rate–distance product BL to values below 10 (Gb/s)-km for systems operating near 1.55 μm (see Fig. 2.13). The BL product can be increased by designing lasers oscillating in a single longitudinal mode. Such lasers are discussed in Section 3.3.5.

3.3.3 Laser Structures

The simplest structure of a semiconductor laser consists of a thin active layer (thickness $\sim 0.1\ \mu$m) sandwiched between p-type and n-type cladding layers of another semiconductor with a higher bandgap. The resulting p–n hetero-junction is forward-biased through metallic contacts. Such lasers are called *broad-area semiconductor lasers* since the current is injected over a relatively broad area covering the entire width of the laser chip ($\sim 100\ \mu$m). Figure 3.13 shows such a structure. The laser light is emitted from the two cleaved facets

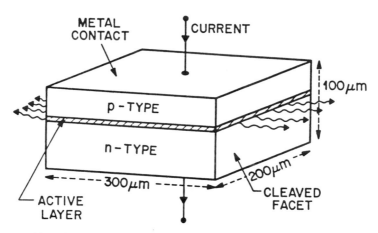

Figure 3.13 Broad-area semiconductor laser. The active layer (hatched region) is sandwiched between p-type and n-type cladding layers of a higher-bandgap material.

in the form of an elliptic spot of dimensions $\sim 1 \times 100$ μm^2. In the direction perpendicular to the junction plane, the spot size is ~ 1 μm because of the heterostructure design of the laser. As discussed in Section 3.1.2, the active layer acts as a planar waveguide because its refractive index is larger than that of the surrounding cladding layers ($\Delta n \approx 0.3$). Similar to the case of optical fibers, it supports a certain number of modes, known as transverse modes. In practice, the active layer is thin enough (~ 0.1 μm) that the planar waveguide supports a single transverse mode. However, there is no such light-confinement mechanism in the lateral direction parallel to the junction plane. Consequently, the light generated spreads over the entire width of the laser. Broad-area semiconductor lasers suffer from a number of deficiencies and are rarely used in optical communication systems. The major drawbacks are a relatively high threshold current and a spatial pattern that is highly elliptical and that changes in an uncontrollable manner with the current. These problems can be solved by introducing a mechanism for light confinement in the lateral direction. The resulting semiconductor lasers are classified into two broad categories, discussed next.

Gain-Guided Semiconductor Lasers

A simple scheme solves the light-confinement problem by limiting current injection over a narrow stripe. Such lasers are known as *stripe-geometry semiconductor lasers*. Figure 3.14 shows two laser structures schematically. In one approach [31], a dielectric (SiO_2) layer is deposited on top of the p-layer with a central opening through which the current is injected. In another [32], an n-type layer is deposited on top of the p-layer. Diffusion of Zn over the central region converts the n-region into p-type. Current flows only through the central region and is blocked elsewhere because of the reverse-biased nature of the p–n junction. Many other variations exist [10]. In all designs, current injection over

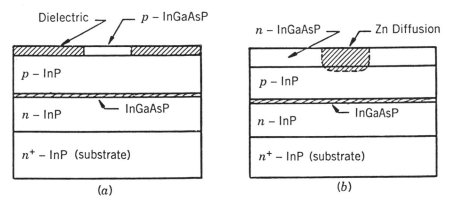

Figure 3.14 Cross section of two stripe-geometry laser structures used to design gain-guided semiconductor lasers and referred to as (a) oxide stripe and (b) junction stripe.

a narrow central stripe (~ 5 μm width) leads to a spatially varying distribution of the carrier density (governed by carrier diffusion) in the lateral direction. The optical gain also peaks at the center of the stripe. Since the active layer exhibits large absorption losses in the region beyond the central stripe, light is confined to the stripe region. As the confinement of light is aided by gain, such lasers are called *gain-guided semiconductor lasers*. Their threshold current is typically in the range 50–100 mA, and light is emitted in the form of an elliptic spot of dimensions $\sim 1 \times 5$ μm^2. The major drawback is that the spot size is not stable as the laser power is increased [10]. Such lasers are rarely used in optical communication systems because of mode-stability problems.

Index-Guided Semiconductor Lasers

The light-confinement problem is solved in *index-guided semiconductor lasers* by introducing an index step Δn_L in the lateral direction so that a waveguide is formed in a way similar to the waveguide formed in the transverse direction by the heterostructure design. Such lasers can be subclassified as weakly and strongly index-guided semiconductor lasers, depending on the magnitude of Δn_L. Figure 3.15 shows examples of the two kinds of lasers. In a specific design known as the *ridge-waveguide laser*, a ridge is formed by etching parts of the p-layer [10]. A SiO$_2$ layer is then deposited to block the current flow and to induce weak index guiding. Since the refractive index of SiO$_2$ is considerably lower than the central p-region, the effective index of the transverse mode is different in the two regions [33], resulting in an index step $\Delta n_L \sim 0.01$. This index step confines the generated light to the ridge region. The magnitude of the index step is sensitive to many fabrication details, such as the ridge width and the proximity of the SiO$_2$ layer to the active layer. However, the relative simplicity of the ridge-waveguide design and the resulting low cost make such lasers attractive for some applications.

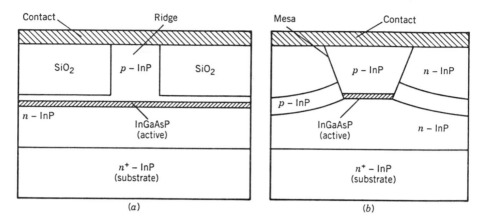

Figure 3.15 Cross section of two index-guided semiconductor lasers: (a) ridge-waveguide structure for weak index guiding; (b) etched-mesa buried heterostructure for strong index guiding.

In strongly index-guided semiconductor lasers, the active region of dimensions $\sim 0.1 \times 1$ μm^2 is buried on all sides by several layers of lower refractive index. For this reason, such lasers are often called *buried heterostructure* (BH) lasers. Several different kinds of BH lasers have been developed. They are known under names such as etched-mesa BH, planar BH, double-channel planar BH, and V-grooved or channeled substrate BH lasers, depending on the fabrication method used to realize the laser structure [10]. They all allow a relatively large index step ($\Delta n_L \sim 0.1$) in the lateral direction and, as a result, permit strong mode confinement. Because of a large built-in index step, the spatial distribution of the emitted light is inherently stable, provided that the laser is designed to support a single spatial mode. Most lightwave systems employ strongly index-guided semiconductor lasers because of their superior performance.

3.3.4   Laser Modes

In this section we discuss the spatial modes of a BH laser together with the condition under which such a laser would support a single spatial mode [10]. The results also provide spot-size dimensions and divergence angles in the lateral and transverse directions. The spatial characteristics of the emitted light are independent of the injected current for BH lasers, a feature that makes such lasers attractive for optical communication systems.

The starting point is the wave equation obtained in Section 2.2.1,

$$\nabla^2 \mathbf{E} + n^2 k_0^2 \mathbf{E} = 0, \tag{3.3.6}$$

where $k_0 = \omega/c = 2\pi/\lambda$ and the refractive index n equals n_1 inside the active region and n_2 outside it. The main difference from the case of fiber modes (see Section 2.2.2) is related to the boundary conditions. Because of a rectangular cross section of the active region, one needs to solve a two-dimensional wave-guide problem without the simplification of the cylindrical geometry. The exact analysis is quite complicated. However, the problem can be simplified consider-ably by using the *effective-index approximation*, which amounts to solving two one-dimensional waveguide problems. This approximation is justified for prac-tical BH lasers with typical dimensions of < 0.2 μm in the transverse y direction and > 1 μm in the lateral x direction. The waveguide problem is first solved in the y direction for each fixed x and then solved in the x direction. The electric field is assumed to be of the form

$$\mathbf{E} = \hat{\mathbf{e}}\,\phi(y, x)\psi(x)\exp(i\beta z), \tag{3.3.7}$$

where $\hat{\mathbf{e}}$ is the polarization unit vector and β is the propagation constant. The laser modes can be classified as TE and TM modes for light polarized along the x and y axes, respectively. By substituting Eq. (3.3.7) in Eq. (3.3.6) and using the method of separation of variables, we obtain

$$\partial^2\phi/\partial y^2 + [n^2(x, y)k_0^2 - \beta_{\text{eff}}^2(x)]\phi = 0, \tag{3.3.8}$$

$$d^2\psi/dx^2 + [\beta_{\text{eff}}^2(x) - \beta^2]\psi = 0, \tag{3.3.9}$$

where the *effective index*, defined as $n_{\text{eff}} = \beta_{\text{eff}}/k_0$, is x dependent. The formal-ism can be applied to both weakly and strongly index-guided lasers.

In the case of a BH laser with an active region of width w and thickness d, Eq. (3.3.8) reduces to a planar-waveguide problem for $|x| \leq w/2$ since $n = n_1$ for $|y| \leq d/2$ and n_2 otherwise. The general solution of Eq. (3.3.8) is given by

$$\phi(y, x) = \begin{cases} A_1\cos(\kappa y) + B_1\sin(\kappa y) & : & |y| \leq d/2, \\ A_2\exp[-\gamma(|y| - d/2)] & : & |y| > d/2, \end{cases} \tag{3.3.10}$$

where κ and γ are defined similar to the case of optical fibers as

$$\kappa^2 = n_1^2 k_0^2 - \beta_{\text{eff}}^2 \quad \text{and} \quad \gamma^2 = \beta_{\text{eff}}^2 - n_2^2 k_0^2. \tag{3.3.11}$$

One can separately consider the even and odd modes defined according to whether $\phi(y, x)$ is an even or odd function of y. In Eq. (3.3.10), $B_1 = 0$ for even modes, and $A_1 = 0$ for odd modes. Consider first the case of even TE modes. By requiring the continuity of the tangential component of $\mathbf{E}$ and $\mathbf{H}$ at the interface $|y| = d/2$ (it amounts to making ϕ and $\partial\phi/\partial y$ continuous across the interface), one obtains a set of two homogeneous equations for A_1 and A_2. The nontrivial solutions of this set exist only for certain values of β_{eff} determined from the eigenvalue equation

$$\gamma = \kappa\tan(\kappa d/2). \tag{3.3.12}$$

The procedure is similar to the case of optical fibers discussed in Section 2.2.2 (but much simpler). The odd TE modes can be treated in an identical manner and are governed by the eigenvalue equation

$$\gamma = -\kappa\cot(\kappa d/2). \tag{3.3.13}$$

Multiple solutions of Eq. (3.3.12) and (3.3.13) correspond to different TE modes. A similar procedure can be carried out for the TM modes. By following the analysis of Section 2.2.2, the condition under which the waveguide supports a single mode is found to be [10]

$$D = k_0 d(n_1^2 - n_2^2)^{1/2} < \pi, \tag{3.3.14}$$

where D is defined in a manner similar to the fiber V parameter [see Eq. (2.2.38)]. Typically, $n_1 = 3.55$, $n_2 = 3.2$, $d = 0.1$ μm, and $D < \pi$ is easily satisfied in the entire wavelength range 0.8–1.6 μm.

The *lateral modes* are obtained in a similar way by solving Eq. (3.3.9) together with

$$\beta_{\text{eff}}(x) = \begin{cases} n_{\text{eff}} k_0 & : \quad |x| \leq w/2, \\ n_2 k_0 & : \quad |x| > w/2. \end{cases} \tag{3.3.15}$$

In fact, the solution takes an identical form if we make the identification $n_1 = n_{\text{eff}}$ and $\beta_{\text{eff}} = \beta$ in Eq. (3.3.11). The single-mode condition (3.3.14) can be written as [10]

$$W = k_0 w(n_{\text{eff}}^2 - n_2^2)^{1/2} < \pi, \tag{3.3.16}$$

where W is the normalized width. The effective index n_{eff} is obtained from Eqs. (3.3.11) and (3.3.12) and depends on the active layer thickness d. Typically, $n_{\text{eff}} = 3.4$, $n_2 = 3.2$, and $w < 2$ μm. The single-mode condition (3.3.16) is not always satisfied for the lateral modes, and some BH lasers may emit light in higher-order lateral modes, particularly at high operating powers.

Most of the important laser parameters can be expressed in terms of the dimensionless parameters D and W. An important parameter is the mode index $\bar{n}$, defined as $\bar{n} = \beta/k_0$. It can be approximated by [34]

$$\bar{n}^2 = n_2^2 + \Gamma_L(n_{\text{eff}}^2 - n_2^2), \tag{3.3.17}$$

$$n_{\text{eff}}^2 = n_2^2 + \Gamma_T(n_1^2 - n_2^2), \tag{3.3.18}$$

where Γ_L and Γ_T are the *mode-confinement factors* in the lateral and transverse directions, respectively. They are defined as

$$\Gamma_L = \frac{\int_{-w/2}^{w/2} |\psi|^2 \, dx}{\int_{-\infty}^{\infty} |\psi|^2 \, dx}, \qquad \Gamma_T = \frac{\int_{-d/2}^{d/2} |\phi|^2 \, dy}{\int_{-\infty}^{\infty} |\phi|^2 \, dy}. \tag{3.3.19}$$

The approximate expressions for Γ_L and Γ_T are given by [34]

$$\Gamma_L = W^2/(2 + W^2) \qquad \text{and} \qquad \Gamma_T = D^2/(2 + D^2). \tag{3.3.20}$$

The total confinement factor $\Gamma = \Gamma_L \Gamma_T$ plays an important role, as it represents the fraction of mode energy contained inside the active region. Typically, $W \approx 3$, $D \approx 1$, and Γ is in the range 0.2–0.3. For a single-quantum-well laser, Γ is quite small ($\Gamma < 0.01$) but can be increased by using a MQW design.

The near-field and far-field distributions of the fundamental mode can be obtained by using Eqs. (3.3.10)–(3.3.12). Similar to the case of optical fibers discussed in Section 2.2.3, it is useful to fit the near-field distribution to a Gaussian

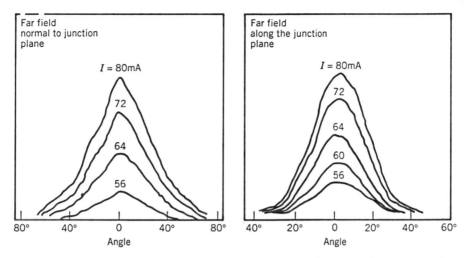

Figure 3.16 Far-field scans in a plane parallel and perpendicular to the junction plane for a 1.3-μm BH laser operating at various current levels. (After Ref. [10]. ©1993 Van Nostrand Reinhold. Reprinted with permission.)

profile. The near field is then characterized by the full width at half maximum (FWHM) of the Gaussian profile in the lateral and transverse directions. The approximate expressions for the FWHM are given by [35]

$$w_L = w(2\ln 2)^{1/2}(0.321 + 2.1W^{-3/2} + 4W^{-6}), \quad (3.3.21)$$
$$w_T = d(2\ln 2)^{1/2}(0.321 + 2.1D^{-3/2} + 4D^{-6}). \quad (3.3.22)$$

These relations are reasonably accurate for values of W and D in the range 1.8–6. Typically, $w_L = 1$–2 μm and $w_T = 0.5$–1 μm, depending on the actual active region dimensions of the BH laser.

The far-field distribution is obtained by taking the Fourier transform of the near field and is a measure of the angular divergence of emitted light. In the lateral direction, the angular distribution is obtained by using

$$S_L^{FF}(\theta) = \cos^2\theta \left| \int_{-\infty}^{\infty} \psi(x)\exp(ik_0 x\sin\theta)\,dx \right|^2. \quad (3.3.23)$$

The angular distribution in the transverse direction is given by a similar relation. Figure 3.16 shows the far-field scans in the lateral and transverse direction of a 1.3-μm BH laser at several current levels. The angular widths θ_L and θ_T, defined as the FWHM of the corresponding angular distribution, are generally used to characterize the far field. For BH lasers, θ_L and θ_T are usually in the range 30–40° and 40–50°, respectively. Although considerably smaller than LEDs, the angular divergence of semiconductor lasers is quite large compared with other kinds of lasers. The elliptic spot size of the emitted light, together with a large divergence angle, makes it difficult to couple light into the fiber efficiently. Coupling issues are discussed in Section 3.4.1. Typical coupling efficiencies are in the range 30–50% for most optical transmitters.

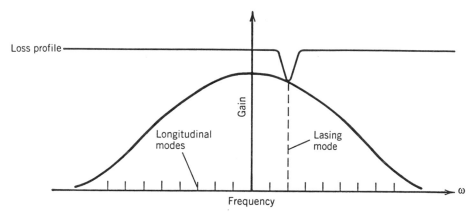

Figure 3.17 Gain and loss profiles of a semiconductor laser oscillating predominantly in a single longitudinal mode.

3.3.5 Single-Longitudinal-Mode Operation

We have seen that BH semiconductor lasers can be designed to emit light into a single spatial mode by controlling the width and the thickness of the active layer. However, as discussed in Section 3.3.2, such lasers oscillate in several longitudinal modes simultaneously because of a relatively small gain difference (~ 0.1 cm^{-1}) between neighboring modes of the FP cavity. The resulting spectral width (2–4 nm) is acceptable for lightwave systems operating near 1.3 μm at bit rates of up to 2.5 Gb/s. However, such multimode lasers cannot be used for systems designed to operate near 1.55 μm with standard telecommunication fibers. Dispersion-shifted fibers offer one solution, as they exhibit minimum loss and minimum dispersion in the same wavelength region. An alternative solution is to design semiconductor lasers [36]–[45] such that they emit light predominantly in a *single longitudinal mode* (SLM).

The SLM semiconductor lasers are designed such that cavity losses are different for different longitudinal modes of the cavity, in contrast with FP lasers, whose losses are mode independent. Figure 3.17 shows the gain and loss profiles schematically for such a laser. The longitudinal mode with the smallest cavity loss reaches threshold first and becomes the dominant mode. Other neighboring modes are discriminated by their higher losses, which prevent their buildup from spontaneous emission. The power carried by these side modes is usually a small fraction ($< 1\%$) of the total emitted power. The performance of a SLM laser is often characterized by the *mode-suppression ratio* (MSR), defined as [39]

$$\text{MSR} = P_{mm}/P_{sm}, \qquad (3.3.24)$$

where P_{mm} is the main-mode power and P_{sm} is the power of the most dominant side mode. The MSR should exceed 1000 (or 30 dB) for a good SLM laser. Several techniques described below can be used for SLM operation.

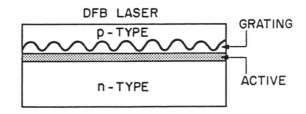

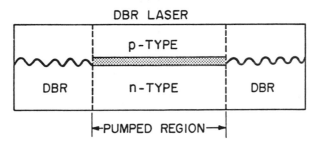

Figure 3.18 DFB and DBR laser structures. The shaded area shows the active region of the device.

Distributed Feedback Semiconductor Lasers

The feedback in *distributed feedback* (DFB) lasers [42], as the name implies, is not localized at the facets but is distributed throughout the cavity length. This is achieved through an internal built-in grating that leads to a periodic variation of the mode index. Feedback occurs by means of *Bragg diffraction*, a phenomenon that couples the waves propagating in the forward and backward directions. Mode selectivity of the DFB mechanism results from the *Bragg condition*; coupling occurs only for wavelengths λ_B satisfying

$$\Lambda = m(\lambda_B/2\bar{n}), \tag{3.3.25}$$

where Λ is the grating period, $\bar{n}$ is the average mode index, and the integer m represents the order of Bragg diffraction. The coupling between the forward and backward waves is strongest for the first-order Bragg diffraction ($m = 1$). For a DFB laser operating at $\lambda_B = 1.55$ μm, Λ is about 235 nm if we use $m = 1$ and $\bar{n} = 3.3$ in Eq. (3.3.25). Such gratings can be made by using holographic techniques.

From the standpoint of device operation, semiconductor lasers employing the DFB mechanism can be classified into two broad categories: DFB lasers and *distributed Bragg reflector* (DBR) lasers. Figure 3.18 shows two kinds of laser structures. Though the feedback occurs throughout the cavity length in DFB lasers, it does not take place inside the active region of a DBR laser. In effect, the end regions of a DBR laser act as mirrors whose reflectivity is maximum for a wavelength λ_B satisfying Eq. (3.3.25). The cavity losses are therefore minimum for the longitudinal mode closest to λ_B and increase substantially for other longitudinal modes (see Fig. 3.17). The MSR is determined by the gain margin defined as the excess gain required by the most dominant side mode to reach threshold. A gain margin of 3–5 cm^{-1} is generally enough to realize an MSR > 30 dB for DFB lasers operating continuously [39]. However, a larger gain margin is needed (> 10 cm^{-1}) when DFB lasers are modulated directly. *Phase-shifted DFB lasers* [38], in which the grating is shifted by $\lambda_B/4$ in the middle of the laser to produce a $\pi/2$ phase shift, are often used, since they are capable of providing much larger gain margin than that of conventional DFB lasers. Another design that has led to improvements in the device performance is known as the *gain-coupled DFB laser* [43]–[45]. In these lasers, both the optical gain and the mode index vary periodically along the cavity length.

Fabrication of DFB semiconductor lasers requires advanced technology with multiple epitaxial growths [42]. The principal difference from FP lasers is that a grating is etched onto one of the cladding layers surrounding the active layer. A thin n-type waveguide layer with a refractive index intermediate to that of active layer and the substrate acts as a grating. The periodic variation of the thickness of the waveguide layer translates into a periodic variation of the mode index $\bar{n}$ along the cavity length and leads to a coupling between the forward and backward propagating waves through Bragg diffraction.

A holographic technique is often used to form a grating with a ~ 0.2-μm periodicity. It works by forming a fringe pattern on a photoresist (deposited on the wafer surface) through interference between two optical beams. In the alternative electron-beam lithographic technique, an electron beam writes the desired pattern on the electron-beam resist. Both methods use chemical etching to form grating corrugations, with the patterned resist acting as a mask. Once the grating has been etched onto the substrate, multiple layers are grown by using an epitaxial growth technique. A second epitaxial regrowth is needed to make a BH device such as that shown in Fig. 3.15(b). Despite the technological complexities, DFB lasers are routinely produced commercially. They are used in nearly all 1.55-μm optical communication systems operating at bit rates of 2.5 Gb/s or more. Since 1992, several transoceanic lightwave systems have been designed by making use of DFB lasers.

Coupled-Cavity Semiconductor Lasers

In the case of *coupled-cavity semiconductor lasers* [10], the SLM operation is realized by coupling the light to an external cavity (see Fig. 3.19). A portion of the reflected light is fed back into the laser cavity. The feedback from the external cavity is not necessarily in phase with the optical field inside the laser cavity because of the phase shift occurring in the external cavity. The in-phase feedback occurs only for those laser modes whose wavelength nearly coincides with one of the longitudinal modes of the external cavity. In effect, the effective reflectivity of the laser facet facing the external cavity becomes wavelength dependent and leads to the loss profile shown in Fig. 3.19. The longitudinal mode that is closest to the gain peak and has the lowest cavity loss becomes the dominant mode. Several kinds of coupled-cavity schemes have been devel-

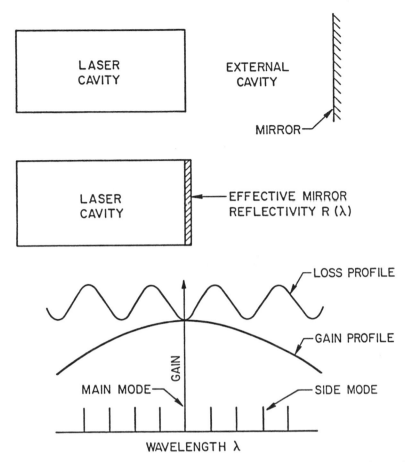

Figure 3.19 Longitudinal-mode selectivity in a coupled-cavity laser. Phase shift in the external cavity makes the effective mirror reflectivity wavelength dependent and results in a periodic loss profile for the laser cavity.

oped; Fig. 3.20 shows three among them. A simple scheme for making SLM lasers couples the light from a semiconductor laser to an external grating [Fig. 3.20(a)]. It is necessary to reduce the natural reflectivity of the cleaved facet facing the grating through an antireflection coating to provide a strong coupling. Such lasers are called external-cavity semiconductor lasers and have attracted considerable attention because of their tunability [36]. The wavelength of the SLM selected by the coupled-cavity mechanism can be tuned over a wide range (typically 50 nm) simply by rotating the grating. Wavelength tunability is a desirable feature for lasers used for multichannel and coherent communication systems (see Chapters 6 and 7). A drawback of the laser shown in Fig. 3.20(a) from the system standpoint is its nonmonolithic nature, which makes it difficult to realize the mechanical stability generally required of optical transmitters.

A monolithic design for coupled-cavity lasers is offered by the cleaved-coupled-

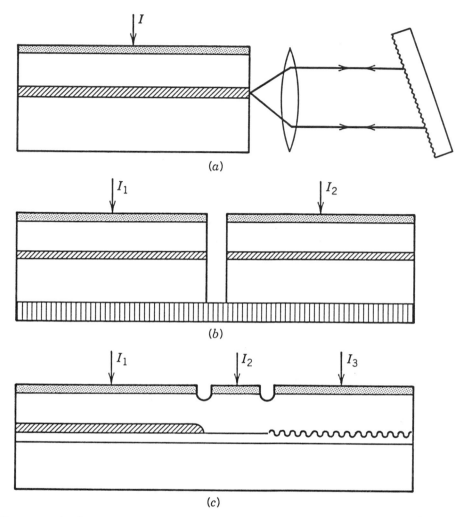

Figure 3.20 Coupled-cavity laser structures: (a) external-cavity laser; (b) C³ laser; (c) multisection DBR laser.

cavity (C³) laser [37] shown in Fig. 3.20(b). Such lasers are made by cleaving a conventional multimode semiconductor laser in the middle so that the laser is divided into two sections of about the same length but separated by a narrow air gap (width ∼ 1 μm). The reflectivity of cleaved facets (∼ 30%) allows enough coupling between the two sections as long as the gap is not too wide. It is even possible to tune the wavelength of a C³ laser over a tuning range ∼ 20 nm by varying the current injected into one of the cavity sections acting as a mode controller. However, tuning is not continuous, since it corresponds to successive mode hops of about 2 nm.

Tunable Semiconductor Lasers

Many lightwave systems (see Chapters 6 and 7) require a stable, narrow-linewidth laser whose wavelength can be tuned over a wide range. Multisection DFB and DBR lasers [46]–[51] have been developed to meet the somewhat conflicting requirements of stability and tunability. Figure 3.20(c) shows a typical laser structure. It consists of three sections, referred to as the active section, the phase-control section, and the Bragg section. Each section can be biased independently by injecting different amounts of currents. The current injected into the Bragg section is used to change the Bragg wavelength ($\lambda_B = 2n\Lambda$) through carrier-induced changes in the refractive index n. The current injected into the phase-control section is used to change the phase of the feedback from the DBR through carrier-induced index changes in that section. The laser wavelength can be tuned continuously over the range 5–7 nm by controlling the currents in the three sections. The laser operates stably because its wavelength is determined by the built-in grating in the Bragg section.

Several other designs of tunable DFB lasers have been developed during the 1990s. In one scheme, the built-in grating is chirped by varying the grating period Λ or the mode index $\bar{n}$ along the cavity length. As seen from Eq. (3.3.25), the Bragg wavelength itself then changes along the cavity length. Since the laser wavelength is determined by the Bragg wavelength, such a laser can be tuned over a wavelength range determined by the grating chirp. In a simple implementation of the basic idea, the grating period remains uniform, but the waveguide is bent to change the effective mode index $\bar{n}$. Such multisection DFB lasers can be tuned over 5–6 nm while maintaining a single longitudinal mode with high side-mode suppression [51].

In another scheme, a *superstructure grating* is used for the DBR section of a multisection laser [52], [53]. A superstructure grating consists of an array of gratings (uniform or chirped) separated by a constant distance. As a result, its reflectivity peaks at several wavelengths whose spacing is determined by the spacing among the individual gratings forming the array. Such multisection DBR lasers can be tuned discretely over a wavelength range exceeding 100 nm. By controlling the current in the phase-control section, a quasicontinuous tuning range of 40 nm has been realized with a super-structure grating [52]. Such widely tunable DBR lasers are likely to find applications in many lightwave systems (see Section 7.1).

Vertical-Cavity Surface-Emitting Lasers

A new class of semiconductor lasers, known as *vertical-cavity surface-emitting lasers* (VCSELs), has emerged with many potential applications [54]–[56]. VCSELs operate in a single longitudinal mode by virtue of an extremely small cavity length ($\sim 1~\mu\text{m}$) for which the mode spacing exceeds the gain bandwidth (see Fig. 3.12). They emit light in a direction normal to the active-layer plane in a manner analogous to that of a surface-emitting LED (see Fig. 3.9). Two high-reflectivity ($> 99.5\%$) DBR mirrors are grown epitaxially on both sides of the active layer to form a microcavity. Operation in a single transverse mode can be realized by reducing the VCSEL diameter to 2–3 μm. Although the output power and the bandwidth of VCSELs are typically lower than those of edge-emitting DFB lasers, VCSELs are likely to be used for local-area network and local-loop applications because of their low-cost packaging. Moreover, VCSEL arrays in which each laser operates at a different wavelength [56] are ideally suited for multichannel lightwave systems (see Chapter 7).

3.3.6 Light-Current Characteristics

The operating characteristics of semiconductor lasers are well described by a set of rate equations that govern the interaction of photons and electrons inside the active region. A rigorous derivation of these *rate equations* generally starts from Maxwell's equations (see Section 2.2.1) together with a quantum-mechanical approach for the induced polarization. However, the rate equations can also be obtained heuristically by considering the physical phenomena through which the number of photons, P, and the number of electrons, N, change with time inside the active region. For a single-mode laser, these equations take the form [10]

$$\frac{dP}{dt} = GP + R_{sp} - \frac{P}{\tau_p}, \tag{3.3.26}$$

$$\frac{dN}{dt} = \frac{I}{q} - \frac{N}{\tau_c} - GP, \tag{3.3.27}$$

where

$$G = \Gamma v_g g = G_N(N - N_0). \tag{3.3.28}$$

G is the net rate of stimulated emission and R_{sp} is the rate of spontaneous emission into the lasing mode. Note that R_{sp} is much smaller than the total spontaneous-emission rate. This can be understood by noting that spontaneous emission occurs in all directions over a wide spectral range ($\sim$ 30–40 nm) whereas only a small fraction of it, propagating along the cavity axis and emitted at the laser frequency, actually contributes to Eq. (3.3.26). In fact, R_{sp} and G are related by $R_{sp} = n_{sp}G$, where n_{sp} is known as the *spontaneous-emission factor* and is about 2 for semiconductor lasers [10]. Although the same notation is used for convenience, the variable N in the rate equations represents the number of electrons rather than the carrier density; the two are related by the active volume V. In Eq. (3.3.28), v_g is the group velocity, Γ is the confinement factor, and g is the optical gain at the mode frequency. By using Eq. (3.3.1), G varies linearly with N with $G_N = \Gamma v_g \sigma_g / V$ and $N_0 = N_T V$.

The last term in Eq. (3.3.26) takes into account the loss of photons inside the cavity. The parameter τ_p is referred to as the *photon lifetime*. It is related to the *cavity loss* α_{cav} introduced in Eq. (3.3.4) as

$$\tau_p^{-1} = v_g \alpha_{\text{cav}} = v_g(\alpha_{\text{mir}} + \alpha_{\text{int}}). \tag{3.3.29}$$

The three terms in Eq. (3.3.27) indicate the rates at which electrons are created or destroyed inside the active region. This equation is similar to Eq. (3.2.8)

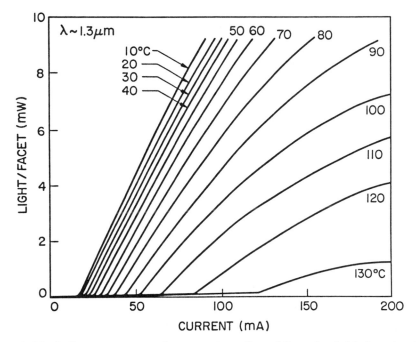

Figure 3.21 *L–I* curves at several temperatures for a 1.3-μm buried heterostructure laser. (After Ref. [10]. ©1993 Van Nostrand Reinhold. Reprinted with permission.)

except for the addition of the last term, which governs the rate of electron–hole recombination through stimulated emission. The carrier lifetime τ_c includes the loss of electrons due to both spontaneous emission and nonradiative recombination, as indicated in Eq. (3.1.18).

The *L–I* curve characterizes the emission properties of a semiconductor laser, as it indicates not only the threshold level but also the current that needs to be applied to obtain a certain amount of power. Figure 3.21 shows the *L–I* curves of a 1.3-μm InGaAsP laser at temperatures in the range 10–130°C. At room temperature the threshold is reached at about 20 mA and the laser can emit about 10 mW of output power from each facet at 100 mA of applied current. The laser performance degrades at high temperatures. The threshold current is found to increase exponentially with temperature, i.e.,

$$I_{th}(T) = I_0 \exp(T/T_0), \tag{3.3.30}$$

where I_0 is a constant and T_0 is a *characteristic temperature* often used to express the temperature sensitivity of threshold current. For InGaAsP lasers T_0 is typically in the range 50–70 K. By contrast, T_0 exceeds 120 K for GaAs lasers. Because of the temperature sensitivity of InGaAsP lasers, it is often necessary to control their temperature through a built-in thermoelectric cooler. InGaAsP lasers typically cannot emit light beyond 100°C.

The rate equations can be used to understand most of the features seen in Fig. 3.21. In the case of CW operation at a constant current I, the time derivatives in Eqs. (3.3.26) and (3.3.27) can be set to zero. The solution takes a particularly simple form if spontaneous emission is neglected by setting $R_{sp} = 0$. For currents such that $G\tau_p < 1$, $P = 0$ and $N = \tau_c I/q$. The threshold is reached at a current for which $G\tau_p = 1$. The carrier population is then clamped to the threshold value $N_{th} = N_0 + (G_N\tau_p)^{-1}$. The threshold current is given by

$$I_{th} = \frac{qN_{th}}{\tau_c} = \frac{q}{\tau_c}\left(N_0 + \frac{1}{G_N\tau_p}\right). \qquad (3.3.31)$$

For $I > I_{th}$, the photon number P increases linearly with I as

$$P = (\tau_p/q)(I - I_{th}). \qquad (3.3.32)$$

The emitted power P_e is related to P by the relation

$$P_e = \tfrac{1}{2}(v_g\alpha_{\mathrm{mir}})\hbar\omega P. \qquad (3.3.33)$$

The derivation of Eq. (3.3.33) is intuitively obvious if we note that $v_g\alpha_{\mathrm{mir}}$ is the rate at which photons of energy $\hbar\omega$ escape from the two facets. The factor of $1/2$ makes P_e the power emitted from each facet for a FP laser with equal facet reflectivities. For FP lasers with coated facets or for DFB lasers, Eq. (3.3.33) needs to be suitably modified [10]. By using Eqs. (3.3.29) and (3.3.32) in Eq. (3.3.33), the emitted power is given by

$$P_e = \frac{\hbar\omega}{2q}\frac{\eta_{\mathrm{int}}\alpha_{\mathrm{mir}}}{\alpha_{\mathrm{mir}} + \alpha_{\mathrm{int}}}(I - I_{th}), \qquad (3.3.34)$$

where the internal quantum efficiency η_{int} is introduced phenomenologically to indicate the fraction of injected electrons that is converted into photons through stimulated emission. In the above-threshold regime, η_{int} is almost 100% for most semiconductor lasers. Equation (3.3.34) should be compared with Eq. (3.2.2) obtained for an LED.

A quantity of practical interest is the slope of the L–I curve for $I > I_{th}$, called the *slope efficiency* and defined as

$$\frac{dP_e}{dI} = \frac{\hbar\omega}{2q}\eta_d \qquad \text{with} \qquad \eta_d = \frac{\eta_{\mathrm{int}}\alpha_{\mathrm{mir}}}{\alpha_{\mathrm{mir}} + \alpha_{\mathrm{int}}}. \qquad (3.3.35)$$

The quantity η_d is called the *differential quantum efficiency*, as it is a measure of the efficiency with which light output increases with an increase in the injected current. One can define the external quantum efficiency η_{ext} as

$$\eta_{\mathrm{ext}} = \frac{\text{photon-emission rate}}{\text{electron-injection rate}} = \frac{2P_e/\hbar\omega}{I/q} = \frac{2q}{\hbar\omega}\frac{P_e}{I}. \qquad (3.3.36)$$

By using Eqs. (3.3.34)–(3.3.36), η_{ext} and η_d are found to be related by

$$\eta_{\mathrm{ext}} = \eta_d(1 - I_{th}/I). \qquad (3.3.37)$$

Generally, $\eta_{\text{ext}} < \eta_d$ but becomes nearly the same for $I \gg I_{th}$. Similar to the case of LEDs, one can define the total quantum efficiency (or wall-plug efficiency) as $\eta_{\text{tot}} = 2P_e/(V_0 I)$, where V_0 is the applied voltage. It is related to η_{ext} as

$$\eta_{\text{tot}} = \frac{\hbar\omega}{qV_0}\, \eta_{\text{ext}} \approx \frac{E_g}{qV_0}\, \eta_{\text{ext}}, \qquad (3.3.38)$$

where E_g is the bandgap energy. Note that both η_{ext} and η_{tot} are defined by taking into account the total power emitted from both facets. Generally, $\eta_{\text{tot}} < \eta_{\text{ext}}$ since the applied voltage exceeds E_g/q. For GaAs lasers, η_d can exceed 80% and η_{tot} can approach 50%. By contrast, InGaAsP lasers are less efficient with $\eta_d \sim 50\%$ and $\eta_{\text{tot}} \sim 20\%$.

The exponential increase in the threshold current with temperature can be understood from Eq. (3.3.31). The carrier lifetime τ_c is generally N dependent because of Auger recombination and decreases with N as N^2. The rate of Auger recombination increases exponentially with temperature and is responsible for the temperature sensitivity of InGaAsP lasers. Figure 3.21 also shows that the slope efficiency decreases with an increase in the output power (bending of the L–I curves). This decrease can be attributed to junction heating under CW operation. It can also result from an increase in the internal loss or in the current leakage at high operating powers. Despite these problems, the performance of semiconductor lasers improved substantially during the 1990s [16]. By 1996, DFB lasers capable of delivering more than 100 mW of power were available.

3.3.7 Modulation Response

The modulation response of semiconductor lasers is studied by solving the rate equations (3.3.26) and (3.3.27) with a time-dependent current of the form

$$I(t) = I_b + I_m f_p(t), \qquad (3.3.39)$$

where I_b is the bias current, I_m is the current, and $f_p(t)$ represents the shape of the current pulse. Two changes are necessary for a realistic description. First, Eq. (3.3.28) for the gain G must be modified to become [57], [58]

$$G = G_N(N - N_0)(1 - \epsilon_{NL}P), \qquad (3.3.40)$$

where ϵ_{NL} is a nonlinear-gain parameter that leads to a slight reduction in G as P increases. The physical mechanism behind this reduction can be attributed to several phenomena, such as spatial hole burning, spectral hole burning, carrier heating, and two-photon absorption [59]–[62]. Typical values of ϵ_{NL} are $\sim 10^{-7}$. Equation (3.3.40) is valid as long as $\epsilon_{NL}P \ll 1$. The factor $1 - \epsilon_{NL}P$ should be replaced by $(1 + P/P_s)^{-b}$, where P_s is a material parameter, when the laser power exceeds far above 10 mW. The exponent b equals $1/2$ for spectral hole burning [60] but can vary over the range 0.2–1.0 because of the contribution of carrier heating [62].

The second change is related to an important property of semiconductor lasers. It turns out that whenever the optical gain changes as a result of changes in the carrier population N, the refractive index also changes. From a physical standpoint, amplitude modulation in semiconductor lasers is always accompanied by phase modulation because of carrier-induced changes in the mode index $\bar{n}$. Phase modulation can be included through the equation [10]

$$\frac{d\phi}{dt} = \frac{1}{2}\beta_c \left(G_N(N - N_0) - \frac{1}{\tau_p} \right), \tag{3.3.41}$$

where β_c is the *amplitude-phase coupling* parameter, commonly called the *linewidth enhancement factor*, as it leads to an enhancement of the spectral width associated with a single longitudinal mode (see Section 3.3.8). Typical values of β_c for InGaAsP lasers are in the range 4–8, depending on the operating wavelength [63]. Lower values of β_c occur in MQW lasers, especially for strained quantum wells [20].

Small-Signal Modulation

In general, the nonlinear nature of the rate equations makes it necessary to solve them numerically. A useful analytic solution can be obtained for the case of small-signal modulation in which the laser is biased above threshold ($I_b > I_{th}$) and modulated such that $I_m \ll I_b - I_{th}$. The rate equations can be linearized in that case and solved analytically, using the Fourier-transform technique, for an arbitrary form of $f_p(t)$. The small-signal modulation bandwidth can be obtained by considering the response of semiconductor lasers to sinusoidal modulation at the frequency ω_m so that $f_p(t) = \sin(\omega_m t)$. The laser output is also modulated sinusoidally. The general solution of Eqs. (3.3.26) and (3.3.27) is given by

$$
\begin{align}
P(t) &= P_b + |p_m| \sin(\omega_m t + \theta_m), \tag{3.3.42}\\
N(t) &= N_b + |n_m| \sin(\omega_m t + \psi_m), \tag{3.3.43}
\end{align}
$$

where P_b and N_b are the steady-state values at the bias current I_b, $|p_m|$ and $|n_m|$ are small changes occurring because of current modulation, and θ_m and ψ_m govern the phase lag associated with the small-signal modulation. In particular, $p_m = |p_m| \exp(i\theta_m)$ is given by [10]

$$p_m(\omega_m) = \frac{P_b G_N I_m/q}{(\Omega_R + \omega_m - i\Gamma_R)(\Omega_R - \omega_m + i\Gamma_R)}, \tag{3.3.44}$$

where

$$
\begin{align}
\Omega_R &= [GG_N P_b - (\Gamma_P - \Gamma_N)^2/4]^{1/2}, & \Gamma_R &= (\Gamma_P + \Gamma_N)/2, \tag{3.3.45}\\
\Gamma_P &= R_{sp}/P_b + \epsilon_{NL} GP_b, & \Gamma_N &= \tau_c^{-1} + G_N P_b. \tag{3.3.46}
\end{align}
$$

Ω_R and Γ_R are the frequency and the damping rate of *relaxation oscillations*. These two parameters play an important role in governing the dynamic response of semiconductor lasers. In particular, the efficiency is reduced when the modulation frequency exceeds Ω_R frequency by a large amount.

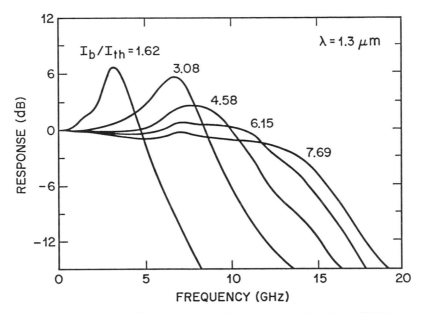

Figure 3.22 Measured small-signal modulation response of a 1.3-μm DFB laser as a function of modulation frequency at several bias levels. (After Ref. [64]. ©1987 IEEE. Reprinted with permission.)

Similar to the case of LEDs, one can introduce the *transfer function* as

$$H(\omega_m) = \frac{p_m(\omega_m)}{p_m(0)} = \frac{\Omega_R^2 + \Gamma_R^2}{(\Omega_R + \omega_m - i\Gamma_R)(\Omega_R - \omega_m + i\Gamma_R)}. \qquad (3.3.47)$$

The modulation response is flat [$H(\omega_m) \approx 1$] for frequencies such that $\omega_m \ll \Omega_R$, peaks at $\omega_m = \Omega_R$, and then drops sharply for $\omega_m \gg \Omega_R$. These features are observed experimentally for all semiconductor lasers. Figure 3.22 shows the modulation response of a 1.3-μm DFB laser at several bias levels [64]. The 3-dB modulation bandwidth, $f_{3\,\mathrm{dB}}$, is defined as the frequency at which $|H(\omega_m)|$ is reduced by 3 dB (by a factor of 2) compared with its direct-current (dc) value. Equation (3.3.47) provides the following analytic expression for $f_{3\,\mathrm{dB}}$:

$$f_{3\,\mathrm{dB}} = \frac{1}{2\pi}\left[\Omega_R^2 + \Gamma_R^2 + 2(\Omega_R^4 + \Omega_R^2\Gamma_R^2 + \Gamma_R^4)^{1/2}\right]^{1/2}. \qquad (3.3.48)$$

For most lasers, $\Gamma_R \ll \Omega_R$, and $f_{3\,\mathrm{dB}}$ can be approximated by

$$f_{3\,\mathrm{dB}} \approx \frac{\sqrt{3}\,\Omega_R}{2\pi} \approx \left(\frac{3G_N P_b}{4\pi^2\tau_p}\right)^{1/2} = \left[\frac{3G_N}{4\pi^2 q}(I_b - I_{th})\right]^{1/2}, \qquad (3.3.49)$$

where Ω_R was approximated by $(GG_N P_b)^{1/2}$ in Eq. (3.3.45) and G was replaced by $1/\tau_p$ since gain equals loss in the above-threshold regime. The last expression was obtained by using Eq. (3.3.32) at the bias level. Equation (3.3.49) provides

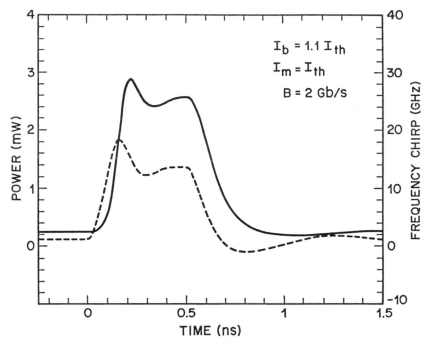

Figure 3.23 Calculated large-signal modulation response of a semiconductor laser to a 500-ps rectangular current pulse. Solid curve shows the shape of the emitted optical pulse; the dashed curve shows the frequency chirp imposed on the pulse by carrier-induced index changes ($\beta_c = 5$).

a remarkably simple expression for the modulation bandwidth. It shows that $f_{3\,\mathrm{dB}}$ increases with an increase in the bias level as $\sqrt{P_b}$ or as $(I_b - I_{th})^{1/2}$. This square-root dependence has been verified for many semiconductor lasers. Figure 3.22 shows how $f_{3\,\mathrm{dB}}$ can be increased to about 14 GHz for a DFB laser biased 7.7 times above threshold [64]. A modulation bandwidth of 25 GHz has been realized for a packaged 1.55-μm InGaAsP laser specifically designed for high-speed response [65]. However, it should be stressed that most semiconductor lasers are limited in practice to a bandwidth below 10 GHz because of electrical parasitics. In spite of this limitation, semiconductor lasers represent a considerable advance over the LEDs, whose modulation bandwidth is often limited to 200 MHz or less.

Large-Signal Modulation

The small-signal analysis, although useful for a qualitative understanding of the modulation response, is not generally applicable to optical communication systems where the laser is typically biased close to threshold and modulated considerably above threshold to obtain optical pulses representing digital bits. In this case of large-signal modulation, the rate equations should be solved numerically. Figure 3.23 shows, as an example, the shape of the emitted optical pulse for a laser biased at $I_b = 1.1 I_{th}$ and modulated at 2 Gb/s using rectangular current pulses of duration 500 ps and amplitude $I_m = I_{th}$. The optical pulse does not exhibit sharp leading and trailing edges because of a limited modulation bandwidth and exhibits a rise time ~ 100 ps and a fall time ~ 300 ps. The initial overshoot near the leading edge is a manifestation of relaxation oscillations. Even though the optical pulse is not an exact replica of the applied electrical pulse, deviations are small enough that semiconductor lasers can be used for data transmission up to bit rates of about 10 Gb/s.

As mentioned before, amplitude modulation in semiconductor lasers is accompanied by phase modulation governed by Eq. (3.3.41). A time-varying phase is equivalent to transient changes in the mode frequency from its steady-state value ν_0. Such a pulse is called chirped. The *frequency chirp* $\delta\nu(t)$ is obtained by using Eq. (3.3.41) and is given by

$$\delta\nu(t) = \frac{1}{2\pi}\frac{d\phi}{dt} = \frac{\beta_c}{4\pi}\left[G_N(N - N_0) - \frac{1}{\tau_p}\right]. \qquad (3.3.50)$$

The dashed curve in Fig. 3.23 shows the frequency chirp across the optical pulse. The mode frequency shifts toward the blue side near the leading edge and toward the red side near the trailing edge of the optical pulse [66]. Such a frequency shift implies that the pulse spectrum is considerably broader than that expected in the absence of frequency chirp.

It was seen in Section 2.4 that the frequency chirp can limit the performance of optical communication systems, especially when $\beta_2 C > 0$, where β_2 is the dispersion parameter and C is the chirp parameter. Even though optical pulses emitted from semiconductors are generally not Gaussian, the analysis of Section 2.4 can be used to study chirp-induced pulse broadening [67] if we identify C with $-\beta_c$ in Eq. (2.4.23). It turns out that 1.55-μm lightwave systems with a transmission distance of 80–100 km are limited to a bit rate below 2 Gb/s because of the frequency chirp [66] when conventional fibers are used ($\beta_2 \approx -20$ ps^2/km). Higher bit rates can be achieved only by operating the system near the zero-dispersion wavelength to minimize the effect of frequency chirp [68], [69]. Chirp limitations are discussed in Section 5.4.4.

Since frequency chirp is often the limiting factor for lightwave systems operating near 1.55 μm, several methods have been used to reduce its magnitude. One scheme requires careful tailoring of the shape of applied current pulse [70], [71]. Another makes use of injection locking [72]. Chirp is also reduced for coupled-cavity lasers [73], [74], discussed in Section 3.3.5. Chirp reduction by a factor of 2 has been observed in C^3 lasers [75]. An even higher reduction has been realized by coupling a semiconductor laser to an external Bragg reflector [76]. A direct way to reduce the frequency chirp is to design semiconductor lasers with small values of the linewidth enhancement factor β_c. The use of quantum-well design reduces β_c by about a factor of 2 [77]. A further reduction occurs for strained quantum wells [78]. Indeed, $\beta_c \approx 1$ has been measured in *modulation-doped* strained MQW lasers [79]. Such lasers exhibit low chirp under direct modulation. The frequency chirp resulting from current modulation can be avoided altogether if the laser is continuously operated, and an external modulator is used to modulate the laser output. In a 1996 system experiment, transmission at 10 Gb/s over 100 km of standard fiber was realized by using a gain-coupled DFB laser with a monolithically integrated modulator [80].

Soliton communication systems, discussed in Chapter 10, require lasers that generate short optical pulses (width ∼ 10 ps) at a high repetition rate equal to the bit rate. An external or integrated modulator then converts the periodic pulse train into an optical signal in the RZ format. Techniques such as *gain switching* and *mode locking* can be used to generate short pulses from a semiconductor laser. A mode-locked fiber laser can also be used for this purpose. Such optical sources are discussed in Section 10.2.3.

3.3.8 Laser Noise

The output of a semiconductor laser exhibits fluctuations in its intensity, phase, and frequency even when the laser is biased at a constant current with negligible current fluctuations. The two fundamental noise mechanisms are *spontaneous emission* and *electron–hole recombination* (shot noise). Noise in semiconductor lasers is dominated by spontaneous emission. Each spontaneously emitted photon adds to the coherent field (established by stimulated emission) a small field component whose phase is random, and thus perturbs both amplitude and phase in a random manner. Moreover, such spontaneous-emission events occur randomly at a high rate ($\sim 10^{12}$ s^{-1}) because of a relatively large value of R_{sp} in semiconductor lasers. The net result is that the intensity and the phase of the emitted light exhibit fluctuations over a time scale as short as 100 ps. Intensity fluctuations lead to a limited *signal-to-noise ratio* (SNR), whereas phase fluctuations lead to a finite spectral linewidth when semiconductor lasers are operated at a constant current. Since such fluctuations can affect the performance of lightwave systems, it is important to estimate their magnitude [81].

The rate equations can be used to study laser noise by adding a noise term, known as the *Langevin force*, to each of them [82], [83]. Equations (3.3.26), (3.3.27), and (3.3.41) then become

$$\frac{dP}{dt} = \left(G - \frac{1}{\tau_p}\right)P + R_{sp} + F_P(t), \tag{3.3.51}$$

$$\frac{dN}{dt} = \frac{I}{q} - \frac{N}{\tau_c} - GP + F_N(t), \tag{3.3.52}$$

$$\frac{d\phi}{dt} = \frac{1}{2}\beta_c \left(G_N(N - N_0) - \frac{1}{\tau_p}\right) + F_\phi(t), \tag{3.3.53}$$

where $F_P(t)$, $F_N(t)$, and $F_\phi(t)$ are the Langevin forces. They are assumed to be Gaussian random processes with zero mean and to have a correlation function of the form (the *Markoffian approximation*)

$$\langle F_i(t)F_j(t')\rangle = 2D_{ij}\delta(t - t'), \tag{3.3.54}$$

where $i, j = P$, N, or ϕ, angle brackets denote the ensemble average, and D_{ij} is called the *diffusion coefficient* [82]. The dominant contribution to laser noise comes from only two diffusion coefficients $D_{PP} = R_{sp}P$ and $D_{\phi\phi} = R_{sp}/4P$; others can be assumed to be nearly zero [83].

Intensity Noise

The intensity-autocorrelation function is defined as

$$C_{pp}(\tau) = \langle \delta P(t) \delta P(t+\tau) \rangle / \bar{P}^2, \tag{3.3.55}$$

where $\bar{P}$ is the average value and $\delta P = P - \bar{P}$ represents a fluctuation. The Fourier transform of $C_{pp}(\tau)$ is known as the *relative-intensity-noise* (RIN) spectrum and is given by

$$\text{RIN}(\omega) = \int_{-\infty}^{\infty} C_{pp}(\tau) \exp(-i\omega t)\, dt. \tag{3.3.56}$$

The RIN can be calculated by linearizing Eqs. (3.3.51) and (3.3.52) in δP and δN, solving the linearized equations in the frequency domain, and performing the average with the help of Eq. (3.3.54). It is given approximately by [10]

$$\text{RIN}(\omega) = \frac{2R_{sp}\{(\Gamma_N^2 + \omega^2) + G_N \bar{P}[G_N \bar{P}(1 + N/\tau_c R_{sp}\bar{P}) - 2\Gamma_N]\}}{\bar{P}[(\Omega_R - \omega)^2 + \Gamma_R^2][(\Omega_R + \omega)^2 + \Gamma_R^2]}, \tag{3.3.57}$$

where Ω_R and Γ_R are the frequency and the damping rate of relaxation oscillations. They are given by Eq. (3.3.45), with P_b replaced by $\bar{P}$.

Figure 3.24 shows the calculated RIN spectra at several power levels for a typical 1.55-μm InGaAsP laser. The RIN is considerably enhanced near the relaxation-oscillation frequency Ω_R but decreases rapidly for $\omega \gg \Omega_R$, since the laser is not able to respond to fluctuations at such high frequencies. In essence, the semiconductor laser acts as a bandpass filter of bandwidth Ω_R to spontaneous-emission fluctuations. At a given frequency, RIN decreases with an increase in the laser power as P^{-3} at low powers, but this behavior changes to P^{-1} dependence at high powers.

The autocorrelation function $C_{pp}(\tau)$ can be calculated by using Eqs. (3.3.56) and (3.3.57). The calculation shows that $C_{pp}(\tau)$ follows relaxation oscillations and approaches zero for $\tau > \Gamma_R^{-1}$ [84]. This behavior indicates that intensity fluctuations do not remain correlated for times longer than the damping time of relaxation oscillations. The quantity of practical interest is the SNR defined as $\bar{P}/\sigma_p$, where σ_p is the root-mean-square (RMS) noise. From Eq. (3.3.55), SNR $= [C_{pp}(0)]^{-1/2}$. At power levels above a few milliwatts, the SNR exceeds 20 dB and improves linearly with the power as

$$\text{SNR} = \left(\frac{\epsilon_{NL}}{R_{sp}\tau_p}\right)^{1/2} \bar{P}. \tag{3.3.58}$$

The presence of ϵ_{NL} indicates that the nonlinear form of the gain in Eq. (3.3.40) plays a crucial role. This form needs to be modified at high powers. Indeed,

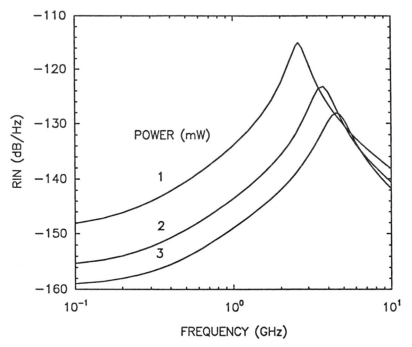

Figure 3.24 Calculated RIN spectra at several power levels for a typical 1.55-μm semiconductor laser.

a more accurate treatment shows [84] that the SNR eventually saturates at a value of about 30 dB and becomes power independent.

So far, the laser has been assumed to oscillate in a single longitudinal mode. In practice, even DFB lasers are accompanied by one or more side modes. Even though side modes remain suppressed by more than 20 dB on the basis of the average power, their presence can affect the RIN significantly. In particular, the main and side modes can fluctuate in such a way that individual modes exhibit large intensity fluctuations, but the total intensity remains relatively constant. This phenomenon is called *mode-partition noise* (MPN) and occurs due to an anticorrelation between the main and side modes [10]. It manifests through the enhancement of RIN for the main mode by 20 dB or more in the low-frequency range 0–1 GHz; the exact value of the enhancement factor depends on the MSR [85]. In the absence of fiber dispersion, MPN would be harmless for optical communication systems, as all modes would remain synchronized during transmission and detection. However, in practice all modes do not arrive simultaneously at the receiver because they travel at slightly different speeds. Such a desynchronization not only degrades the SNR of the received signal but also leads to intersymbol interference. The effect of MPN on the system performance is discussed in Section 5.4.3.

Spectral Linewidth

The spectrum of emitted light is related to the field-autocorrelation function $\Gamma_{EE}(\tau)$ through a Fourier-transform relation similar to Eq. (3.3.56), i.e.,

$$S(\omega) = \int_{-\infty}^{\infty} \Gamma_{EE}(t) \exp[-i(\omega - \omega_0)\tau] \, d\tau, \qquad (3.3.59)$$

where $\Gamma_{EE}(t) = \langle E^*(t)E(t+\tau)\rangle$ and $E(t) = \sqrt{P}\exp(i\phi)$ is the optical field. If intensity fluctuations are neglected, $\Gamma_{EE}(t)$ is given by

$$\Gamma_{EE}(t) = \langle \exp[i\Delta\phi(t)]\rangle = \exp[-\langle\Delta\phi^2(\tau)\rangle/2], \qquad (3.3.60)$$

where the phase fluctuation $\Delta\phi(\tau) = \phi(t+\tau) - \phi(t)$ is taken to be a Gaussian random process. The phase variance $\langle\Delta\phi^2(\tau)\rangle$ can be calculated by linearizing Eqs. (3.3.51)–(3.3.53) and solving the resulting set of linear equations. The result is [83]

$$\langle\Delta\phi^2(\tau)\rangle = \frac{R_{sp}}{2\bar{P}}\left[(1+\beta_c^2 b)\tau + \frac{\beta_c^2 b}{2\Gamma_R\cos\delta}[\cos(3\delta) - \exp(-\Gamma_R\tau)\cos(\Omega_R\tau - 3\delta)]\right],$$
$$(3.3.61)$$

where

$$b = \Omega_R/(\Omega_R^2 + \Gamma_R^2)^{1/2} \qquad \text{and} \qquad \delta = \tan^{-1}(\Gamma_R/\Omega_R). \qquad (3.3.62)$$

The spectrum is obtained by using Eqs. (3.3.59)–(3.3.61). It is found to consist of a dominant central peak located at ω_0 and multiple satellite peaks located at $\omega = \omega_0 \pm m\Omega_R$, where m is an integer. The amplitude of satellite peaks is typically less than 1% of that of the central peak. The physical origin of the satellite peaks is related to relaxation oscillations, which are responsible for the term proportional to b in Eq. (3.3.61). If this term is neglected, the autocorrelation function $\Gamma_{EE}(\tau)$ decays exponentially with τ. The integral in Eq. (3.3.59) can then be performed analytically, and the spectrum is found to be Lorentzian. The spectral linewidth $\Delta\nu$ is defined as the full width at half maximum (FWHM) of this Lorentzian line and is given by [83]

$$\Delta\nu = R_{sp}(1+\beta_c^2)/(4\pi\bar{P}), \qquad (3.3.63)$$

where $b = 1$ was assumed as $\Gamma_R \ll \Omega_R$ under typical operating conditions. The linewidth is enhanced by a factor of $1 + \beta_c^2$ as a result of the amplitude-phase coupling governed by β_c in Eq. (3.3.53); β_c is called the linewidth enhancement factor for this reason.

Equation (3.3.63) shows that $\Delta\nu$ should decrease as $\bar{P}^{-1}$ with an increase in the laser power. Such an inverse dependence is observed experimentally at low power levels (< 10 mW) for most semiconductor lasers. However, often the linewidth is found to saturate to a value in the range 1–10 MHz at a power level above 10 mW. Figure 3.25 shows such linewidth-saturation behavior for several 1.55-μm DFB lasers [86]. It also shows that the linewidth can be reduced

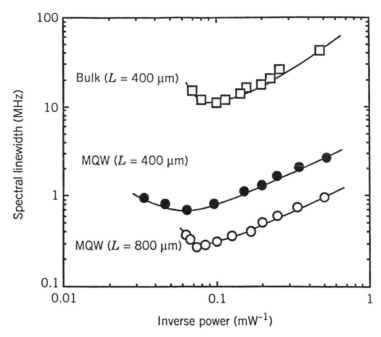

Figure 3.25 Measured spectral linewidth as a function of the optical power for several 1.55-μm DFB lasers. The upper curve corresponds to a conventional laser with an active layer about 100 nm thick. The lower two curves correspond to DFB lasers whose active region consists of multiple quantum wells about 10 nm thick. (After Ref. [86]. ©1991 IEEE. Reprinted with permission.)

considerably by using a MQW design for the DFB laser. The reduction is due to a smaller value of the parameter β_c realized by such a design. The linewidth can also be reduced by increasing the cavity length L, since R_{sp} decreases and P increases at a given output power as L is increased. Although not obvious from Eq. (3.3.63), $\Delta\nu$ can be shown to vary as L^{-2} when the length dependence of R_{sp} and P is incorporated. As seen in Fig. 3.25, $\Delta\nu$ is reduced by about a factor of 4 when the cavity length is doubled. The 800-μm-long MQW-DFB laser is found to exhibit a linewidth as small as 270 kHz at a power output of 13.5 mW [86]. It is further reduced in strained MQW lasers because of relatively low values of β_c, and a value of about 100 kHz has been measured in lasers with $\beta_c \approx 1$ [79]. It should be stressed, however, that the linewidth of most DFB lasers is typically 5–10 MHz when operating at a power level of 10 mW.

Figure 3.25 shows that as the laser power increases, the linewidth not only saturates but begins to rebroaden. Several mechanisms have been invoked to explain such behavior; a few of them are current noise [87], $1/f$-noise [88], non-linear gain [89], side-mode interaction [90]–[92], and index nonlinearity [93], [94]. The issue of laser linewidth is of considerable importance for lightwave systems. Considerable effort has been directed [46]–[52] toward developing semiconductor lasers that not only operate in a single longitudinal mode but whose wavelength can also be tuned over a considerable range while maintaining a narrow linewidth (1 MHz or less). DBR lasers with a superstructure grating can be tuned over 40 nm while maintaining a linewidth below 400 kHz [52].

3.4 TRANSMITTER DESIGN

So far this chapter has focused on the properties of optical sources. Although an optical source is a major component of optical transmitters, it is not the only component. Other components include a modulator for converting electrical data into optical form, a coupler for launching the optical signal into the fiber, and an electrical driving circuit for supplying current to the optical source (see Fig. 1.8). An external modulator is not always necessary and is rarely used in practice, since both LEDs and semiconductor lasers can be modulated directly. This section covers the design of optical transmitters with emphasis on the packaging issues [95]–[105]. Transmitters with a monolithically integrated external modulator are also discussed.

3.4.1 Source–Fiber Coupling

The design objective for any transmitter is to couple as much light as possible into the optical fiber to increase the launched power. In practice, the coupling efficiency depends on the type of optical source (LED or semiconductor laser) as well as on the type of fiber (multimode versus single mode). The coupling can be very inefficient when light from an LED is coupled into a single-mode fiber. As discussed briefly in Section 3.2.1, the coupling efficiency for an LED changes with the numerical aperture, and can become < 1% in the case of single-mode fibers. By contrast, the coupling efficiency for semiconductor lasers is typically 40–50%. A small piece of fiber (known as a pigtail) is included with the transmitter so that the coupling efficiency can be maximized during packaging; a splice or connector is used to join the pigtail with the fiber cable.

Two approaches have been used for source–fiber coupling. In one approach, known as direct or *butt coupling*, the fiber is brought close to the source and held in place by epoxy. In the other, known as *lens coupling*, a lens is used to maximize the coupling efficiency. Each approach has its own merits, and the choice generally depends on the design objectives. An important criterion is that the coupling efficiency should not change with time; mechanical stability of the coupling scheme is therefore a necessary requirement.

An example of butt coupling is shown in Fig. 3.26(a), where the fiber is brought in contact with a surface-emitting LED. The coupling efficiency for a fiber of numerical aperture NA is given by [96]

$$n_c = (1 - R_f)(\text{NA})^2, \qquad (3.4.1)$$

where R_f is the reflectivity at the fiber front end. R_f is about 4% if an air gap exists between the source and the fiber but can be reduced to nearly zero by placing an index-matching liquid. The coupling efficiency is about 1% for

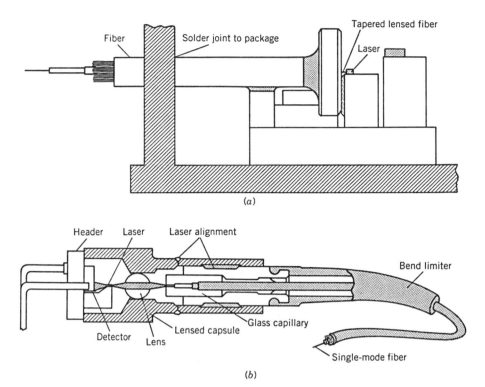

Figure 3.26 Transmitters employing (a) butt-coupling and (b) lens-coupling designs. (After Ref. [97]. ©1989 AT&T. Reprinted with permission.)

a surface-emitting LED and roughly 10% for an edge-emitting LED. Some improvement is possible in both cases by using fibers that are tapered or have a lensed tip. An external lens also improves the coupling efficiency but only at the expense of reduced mechanical tolerance.

The coupling of a semiconductor laser to a single-mode optical fiber is more efficient than that of an LED. The butt coupling provides only about 10% efficiency, as it makes no attempt to match the mode sizes of the laser and the fiber. Typically, index-guided InGaAsP lasers have a mode size of about 1 μm, whereas the mode size of a single-mode fiber is in the range 6–9 μm. The coupling efficiency can be improved by tapering the fiber end and forming a lens at the fiber tip. Figure 3.26(a) shows such a butt-coupling scheme for a commercial transmitter (AT&T Astrotec laser package). The fiber is attached to a jewel, and the jewel is attached to the laser submount by using an epoxy [97]. The fiber tip is aligned with the emitting region of the laser to maximize the coupling efficiency (typically 40%). The use of a lensed fiber [98]–[100] can improve the coupling efficiency, and values close to 100% have been realized with an optimum design [98].

Figure 3.26(b) shows a lens-coupling approach for transmitter design. The coupling efficiency can exceed 70% for such a confocal design in which a sphere is used to collimate the laser light and focus it onto the fiber core. The alignment of the fiber core is less critical for the confocal design because the spot size is magnified to match the fiber's mode size. The mechanical stability of the package is ensured by soldering the fiber into a ferrule which is secured to the body by two sets of laser alignment welds. One set of welds establishes proper axial alignment, while the other set provides transverse alignment.

The lens–fiber coupling issue remains important, and several new schemes have been developed during the 1990s [101]–[105]. In one approach, a *silicon optical bench* is used to align the laser and the fiber [101]–[103]. In another, a *silicon micromirror*, fabricated by using the micro-machining technology, is used for optical alignment [104]. In a different approach, a directional coupler is used as the *spot-size converter* for maximizing the coupling efficiency [105]. The suitability of a specific technique depends on the application and is dictated by cost and performance issues.

3.4.2   Optical Feedback

An important problem that needs to be addressed in designing an optical transmitter is related to the extreme sensitivity of semiconductor lasers to optical feedback [10], [11]. Even a relatively small amount of feedback ($< 0.1\%$) can destabilize the laser and affect the system performance through phenomena such as linewidth broadening, mode hopping, and RIN enhancement [106]–[110]. Attempts are made to reduce the feedback into the laser cavity by using anti-reflection coatings. Feedback can also be reduced by cutting the fiber tip at a slight angle so that the reflected light does not hit the active region of the laser. Such precautions are generally enough to reduce the feedback to a tolerable level. However, it becomes necessary to use an *optical isolator* between the laser and the fiber in transmitters designed for more demanding applications. One such application corresponds to lightwave systems operating at high bit rates and requiring a narrow-linewidth DFB laser.

Most optical isolators make use of the *Faraday effect*, which governs the rotation of the plane of polarization of an optical beam in the presence of a magnetic field: The rotation is in the same direction for light propagating parallel or antiparallel to the magnetic field direction. Optical isolators consist of a rod of Faraday material such as yttrium iron garnet (YIG), whose length is chosen to provide 45° rotation. The YIG rod is sandwiched between two polarizers whose axes are tilted by 45° with respect to each other. Light propagating in one direction passes through the second polarizer because of the Faraday rotation. By contrast, light propagating in the opposite direction is blocked by the first polarizer. Desirable characteristics of optical isolators are low insertion loss, high isolation (> 30 dB), compact size, and a wide spectral bandwidth of operation. A very compact isolator can be designed if the lens in Fig. 3.26(b) is replaced by a YIG sphere so that it serves a dual purpose [111]. Since light from a semiconductor laser is already polarized, a signal polarizer placed between the YIG sphere and the fiber can reduce the feedback by more than 30 dB.

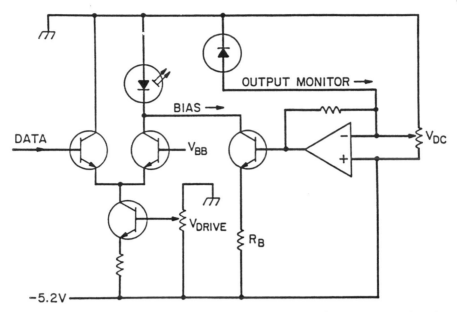

Figure 3.27 Driving circuit for a laser transmitter with feedback control to keep the average optical power constant. A photodiode monitors the output power and provides the control signal. (After Ref. [95]. ©1988 Academic Press. Reprinted with permission.)

3.4.3 Driving Circuitry

The purpose of driving circuitry is to provide electrical power to the optical source and to modulate the light output in accordance with the signal that is to be transmitted. Driving circuits are relatively simple for LED transmitters but become increasingly complicated for high-bit-rate optical transmitters employing semiconductor lasers as an optical source [95]. As discussed in Section 3.3.7, semiconductor lasers are biased near threshold and then modulated through an electrical time-dependent signal. Thus the driving circuit is designed to supply a constant bias current as well as modulated electrical signal. Furthermore, a servo loop is often used to keep the average optical power constant.

Figure 3.27 shows a simple driving circuit that controls the average optical power through a feedback mechanism. A photodiode (see Section 4.2) monitors the laser output and generates the control signal that is used to adjust the laser bias level. The rear facet of the laser is generally used for the monitoring purpose [see Fig. 3.26(b)]. In some transmitters a front-end tap is used to divert a small fraction of the output power to the detector. The bias-level control is essential, since the laser threshold is sensitive to the operating temperature. The threshold current also increases with aging of the transmitter because of gradual degradation of the semiconductor laser.

The driving circuit shown in Fig. 3.27 adjusts the bias level dynamically but leaves the modulation current unchanged. Such an approach is acceptable if the slope efficiency of the laser does not change with aging. As discussed in Section 3.3.6 and seen in Fig. 3.21, the slope efficiency of the laser generally decreases with an increase in temperature. A thermoelectric cooler is often used to stabilize the laser temperature. An alternative approach consists of designing driving circuits that use *dual-loop feedback* circuits and adjust both the bias current and the modulation current automatically [112].

3.4.4 Optoelectronic Integration

The electrical components used in the driving circuit determine the rate at which the transmitter output can be modulated. For lightwave transmitters operating at bit rates above 1 Gb/s, electrical parasitics associated with various transistors and other components often limit the transmitter performance. The performance of high-speed transmitters can be improved considerably by using monolithic integration of the laser with the driver. Since optical and electrical devices are fabricated on the same chip, such monolithic transmitters are referred to as *optoelectronic integrated-circuit* (OEIC) transmitters. The OEIC approach was first applied to integration of GaAs lasers, since the technology for fabrication of GaAs electrical devices is relatively well established [113]–[116]. The technology for fabrication of InP OEICs has evolved rapidly during the 1990s [117]–[122]. A 1.5-μm OEIC transmitter capable of operating at 5 Gb/s was demonstrated in 1988 [117]. By 1992, 10-Gb/s laser transmitters were fabricated by integrating 1.55-μm DFB lasers with field-effect transistors made with the InGaAs/InAlAs material system. Since then, OEIC transmitters with multiple lasers on the same chip have been developed for multichannel applications (see Chapter 7).

A related approach to OEIC integrates the semiconductor laser with a photodetector [123]–[125] and/or with a modulator [126]–[130]. The photodetector is generally used for monitoring and stabilizing the output power of the laser. The role of the modulator is to reduce the dynamic chirp occurring when a semiconductor laser is modulated directly (see Section 3.3.7). Photodetectors can be fabricated by using the same material as that used for the laser (see Chapter 4). However, the integration of a modulator requires a different approach.

A commonly used technique makes use of the Franz–Keldysh effect, according to which the bandgap of a semiconductor decreases when an electric field is applied across it. Thus, a transparent semiconductor layer begins to absorb light when its bandgap is reduced electronically by applying an external voltage. Such modulators are referred to as *electroabsorption modulators*. An extinction ratio of 15 dB or more for an applied reverse bias of 2 V can be realized at a bit rate of a few Gb/s. Low-chirp transmission at a bit rate of 5 Gb/s was demonstrated in 1994 by integrating an electroabsorption modulator with a DBR laser [128]. Such integrated transmitters can be operated at bit rates as high as 20 Gb/s. They can also be used to generate ultrashort pulses suitable for soliton communication systems (see Chapter 10). A strained MQW DFB laser, integrated monolithically with a MQW modulator, was used in 1993 to generate a 20-GHz pulse train [127]. The 7-ps output pulses were nearly transform-limited because of an extremely low chirp associated with the modulator.

A Mach–Zehnder interferometer (MZ) can also be used for intensity modulation (see Section 6.2). Its integration requires splitting of the laser light into two waveguides, voltage-dependent phase shifts in each arm, and coherent addition of the resulting output. In one implementation of this structure, a 1.55-μm DBR laser was integrated with a MZ modulator by using two Y-branch couplers and two MQW waveguides [130]. The differential phase shifts between the two waveguides is induced through electro-refraction. Such devices can be used at bit rates $\sim$ 10 Gb/s. In a 1996 demonstration, transmission at 10 Gb/s over 100 km of standard fiber was demonstrated by using a gain-coupled DFB laser integrated with a MZ modulator [80].

The concept of monolithic integration can be extended to build single-chip transmitters by adding all functionality on the same chip. Considerable effort has been directed toward developing such OEICs, often called *photonic integrated circuits* [15], which integrate on the same chip multiple optical components, such as lasers, detectors, modulators, amplifiers, filters, and waveguides [131]–[136]. Such integrated circuits should prove quite beneficial to lightwave technology.

3.4.5   Reliability and Packaging

An optical transmitter should operate reliably over a relatively long period of time (10 years or more) in order to be useful as a major component of lightwave systems. The reliability requirements are quite stringent for undersea lightwave systems, for which repairs and replacement are prohibitively expensive. By far the major reason for failure of optical transmitters is the optical source itself. Considerable testing is performed during assembly and manufacture of transmitters to ensure a reasonable lifetime for the optical source. It is common to quantify the lifetime by a parameter t_F known as *mean time to failure* (MTTF) [95]. Its use is based on the assumption of an exponential failure probability [$P_F = \exp(-t/t_F)$]. Typically, t_F should exceed 10^5 hours (about 11 years) for the optical source. Reliability of semiconductor lasers has been studied extensively to ensure their operation under realistic operating conditions [137]–[141].

Both LEDs and semiconductor lasers can stop operating suddenly (catastrophic degradation) or may exhibit a gradual mode of degradation in which the device efficiency degrades with aging [138]. Attempts are made to identify devices that are likely to degrade catastrophically. A common method is to operate the device at high temperatures and high current levels. This technique is referred to as burn-in or *accelerated aging* [137] and is based on the assumption that under high-stress conditions weak devices will fail, while others will stabilize after an initial period of rapid degradation. The change in the operating current at a constant power is used as a measure of device degradation. Figure 3.28 shows the change in the operating current of a 1.3-μm InGaAsP laser aged at 60°C under a constant output power of 5 mW from each facet. The operating

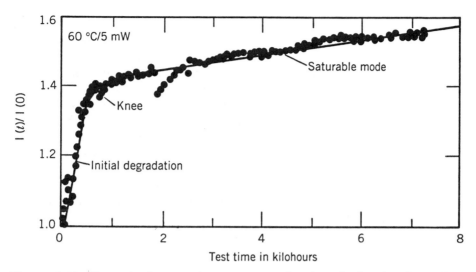

Figure 3.28 Change in the operating current as a function of aging time for a 1.3-μm InGaAsP laser aged at 60°C with 5 mW of output power. (After Ref. [139]. ©1985 AT&T. Reprinted with permission.)

current for this laser increases by 40% in the first 400 hours but then stabilizes and increases at a much reduced rate indicative of gradual degradation. The degradation rate can be used to estimate the laser lifetime and the MTTF at the elevated temperature. The MTTF at the normal operating temperature is then extrapolated by using an Arrhenius-type relation $t_F = t_0 \exp(-E_a/k_B T)$, where t_0 is a constant and E_a is the activation energy with a typical value of about 1 eV [138]. Physically, gradual degradation is due to the generation of various kinds of defects (dark-line defects, dark-spot defects) within the active region of the laser or LED [10].

Extensive tests have shown that LEDs are normally more reliable than semiconductor lasers under the same operating conditions. The MTTF for GaAs LEDs easily exceeds 10^6 hours and can be > 10^7 hours at 25°C [138]. The MTTF for InGaAsP LEDs is even larger, approaching a value $\sim 10^9$ hours. By contrast, the MTTF for InGaAsP lasers is generally limited to 10^6 hours at 25°C [139]–[141]. Nonetheless, this value is large enough that semiconductor lasers can be used in undersea optical transmitters designed to operate reliably for a period of 25 years. Because of the adverse effect of high temperatures on device reliability, most transmitters use a thermoelectric cooler to maintain the source temperature near 20°C even when the outside temperature may be as high as 80°C.

Even with a reliable optical source, a transmitter may fail in an actual system if the coupling between the source and the fiber degrades with aging. Coupling stability is an important issue in the design of reliable optical transmitters. It depends ultimately on the packaging of transmitters. Although LEDs are often packaged nonhermetically, an hermetic environment is essential for semiconductor lasers. It is common to package the laser separately so that it is isolated from other transmitter components. Figure 3.26 showed two examples of laser packages. In the butt-coupling scheme [Fig. 3.26(a)], an epoxy is used to hold the laser and fiber in place. Coupling stability in this case depends on how epoxy changes with aging of the transmitter. In the lens-coupling scheme [Fig. 3.26(b)], laser welding is used to hold various parts of the assembly together. The laser package becomes a part of the transmitter package, which includes other electrical components associated with the driving circuit. The choice of transmitter package depends on the type of application; a dual-in-line package or a butterfly housing with multiple pins is typically used.

Testing and packaging of optical transmitters are two important parts of the manufacturing process [140], and both of them add considerably to the cost of a transmitter. The development of low-cost packaged transmitters is necessary, especially for local-area and local-loop applications.

PROBLEMS

3.1 Show that the external quantum efficiency of a planar LED is given approximately by $\eta_{\text{ext}} = n^{-1}(n+1)^{-2}$, where n is the refractive index of the semiconductor–air interface. Consider Fresnel reflection and total internal reflection at the output facet. Assume that the internal radiation is uniform in all directions.

3.2 Prove that the 3-dB optical bandwidth of a LED is related to the 3-dB electrical bandwidth by the relation $f_{3\,\text{dB}}(\text{optical}) = \sqrt{3} f_{3\,\text{dB}}(\text{electrical})$.

3.3 Find the composition of the quaternary alloy InGaAsP for making semiconductor lasers operating at 1.3- and 1.55-μm wavelengths.

3.4 The active region of a 1.3-μm InGaAsP laser is 0.1 μm thick, 1 μm wide, and 250 μm long. Calculate the confinement factor and the modal index of the fundamental waveguide mode supported by the laser. Assume that $n_1 = 3.5$ and $n_2 = 3.2$.

3.5 Determine the active-region gain required for the laser of Problem 3.4 to reach threshold. Assume an internal loss of 30 cm^{-1} for the laser cavity.

3.6 Derive the eigenvalue equation for the transverse-electric (TE) modes of a planar waveguide of thickness d and refractive index n_1 sandwiched between two cladding layers of refractive index n_2. Assume infinitely thick cladding layers.

3.7 Calculate the maximum allowed thickness of the active layer for single-mode operation of a 1.3-μm semiconductor laser. How does this value change if the laser operates at 1.55 μm? Assume that $n_1 = 3.5$ and $n_2 = 3.2$.

3.8 Estimate the spot size (FWHM) in the lateral and transverse directions for a 1.3-μm semiconductor laser whose active layer is 0.3 μm thick and

1 μm wide. Also calculate the angular width (FWHM) of the far field in the two directions. Assume that $n_1 = 3.5$ and $n_2 = 3.2$.

3.9 Solve the rate equations in the steady state and obtain the analytic expressions for P and N as a function of the injection current I. Neglect spontaneous emission for simplicity.

3.10 A semiconductor laser is operating continuously at a certain current. Its output power changes slightly because of a transient current fluctuation. Show that the laser power will attain its original value through an oscillatory approach. Obtain the frequency and the damping time of such relaxation oscillations.

3.11 A 250-μm-long InGaAsP laser has an internal loss of 40 cm^{-1}. It operates in a single mode with the modal index 3.3 and the group index 3.4. Calculate the photon lifetime. What is the threshold value of the electron population, N_{th}? Assume that the gain varies as $G = G_N(N - N_0)$ with $G_N = 6 \times 10^3$ s^{-1} and $N_0 = 1 \times 10^8$.

3.12 Determine the threshold current for the semiconductor laser of Problem 3.11 by taking 2 ns as the carrier lifetime. How much power is emitted from one facet when the laser is operated twice above threshold?

3.13 Consider the laser of Problem 3.11 operating twice above threshold. Calculate the differential quantum efficiency and the external quantum efficiency for the laser. What is the device (wall-plug) efficiency if the external voltage is 1.5 V? Assume that the internal quantum efficiency is 90%.

3.14 Calculate the frequency (in GHz units) and the damping time of the relaxation oscillations for the laser of Problem 3.11 operating twice above threshold. Assume that $G_P = -4 \times 10^4$ s^{-1}, where G_P is the derivative of G with respect to P. Also assume that $R_{sp} = 2/\tau_p$.

3.15 Determine the 3-dB modulation bandwidth for the laser of Problem 3.11 biased to operate twice above threshold. What is the corresponding 3-dB electrical bandwidth?

3.16 The threshold current of a semiconductor laser doubles when the operating temperature is increased by 50°C. What is the characteristic temperature of the laser?

3.17 Derive an expression for the 3-dB modulation bandwidth by assuming that the gain G in the rate equations varies with N and P as

$$G(N, P) = G_N(N - N_0)(1 + P/P_s)^{-1/2}.$$

Show that the bandwidth saturates at high operating powers.

3.18 Solve the rate equations (3.3.26) and (3.3.27) numerically by using $I(t) = I_b + I_m f_p(t)$, where $f_p(t)$ represents a rectangular pulse of 200-ps duration. Assume that $I_b/I_{th} = 0.8$, $I_m/I_{th} = 3$, $\tau_p = 3$ ps, $\tau_c = 2$ ns, and $R_{sp} = 2/\tau_p$. Use Eq. (3.3.40) for the gain G with $G_N = 10^4$ s^{-1}, $N_0 = 10^8$, and $\epsilon_{NL} = 10^{-7}$. Plot the optical pulse shape and the frequency chirp. Why is the optical pulse much shorter than the applied current pulse?

3.19 Complete the derivation of Eq. (3.3.57) for the RIN. How does this expression change if the gain G is assumed of the form of Problem 3.17?

3.20 Calculate the autocorrelation $C_{pp}(\tau)$ by using Eqs. (3.3.56) and (3.3.57). Use it to derive an expression for the SNR of the laser output.

REFERENCES

[1] R. N. Hall, G. E. Fenner, J. D. Kingsley, T. J. Soltys, and R. O. Carlson, *Phys. Rev. Lett.* **9**, 366 (1962).

[2] M. I. Nathan, W. P. Dumke, G. Burns, F. H. Dill, Jr., and G. Lasher, *Appl. Phys. Lett.* **1**, 62 (1962).

[3] T. M. Quist, R. H. Rediker, R. J. Keyes, W. E. Krag, B. Lax, A. L. McWhorter, and H. J. Zeiger, *Appl. Phys. Lett.* **1**, 91 (1962).

[4] N. Holonyak, Jr. and S. F. Bevacqua, *Appl. Phys. Lett.* **1**, 82 (1962).

[5] I. Hayashi, M. B. Panish, P. W. Foy, and S. Sumski, *Appl. Phys. Lett.* **17**, 109 (1970).

[6] Zh. I. Alferov, V. M. Andreev, D. Z. Garbuzov, Yu. V. Zhilyaev, E. P. Morozov, E. L. Portnoi, and V. G. Trofim, *Sov. Phys. Semicond.* **4**, 1573 (1971).

[7] H. Kressel and J. K. Butler, *Semiconductor Lasers and Heterojunction LEDs*, Academic Press, San Diego, CA, 1977.

[8] H. C. Casey, Jr. and M. B. Panish, *Heterostructure Lasers*, Parts A and B, Academic Press, San Diego, CA, 1978.

[9] G. H. B. Thompson, *Physics of Semiconductor Laser Devices*, Wiley, New York, 1980.

[10] G. P. Agrawal and N. K. Dutta, *Semiconductor Lasers*, 2nd ed., Van Nostrand Reinhold, New York, 1993.

[11] K. Petermann, *Laser Diode Modulation and Noise*, 2nd ed., Kluwer Academic, Boston, 1991.

[12] P. S. Zory, Jr., Ed., *Quantum Well Lasers*, Academic Press, San Diego, CA, 1993.

[13] W. W. Chow, S. W. Koch, and M. Sargent, *Semiconductor-Laser Physics*, Springer-Verlag, New York, 1994.

[14] P. Vasil'ev, *Ultrafast Diode Lasers: Fundamentals and Applications*, Artech House, Boston, 1995.

[15] L. A. Coldren and S. W. Corzine, *Diode Lasers and Photonic Integrated Circuits*, Wiley, New York, 1995.

[16] G. P. Agrawal, Ed., *Semiconductor Lasers: Past, Present, and Future*, AIP Press, Woodbury, NY, 1995.

[17] B. Saleh and M. Teich, *Fundamental of Photonics*, Wiley, New York, 1991, Chaps. 15 and 16.

[18] J. Gower, *Optical Communication Systems*, 2nd ed., Prentice-Hall, Upper Saddle River, NJ, 1993.

[19] A. Einstein, *Phys. Z.* **18**, 121 (1917).

[20] S. L. Chuang, *Physics of Optoelectronic Devices*, Wiley, New York, 1995.

[21] M. G. A. Bernard and G. Duraffourg, *Phys. Status Solidi* **1**, 699 (1961).

[22] H. Temkin, G. V. Keramidas, M. A. Pollack, and W. R. Wagner, *J. Appl. Phys.*
      52, 1574 (1981); H. Temkin et al., *Bell Syst. Tech. J.* **62**, 1 (1983).

[23] C. A. Burrus and R. W. Dawson, *Appl. Phys. Lett.* **17**, 97 (1970).

[24] R. C. Goodfellow, A. C. Carter, I. Griffith, and R. R. Bradley, *IEEE Trans.*
      Electron. Dev. **26**, 1215 (1979).

[25] O. Wada, S. Yamakoshi, A. Masayuki, Y. Nishitani, and T. Sakurai, *IEEE J.*
      Quantum Electron. **17**, 174 (1981).

[26] D. Marcuse, *IEEE J. Quantum Electron.* **13**, 819 (1977).

[27] D. Botez and M. Ettenburg, *IEEE Trans. Electron. Dev.* **26**, 1230 (1979).

[28] S. T. Wilkinson, N. M. Jokerst, and R. P. Leavitt, *Appl. Opt.* **34**, 8298 (1995).

[29] M. C. Larson and J. S. Harris, Jr., *IEEE Photon. Technol. Lett.* **7**, 1267 (1995).

[30] I. J. Fritz, J. F. Klem, M. J. Hafich, A. J. Howard, and H. P. Hjalmarson, *IEEE*
      Photon. Technol. Lett. **7**, 1270 (1995).

[31] J. C. Dyment, *Appl. Phys. Lett.* **10**, 84 (1967).

[32] K. Oe, S. Ando, and K. Sugiyama, *J. Appl. Phys.* **51**, 43 (1980).

[33] G. P. Agrawal, *J. Lightwave Technol.* **2**, 537 (1984).

[34] D. Botez, *IEEE J. Quantum Electron.* **17**, 178 (1981).

[35] D. Botez and M. Ettenberg, *IEEE J. Quantum Electron.* **14**, 827 (1978).

[36] R. Wyatt and W. J. Devlin, *Electron. Lett.* **19**, 110 (1983).

[37] W. T. Tsang, in *Semiconductors and Semimetals*, vol. 22B, W. T. Tsang, Ed.,
      Academic Press, San Diego, CA, 1985, Chap. 4.

[38] S. Akiba, M. Usami, and K. Utaka, *J. Lightwave Technol.* **5**, 1564 (1987).

[39] G. P. Agrawal, in *Progress in Optics*, Vol. 26, E. Wolf, Ed., North-Holland,
      Amsterdam, 1988, Chap. 3.

[40] J. Buus, *Single Frequency Semiconductor Lasers*, SPIE Press, Bellingham, WA,
      1991.

[41] M. Ohtsu, *Highly Coherent Semiconductor Lasers*, Artech House, Boston, 1992.

[42] N. Chinone and M. Okai, in *Semiconductor Lasers: Past, Present, and Future*,
      G. P. Agrawal, Ed., AIP Press. Woodbury, NY, 1995, Chap. 2.

[43] G. P. Li, T. Makino, R. Moor, N. Puetz, K. W. Leong, and H. Lu, *IEEE J.*
      Quantum Electron. **29**, 1736 (1993).

[44] H. Lu, C. Blaauw, B. Benyon, G. P. Li, and T. Makino, *IEEE J. Sel. Topics*
      Quantum Electron. **1**, 375 (1995).

[45] C.-Y. Wang, Z.-M. Chuang, W. Lin, Y.-K. Tu, and C.-T. Lee, *IEEE Photon.*
      Technol. Lett. **8**, 331 (1996).

[46] K. Kobayashi and I. Mito, *J. Lightwave Technol.* **6**, 1623 (1988).

[47] Y. Kotaki, M. Matsuda, H. Ishikawa, and H. Imai, *Electron. Lett.* **24**, 503 (1988).

[48] T. L. Koch, U. Koren, and B. J. Miller, *Appl. Phys. Lett.* **53**, 1036 (1988).

[49] T. L. Koch and U. Koren, *J. Lightwave Technol.* **8**, 274 (1990).

[50] Y. Kotaki and H. Ishikawa, *IEE Proc.* **138**, Pt. J, 171 (1991).

[51] H. Hillmer, A. Grabmaier, S. Hansmann, H.-L. Zhu, H. Burkhard, and K. Magari, *IEEE J. Sel. Topics Quantum Electron.* **1**, 356 (1995).

[52] H. Ishii, F. Kano, Y. Tohmori, Y. Kondo, T. Tamamura, and Y. Yoshikuni, *IEEE J. Sel. Topics Quantum Electron.* **1**, 401 (1995).

[53] P.-J. Rigole, S. Nilsson, I. Bäckbom, T. Klinga, J. Wallin, B. Stalnacke, E. Berglind, and B. Stoltz, *IEEE Photon. Technol. Lett.* **7**, 697 (1995); **7**, 1249 (1995).

[54] T. E. Sale, *Vertical-Cavity Surface-Emitting Lasers*, Wiley, New York, 1995.

[55] C. J. Chang-Hasnain, in *Semiconductor Lasers: Past, Present, and Future*, G. P. Agrawal, Ed., AIP Press. Woodbury, NY, 1995, Chap. 5.

[56] L. E. Eng, K. Bacher, W. Yuen, J. S. Harris, Jr., and C. J. Chang-Hasnain, *IEEE J. Sel. Topics Quantum Electron.* **1**, 624 (1995).

[57] R. S. Tucker, *J. Lightwave Technol.* **3**, 1180 (1985).

[58] J. Manning, R. Olshansky, D. M. Fye, and W. Powazinik, *Electron. Lett.* **21**, 496 (1985).

[59] G. P. Agrawal, *IEEE J. Quantum Electron.* **23**, 860 (1987).

[60] G. P. Agrawal, *IEEE J. Quantum Electron.* **26**, 1901 (1990).

[61] G. P. Agrawal and G. R. Gray, *Proc. SPIE* **1497**, 444 (1991).

[62] C. Z. Ning and J. V. Moloney, *Appl. Phys. Lett.* **66**, 559 (1995).

[63] M. Osinski and J. Buus, *IEEE J. Quantum Electron.* **23**, 9 (1987).

[64] H. Ishikawa, H. Soda, K. Wakao, K. Kihara, K. Kamite, Y. Kotaki, M. Matsuda, H. Sudo, S. Yamakoshi, S. Isozumi, and H. Imai, *J. Lightwave Technol.* **5**, 848 (1987).

[65] P. A. Morton, T. Tanbun-Ek, R. A. Logan, N. Chand, K. W. Wecht, A. M. Sergent, and P. F. Sciortino, Jr., *Electron. Lett.* **30**, 2044 (1994).

[66] R. A. Linke, *Electron. Lett.* **20**, 472 (1984); *IEEE J. Quantum Electron.* **21**, 593 (1985).

[67] G. P. Agrawal and M. J. Potasek, *Opt. Lett.* **11**, 318 (1986).

[68] S. Fujita, M. Kitamura, N. Torikai, N. Henmi, H. Yamada, T. Suzuki, I. Takano, and M. Shikada, *Electron. Lett.* **25**, 702 (1989).

[69] A. H. Gnauck, C. A. Burrus, S.-J. Wang, and N. K. Dutta, *Electron. Lett.* **25**, 1356 (1989).

[70] R. Olshansky and D. Fye, *Electron. Lett.* **20**, 928 (1984).

[71] L. Bickers and L. D. Westbrook, *Electron. Lett.* **21**, 103 (1985).

[72] C. Lin, J. K. Anderson, and F. Mengel, *Electron. Lett.* **21**, 80 (1985).

[73] T. Fujita, J. Ohya, S. Ishizuka, K. Fujito, and H. Sato, *Electron. Lett.* **20**, 416 (1984).

[74] G. P. Agrawal, *Opt. Lett.* **10**, 10 (1985).

[75] G. P. Agrawal, N. A. Olsson, and N. K. Dutta, *Appl. Phys. Lett.* **45**, 119 (1984).

[76] N. A. Olsson, C. H. Henry, R. F. Kazarinov, H. J. Lee, and K. J. Orlowsky, *IEEE J. Quantum Electron.* **24**, 143 (1988).

[77] C. A. Green, N. K. Dutta, and W. Watson, *Appl. Phys. Lett.* **50**, 1409 (1987).

[78] H. D. Summers and I. H. White, *Electron. Lett.* **30**, 1140 (1994).

[79] F. Kano, T. Yamanaka, N. Yamamoto, H. Mawatan, Y. Tohmori, and Y. Yoshikuni, *IEEE J. Quantum Electron.* **30**, 533 (1994).

[80] D. M. Adams, C. Rolland, N. Puetz, R. S. Moore, F. R. Shepard, H. B. Kim, and S. Bradshaw, *Electron. Lett.* **32**, 485 (1996).

[81] G. P. Agrawal, *Proc. SPIE* **1376**, 224 (1991).

[82] M. Lax, *Rev. Mod. Phys.* **38**, 541 (1966); *IEEE J. Quantum Electron.* **3**, 37 (1967).

[83] C. H. Henry, *IEEE J. Quantum Electron.* **18**, 259 (1982); **19**, 1391 (1983); *J. Lightwave Technol.* **4**, 298 (1986).

[84] G. P. Agrawal, *Electron. Lett.* **27**, 232 (1991).

[85] G. P. Agrawal, *Phys. Rev. A* **37**, 2488 (1988).

[86] M. Aoki, K. Uomi, T. Tsuchiya, S. Sasaki, M. Okai, and N. Chinone, *IEEE J. Quantum Electron.* **27**, 1782 (1991).

[87] G. P. Agrawal and R. Roy, *Phys. Rev. A* **37**, 2495 (1988).

[88] K. Kikuchi, *Electron. Lett.* **24**, 1001 (1988); *IEEE J. Quantum Electron.* **25**, 684 (1989).

[89] G. P. Agrawal, *IEEE Photon. Technol. Lett.* **1**, 212 (1989).

[90] U. Kruger and K. Petermann, *IEEE J. Quantum Electron.* **24**, 2355 (1988); 26, 2058 (1990).

[91] S. E. Miller, *IEEE J. Quantum Electron.* **24**, 750 (1988); **24**, 1873 (1988).

[92] G. R. Gray and G. P. Agrawal, *IEEE Photon. Technol. Lett.* **3**, 204 (1991).

[93] G. P. Agrawal, G.-H. Duan, and P. Gallion, *Electron. Lett.* **28**, 1773 (1992).

[94] F. Girardin, G.-H. Duan, and P. Gallion, *IEEE Photon. Technol. Lett.* **8**, 334 (1996).

[95] P. W. Shumate, in *Optical Fiber Telecommunications II*, S. E. Miller and I. P. Kaminow, Eds., Academic Press, San Diego, CA, 1988, Chap. 19.

[96] T. P. Lee, C. A. Burrus, and R. H. Saul, in *Optical Fiber Telecommunications II*, S. E. Miller and I. P. Kaminow, Eds., Academic Press, San Diego, CA, 1988, Chap. 12.

[97] D. S. Alles and K. J. Brady, *AT&T Tech. J.* **68**, 183 (1989).

[98] H. M. Presby and C. A. Edwards, *Electron. Lett.* **28**, 582 (1992).

[99] R. A. Modavis and T. W. Webb, *IEEE Photon. Technol. Lett.* **7**, 798 (1995).

[100] K. Shiraishi, N. Oyama, K. Matsumura, I. Ohisi, and S. Suga, *J. Lightwave Technol.* **13**, 1736 (1995).

[101] J. P. Schmidt, A. Cordes, J. Müller, and H. Burkhardt, *Proc. SPIE* **2449**, 176 (1995).

[102] P. C. Chen and T. D. Milster, *Laser Diode Chip and Packaging Technology*, Vol. 2610, SPIE Press, Bellingham, WA, 1996.

[103] I. P. Hall, *Microelectron. Int.* **39**, 6 (1996).

[104] M. J. Daneman, O. Sologaard, N. C. Tien, K. Y. Lau, and R. S. Muller, *IEEE Photon. Technol. Lett.* **8**, 396 (1996).

[105] B. M. A. Rahman, M. Rajarajan, T. Wongcharoen, and K. T. V. Grattan, *IEEE Photon. Technol. Lett.* **8**, 557 (1996).

[106] G. P. Agrawal, *IEEE J. Quantum Electron.* **20**, 468 (1984).

[107] G. P. Agrawal and T. M. Shen, *J. Lightwave Technol.* **4**, 58 (1986).

[108] A. T. Ryan, G. P. Agrawal, G. R. Gray, and E. C. Gage, *IEEE J. Quantum Electron.* **30**, 668 (1994).

[109] G. H. M. van Tartwijk and D. Lenstra, *Quantum Semiclass. Opt.* **7**, 87 (1995).

[110] K. Petermann, *IEEE J. Sel. Topics Quantum Electron.* **1**, 480 (1995).

[111] T. Sugie and M. Saruwatari, *Electron. Lett.* **18**, 1026 (1982).

[112] F. S. Chen, *Electron. Lett.* **16**, 7 (1980).

[113] O. Wada, T. Sakurai, and T. Nakagami, *IEEE J. Quantum Electron.* **22**, 805 (1986).

[114] S. R. Forrest, *Proc. IEEE* **75**, 1488 (1987).

[115] T. Horimatsu and M. Sasaki, *J. Lightwave Technol.* **7**, 1613 (1989).

[116] M. Dagenais, R. F. Leheny, H. Temkin, and P. Battacharya, *J. Lightwave Technol.* **8**, 846 (1990).

[117] N. Suzuki, H. Furuyama, Y. Hirayama, M. Morinaga, K. Eguchi, M. Kushibe, M. Funamizu, and M. Nakamura, *Electron. Lett.* **24**, 467 (1988).

[118] M. Banu, B. Jalali, R. Nottenburg, D. A. Humphrey, R. K. Montgomery, R. A. Hamm, and M. B. Panish, *Electron. Lett.* **27**, 278 (1991).

[119] Y. H. Lo, R. Bhat, P. S. D. Lin, and T. P. Lee, *Proc. SPIE* **1582**, 60 (1991).

[120] P. N. Woolnough, P. Birdsall, P. J. O'Sullivan, A. J. Cockburn, and M. J. Harlow, *Electron. Lett.* **29**, 1388 (1993).

[121] K. Pedrotti, *Proc. SPIE* **2149**, 178 (1994).

[122] O. Calliger, A. Clei, D. Robein, R. Azoulay, B. Pierre, S. Biblemont, and C. Kazmierski, *IEE Proc.* **142**, Pt. J, 13 (1995).

[123] K. Sato, I. Katoka, K. Wakita, Y. Kondo, and M. Yamamoto, *Electron. Lett.* **29**, 1087 (1993).

[124] D. Hofstetter, H. P. Zappe, J. E. Epler, and P. Riel, *IEEE Photon. Technol. Lett.* **7**, 1022 (1995).

[125] U. Koren, B. I. Miller, M. G. Young, M. Chien, K. Dreyer, R. Ben-Michael, and R. J. Capik, *IEEE Photon. Technol. Lett.* **8**, 364 (1996).

[126] M. Aoki, M. Suzuki, H. Sano, T. Kawano, T. Ido, T. Taniwatari, K. Uomi, and A. Takai, *IEEE J. Quantum Electron.* **29**, 2088 (1993).

[127] K. Wakita, K. Sato, I. Kotaka, M. Yamamoto, and M. Asobe, *IEEE Photon. Technol. Lett.* **5**, 899 (1993).

[128] G. Raybon, U. Koren, M. G. Young, B. I. Miller, M. Chien, T. H. Wood, and H. M. Presby, *Electron. Lett.* **30**, 1330 (1994).

[129] A. Ramdane, P. Krauz, E. V. K. Rao, A. Hamoudi, A. Ougazzaden, D. Robein, A. Gloukhian, and M. Carré, *IEEE Photon. Technol. Lett.* **7**, 1016 (1995).

[130] T. Tanbun-Ek, P. F. Sciortino, A. M. Sergent, K. W. Wecht, P. Wisk, Y. K. Chen, C, G. Bethea, and S. K. Sputz, *IEEE Photon. Technol. Lett.* **7**, 1019 (1995).

[131] T. L. Koch and U. Koren, *IEE Proc.* **138**, Pt. J, 139 (1991); *IEEE J. Quantum Electron.* **27**, 641 (1991).

[132] P. J. Williams and A. C. Carter, *GEC J. Res.* **10**, 91 (1993).

[133] R. Matz, J. G. Bauer, P. C. Clemens, G. Heise, H. F. Mahlein, W. Metzger, H. Michel, and G. Schulte-Roth, *IEEE Photon. Technol. Lett.* **6**, 1327 (1994).

[134] Y. Sasaki, Y. Sakata, T. Morimoto, Y. Inomoto, and T. Murakami, *NEC Tech. J.* **48**, 219 (1995).

[135] M. N. Armenise and K.-K. Wong, Eds., *Functional Photonic Integrated Circuits*, SPIE Proc. Series, Vol. 2401, SPIE Press, Bellingham, WA, 1995.

[136] W. Metzger, J. G. Bauer, P. C. Clemens, G. Heise, M. Klein, H. F. Mahlein, R. Matz, H. Michel, and J. Rieger, *Opt. Quantum Electron.* **28**, 51 (1996).

[137] F. R. Nash, W. B. Joyce, R. L. Hartman, E. I. Gordon, and R. W. Dixon, *AT&T Tech. J.* **64**, 671 (1985).

[138] N. K. Dutta and C. L. Zipfel, in *Optical Fiber Telecommunications II*, S. E. Miller and I. P. Kaminow, Eds., Academic Press, San Diego, CA, 1988, Chap. 17.

[139] B. W. Hakki, P. E. Fraley, and T. F. Eltringham, *AT&T Tech. J.* **64**, 771 (1985).

[140] M. Fallahi and S. C. Wang, Eds., *Fabrication, Testing, and Reliability of Semiconductor Lasers*, SPIE Proc. Series, Vol. 2863, SPIE Press, Bellingham, WA, 1995.

[141] O. Ueda, *Reliability and Degradation of III–V Optical Devices*, Artec House, Boston, 1996.

Chapter 4

OPTICAL RECEIVERS

The role of an optical receiver is to convert the optical signal back into electrical form and recover the data transmitted through the lightwave system. Its main component is a photodetector that converts light into electricity through the photoelectric effect. The requirements for a photodetector are similar to those of an optical source. It should have high sensitivity, fast response, low noise, low cost, and high reliability. Its size should be compatible with the fiber-core size. These requirements are best met by photodetectors made of semiconductor materials. Photodetectors and optical receivers have been discussed extensively in the literature on optical communications [1]–[8]. This chapter covers the main topics only. Section 4.1 introduces the basic concepts behind the photodetection process. In Section 4.2 we discuss several kinds of photodetectors commonly used for optical receivers. The components of an optical receiver are described in Section 4.3 with emphasis on the role played by each component. Section 4.4 deals with various noise sources that limit the signal-to-noise ratio in optical receivers. Section 4.5 considers receiver sensitivity, and Section 4.6 covers the degradation of sensitivity under nonideal conditions. The performance of optical receivers in actual transmission experiments is discussed in the final section.

4.1 BASIC CONCEPTS

The fundamental mechanism behind the photodetection process is optical absorption. Consider the semiconductor slab shown schematically in Fig. 4.1. If the energy $h\nu$ of incident photons exceeds the bandgap energy, an electron–hole pair is generated each time a photon is absorbed by the semiconductor. Under the influence of an electric field set up by an applied voltage, electrons and holes are swept across the semiconductor, resulting in a flow of electric current. The photocurrent I_p is directly proportional to the incident optical power P_{in}, i.e.,

$$I_p = RP_{in}, \qquad (4.1.1)$$

where R is the *responsivity* of the photodetector (in units of A/W).

138

Incident light

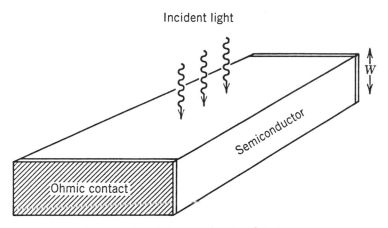

Figure 4.1 Photoconductive detector.

The responsivity R can be expressed in terms of a fundamental quantity η, called the *quantum efficiency* and defined as

$$\eta = \frac{\text{electron generation rate}}{\text{photon incidence rate}} = \frac{I_p/q}{P_{in}/h\nu} = \frac{h\nu}{q}R, \qquad (4.1.2)$$

where Eq. (4.1.1) was used. The responsivity R is thus given by

$$R = \frac{\eta q}{h\nu} \approx \frac{\eta\lambda}{1.24}, \qquad (4.1.3)$$

where $\lambda = c/\nu$ is expressed in micrometers. The responsivity of a photodetector increases with the wavelength λ simply because more photons are present for the same optical power. Such a linear dependence on λ is not expected to continue forever, since eventually the photon energy becomes too small to generate electrons. In semiconductors, this happens for $h\nu < E_g$, where E_g is the bandgap. The quantum efficiency η then drops to zero.

The dependence of η on λ enters through the absorption coefficient α. If the facets of the semiconductor slab in Fig. 4.1 are assumed to have an anti-reflection coating, the power transmitted through the slab of width W is $P_{tr} = \exp(-\alpha W)P_{in}$. The absorbed power is thus given by

$$P_{abs} = P_{in} - P_{tr} = [1 - \exp(-\alpha W)]P_{in}. \qquad (4.1.4)$$

Since each absorbed photon creates an electron–hole pair, the quantum efficiency η is given by

$$\eta = P_{abs}/P_{in} = 1 - \exp(-\alpha W). \qquad (4.1.5)$$

As expected, η becomes zero when $\alpha = 0$. On the other hand, η approaches 1 if $\alpha W \gg 1$.

Figure 4.2 shows the wavelength dependence of α for several semiconductor materials commonly used to make photodetectors for lightwave systems. The

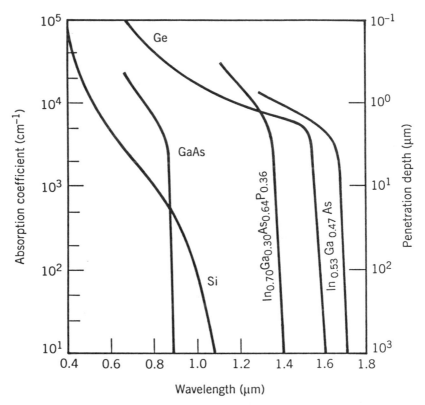

Figure 4.2 Wavelength dependence of the absorption coefficient for several semiconductor materials. (After Ref. [2]. ©1979 Academic Press. Reprinted with permission.)

wavelength λ_c at which α becomes zero is called the cutoff wavelength, since that material can be used as a photodetector only for $\lambda < \lambda_c$. As seen in Fig. 4.2, indirect-bandgap semiconductors such as Si and Ge can be used to make photodetectors even though the absorption edge is not as sharp as for direct-bandgap materials. Large values of α ($\sim 10^4$ cm^{-1}) can be realized for most semiconductors, and η can approach 100% for $W \sim 10$ μm. This feature illustrates the efficiency of semiconductors for the purpose of photodetection.

The *bandwidth* of a photodetector is determined by the speed with which it responds to variations in the incident optical power. It is useful to introduce the concept of *rise time* T_r, defined as the time over which the current builds up from 10% to 90% of its final value when the incident optical power is changed abruptly. The rise time can be written as (see Section 5.3.2)

$$T_r = (\ln 9)(\tau_{tr} + \tau_{RC}), \qquad (4.1.6)$$

where τ_{tr} is the *transit time* and τ_{RC} is the time constant of the equivalent RC circuit. The transit time is added to τ_{RC} because it takes some time before the carriers are collected after their generation through absorption of photons. The maximum collection time is just equal to the time an electron takes to traverse the absorption region. Clearly, τ_{tr} can be reduced by decreasing W. However, as seen from Eq. (4.1.5), the quantum efficiency η begins to decrease significantly for $\alpha W < 3$. Thus, there is a trade-off between the bandwidth and the responsivity (speed versus sensitivity) of a photodetector. Often, the RC time constant τ_{RC} limits the bandwidth because of electrical parasitics. The numerical values of τ_{tr} and τ_{RC} depend on the detector design and can vary over a wide range. The bandwidth Δf is related to them as (see Section 5.3.2)

$$\Delta f = [2\pi(\tau_{tr} + \tau_{RC})]^{-1}. \tag{4.1.7}$$

Together with the bandwidth and the responsivity, the dark current I_d of a photodetector is the third important parameter. I_d is the current generated in a photodetector in the absence of any optical signal and originates from stray light or from thermally generated electron–hole pairs. For a good photodetector, the dark current should be negligible ($I_d < 10$ nA).

4.2 PHOTODETECTOR DESIGN

Photodetectors can be broadly classified into two categories: photoconductive and photovoltaic. A homogeneous semiconductor slab with ohmic contacts (see Fig. 4.1) acts as a simple kind of photoconductive detector. Little current flows when no light is incident because of low conductivity of semiconductors. Incident light increases conductivity through electron–hole generation, and allows current to flow in proportion to the optical power. Photovoltaic detectors (e.g., solar cells) operate by producing a voltage in the presence of light. This chapter focuses on photoconductive detectors. Reverse-biased p–n junctions fall in this category and are commonly used for lightwave system. This section is devoted largely to them. Metal–semiconductor–metal (MSM) photodetectors are also discussed briefly.

4.2.1 p–n Photodiodes

A reverse-biased p–n junction consists of a region, known as the depletion region, that is essentially devoid of free charge carriers and where a large built-in electric field opposes flow of electrons from the n-side to the p-side (and of holes from p to n). When such a p–n junction is illuminated with light on one side, say the p-side (see Fig. 4.3), electron–hole pairs are created through absorption. Because of the large built-in electric field, electrons and holes generated inside the depletion region accelerate in opposite directions and drift to the n- and p-sides, respectively. The resulting flow of current is proportional to the incident optical power. Thus a reverse-biased p–n junction acts as a photodetector and is referred to as the p–n photodiode.

Figure 4.3(a) shows the structure of a p–n photodiode. As shown in Fig. 4.3(b), incident light is absorbed mostly inside the *depletion region*. The electron–hole pairs generated experience a large electric field and drift rapidly toward the

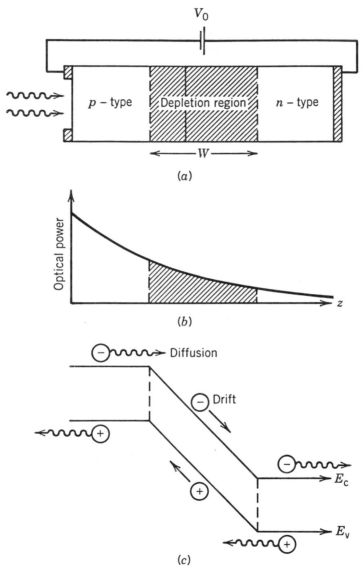

Figure 4.3 (a) A p–n photodiode under reverse bias; (b) variation of optical power inside the photodiode; (c) energy-band diagram showing carrier movement through drift and diffusion.

p- or n-side, depending on the electric charge [Fig. 4.3(c)]. The resulting current flow constitutes the photodiode response to the incident optical power in accordance with Eq. (4.1.1). The responsivity of a photodiode is quite high ($R \sim 1$ A/W) because of a high quantum efficiency.

The bandwidth of a p–n photodiode is often limited by the transit time τ_{tr} in Eq. (4.1.7). If W is the width of the depletion region and v_d is the drift

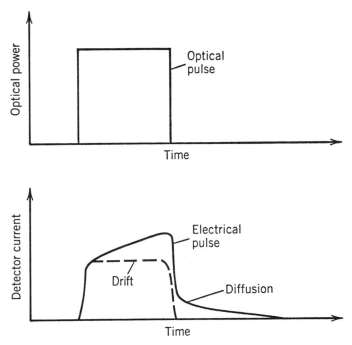

Figure 4.4 Response of a p–n photodiode to a rectangular optical pulse when both drift and diffusion contribute to the detector current.

velocity, the transit time is given by

$$\tau_{tr} = W/v_d. \tag{4.2.1}$$

Typically, $W \sim 10~\mu$m, $v_d \sim 10^5$ m/s, and $\tau_{tr} \sim 100$ ps. Both W and v_d can be optimized to minimize τ_{tr}. The depletion-layer width depends on the acceptor and donor concentrations and can be controlled through them. The velocity v_d depends on the applied voltage but attains a maximum value (called the *saturation velocity*) $\sim 10^5$ m/s that depends on the material used for the photodiode. The RC time constant τ_{RC} can be written as

$$\tau_{RC} = (R_L + R_s)C_p, \tag{4.2.2}$$

where R_L is the external load resistance, R_s is the internal series resistance, and C_p is the parasitic capacitance. Typically, $\tau_{RC} \sim 100$ ps, although lower values are possible with proper design. This value is small enough that p–n photodiodes are capable of operating at bit rates of about 1 Gb/s.

The limiting factor for the bandwidth of p–n photodiodes is the presence of a diffusive component in the photocurrent. The physical origin of the diffusive component is related to the absorption of incident light outside the depletion region. Electrons generated in the p-region have to diffuse to the depletion-region boundary before they can drift to the n-side; similarly, holes generated in the n-region must diffuse to the depletion-region boundary. Diffusion is an inherently

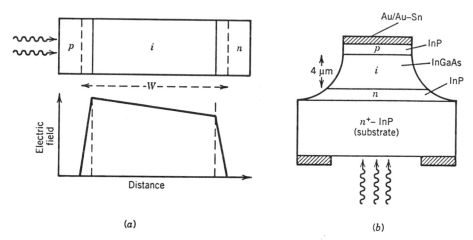

Figure 4.5 (a) A p–i–n photodiode together with the electric-field distribution under reverse bias; (b) design of an InGaAs p–i–n photodiode.

slow process; carriers take a nanosecond or longer to diffuse over a distance of about 1 μm. Figure 4.4 shows how the presence of diffusive component can distort the temporal response of a photodiode. In practice, the diffusion contribution depends on the bit rate and becomes negligible if the optical pulse is much shorter than the diffusion time. It can also be reduced by decreasing the widths of p- and n-regions and increasing the depletion-region width so that most of the incident optical power is absorbed inside it. This is the approach adopted for p–i–n photodiodes, discussed next.

4.2.2 p–i–n Photodiodes

A simple way to increase the depletion-region width is to insert a layer of undoped (or lightly doped) semiconductor material between the p–n junction. Since the middle layer consists of nearly intrinsic material, such a structure is referred to as the p–i–n photodiode. Figure 4.5(a) shows the device structure together with the electric-field distribution inside it under reverse-bias operation. Because of its intrinsic nature, the middle i-layer offers a high resistance, and most of the voltage drop occurs across it. As a result, a large electric field exists in the i-layer. In essence, the depletion region extends throughout the i-region, and its width W can be controlled by changing the middle-layer thickness. The main difference from the p–n photodiode is that the drift component of the photocurrent dominates over the diffusion component simply because most of the incident power is absorbed inside the i-region of a p–i–n photodiode.

Since the depletion width W can be tailored in p–i–n photodiodes, a natural question is how large W should be. As discussed in Section 4.1, the optimum value of W depends on a compromise between speed and sensitivity. The responsivity can be increased by increasing W so that the quantum efficiency η approaches 100% [see Eq. (4.1.5)]. However, the response time also increases,

Table 4.1 Characteristics of common p–i–n photodiodes

Parameter	Symbol	Unit	Si	Ge	InGaAs
Wavelength	λ	μm	0.4–1.1	0.8–1.8	1.0–1.7
Responsivity	R	A/W	0.4–0.6	0.5–0.7	0.6–0.9
Quantum efficiency	η	%	75–90	50–55	60–70
Dark Current	I_d	nA	1–10	50–500	1–20
Rise time	T_r	ns	0.5–1	0.1–0.5	0.05–0.5
Bandwidth	Δf	GHz	0.3–0.6	0.5–3	1–5
Bias voltage	V_b	V	50–100	6–10	5–6

as it takes longer for carriers to drift across the depletion region. For indirect-bandgap semiconductors such as Si and Ge, typically W must be in the range 20–50 μm to ensure a reasonable quantum efficiency. The bandwidth of such photodiodes is then limited by a relatively long transit time ($\tau_{tr} > 200$ ps). By contrast, W can be as small as 3–5 μm for photodiodes that use direct bandgap semiconductors, such as InGaAs. The transit time for such photodiodes is in the range $\tau_{tr} = 30$–50 ps if we use $v_d = 1 \times 10^5$ m/s for the saturation velocity. Such values of τ_{tr} correspond to a detector bandwidth $\Delta f = 3$–5 GHz if we use Eq. (4.1.7) with $\tau_{tr} \gg \tau_{RC}$.

The performance of p–i–n photodiodes can be improved considerably by using a double-heterostructure design. Similar to the case of semiconductor lasers, the middle i-type layer is sandwiched between the p-type and n-type layers of a different semiconductor whose bandgap is chosen such that light is absorbed only in the middle i-layer. A p–i–n photodiode commonly used for lightwave applications uses InGaAs for the middle layer and InP for the surrounding p-type and n-type layers [10], [11]. Figure 4.5(b) shows such an InGaAs p–i–n photodiode. Since the bandgap of InP is 1.35 eV, InP is transparent for light whose wavelength exceeds 0.92 μm. By contrast, the bandgap of lattice-matched $In_{1-x}Ga_xAs$ material with $x = 0.47$ is about 0.75 eV (see Section 3.1.4), a value that corresponds to a cutoff wavelength of 1.65 μm. The middle InGaAs layer thus absorbs strongly in the wavelength region 1.3–1.6 μm. The diffusive component of the detector current is eliminated completely in such a heterostructure photodiode simply because photons are absorbed only inside the depletion region. The front facet is often coated using suitable dielectric layers to minimize reflections. The quantum efficiency η can be made almost 100% by using an InGaAs layer 4–5 μm thick. InGaAs photodiodes are quite useful for lightwave systems and are often used in practice. Table 4.1 lists the operating characteristics of three common p–i–n photodiodes.

Considerable effort was directed during the 1990s toward developing high-speed p–i–n photodiodes capable of operating at bit rates exceeding 10 Gb/s [10]–[19]. Bandwidths as high as 70 GHz were realized as early as 1986 by using a thin absorption layer (< 1 μm) and by reducing the parasitic capacitance C_p with a small size, but only at the expense of a lower quantum efficiency and responsivity [11]. By 1995, p–i–n photodiodes exhibited a bandwidth of 110 GHz for devices designed to reduce τ_{RC} to $\sim$ 1 ps [12].

Several techniques have been developed to improve the efficiency of high-speed photodiodes. In one approach, a Fabry–Perot (FP) cavity is formed around the p–i–n structure to enhance the quantum efficiency [13]–[15], resulting in a laserlike structure. As discussed in Section 3.3.2, a FP cavity has a set of longitudinal modes at which the internal optical field is resonantly enhanced through constructive interference. As a result, when the incident wavelength is close to a longitudinal mode, such a photodiode exhibits high sensitivity. The wavelength selectivity can even be used to advantage in some WDM applications (see Chapter 7). A nearly 100% quantum efficiency was realized in a photodiode in which one mirror of the FP cavity was formed by using the Bragg reflectivity of a stack of AlGaAs/AlAs layers [14]. This approach was extended to InGaAs photodiodes by inserting a 90-nm thick InGaAs absorbing layer into a microcavity composed of a GaAs/AlAs Bragg mirror and a dielectric mirror. The device exhibited 94% quantum efficiency at the cavity resonance with a bandwidth of 14 nm [15]. By using an air-bridged metal waveguide together with an undercut mesa structure, a bandwidth of 120 GHz has been realized recently [16]. The use of such a structure within a FP cavity should provide a p–i–n photodiode with a high bandwidth ($>$ 100 GHz) and high efficiency.

Another approach to realize efficient high-speed photodiodes makes use of an optical waveguide into which the optical signal is edge-coupled [17]–[19]. Such a structure resembles an unpumped semiconductor laser except that various epitaxial layers are optimized differently. In contrast with a semiconductor laser, the waveguide can be made wide to support multiple transverse modes in order to improve the coupling efficiency [17]. Since absorption takes place along the length of the optical waveguide ($\sim$ 10 μm), the quantum efficiency can be nearly 100% even for an ultrathin absorption layer. The bandwidth of such *waveguide photodiodes* is limited by τ_{RC} in Eq. (4.1.7), which can be decreased by controlling the waveguide cross-section area. Indeed, a 50-GHz bandwidth was realized in 1992 for a waveguide photodiode [17]. The bandwidth could be increased to 110 GHz by adopting the mushroom-mesa waveguide structure [18] shown schematically in Fig. 4.6. In this structure the width of the i-type absorbing layer was reduced to 1.5 μm while the p- and n-type cladding layers were made 6 μm wide. In this way, both the parasitic capacitance and the internal series resistance were minimized, reducing τ_{RC} to about 1 ps. The frequency response of such a device at the 1.55-μm wavelength is also shown in Fig. 4.6. It was measured [18] by using a spectrum analyzer (circles) as well as by taking the Fourier transform of the short-pulse response (solid curve). Clearly, waveguide p–i–n photodiodes can provide both a high responsivity and a large bandwidth. A packaged module in which a waveguide photodiode was coupled to a single-mode fiber pigtail through microlenses had a 50-GHz bandwidth with 80% quantum efficiency [19].

The performance of waveguide photodiodes can be improved further by adopting an electrode structure designed to support traveling electrical waves with matching impedance to avoid reflections. Such photodiodes are called

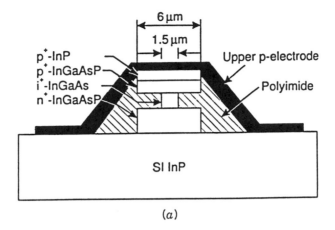

(a)

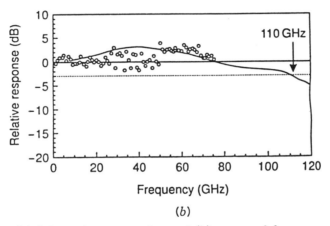

(b)

Figure 4.6 (a) Schematic cross section and (b) measured frequency response of a mushroom-mesa waveguide photodiode. (After Ref. [18]. ©1994 IEEE. Reprinted with permission.)

traveling-wave photodetectors [20]. In a GaAs-based implementation of this idea, a bandwidth of 172 GHz with 45% quantum efficiency was realized in a traveling-wave photodetector designed with a 1-μm-wide waveguide and a characteristic impedance of 50 Ω [21].

4.2.3   Avalanche Photodiodes

All detectors require a certain minimum current to operate reliably. The current requirement translates into a minimum power requirement through $P_{in} = I_p/R$ [see Eq. (4.1.1)]. Detectors with a large responsivity R are preferred since they require less optical power. The responsivity of p–i–n photodiodes is limited by Eq. (4.1.3) and takes its maximum value $R = q/h\nu$ for $\eta = 1$. APDs can have

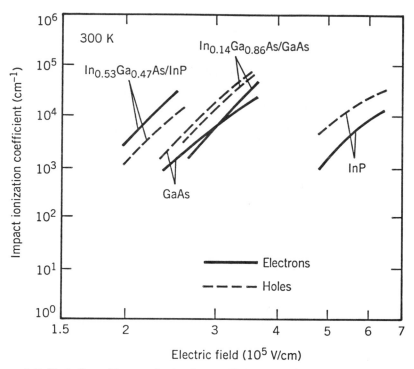

Figure 4.7 Variation of impact-ionization coefficient as a function of the electric field for electrons (solid line) and holes (dashed line) for several semiconductors. (After Ref. [23]. ©1977 Elsevier. Reprinted with permission.)

much larger values of R, as they are designed to provide an internal current gain in a way similar to photomultiplier tubes. They are used when the amount of optical power that can be spared for the receiver is limited.

The physical phenomenon behind the internal current gain is known as *impact ionization* [22], [23]. Under certain conditions, an accelerating electron can acquire sufficient energy to generate a new electron–hole pair. In the band picture (see Fig. 3.2) the energetic electron gives a part of its kinetic energy to another electron in the valence band that ends up in the conduction band, leaving behind a hole. The net result of impact ionization is that a single primary electron, generated through absorption of a photon, creates many secondary electrons and holes, all of which contribute to the photodiode current. Of course, the primary hole can also generate secondary electron–hole pairs that contribute to the current. The generation rate is governed by two parameters, α_e and α_h, known as the *impact-ionization coefficients* of electrons and holes, respectively. The numerical value of α_e and α_h depends on the semiconductor material and on the electric field that accelerates electrons and holes. Figure 4.7 shows α_e and α_h for several semiconductors. Values $\sim 1 \times 10^4$ cm^{-1} are obtained for electric fields in the range 2–4×10^5 V/cm. Such large fields can be realized by applying a high voltage (~ 100 V) to the APD.

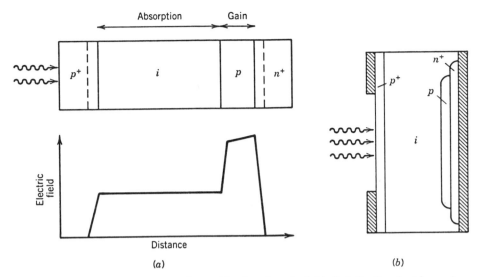

Figure 4.8 (a) An APD together with the electric-field distribution inside various layers under reverse bias; (b) design of a silicon reach-through APD.

APDs differ in their design from that of p–i–n photodiodes mainly in one respect: an additional layer is added in which secondary electron–hole pairs are generated through impact ionization. Figure 4.8(a) shows the APD structure together with the variation of electric field in various layers. Under reverse bias, a high electric field exists in the p-type layer sandwiched between i-type and n$^+$-type layers. This layer is referred to as the *multiplication layer*, since secondary electron–hole pairs are generated here through impact ionization. The i-layer still acts as the depletion region in which most of the incident photons are absorbed and primary electron–hole pairs are generated. Electrons generated in the i-region cross the gain region and generate secondary electron–hole pairs responsible for the current gain.

The current gain for APDs can be calculated by using two equations governing current flow within the multiplication layer [22]:

$$\frac{di_e}{dx} = \alpha_e i_e + \alpha_h i_h, \tag{4.2.3}$$

$$-\frac{di_h}{dx} = \alpha_e i_e + \alpha_h i_h, \tag{4.2.4}$$

where i_e is the electron current and i_h is the hole current. The minus sign in Eq. (4.2.4) is due to the opposite direction of the hole current. The total current,

$$I = i_e(x) + i_h(x), \tag{4.2.5}$$

remains constant at every point inside the multiplication region. If we replace i_h in Eq. (4.2.3) by $I - i_e$, we obtain

$$di_e/dx = (\alpha_e - \alpha_h)i_e + \alpha_h I. \tag{4.2.6}$$

In general, α_e and α_h are x dependent if the electric field across the gain region is nonuniform. The analysis is considerably simplified if we assume a uniform electric field and treat α_e and α_h as constants. We also assume that $\alpha_e > \alpha_h$. The avalanche process is initiated by electrons that enter the gain region of thickness d at $x = 0$. By using the condition $i_h(d) = 0$ (only electrons cross the boundary to enter the n-region), the boundary condition for Eq. (4.2.6) is $i_e(d) = I$. The *multiplication factor* M is defined as $M = i_e(d)/i_e(0)$ and is given by

$$M = \frac{1 - k_A}{\exp[-(1 - k_A)\alpha_e d] - k_A}, \qquad (4.2.7)$$

where $k_A = \alpha_h/\alpha_e$. The APD gain is quite sensitive to the ratio of the impact-ionization coefficients. When $\alpha_h = 0$ so that only electrons participate in the avalanche process, $M = \exp(\alpha_e d)$, and the APD gain increases exponentially with d. On the other hand, when $\alpha_h = \alpha_e$, so that $k_A = 1$ in Eq. (4.2.7), $M = (1 - \alpha_e d)^{-1}$. The APD gain then becomes infinite for $\alpha_e d = 1$, a condition known as the *avalanche breakdown*. Although higher APD gain can be realized with a smaller gain region when α_e and α_h are comparable, the performance is better in practice for APDs in which either $\alpha_e \gg \alpha_h$ or $\alpha_h \gg \alpha_e$ so that the avalanche process is dominated by only one type of charge carrier. The reason behind this requirement is discussed in Section 4.4, where issues related to the receiver noise are considered.

Because of the current gain, the responsivity of an APD is enhanced by the multiplication factor M and is given by

$$R_{\text{APD}} = MR = M(\eta q/h\nu), \qquad (4.2.8)$$

where Eq. (4.1.3) was used. It should be mentioned that the avalanche process in APDs is intrinsically noisy and results in a gain factor that fluctuates around an average value. The quantity M in Eq. (4.2.8) refers to the average APD gain. The noise characteristics of APDs are considered in Section 4.4.

The intrinsic bandwidth of an APD depends on the multiplication factor M. This is easily understood by noting that the transit time τ_{tr} for an APD is no longer given by Eq. (4.2.1) but increases considerably simply because generation and collection of secondary electron–hole pairs take additional time. The APD gain decreases at high frequencies because of such an increase in the transit time and limits the bandwidth. The decrease in $M(\omega)$ can be written as [23]

$$M(\omega) = M_0[1 + (\omega\tau_e M_0)^2]^{-1/2}, \qquad (4.2.9)$$

where $M_0 = M(0)$ is the low-frequency gain and τ_e is the effective transit time that depends on the ionization coefficient ratio $k_A = \alpha_h/\alpha_e$. For the case $\alpha_h < \alpha_e$, $\tau_e = c_A k_A \tau_{tr}$, where c_A is a constant ($c_A \sim 1$). Assuming that $\tau_{RC} \ll \tau_e$, the APD bandwidth is given approximately by $\Delta f = (2\pi\tau_e M_0)^{-1}$. This relation shows the *trade-off* between the APD gain M_0 and the bandwidth Δf (speed versus sensitivity). It also shows the advantage of using a semiconductor material for which $k_A \ll 1$.

Table 4.2 Characteristics of common APDs

Parameter	Symbol	Unit	Si	Ge	InGaAs
Wavelength	λ	μm	0.4–1.1	0.8–1.8	1.0–1.7
Responsivity	R_{APD}	A/W	80–130	3–30	5–20
APD gain	M	—	100–500	50–200	10–40
k-factor	k_A	—	0.02–0.05	0.7–1.0	0.5–0.7
Dark Current	I_d	nA	0.1–1	50–500	1–5
Rise time	T_r	ns	0.1–2	0.5–0.8	0.1–0.5
Bandwidth	Δf	GHz	0.2–1.0	0.4–0.7	1–3
Bias voltage	V_b	V	200–250	20–40	20–30

Table 4.2 compares the operating characteristics Si, Ge, and InGaAs APDs. As $k_A \ll 1$ for Si, silicon APDs can be designed to provide high performance and are useful for lightwave system operating near 0.8 μm at bit rates $\sim$ 100 Mb/s. A particularly useful design, shown in Fig. 4.8(b), is known as reach-through APD because the depletion layer reaches to the contact layer through the absorption and multiplication regions. It can provide high gain ($M \approx 100$) with low noise and a relatively large bandwidth. For lightwave systems operating in the wavelength range 1.3–1.6 μm, Ge or InGaAs APDs must be used. The improvement in sensitivity for such APDs is limited to a factor below 10 because of a relatively low APD gain ($M \sim 10$) that must be used to reduce the noise (see Section 4.4.3).

Considerable progress has been made in improving the performance of In-GaAs APDs through suitable design modifications to the basic APD structure shown in Fig. 4.8. The main reason for a relatively poor performance of InGaAs APDs is related to the comparable numerical values of the impact-ionization coefficients α_e and α_h (see Fig. 4.7). As a result, the bandwidth is considerably reduced, and the noise is also relatively high (see Section 4.4). Furthermore, because of a relatively narrow bandgap, InGaAs undergoes tunneling breakdown at electric fields of about 1×10^5 V/cm, a value that is below the threshold for avalanche multiplication. This problem can be solved in heterostructure APDs by using an InP layer for the gain region because quite high electric fields ($> 5 \times 10^5$ V/cm) can exist in InP without tunneling breakdown. Since the absorption region (*i*-type InGaAs layer) and the multiplication region (*n*-type InP layer) are separate in such a device, this structure is known as SAM, where SAM stands for *separate absorption and multiplication* regions. Since $\alpha_h > \alpha_e$ for InP (see Fig. 4.7), the APD is designed such that holes initiate the avalanche process in an *n*-type InP layer, and k_A is defined as $k_A = \alpha_e/\alpha_h$. Figure 4.9(a) shows a mesa-type SAM APD structure.

One problem with the SAM APD is related to the large bandgap difference between InP ($E_g = 1.35$ eV) and InGaAs ($E_g = 0.75$ eV). Because of a valence-band step of about 0.4 eV, holes generated in the InGaAs layer are trapped at the heterojunction interface and are considerably slowed before they reach

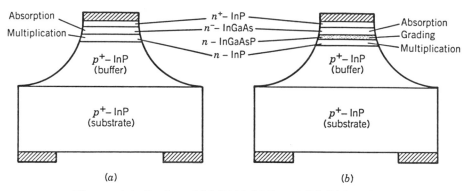

Figure 4.9 Design of (a) SAM APD and (b) SAGM APD.

the multiplication region (InP layer). Such an APD has an extremely slow response and a relatively small bandwidth. The problem can be solved by using another layer between the absorption and multiplication regions whose bandgap is intermediate to those of InP and InGaAs layers. The quaternary material InGaAsP, the same material used for semiconductor lasers, can be tailored to have a bandgap anywhere in the range 0.75–1.35 eV and is ideal for this purpose. It is even possible to grade the composition of InGaAsP over a region of 10–100 nm thickness. Such APDs are called SAGM APDs, where SAGM indicates *separate absorption, grading, and multiplication* regions [24]. Figure 4.9(b) shows the design of an InGaAs APD with the SAGM structure. The use of InGaAsP grading layer improves the bandwidth considerably. As early as 1987, a SAGM APD exhibited a gain–bandwidth product $M\Delta f = 70$ GHz for $M > 12$ [25]. A value of $M\Delta f = 100$ GHz was demonstrated in 1991 by using a charge region between the grading and multiplication regions [26]. In such SAGCM APDs, the InP multiplication layer is undoped, while the InP charge layer is heavily n-doped. Holes accelerate in the charge layer because of a strong electric field, but the generation of secondary electron–hole pairs takes place in the undoped InP layer. SAGCM APDs have been studied extensively during the 1990s [27]–[31].

A different approach to the design of high-performance APDs makes use of a superlattice structure [32]–[38]. The major limitation of InGaAs APDs results from comparable values of α_e and α_h. A superlattice design offers the possibility of reducing the ratio $k_A = \alpha_h/\alpha_e$ from its standard value of nearly unity. In one scheme, the absorption and multiplication regions alternate and consist of thin layers (~ 10 nm) of semiconductor materials with different bandgaps. This approach was first demonstrated for GaAs/AlGaAs MQW APDs and resulted in a considerable enhancement of the impact-ionization coefficient for electrons [32]. Its use is less successful for the InGaAs/InP material system. Nonetheless, considerable progress has been made through the so-called *staircase* APDs, in which the InGaAsP layer is compositionally graded to form a sawtooth kind of structure in the energy-band diagram that looks like a staircase under reverse bias [32]. Another scheme for making high-speed APDs uses alternate layers of

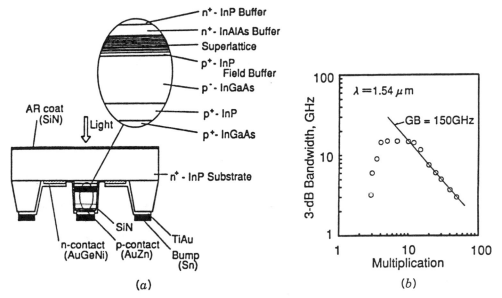

Figure 4.10 (a) Device structure and (b) measured 3-dB bandwidth as a function of the multiplication factor for a superlattice APD. Solid line corresponds to a gain–bandwidth product of 150 GHz. (After Ref. [38]. ©1996 IEEE. Reprinted with permission.)

InP and InGaAs for the grading region [32]. However, the ratio of the widths of the InP to InGaAs layers varies from zero near the absorbing region to almost infinity near the multiplication region. Since the effective bandgap of a quantum well depends on the quantum-well width (InGaAs layer thickness), a graded "pseudo-quaternary" compound is formed as a result of variation in the layer thickness.

The most successful design for InGaAs APDs uses a superlattice structure for the multiplication region of a SAM APD. Figure 4.10 shows the device structure schematically together with the 3-dB bandwidth measured with a 25-Ω load resistor as a function of the APD gain. In this device, the multiplication region is 231 nm thick and consists of 11 periods of 9-nm-thick InAlGaAs quantum wells, separated by 12-nm-thick InAlAs barrier layers. An InP field-buffer layer separates the InGaAs absorption region from the superlattice multiplication region. The thickness of this buffer layer is quite critical for APD performance. For a 52-nm-thick field-buffer layer, the gain–bandwidth product was limited to $M\Delta f = 120$ GHz [34] but increased to 150 GHz when the thickness was reduced to 33.4 nm [38]. Moreover, as seen in Fig. 4.10, this APD has a bandwidth of 15 GHz for $M = 10$ (with a dark current of only 56 nA). Such a photodiode can be used for lightwave systems operating at bit rates as high as 30 Gb/s and is at least 10 times more sensitive than comparable-bandwidth p–i–n photodiodes.

The gain-bandwidth limitation of InGaAs APDs results primarily from using the InP material system for the generation of secondary electron–hole pairs. A hybrid approach in which a Si multiplication layer is incorporated next to an InGaAs absorption layer may be useful provided the heterointerface problems can be overcome. In a 1997 experiment, a gain-bandwidth product of more than 300 GHz was realized by using such a hybrid approach [39]. The APD exhibited a 3-dB bandwidth of over 9 GHz for values of M as high as 35 while maintaining a 60% quantum efficiency.

Most APDs use an absorbing layer thick enough (about 1 μm) that the quantum efficiency exceeds 50%. The thickness of the absorbing layer affects both the transit time τ_{nr} and the bias voltage V_b. In fact, both of them can be reduced significantly by using a thin absorbing layer ($\sim$ 0.1 μm), resulting in improved APDs provided that a high quantum efficiency can be maintained. Two approaches have been used to meet these somewhat conflicting design requirements. In one design a FP cavity is used to enhance the absorption within a thin layer through multiple round trips [40]. An external quantum efficiency of 77% was achieved in a resonant-cavity-enhanced SAM APD having only a 50-nm-thick absorbing layer [41]. In another approach, an optical waveguide of about 40 μm length, 4 μm width, and 0.1 μm thickness is used into which the incident light is edge-coupled [42]. Both of these approaches reduce the bias voltage to near 10 V, maintain high efficiency, and reduce the transit time to $\sim$ 1 ps. Such APDs can have a high bandwidth provided that τ_{RC} can also be reduced by controlling the electrical parasitics.

4.2.4 MSM Photodetectors

In MSM photodetectors, a semiconductor absorbing layer is sandwiched between two metals, forming a Schottky barrier at each metal–semiconductor interface that prevents flow of electrons from metal to the semiconductor. Similar to a p–i–n photodiode, electron–hole pairs generated through photoabsorption flow toward the metal contacts, resulting in a photocurrent that is a measure of the incident optical power, as indicated in Eq. (4.1.1). For practical reasons, the two metal contacts are made on the same (top) side of the epitaxially grown absorbing layer by using an *interdigited* electrode structure with a finger spacing of about 1 μm [43]. This scheme results in a planer structure with an inherently low parasitic capacitance that allows high-speed operation (up to 300 GHz) of MSM photodetectors. If the light is incident from the electrode side, the responsivity of a MSM photodetector is reduced because of its blockage by the opaque electrodes. This problem can be solved by back illumination if the substrate is transparent to the incident light.

GaAs-based MSM photodetectors were developed throughout the 1980s and exhibit excellent operating characteristics [43]. The development of InGaAs-based MSM photodetectors, suitable for lightwave systems operating in the range 1.3–1.6 μm, started in the late 1980s, with most progress made during the 1990s [44]–[51]. The major problem with InGaAs is its relatively low *Schottky-barrier height* (about 0.2 eV). This problem was solved by introducing a thin layer of InP or InAlAs between the InGaAs layer and the metal contact. Such a layer, called the *barrier-enhancement layer*, improves the performance

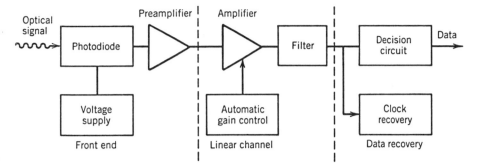

Figure 4.11 Diagram of a digital optical receiver showing various components. Vertical dashed lines group receiver components into three sections.

of InGaAs MSM photodetectors drastically. The use of a 20-nm-thick InAlAs barrier-enhancement layer resulted in 1992 in 1.3-μm MSM photodetectors exhibiting 92% quantum efficiency (through back illumination) with a low dark current [45]. A packaged device had a bandwidth of 4 GHz despite a large 150-μm diameter. If top illumination is desirable for processing or packaging reasons, the responsivity can be enhanced by using semitransparent metal contacts. In one experiment, the responsivity at 1.55 μm increased from 0.4 A/W to 0.7 A/W when the thickness of gold contact was reduced from 100 nm to 10 nm [46]. In another approach, the structure is separated from the host substrate and bonded to a silicon substrate with the interdigited contact on bottom. Such an "inverted" MSM photodetector then exhibits high responsivity when illuminated from the top [47].

The temporal response of MSM photodetectors is generally different under back and top illuminations [48]. In particular, the bandwidth Δf is larger by about a factor of 2 for top illumination, although the responsivity is reduced because of metal shadowing. Bandwidths as high as 20 GHz were measured under top illumination. The performance of a MSM photodetector can be further improved by using a graded superlattice structure. Such devices exhibit a low dark-current density of 0.75 pA/μm^2, a responsivity of about 0.6 A/W at 1.3 μm, and a rise time of about 16 ps [51]. The planar structure of MSM photodetectors is most suitable for monolithic integration, an issue covered in the next section.

4.3   RECEIVER DESIGN

The design of an optical receiver depends largely on the modulation format used by the transmitter. In particular, it depends on whether the signal is transmitted in an analog or digital format. Since most lightwave systems employ the digital format, we focus in this chapter on digital optical receivers. Figure 4.11 shows a block diagram of such a receiver. Its components can be arranged into three

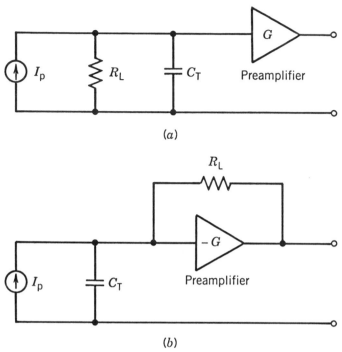

Figure 4.12 Equivalent circuit for (a) high-impedance and (b) transimpedance front ends in optical receivers. The photodiode is modeled as a current source in both cases.

groups: the front end, the linear channel, and the data-recovery section. In this section we discuss each group separately.

4.3.1   Front End

The front end of a receiver consists of a photodiode followed by a preamplifier. The optical signal is coupled onto the photodiode by using a coupling scheme similar to that used for optical transmitters (see Section 3.4.1); butt coupling is often used in practice. The photodiode converts the optical bit stream into an electrical time-varying signal. The role of the preamplifier is to amplify the electrical signal for further processing.

The design of the front end requires a trade-off between speed and sensitivity. Since the input voltage to the preamplifier can be increased by using a large load resistor R_L, a *high-impedance front end* is often used [see Fig. 4.12(a)]. Furthermore, as discussed in Section 4.4, a large R_L reduces the thermal noise and improves the receiver sensitivity. The main drawback of high-impedance front end is its low bandwidth given by $\Delta f = (2\pi R_L C_T)^{-1}$, where $R_s \ll R_L$ is assumed in Eq. (4.2.2) and $C_T = C_p + C_A$ is the total capacitance, which includes the contributions from the photodiode (C_p) and the transistor used for amplification (C_A). The receiver bandwidth is limited by its slowest component. A high-impedance front end cannot be used if Δf is considerably less than the bit rate. An equalizer is sometimes used to increase the bandwidth. The equalizer acts as a filter that attenuates low-frequency components of the signal more than the high-frequency components, thereby effectively increasing the front-end bandwidth. If the receiver sensitivity is not of concern, one can simply decrease R_L to increase the bandwidth. Such front ends are called *low-impedance front ends*.

Transimpedance front ends provide a configuration that has high sensitivity together with a large bandwidth. Its dynamic range is also improved compared with high-impedance front ends. As seen in Fig. 4.12(b), the load resistor is connected as a feedback resistor around an inverting amplifier. Even though R_L is large, the *negative feedback* reduces the effective input impedance by a factor of G, where G is the amplifier gain. The bandwidth is thus enhanced by a factor of G compared with high-impedance front ends. Transimpedance front ends are often used in optical receivers because of their improved characteristics. A major design issue is related to the stability of the feedback loop. More details can be found in Refs. [5]–[8].

4.3.2 Linear Channel

The linear channel in optical receivers consists of a high-gain amplifier (the main amplifier) and a low-pass filter. An equalizer is sometimes included just before the amplifier to correct for the limited bandwidth of the front end. The amplifier gain is controlled automatically to limit the average output voltage to a fixed level irrespective of the incident average optical power at the receiver. The low-pass filter shapes the voltage pulse. Its purpose is to reduce the noise without introducing much *intersymbol interference* (ISI). As discussed in Section 4.4, the receiver noise is proportional to the receiver bandwidth and can be reduced by using a low-pass filter whose bandwidth Δf is smaller than the bit rate. Since other components of the receiver are designed to have a bandwidth larger than the filter bandwidth, the receiver bandwidth is determined by the low-pass filter used in the linear channel. For $\Delta f < B$, the electrical pulse spreads beyond the allocated bit slot. Such a spreading can interfere with the detection of neighboring bits, a phenomenon referred to as ISI.

It is possible to design a low-pass filter in such a way that ISI is minimized [1]. Since the combination of preamplifier, main amplifier, and the filter acts as a linear system (hence the name *linear channel*), the output voltage can be written as

$$V_{\text{out}}(t) = \int_{-\infty}^{\infty} z_T(t - t') I_p(t') \, dt', \qquad (4.3.1)$$

where $I_p(t)$ is the photocurrent generated in response to the incident optical power ($I_p = RP_{in}$). In the frequency domain,

$$\tilde{V}_{\text{out}}(\omega) = Z_T(\omega) \tilde{I}_p(\omega), \qquad (4.3.2)$$

where Z_T is the total impedance at the frequency ω and a tilde represents the Fourier transform. $Z_T(\omega)$ is determined by the transfer functions associated with various receiver components and can be written as [3]

$$Z_T(\omega) = G_p(\omega)G_A(\omega)H_F(\omega)/Y_{\text{in}}(\omega), \qquad (4.3.3)$$

where $Y_{\text{in}}(\omega)$ is the input admittance and $G_p(\omega)$, $G_A(\omega)$, and $H_F(\omega)$ are transfer functions of the preamplifier, the main amplifier, and the filter. It is useful to isolate the frequency dependence of $\tilde{V}_{\text{out}}(\omega)$ and $\tilde{I}_p(\omega)$ through normalized spectral functions $H_{\text{out}}(\omega)$ and $H_p(\omega)$, which are related to the Fourier transform of the output and input pulse shapes, respectively, and write Eq. (4.3.2) as

$$H_{\text{out}}(\omega) = H_T(\omega)H_p(\omega), \qquad (4.3.4)$$

where $H_T(\omega)$ is the total transfer function of the linear channel and is related to the total impedance as $H_T(\omega) = Z_T(\omega)/Z_T(0)$. If the amplifiers have a much larger bandwidth than the low-pass filter, $H_T(\omega)$ can be approximated by $H_F(\omega)$.

The ISI is minimized when $H_{\text{out}}(\omega)$ corresponds to the transfer function of a *raised-cosine filter* and is given by [3]

$$H_{\text{out}}(f) = \begin{cases} \frac{1}{2}[1 + \cos(\pi f/B)] & : \ f < B, \\ 0 & : \ f \geq B, \end{cases} \qquad (4.3.5)$$

where $f = \omega/2\pi$ and B is the bit rate. The impulse response, obtained by taking the Fourier transform of $H_{\text{out}}(f)$, is given by

$$h_{\text{out}}(t) = \frac{\sin(2\pi Bt)}{2\pi Bt}\frac{1}{1 - (2Bt)^2}. \qquad (4.3.6)$$

The functional form of $h_{\text{out}}(t)$ corresponds to the shape of the voltage pulse $V_{\text{out}}(t)$ received by the decision circuit. At the decision instant $t = 0$, $h_{\text{out}}(t) = 1$, and the signal is maximum. At the same time, $h_{\text{out}}(t) = 0$ for $t = m/B$, where m is an integer. Since $t = m/B$ corresponds to the decision instant of the neighboring bits, the voltage pulse of Eq. (4.3.6) does not interfere with the neighboring bits.

The linear-channel transfer function $H_T(\omega)$ that will result in output pulse shapes of the form (4.3.6) is obtained from Eq. (4.3.4) and is given by

$$H_T(f) = H_{\text{out}}(f)/H_p(f). \qquad (4.3.7)$$

For an ideal bit stream in the NRZ format (rectangular input pulses of duration $T_B = 1/B$), $H_p(f) = B\sin(\pi f/B)/\pi f$, and $H_T(f)$ becomes

$$H_T(f) = (\pi f/2B)\cot(\pi f/2B). \qquad (4.3.8)$$

Equation (4.3.8) determines the frequency response of the linear channel that would produce output pulse shape given by Eq. (4.3.6) under ideal conditions. In practice, the input pulse shape is far from being rectangular. The output pulse shape also deviates from Eq. (4.3.6), and some ISI invariably occurs.

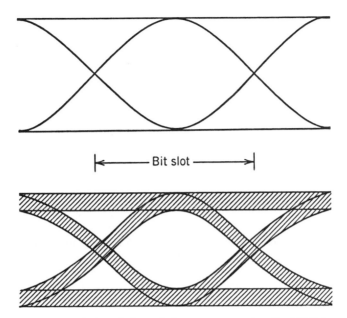

Figure 4.13 Ideal and degraded eye patterns for the NRZ format.

4.3.3   Data Recovery

The data-recovery section of optical receivers consists of a decision circuit and a clock-recovery circuit. The purpose of the latter is to isolate a spectral component at $f = B$ from the received signal. This component provides information about the bit slot ($T_B = 1/B$) to the decision circuit and helps to synchronize the decision process. In the case of RZ (return-to-zero) format, a spectral component at $f = B$ is present in the received signal; a narrow-bandpass filter such as a surface-acoustic-wave filter can isolate this component easily. Clock recovery is more difficult in the case of NRZ format because the signal received lacks a spectral component at $f = B$. A commonly used technique generates such a component by squaring and rectifying the spectral component at $f = B/2$ that can be obtained by passing the received signal through a high-pass filter.

The decision circuit compares the output from the linear channel to a threshold level, at sampling times determined by the clock-recovery circuit, and decides whether the signal corresponds to bit 1 or bit 0. The best sampling time corresponds to the situation in which the signal level difference between 1 and 0 bits is maximum. It can be determined from the *eye diagram* formed by superposing 2–3-bit-long electrical sequences in the bit stream on top of each other. The resulting pattern is called an eye diagram because of its appearance. Figure 4.13 shows an ideal eye diagram together with a degraded one in which the noise and the timing jitter lead to a partial closing of the eye. The best sampling time corresponds to maximum opening of the eye.

Because of noise inherent in any receiver, there is always a finite probability that a bit would be identified incorrectly by the decision circuit. Digital receivers

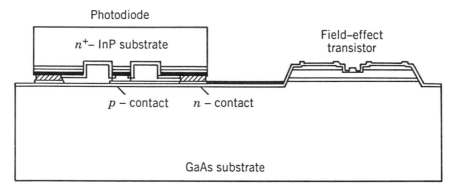

Figure 4.14 Flip-chip OEIC technology for integrated receivers. The InGaAs photodiode is fabricated on an InP substrate and then bonded to the GaAs chip through common electrical contacts. (After Ref. [56]. ©1988 IEE. Reprinted with permission.)

are designed to operate in such a way that the error probability is quite small (typically $< 10^{-9}$). Issues related to receiver noise and decision errors are discussed in Sections 4.4 and 4.5. The eye diagram provides a visual way of monitoring the receiver performance: Closing of the eye is an indication that the receiver is not performing properly.

4.3.4 Integrated Receivers

All receiver components shown in Fig. 4.11, with the exception of the photodiode, are standard electrical components and can be easily integrated on the same chip by using the integrated-circuit (IC) technology developed for microelectronic devices. Integration is particularly necessary for receivers operating at high bit rates. By 1988, both Si and GaAs IC technologies have been used to make integrated receivers up to a bandwidth of more than 2 GHz [52]. Since then, the bandwidth has been extended to 10 GHz.

Considerable effort has been directed at developing monolithic optical receivers that integrate all components, including the photodetector, on the same chip by using the *optoelectronic integrated-circuit* (OEIC) technology [53]–[76]. Such a complete integration is relatively easy for GaAs receivers, and the technology behind GaAs-based OEICs is quite advanced. The use of MSM photodiodes has proved especially useful as they are structurally compatible with the well-developed *field-effect-transistor* (FET) technology. It was used in 1986 to demonstrate a four-channel OEIC receiver chip [55].

For lightwave systems operating in the wavelength range 1.3–1.6 μm, InP-based OEIC receivers are needed. Since the IC technology for GaAs is much more mature than for InP, a hybrid approach is sometimes used for InGaAs receivers. In this approach, called *flip-chip OEIC technology* [56], the electronic components are integrated on a GaAs chip, whereas the photodiode is made on top of an InP chip. The two chips are then connected by flipping the InP chip on the GaAs chip, as shown in Fig. 4.14. The advantage of the flip-chip technique is that the photodiode and the electrical components of the receiver can be independently optimized while keeping the parasitics (e.g., effective input capacitance) to a bare minimum.

The InP-based IC technology has advanced considerably during the 1990s, making it possible to develop InGaAs OEIC receivers [57]–[76]. Several kinds of transistors have been used for this purpose. In one approach, a p–i–n photodiode is integrated with the FETs or high-electron-mobility transistors (HEMTs) side by side on an InP substrate [58]–[63]. By 1993, HEMT-based receivers were capable of operating at 10 Gb/s with high sensitivity [61]. The bandwidth of such receivers has been increased to 22 GHz, making it possible to use them at bit rates above 20 Gb/s [62] A waveguide p–i–n photodiode has also been integrated with HEMTs to develop a two channel OEIC receiver array [63].

In another approach [64]–[72], the heterojunction-bipolar transistor (HBT) technology is used to fabricate the p–i–n photodiode within the HBT structure itself through a common-collector configuration. Such transistors are often called *heterojunction phototransistors*. Such OEIC receivers operating at 5 Gb/s (bandwidth $\Delta f = 3$ GHz) were demonstrated [64] in 1993. The bandwidth can be increased to 7.1 GHz by using a transimpedance front end for the OEIC receiver [66]. By 1995, OEIC receivers making use of the HBT technology exhibited a bandwidth of up to 16 GHz, together with a high gain, when their design was suitably optimized [67], [68]. Such receivers can be used at bit rates of more than 20 Gb/s. Indeed, a high-sensitivity OEIC receiver module has been used at a bit rate of 20 Gb/s in a 1.55-μm lightwave system [69]. Even a decision circuit can be integrated within the OEIC receiver by using the HBT technology, and its operation at 13 Gb/s has been demonstrated [70].

A third approach to InP-based OEIC receivers integrates a MSM photodetector with an HEMT or a modulation-doped FET [73]–[76]. By 1995, a bandwidth of 15 GHz was realized for such an OEIC by using modulation-doped FETs [75]. It was later increased to 18.5 GHz by using a transimpedance amplifier circuit with active tunable feedback [76]. Figure 4.15 shows the circuit diagram together with the epitaxial-layer structure of such an OEIC receiver. The MSM photodetector was grown on top of the modulation-doped FET, and a Fe-doped InP barrier layer separated the two. The active area of the MSM photodetector was 25×25 μm^2. The responsivity at 1.55 μm was 0.26 A/W, with 70-nA dark current at the 8-V operating voltage. The threshold voltage of the FET was only -0.7 V. The bandwidth of such an OEIC receiver was measured to be 18.5 GHz at a transimpedance of 80 Ω.

Since the lightwave communication technology is moving toward high bit rates through wavelength-division multiplexing, considerable effort is directed toward developing devices that can detect several channels by using a single OEIC receiver array. Such devices are covered in Chapter 7.

Similar to the case of optical transmitters (Section 3.4), packaging of optical receivers is also an important issue [77]–[80]. The fiber–detector coupling issue is even more critical at the receiver than the fiber–source coupling issue at the transmitter (Section 3.4.1) since only a small amount of optical power is typically available at the photodetector. The optical-feedback issue is also important since

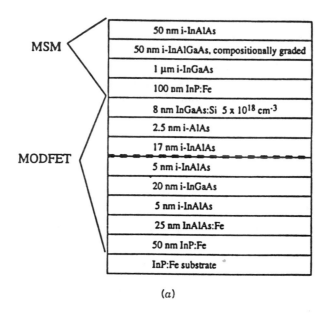

(a)

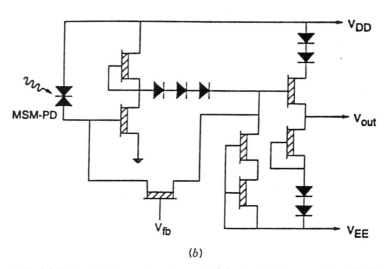

(b)

Figure 4.15 (a) Epitaxial-layer structure and (b) circuit diagram of an OEIC receiver in which a MSM photodetector (MSM-PD) is grown on top of the modulation-doped FET (MODFET). (After Ref. [76]. ©1996 IEEE. Reprinted with permission.)

unintentional reflections fed back into the transmission fiber can affect system performance and should be minimized. In practice, the fiber tip is cut at an angle to reduce the optical feedback. Several different techniques have been used to produce packaged optical receivers capable of operating at bit rates as high as 10 Gb/s. In one approach, an InGaAs APD was bonded to the Si-based IC by using the flip-chip technique [77]. Efficient fiber–APD coupling was realized by using a *slant-ended fiber* and a microlens monolithically fabricated on the photodiode. The fiber ferrule was directly laser-welded to the package wall with a double-ring structure for mechanical stability. The resulting receiver module withstood shock and vibration tests and had a bandwidth of 10 GHz.

Another hybrid approach makes use of a *planar-lightwave-circuit* platform containing silica waveguides on a silicon substrate. In one experiment [78], an InP-based OEIC receiver with two channels was flip-chip bonded to the platform. The resulting module could detect two 10-Gb/s channels with negligible crosstalk. GaAs ICs have also been used to fabricate a compact receiver module capable of operating at a bit rate of 10 Gb/s [79]. For local-loop and local-area applications, a low-cost package is needed, and the receiver should be able to operate over a wide temperature range (-40–$85°$C). Recently, a plastic-molded receiver module, capable of operating at 155 Mb/s with only 300 nW of optical power, has been developed for such applications [80].

4.4 RECEIVER NOISE

Optical receivers convert incident optical power P_{in} into electric current through a photodiode. The relation $I_p = RP_{in}$ in Eq. (4.1.1) assumes that such a conversion is noise free. However, this is not the case even for a perfect receiver. Two fundamental noise mechanisms, shot noise and thermal noise [81]–[83], lead to fluctuations in the current even when the incident optical signal has a constant power. The relation $I_p = RP_{in}$ still holds if we interpret I_p as the average current. However, electrical noise induced by current fluctuations affects the receiver performance. The objective of this section is to review the noise mechanisms and then discuss the signal-to-nose ratio (SNR) in optical receivers. The p–i–n and APD receivers are considered in separate subsections, as the SNR is also affected by the avalanche gain mechanism in APDs.

4.4.1   Noise Mechanisms

As mentioned above, shot noise and thermal noise are the two fundamental noise mechanisms responsible for current fluctuations in all optical receivers even when the incident optical power P_{in} is constant. Of course, additional noise is generated if P_{in} is itself fluctuating because of intensity noise associated with the transmitter. This subsection considers only the noise generated at the receiver; intensity noise is discussed in Section 4.6.2.

Shot Noise

Shot noise is a manifestation of the fact that the electric current consists of a stream of electrons that are generated at random times. It was first studied by Schottky [84] in 1918 and has been thoroughly investigated since then [81]–[83]. The photodiode current generated in response to a constant optical signal can be written as

$$I(t) = I_p + i_s(t), \qquad (4.4.1)$$

where $I_p = RP_{in}$ is the average current and $i_s(t)$ is a current fluctuation related to shot noise. Mathematically, $i_s(t)$ is a stationary random process with *Poisson statistics*, which in practice can be approximated by *Gaussian statistics*. The autocorrelation function of $i_s(t)$ is related to the spectral density $S_s(f)$ by the *Wiener–Khinchin theorem* [83]

$$\langle i_s(t)i_s(t+\tau)\rangle = \int_{-\infty}^{\infty} S_s(f)\exp(2\pi i f \tau)\,df, \qquad (4.4.2)$$

where angle brackets denote an ensemble average over fluctuations. The spectral density of shot noise is constant and is given by $S_s(f) = qI_p$ (an example of *white noise*). Note that $S_s(f)$ is the *two-sided* spectral density, as negative frequencies are included in Eq. (4.4.2). If only positive frequencies are considered by changing the lower limit of integration to zero, the *one-sided* spectral density becomes $2qI_p$.

The noise variance is obtained by setting $\tau = 0$ in Eq. (4.4.2), i.e.,

$$\sigma_s^2 = \langle i_s^2(t)\rangle = \int_{-\infty}^{\infty} S_s(f)\,df = 2qI_p\,\Delta f, \qquad (4.4.3)$$

where Δf is the *effective noise bandwidth* of the receiver. The actual value of Δf depends on receiver design. It corresponds to the intrinsic photodetector bandwidth if fluctuations in the photocurrent are measured. In practice, a decision circuit may use voltage or some other quantity (e.g., signal integrated over the bit slot). One then has to consider the transfer functions of other receiver components such as the preamplifier and the low-pass filter. It is common to consider current fluctuations, and include the total transfer function $H_T(f)$ by modifying Eq. (4.4.3) as

$$\sigma_s^2 = 2qI_p \int_0^{\infty} |H_T(f)|^2\,df = 2qI_p\,\Delta f, \qquad (4.4.4)$$

where $\Delta f = \int_0^{\infty} |H_T(f)|^2 df$, and $H_T(f)$ is given by Eq. (4.3.7). Since the dark current I_d also generates shot noise, its contribution is included in Eq. (4.4.4) by replacing I_p by $I_p + I_d$. The total shot noise is then given by

$$\sigma_s^2 = 2q(I_p + I_d)\Delta f. \qquad (4.4.5)$$

The quantity σ_s is the root-mean-square (RMS) value of the noise current induced by shot noise.

Thermal Noise

At a finite temperature, electrons move randomly in any conductor. Random thermal motion of electrons in a resistor manifests as a fluctuating current even in the absence of an applied voltage. The load resistor in the front end of an optical receiver (see Fig. 4.12) adds such fluctuations to the current generated by the photodiode. This additional noise component is referred to as thermal noise. It is also called *Johnson noise* [85] or *Nyquist noise* [86] after the two scientists who first studied it experimentally and theoretically. Thermal noise can be included by modifying Eq. (4.4.1) as

$$I(t) = I_p + i_s(t) + i_T(t), \qquad (4.4.6)$$

where $i_T(t)$ is a current fluctuation induced by thermal noise. Mathematically, $i_T(t)$ is modeled as a stationary Gaussian random process with a spectral density that is frequency independent up to $f \sim 1$ THz (nearly white noise) and is given by

$$S_T(f) = 2k_B T / R_L, \qquad (4.4.7)$$

where k_B is the *Boltzmann constant*, T is the absolute temperature, and R_L is the load resistor. As mentioned before, $S_T(f)$ is the two-sided spectral density.

The autocorrelation function of $i_T(t)$ is given by Eq. (4.4.2) if we replace the subscript s by T. The noise variance is obtained by setting $\tau = 0$ and becomes

$$\sigma_T^2 = \langle i_T^2(t) \rangle = \int_{-\infty}^{\infty} S_T(f) \, df = (4k_B T / R_L) \Delta f, \qquad (4.4.8)$$

where Δf is the effective noise bandwidth. The same bandwidth appears in the case of both shot and thermal noises. Note that σ_T^2 does not depend on the average current I_p, whereas σ_s^2 does.

Equation (4.4.8) includes thermal noise generated in the load resistor. An actual receiver contains many other electrical components, some of which add additional noise. For example, noise is invariably added by the pre- and main amplifiers (see Fig. 4.12). The amount of noise depends on the front-end design and the type of amplifiers used. In particular, the added thermal noise is different for field-effect and bipolar transistors. Considerable work has been done to estimate the amplifier noise for different front-end designs [3], [5]. A simple approach accounts for the amplifier noise by introducing a quantity F_n, referred to as the *amplifier noise figure*, and modifying Eq. (4.4.8) as

$$\sigma_T^2 = (4k_B T / R_L) F_n \Delta f. \qquad (4.4.9)$$

Physically, F_n represents the factor by which thermal noise is enhanced by various resistors used in pre- and main amplifiers.

The total current noise can be obtained by adding the contributions of shot noise and thermal noise. Since $i_s(t)$ and $i_T(t)$ in Eq. (4.4.6) are independent random processes with approximately Gaussian statistics, the total variance of current fluctuations, $\Delta I = I - I_p = i_s + i_T$, can be obtained simply by adding individual variances. The result is

$$\sigma^2 = \langle (\Delta I)^2 \rangle = \sigma_s^2 + \sigma_T^2 = 2q(I_p + I_d)\Delta f + (4k_B T / R_L) F_n \Delta f. \qquad (4.4.10)$$

Equation (4.4.10) can be used to calculate the SNR of the photocurrent.

4.4.2 p–i–n Receivers

The performance of an optical receiver depends on the SNR. The SNR of a receiver with a p–i–n photodiode is considered here; APD receivers are discussed in the following subsection. The SNR of any electrical signal is defined as

$$\text{SNR} = \frac{\text{average signal power}}{\text{noise power}} = \frac{I_p^2}{\sigma^2}, \tag{4.4.11}$$

where we used the fact that electrical power varies as the square of the current. By using Eq. (4.4.10) in Eq. (4.4.11) together with $I_p = RP_{in}$, the SNR is related to the incident optical power as

$$\text{SNR} = \frac{R^2 P_{in}^2}{2q(RP_{in} + I_d)\Delta f + 4(k_B T/R_L)F_n \Delta f}, \tag{4.4.12}$$

where $R = \eta q/h\nu$ is the responsivity of the p–i–n photodiode.

Thermal-Noise Limit

In most cases of practical interest, thermal noise dominates receiver performance $(\sigma_T^2 \gg \sigma_s^2)$. Neglecting the shot-noise term in Eq. (4.4.12), the SNR becomes

$$\text{SNR} = \frac{R_L R^2 P_{in}^2}{4k_B T F_n \Delta f}. \tag{4.4.13}$$

Thus, the SNR varies as P_{in}^2 in the thermal-noise limit. It can also be improved by increasing the load resistance. As discussed in Section 4.3.1, this is the reason why most receivers use a high-impedance or transimpedance front end. The effect of thermal noise is often quantified through a quantity called the *noise-equivalent power* (NEP). The NEP is defined as the minimum optical power per unit bandwidth required to produce SNR = 1 and is given by

$$\text{NEP} = \frac{P_{in}}{\sqrt{\Delta f}} = \left(\frac{4k_B T F_n}{R_L R^2}\right)^{1/2} = \frac{h\nu}{\eta q}\left(\frac{4k_B T F_n}{R_L}\right)^{1/2}. \tag{4.4.14}$$

Another quantity, called *detectivity* and defined as $(\text{NEP})^{-1}$, is also used for this purpose. The advantage of specifying NEP or the detectivity for a p–i–n receiver is that it can be used to estimate the optical power needed to obtain a specific value of SNR if the bandwidth Δf is known. Typical values of NEP are in the range 1–10 pW/Hz$^{1/2}$.

Shot-Noise Limit

Consider the opposite limit in which the receiver performance is dominated by shot noise $(\sigma_s^2 \gg \sigma_T^2)$. Since σ_s^2 increases linearly with P_{in}, the shot-noise limit can be achieved by making the incident power large. The dark current I_d can be neglected in that situation. Equation (4.4.12) then provides the following expression for SNR:

$$\text{SNR} = \frac{RP_{in}}{2q\Delta f} = \frac{\eta P_{in}}{2h\nu\Delta f}. \tag{4.4.15}$$

The SNR increases linearly with P_{in} in the shot-noise limit and depends only on the quantum efficiency η, the bandwidth Δf, and the photon energy $h\nu$. It can be written in terms of the number of photons N_p contained in the "1" bit. If we use $E_p = P_{in} \int_{-\infty}^{\infty} h_p(t)\, dt = P_{in}/B$ for the pulse energy of a bit of duration $1/B$, where B is the bit rate, and note that $E_p = N_p h\nu$, we can write P_{in} as $P_{in} = N_p h\nu B$. By choosing $\Delta f = B/2$ (a typical value for the bandwidth), the SNR is simply given by ηN_p. In the shot-noise limit, a SNR of 20 dB can be realized if $N_p \approx 100$. By contrast, several thousand photons are required to obtain SNR = 20 dB when thermal noise dominates the receiver. As a reference, for a 1.55-μm receiver operating at 10 Gb/s, $N_p = 100$ when $P_{in} \approx 130$ nW.

4.4.3 APD Receivers

Optical receivers that employ an APD generally provide a higher SNR for the same incident optical power. The improvement is due to the internal gain that increases the photocurrent by a multiplication factor M so that

$$I_p = MRP_{in} = R_{\text{APD}} P_{in}, \tag{4.4.16}$$

where R_{APD} is the APD responsivity, enhanced by a factor of M compared with that of p–i–n photodiodes ($R_{\text{APD}} = MR$). The SNR would improve by a factor of M^2 if the receiver noise were unaffected by the internal gain mechanism of APDs. Unfortunately, the noise of APD receivers is also enhanced, and the SNR improvement is considerably reduced.

Shot-Noise Enhancement

Thermal noise remains the same for APD receivers, as it originates in the electrical components that are not part of the APD. This is not the case for shot noise. The APD gain results from generation of secondary electron–hole pairs through the process of impact ionization. Since such pairs are generated at random times, an additional contribution is added to the shot noise associated with the generation of primary electron–hole pairs. In effect, the multiplication factor itself is a random variable, and M appearing in Eq. (4.4.16) represents the average APD gain. Total shot noise can be calculated by using Eqs. (4.2.3) and (4.2.4) and treating i_e and i_h as random variables [87]. The result is

$$\sigma_s^2 = 2qM^2 F_A(RP_{in} + I_d)\Delta f. \tag{4.4.17}$$

where F_A is the *excess noise factor* of the APD and is given by [87]

$$F_A(M) = k_A M + (1 - k_A)(2 - 1/M). \tag{4.4.18}$$

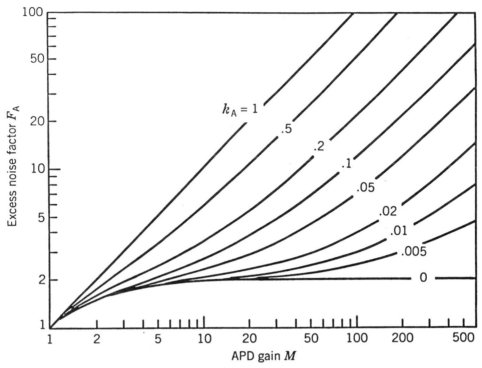

Figure 4.16 Excess noise factor F_A as a function of the average APD gain M for several values of the ionization-coefficient ratio k_A.

The dimensionless parameter $k_A = \alpha_h/\alpha_e$ if $\alpha_h < \alpha_e$, but is defined as $k_A = \alpha_e/\alpha_h$ when $\alpha_h > \alpha_e$. In other words, k_A is in the range $0 < k_A < 1$. Figure 4.16 shows the gain dependence of F_A for several values of k_A. In general, F_A increases with M. However, although F_A is at most 2 for $k_A = 0$, it keeps on increasing linearly $(F_A = M)$ when $k_A = 1$. The ratio k_A should be as small as possible for achieving the best performance from an APD [88].

If the avalanche–gain process were noise free $(F_A = 1)$, both I_p and σ_s would increase by the same factor M, and the SNR would be unaffected as far as the shot-noise contribution is concerned. In practice, the SNR of APD receivers is worse than that of p–i–n receivers when shot noise dominates because of the excess noise generated inside the APD. It is the dominance of thermal noise in practical receivers that makes APDs attractive. In fact, the SNR of APD receivers can be written as

$$\text{SNR} = \frac{I_p^2}{\sigma_s^2 + \sigma_T^2} = \frac{(MRP_{in})^2}{2qM^2F_A(RP_{in} + I_d)\Delta f + 4(k_BT/R_L)F_n\Delta f}, \quad (4.4.19)$$

where Eqs. (4.4.9), (4.4.16), and (4.4.17) were used. In the thermal-noise limit $(\sigma_s \ll \sigma_T)$, the SNR becomes

$$\text{SNR} = (R_L R^2/4k_BTF_n\Delta f)M^2 P_{in}^2, \quad (4.4.20)$$

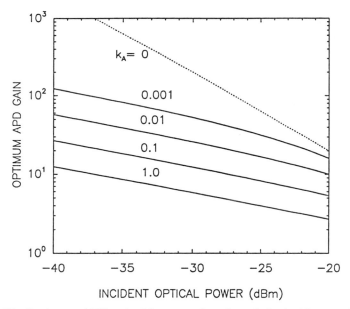

OPTIMUM APD GAIN

INCIDENT OPTICAL POWER (dBm)

Figure 4.17 Optimum APD gain M_{opt} as a function of the incident optical power P_{in} for several values of k_A. Parameter values corresponding to a typical 1.55-μm InGaAs APD receiver were used.

and is improved, as expected, by a factor of M^2 compared with that of p–i–n receivers [see Eq. (4.4.13)]. By contrast, in the shot-noise limit ($\sigma_s \gg \sigma_T$), the SNR is given by

$$\text{SNR} = \frac{RP_{in}}{2qF_A\Delta f} = \frac{\eta P_{in}}{2h\nu F_A\Delta f} \qquad (4.4.21)$$

and is reduced by the excess noise factor F_A compared with that of p–i–n receivers [see Eq. (4.4.15)].

Optimum APD Gain

Equation (4.4.19) shows that for a given P_{in}, the SNR of APD receivers is maximum for an optimum value M_{opt} of the APD gain M. It is easy to show that the SNR is maximum when M_{opt} satisfies the following cubic polynomial:

$$k_A M_{\text{opt}}^3 + (1 - k_A)M_{\text{opt}} = \frac{4k_B T F_n}{qR_L(RP_{in} + I_d)}. \qquad (4.4.22)$$

The optimum value M_{opt} depends on a large number of the receiver parameters, such as the dark current, the responsivity R, and the ionization-coefficient ratio k_A. However, it is independent of receiver bandwidth. The most notable feature of Eq. (4.4.22) is that M_{opt} decreases with an increase in P_{in}. Figure 4.17 shows the variation of M_{opt} with P_{in} for several values of k_A by using typical parameter values $R_L = 1$ kΩ, $F_n = 2$, $R = 1$ A/W, and $I_d = 2$ nA corresponding to a 1.55-μm InGaAs receiver. The optimum APD gain is quite sensitive to the ionization-coefficient ratio k_A. For $k_A = 0$, M_{opt} decreases inversely with P_{in}, as can readily be inferred from Eq. (4.4.22) by noting that the contribution of I_d is negligible in practice. By contrast, M_{opt} varies as $P_{in}^{-1/3}$ for $k_A = 1$, and this form of dependence appears to hold even for k_A as small as 0.01 as long as $M_{opt} > 10$. In fact, by neglecting the second term in Eq. (4.4.22), M_{opt} is well approximated by

$$M_{opt} \approx \left(\frac{4k_B T F_n}{k_A q R_L (R P_{in} + I_d)} \right)^{1/3} \tag{4.4.23}$$

for k_A in the range 0.01–1. This expression shows the critical role played by the ionization-coefficient ratio k_A. M_{opt} can be as large as 100 for Si APDs for which $k_A \ll 1$. By contrast, M_{opt} is in the neighborhood of 10 for InGaAs receivers, since $k_A \approx 0.7$. InGaAs APD receivers are nonetheless useful for optical communication systems simply because of their higher sensitivity. Receiver sensitivity is an important issue in the design of lightwave systems and is discussed next.

4.5 RECEIVER SENSITIVITY

Among a group of optical receivers, a receiver is said to be more sensitive if it achieves the same performance with less optical power incident on it. The performance criterion for digital receivers is governed by the *bit-error rate* (BER), defined as the probability of incorrect identification of a bit by the decision circuit of the receiver. Hence, a BER of 2×10^{-6} corresponds to on average 2 errors per million bits. A commonly used criterion for digital optical receivers requires BER $\leq 1 \times 10^{-9}$. The receiver sensitivity is then defined as the minimum average received power $\bar{P}_{rec}$ required by the receiver to operate at a BER of 10^{-9}. Since $\bar{P}_{rec}$ depends on the BER, let us begin by calculating the BER.

4.5.1 Bit-Error Rate

Figure 4.18(a) shows schematically the fluctuating signal received by the decision circuit, which samples it at the decision instant t_D determined through clock recovery. The sampled value I fluctuates from bit to bit around an average value I_1 or I_0, depending on whether the bit corresponds to 1 or 0 in the bit stream. The decision circuit compares the sampled value with a threshold value I_D and calls it bit 1 if $I > I_D$ or bit 0 if $I < I_D$. An error occurs if $I < I_D$ for bit 1 because of receiver noise. An error also occurs if $I > I_D$ for bit 0. Both sources of errors can be included by defining the *error probability* as

$$\text{BER} = p(1)P(0/1) + p(0)P(1/0), \tag{4.5.1}$$

where $p(1)$ and $p(0)$ are the probabilities of receiving bits 1 and 0, respectively, $P(0/1)$ is the probability of deciding 0 when 1 is received, and $P(1/0)$ is the probability of deciding 1 when 0 is received. Since 1 and 0 bits are equally likely

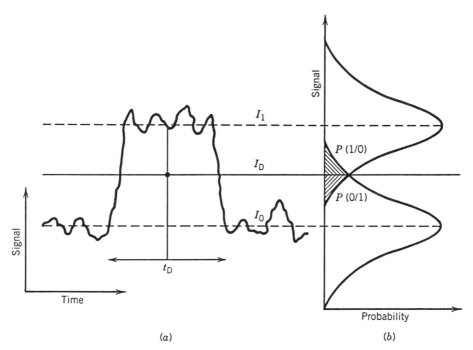

Figure 4.18 (a) Fluctuating signal generated at the receiver. (b) Gaussian probability densities of 1 and 0 bits. The dashed region shows the probability of incorrect identification.

to occur, $p(1) = p(0) = 1/2$, and the BER becomes

$$\text{BER} = \tfrac{1}{2}[P(0/1) + P(1/0)]. \tag{4.5.2}$$

Figure 4.18(b) shows how $P(0/1)$ and $P(1/0)$ depend on the probability density function $p(I)$ of the sampled value I. The functional form of $p(I)$ depends on the statistics of noise sources responsible for current fluctuations. Thermal noise i_T in Eq. (4.4.6) is well described by Gaussian statistics with zero mean and variance σ_T^2. The statistics of shot-noise contribution i_s in Eq. (4.4.6) is also approximately Gaussian for p–i–n receivers but is difficult to obtain for APDs [87]–[91]. A common approximation treats i_s as a Gaussian random variable for both p–i–n and APD receivers but with different variance σ_s^2 given by Eqs. (4.4.5) and (4.4.17), respectively. Since the sum of two Gaussian random variables is also a Gaussian random variable, the sampled value I has a Gaussian probability density function with variance $\sigma^2 = \sigma_s^2 + \sigma_T^2$. However, both the average and the variance are different for 1 and 0 bits since I_p in Eq. (4.4.6) equals I_1 or I_0, depending on the bit received. If σ_1^2 and σ_0^2 are the corresponding variances, the conditional probabilities are given by

$$P(0/1) = \frac{1}{\sigma_1\sqrt{2\pi}} \int_{-\infty}^{I_D} \exp\left(-\frac{(I - I_1)^2}{2\sigma_1^2}\right) dI = \frac{1}{2}\operatorname{erfc}\left(\frac{I_1 - I_D}{\sigma_1\sqrt{2}}\right), \tag{4.5.3}$$

$$P(1/0) = \frac{1}{\sigma_0 \sqrt{2\pi}} \int_{I_D}^{\infty} \exp\left(-\frac{(I - I_0)^2}{2\sigma_0^2}\right) dI = \frac{1}{2} \text{erfc}\left(\frac{I_D - I_0}{\sigma_0 \sqrt{2}}\right), \quad (4.5.4)$$

where erfc stands for the complementary error function, defined as [92]

$$\text{erfc}(x) = \frac{2}{\sqrt{\pi}} \int_x^{\infty} \exp(-y^2)\, dy. \quad (4.5.5)$$

By substituting Eqs. (4.5.3) and (4.5.4) in Eq. (4.5.2), the BER is given by

$$\text{BER} = \frac{1}{4}\left[\text{erfc}\left(\frac{I_1 - I_D}{\sigma_1 \sqrt{2}}\right) + \text{erfc}\left(\frac{I_D - I_0}{\sigma_0 \sqrt{2}}\right)\right]. \quad (4.5.6)$$

Equation (4.5.6) shows that the BER depends on the *decision threshold* I_D. In practice, I_D is optimized to minimize the BER. The minimum occurs when I_D is chosen such that

$$(I_1 - I_D)/\sigma_1 = (I_D - I_0)/\sigma_0 \equiv Q. \quad (4.5.7)$$

An explicit expression for I_D is

$$I_D = \frac{\sigma_0 I_1 + \sigma_1 I_0}{\sigma_0 + \sigma_1}. \quad (4.5.8)$$

When $\sigma_1 = \sigma_0$, $I_D = (I_1 + I_0)/2$, which corresponds to setting the decision threshold in the middle. This is the situation for most p–i–n receivers whose noise is dominated by thermal noise ($\sigma_T \gg \sigma_s$) and is independent of the average current. By contrast, shot noise is larger for bit 1 than for bit 0, since σ_s^2 varies linearly with the average current. In the case of APD receivers, the BER can be minimized by setting the decision threshold in accordance with Eq. (4.5.8).

The BER with the optimum setting of the decision threshold is obtained by using Eqs. (4.5.6) and (4.5.7) and is given by

$$\text{BER} = \frac{1}{2} \text{erfc}\left(\frac{Q}{\sqrt{2}}\right) \approx \frac{\exp(-Q^2/2)}{Q\sqrt{2\pi}}, \quad (4.5.9)$$

where Q is obtained by using Eqs. (4.5.7) and (4.5.8) and is given by

$$Q = \frac{I_1 - I_0}{\sigma_1 + \sigma_0}. \quad (4.5.10)$$

The approximate form of BER is obtained by using the asymptotic expansion [92] of $\text{erfc}(Q/\sqrt{2})$ and is reasonably accurate for $Q > 3$. Figure 4.19 shows how the BER varies with the Q parameter. The BER improves as Q increases and becomes lower than 10^{-12} for $Q > 7$. The receiver sensitivity corresponds to the average optical power for which $Q \approx 6$, since BER $\approx 10^{-9}$ when $Q = 6$. The next subsection provides an explicit expression for the receiver sensitivity.

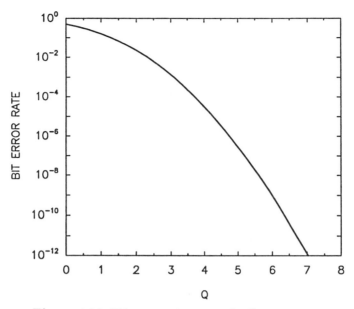

Figure 4.19 Bit-error rate versus the Q parameter.

4.5.2   Minimum Received Power

Equation (4.5.9) can be used to calculate the minimum optical power that a receiver needs to operate reliably with a BER below a specified value. For this purpose Q should be related to the incident optical power. For simplicity, consider the case in which 0 bits carry no optical power so that $P_0 = 0$, and hence $I_0 = 0$. The power P_1 in 1 bits is related to I_1 as

$$I_1 = MRP_1 = 2MR\bar{P}_{\text{rec}}, \tag{4.5.11}$$

where $\bar{P}_{\text{rec}}$ is the *average received power* defined as $\bar{P}_{\text{rec}} = (P_1 + P_0)/2$. The APD gain M is included in Eq. (4.5.11) for generality. The case of p–i–n receivers can be considered by setting $M = 1$.

The RMS noise currents σ_1 and σ_0 include the contributions of both shot noise and thermal noise and can be written as

$$\sigma_1 = (\sigma_s^2 + \sigma_T^2)^{1/2} \quad \text{and} \quad \sigma_0 = \sigma_T, \tag{4.5.12}$$

where σ_s^2 and σ_T^2 are given by Eqs. (4.4.17) and (4.4.9), respectively. Neglecting the contribution of dark current, the noise variances become

$$\sigma_s^2 = 2qM^2 F_A R(2\bar{P}_{\text{rec}})\Delta f, \tag{4.5.13}$$
$$\sigma_T^2 = (4k_B T/R_L)F_n \Delta f. \tag{4.5.14}$$

By using Eqs. (4.5.10)–(4.5.12), the parameter Q is given by

$$Q = \frac{I_1}{\sigma_1 + \sigma_0} = \frac{2MR\bar{P}_{\text{rec}}}{(\sigma_s^2 + \sigma_T^2)^{1/2} + \sigma_T}. \tag{4.5.15}$$

For a specified value of BER, Q is determined from Eq. (4.5.9) and the receiver sensitivity $\bar{P}_{\text{rec}}$ is found from Eq. (4.5.15). A simple analytic expression for $\bar{P}_{\text{rec}}$ is obtained by solving Eq. (4.5.15) for a given value of Q and is given by [3]

$$\bar{P}_{\text{rec}} = \frac{Q}{R}\left(qF_A Q\Delta f + \frac{\sigma_T}{M}\right). \tag{4.5.16}$$

Equation (4.5.16) shows how $\bar{P}_{\text{rec}}$ depends on various receiver parameters and how it can be optimized. Consider first the case of a p–i–n receiver by setting $M = 1$. Since thermal noise σ_T generally dominates for such a receiver, $\bar{P}_{\text{rec}}$ is given by the simple expression

$$(\bar{P}_{\text{rec}})_{\text{pin}} \approx Q\sigma_T/R. \tag{4.5.17}$$

From Eq. (4.5.14), σ_T^2 depends not only on receiver parameters such as R_L and F_n but also on the bit rate through the receiver bandwidth Δf (typically, $\Delta f = B/2$). Thus, $\bar{P}_{\text{rec}}$ increases as $\sqrt{B}$ in the thermal-noise limit. As an example, consider a 1.55-μm p–i–n receiver with $R = 1$ A/W. If we use $\sigma_T = 100$ nA as a typical value and $Q = 6$ corresponding to a BER of 10^{-9}, the receiver sensitivity is given by $\bar{P}_{\text{rec}} = 0.6$ μW or -32.2 dBm.

Equation (4.5.16) shows how receiver sensitivity improves with the use of APD receivers. If thermal noise remains dominant, $\bar{P}_{\text{rec}}$ is reduced by a factor of M, and the received sensitivity is improved by the same factor. However, shot noise increases considerably for APD, and Eq. (4.5.16) should be used in the general case in which shot-noise and thermal-noise contributions are comparable. Similar to the case of SNR discussed in Section 4.4.3, the receiver sensitivity can be optimized by adjusting the APD gain M. By using F_A from Eq. (4.4.18) in Eq. (4.5.16), it is easy to verify that $\bar{P}_{\text{rec}}$ is minimum for an optimum value of M given by [3]

$$M_{\text{opt}} = k_A^{-1/2}\left(\frac{\sigma_T}{Qq\Delta f} + k_A - 1\right)^{1/2} \approx \left(\frac{\sigma_T}{k_A Qq\Delta f}\right)^{1/2}, \tag{4.5.18}$$

and the minimum value is given by

$$(\bar{P}_{\text{rec}})_{\text{APD}} = (2q\Delta f/R)Q^2(k_A M_{\text{opt}} + 1 - k_A). \tag{4.5.19}$$

The improvement in receiver sensitivity obtained by the use of an APD can be estimated by comparing Eqs. (4.5.17) and (4.5.19). It depends on the ionization coefficient ratio k_A and is larger for APDs with a smaller value of k_A. For InGaAs APD receivers, the sensitivity is typically improved by 6–8 dB; such an improvement is sometimes called the APD advantage. Note that $\bar{P}_{\text{rec}}$ for APD receivers increases linearly with the bit rate B ($\Delta f \approx B/2$), in contrast with its $\sqrt{B}$ dependence for p–i–n receivers. The linear dependence of $\bar{P}_{\text{rec}}$ on B is a general feature of shot-noise-limited receivers. For an ideal receiver for which $\sigma_T = 0$, the receiver sensitivity is obtained by setting $M = 1$ in Eq. (4.5.16) and is given by

$$(\bar{P}_{\text{rec}})_{\text{ideal}} = (q\Delta f/R)Q^2. \tag{4.5.20}$$

A comparison of Eqs. (4.5.19) and (4.5.20) shows sensitivity degradation caused by the excess-noise factor in APD receivers.

Alternative measures of receiver sensitivity are sometimes used. For example, the BER can be related to the SNR and to the average number of photons N_p contained within the "1" bit. In the thermal-noise limit $\sigma_0 \approx \sigma_1$. By using $I_0 = 0$, Eq. (4.5.10) provides $Q = I_1/2\sigma_1$. As SNR $= I_1^2/\sigma_1^2$, it is related to Q by the simple relation SNR $= 4Q^2$. Since $Q = 6$ for a BER of 10^{-9}, the SNR must be at least 144 or 21.6 dB for achieving BER $\leq 10^{-9}$. The required value of SNR changes in the shot-noise limit. In the absence of thermal noise, $\sigma_0 \approx 0$, since shot noise is negligible for the "0" bit if the dark-current contribution is neglected. Since $Q = I_1/\sigma_1 = (\text{SNR})^{1/2}$ in the shot-noise limit, an SNR of 36 or 15.6 dB is enough to obtain BER $= 1 \times 10^{-9}$. It was shown in Section 4.4.2 that SNR $\approx \eta N_p$ [see Eq. (4.4.15) and the following discussion] in the shot-noise limit. By using $Q = (\eta N_p)^{1/2}$ in Eq. (4.5.9), the BER is given by

$$\text{BER} = \frac{1}{2}\,\text{erfc}\left(\sqrt{\frac{\eta N_p}{2}}\right). \tag{4.5.21}$$

For a receiver with 100% quantum efficiency ($\eta = 1$), BER $= 1 \times 10^{-9}$ when $N_p = 36$. In practice, most optical receivers require $N_p \sim 1000$ to achieve a BER of 10^{-9}, as their performance is severely limited by thermal noise.

4.5.3 Quantum Limit of Photodetection

The BER expression (4.5.21) obtained in the shot-noise limit is not totally accurate, since its derivation is based on the Gaussian approximation for the receiver noise statistics. For an ideal detector (no thermal noise, no dark current, and 100% quantum efficiency), $\sigma_0 = 0$, as shot noise vanishes in the absence of incident power, and thus the decision threshold can be set quite close to the 0-level signal. Indeed, for such an ideal receiver, 1 bits can be identified without error as long as even one photon is detected. An error is made only if a 1 bit fails to produce even a single electron–hole pair. For such a small number of photons and electrons, shot-noise statistics cannot be approximated by a Gaussian distribution, and the exact Poisson statistics should be used. If N_p is the average number of photons in each 1 bit, the probability of generating m electron–hole pairs is given by the Poisson distribution [93]

$$P_m = \exp(-N_p)N_p^m/m!. \tag{4.5.22}$$

The BER can be calculated by using Eqs. (4.5.2) and (4.5.22). The probability $P(1/0)$ that a 1 is identified when 0 is received is zero since no electron–hole pair is generated when $N_p = 0$. The probability $P(0/1)$ is obtained by setting $m = 0$ in Eq. (4.5.22), since a 0 is decided in that case even though 1 is received. Since $P(0/1) = \exp(-N_p)$, the BER is given by the simple expression

$$\text{BER} = \exp(-N_p)/2. \tag{4.5.23}$$

For BER $< 10^{-9}$, N_p must exceed 20. Since this requirement is a direct result of quantum fluctuations associated with the incoming light, it is referred to as the quantum limit. Each 1 bit must contain at least 20 photons to be detected with a BER $< 10^{-9}$. This requirement can be converted into power by using $P_1 = N_p h\nu B$, where B is the bit rate and $h\nu$ the photon energy. The receiver sensitivity, defined as $\bar{P}_{\text{rec}} = (P_1 + P_0)/2 = P_1/2$, is given by

$$\bar{P}_{\text{rec}} = N_p h\nu B/2 = \bar{N}_p h\nu B. \tag{4.5.24}$$

The quantity $\bar{N}_p$ expresses the receiver sensitivity in terms of the average number of photons/bit and is related to N_p as $\bar{N}_p = N_p/2$ when 0 bits carry no energy. Its use as a measure of receiver sensitivity is quite common. In the quantum limit $\bar{N}_p = 10$. The power can be calculated from Eq. (4.5.24). For example, for a 1.55-μm receiver ($h\nu = 0.8$ eV), $\bar{P}_{\text{rec}} = 13$ nW or -48.9 dBm at $B = 10$ Gb/s. Most receivers operate away from the quantum limit by 20 dB or more. This is equivalent to saying that $\bar{N}_p$ typically exceeds 1000 photons in practical receivers.

4.6   SENSITIVITY DEGRADATION

The sensitivity analysis in Section 4.5 is based on the consideration of receiver noise only. In particular, the analysis assumes that the optical signal incident on the receiver consists of an ideal bit stream such that 1 bits consist of an optical pulse of constant energy while no energy is contained in 0 bits. In practice, the optical signal emitted by a transmitter deviates from this ideal situation. Moreover, it can be degraded during its transmission through the fiber. The minimum average optical power required by the receiver increases because of such nonideal conditions compared with the value $\bar{P}_{\text{rec}}$ derived in Section 4.5 by considering the receiver noise only. This increase in the average received power is referred to as the *power penalty*. Many factors contribute to the power penalty. Some of them occur only when the signal is propagated through the fiber whereas others are present even in the absence of fiber. In this section we discuss the sources of power penalties that can lead to sensitivity degradation even without signal transmission through the fiber. The fiber-related power-penalty mechanisms are discussed in Section 5.4.

4.6.1   Extinction Ratio

A simple source of a power penalty is related to the energy carried by 0 bits. Some power is emitted by most transmitters even in the off-state. In the case of semiconductor lasers, the off-state power P_0 depends on the bias current I_b and the threshold current I_{th}. If $I_b < I_{th}$, the power emitted during 0 bits is due to spontaneous emission, and generally $P_0 \ll P_1$, where P_1 is the on-state power. By contrast, P_0 can be a significant fraction of P_1 if the laser is biased close to but above threshold. The *extinction ratio* is defined as

$$r_{ex} = P_0/P_1. \tag{4.6.1}$$

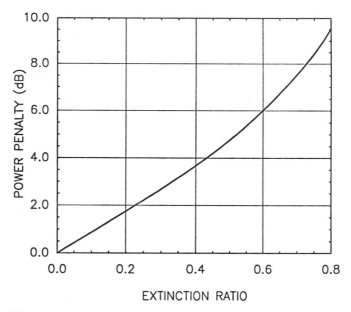

Figure 4.20 Power penalty versus the extinction ratio r_{ex}.

The power penalty can be obtained by using Eq. (4.5.10). For a p–i–n receiver $I_1 = RP_1$ and $I_0 = RP_0$, where R is the responsivity (the APD gain can be included by replacing R by MR). By using the definition of receiver sensitivity $\bar{P}_{rec} = (P_1 + P_0)/2$, the parameter Q is given by

$$Q = \left(\frac{1 - r_{ex}}{1 + r_{ex}}\right) \frac{2R\bar{P}_{rec}}{\sigma_1 + \sigma_0}. \tag{4.6.2}$$

In general, σ_1 and σ_0 depend on $\bar{P}_{rec}$ because of the dependence of the shot-noise contribution on the received optical signal. However, both of them can be approximated by the thermal noise σ_T when receiver performance is dominated by thermal noise. By using $\sigma_1 \approx \sigma_0 \approx \sigma_T$ in Eq. (4.6.2), $\bar{P}_{rec}$ is given by

$$\bar{P}_{rec}(r_{ex}) = \left(\frac{1 + r_{ex}}{1 - r_{ex}}\right) \frac{\sigma_T Q}{R}. \tag{4.6.3}$$

This equation shows that $\bar{P}_{rec}$ increases when $r_{ex} \neq 0$. The power penalty is defined as the ratio $\delta_{ex} = \bar{P}_{rec}(r_{ex})/\bar{P}_{rec}(0)$. It is commonly expressed in decibel (dB) units by using

$$\delta_{ex} = 10 \log_{10}\left(\frac{\bar{P}_{rec}(r_{ex})}{\bar{P}_{rec}(0)}\right) = 10 \log_{10}\left(\frac{1 + r_{ex}}{1 - r_{ex}}\right). \tag{4.6.4}$$

Figure 4.20 shows how the power penalty increases with r_{ex}. A 1-dB penalty occurs for $r_{ex} = 0.12$ and increases to 4.8 dB for $r_{ex} = 0.5$. In practice, for lasers biased below threshold, r_{ex} is typically below 0.05, and the corresponding power penalty (< 0.4 dB) is negligible. Nonetheless, it can become significant if the semiconductor laser is biased above threshold. An expression for $\bar{P}_{\text{rec}}(r_{ex})$ can be obtained [3] for APD receivers by including the APD gain and the shot-noise contribution to σ_0 and σ_1 in Eq. (4.6.2). The optimum APD gain is lower than that in Eq. (4.5.18) when $r_{ex} \neq 0$. The sensitivity is also reduced because of the lower optimum gain. Normally, the power penalty for an APD receiver is larger by about a factor of 2 for the same value of r_{ex}.

4.6.2   Intensity Noise

The noise analysis of Section 4.4 is based on the assumption that the optical power incident on the receiver does not fluctuate. In practice, light emitted by any transmitter exhibits power fluctuations. Such fluctuations, called intensity noise, were discussed in Section 3.3.8 in the context of semiconductor lasers. The optical receiver converts power fluctuations into current fluctuations which add to those resulting from shot noise and thermal noise. As a result, the receiver SNR is degraded and is lower than that given by Eq. (4.4.19). An exact analysis is complicated, as it involves the calculation of photocurrent statistics [94]. A simple approach consists of adding a third term to the current variance given by Eq. (4.4.10), so that

$$\sigma^2 = \sigma_s^2 + \sigma_T^2 + \sigma_I^2, \qquad (4.6.5)$$

where

$$\sigma_I = R\langle(\Delta P_{in}^2)\rangle^{1/2} = RP_{in}r_I. \qquad (4.6.6)$$

The parameter r_I, defined as $r_I = \langle(\Delta P_{in}^2)\rangle^{1/2}/P_{in}$, is a measure of the noise level of the incident optical signal. It is related to the *relative intensity noise* (RIN) of the transmitter as

$$r_I^2 = \frac{1}{2\pi} \int_{-\infty}^{\infty} \text{RIN}(\omega)\, d\omega, \qquad (4.6.7)$$

where $\text{RIN}(\omega)$ is given by Eq. (3.3.57). As discussed in Section 3.3.8, r_I is simply the inverse of the SNR of light emitted by the transmitter. Typically, the transmitter SNR is better than 20 dB, and $r_I < 0.01$.

As a result of the dependence of σ_0 and σ_1 on the parameter r_I, the parameter Q in Eq. (4.5.10) is reduced in the presence of intensity noise, Since Q should be maintained to the same value to maintain the BER, it is necessary to increase the received power. This is the origin of the power penalty induced by intensity noise. To simplify the following analysis, the extinction ratio is assumed to be zero, so that $I_0 = 0$ and $\sigma_0 = \sigma_T$. By using $I_1 = RP_1 = 2R\bar{P}_{\text{rec}}$ and Eq. (4.6.5) for σ_1, Q is given by

$$Q = \frac{2R\bar{P}_{\text{rec}}}{(\sigma_T^2 + \sigma_s^2 + \sigma_I^2)^{1/2} + \sigma_T}, \qquad (4.6.8)$$

where

$$\sigma_s = (4qR\bar{P}_{\text{rec}}\Delta f)^{1/2}, \qquad \sigma_I = 2r_I R\bar{P}_{\text{rec}}, \qquad (4.6.9)$$

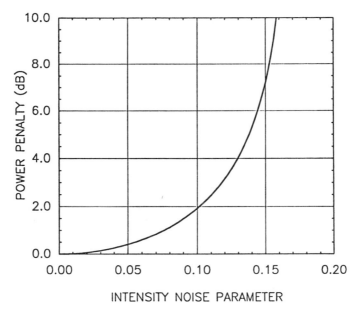

Figure 4.21 Power penalty versus the intensity noise parameter r_I.

and σ_T is given by Eq. (4.4.9). Equation (4.6.8) is easily solved to obtain the following expression for the receiver sensitivity:

$$\bar{P}_{\rm rec}(r_I) = \frac{Q\sigma_T + Q^2 q \Delta f}{R(1 - r_I^2 Q^2)}. \tag{4.6.10}$$

The power penalty, defined as the increase in $\bar{P}_{\rm rec}$ when $r_I \neq 0$, is given by

$$\delta_I = 10\log_{10}[\bar{P}_{\rm rec}(r_I)/\bar{P}_{\rm rec}(0)] = -10\log_{10}(1 - r_I^2 Q^2). \tag{4.6.11}$$

Figure 4.21 shows the power penalty as a function of r_I for maintaining $Q = 6$ corresponding to a BER of 10^{-9}. The penalty is negligible for $r_I < 0.01$ as δ_I is below 0.02 dB. Since this is the case for most optical transmitters, the effect of intensity noise is negligible for digital lightwave receivers. As discussed in Chapter 7, intensity noise becomes a limiting factor for analog lightwave systems. It can become a limiting factor even for digital systems if r_I exceeds 0.1. The power penalty is almost 2 dB for $r_I = 0.1$ and becomes infinite when $r_I = Q^{-1} = 0.167$. An infinite power penalty implies that the receiver cannot operate at the specific BER even if the received optical power is increased indefinitely. In the BER diagram shown in Fig. 4.19, an infinite power penalty corresponds to a saturation of the BER curve above the 10^{-9} level, a feature referred to as the BER floor. In this respect, the effect of intensity noise is qualitatively different than the extinction ratio, for which the power penalty remains finite for all values of r_{ex} such that $r_{ex} < 1$.

The analysis above assumes that the intensity noise at the receiver is the same as at the transmitter. This may not be the case when the optical signal propagates through an optical fiber. For multimode semiconductor lasers, fiber dispersion can lead to considerable degradation of the receiver sensitivity through a phenomenon called mode-partition noise. This effect is discussed in Section 5.4.3. Another phenomenon that can enhance intensity noise is optical feedback from parasitic reflections along the fiber link. The effect of reflection feedback is also considered in the following chapter (Section 5.4.5).

4.6.3   Timing Jitter

The calculation of receiver sensitivity in Section 4.5 is based on the assumption that the signal is sampled at the peak of the voltage pulse. In practice, the decision instant is determined by the clock-recovery circuit (see Fig. 4.11). Because of the noisy nature of the input to the clock-recovery circuit, the sampling time fluctuates from bit to bit. Such fluctuations are called *timing jitter* [95]–[98]. The SNR is degraded because fluctuations in the sampling time lead to additional fluctuations in the signal. This can be understood by noting that if the bit is not sampled at the bit center, the sampled value is reduced by an amount that depends on the timing jitter Δt. Since Δt is a random variable, the reduction in the sampled value is also random. The SNR is reduced as a result of such additional fluctuations, and the receiver performance is degraded compared with that expected in the absence of timing jitter. The SNR can be maintained by increasing the received optical power. This increase is the power penalty induced by timing jitter.

To simplify the following analysis, let us consider a p–i–n receiver dominated by thermal noise σ_T and assume a zero extinction ratio. By using $I_0 = 0$ in Eq. (4.5.10), the parameter Q is given by

$$Q = \frac{I_1 - \langle \Delta i_j \rangle}{(\sigma_T^2 + \sigma_j^2)^{1/2} + \sigma_T}, \tag{4.6.12}$$

where $\langle \Delta i_j \rangle$ is the average value and σ_j is the RMS value of the current fluctuation Δi_j induced by timing jitter Δt. If $h_{\text{out}}(t)$ governs the shape of the current pulse,

$$\Delta i_j = I_1[h_{\text{out}}(0) - h_{\text{out}}(\Delta t)], \tag{4.6.13}$$

where the ideal sampling instant is taken to be $t = 0$.

Clearly, σ_j depends on the shape of the signal pulse at the decision current. A simple choice [95] corresponds to $h_{\text{out}}(t) = \cos^2(\pi B t/2)$, where B is the bit rate. Here Eq. (4.3.6) is used as many optical receivers are designed to provide that pulse shape. Since Δt is likely to be much smaller than the bit period $T_B = 1/B$, it can be approximated as

$$\Delta i_j = (2\pi^2/3 - 4)(B\Delta t)^2 I_1 \tag{4.6.14}$$

by assuming that $B\Delta t \ll 1$. This approximation provides a reasonable estimate of the power penalty as long as the penalty is not too large [95]. This is expected to be the case in practice. To calculate σ_j, the probability density function of

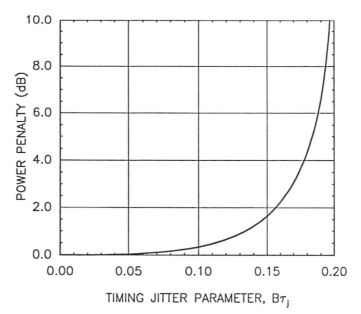

Figure 4.22 Power penalty versus the timing jitter parameter $B\tau_j$.

the timing jitter Δt is assumed to be Gaussian, so that

$$p(\Delta t) = \frac{1}{\tau_j\sqrt{2\pi}}\exp\left(-\frac{\Delta t^2}{2\tau_j^2}\right), \qquad (4.6.15)$$

where τ_j is the RMS value (standard deviation) of Δt. The probability density of Δi_j can be obtained by using Eqs. (4.6.14) and (4.6.15) and noting that Δi_j is proportional to $(\Delta t)^2$. The result is

$$p(\Delta i_j) = \frac{1}{\sqrt{\pi b \Delta i_j I_1}}\exp\left(-\frac{\Delta i_j}{bI_1}\right), \qquad (4.6.16)$$

where

$$b = (4\pi^2/3 - 8)(B\tau_j)^2. \qquad (4.6.17)$$

Equation (4.6.16) is used to calculate $\langle\Delta i_j\rangle$ and $\sigma_j = \langle(\Delta i_j)^2\rangle^{1/2}$. The integration over Δi_j is easily done to obtain

$$\langle\Delta i_j\rangle = bI_1/2, \qquad \sigma_j = bI_1/\sqrt{2}. \qquad (4.6.18)$$

By using Eqs. (4.6.12) and (4.6.18) and noting that $I_1 = 2R\bar{P}_{\text{rec}}$, where R is the responsivity, the receiver sensitivity is given by

$$\bar{P}_{\text{rec}}(b) = \left(\frac{\sigma_T Q}{R}\right)\frac{1 - b/2}{(1 - b/2)^2 - b^2Q^2/2}. \qquad (4.6.19)$$

The power penalty, defined as the increase in $\bar{P}_{rec}$, is given by

$$\delta_j = 10 \log_{10} \left(\frac{\bar{P}_{rec}(b)}{\bar{P}_{rec}(0)} \right) = 10 \log_{10} \left(\frac{1 - b/2}{(1 - b/2)^2 - b^2 Q^2/2} \right). \qquad (4.6.20)$$

Figure 4.22 shows how the power penalty varies with the parameter $B\tau_j$, which has the physical significance of the fraction of the bit period over which the decision time fluctuates (one standard deviation). The power penalty is negligible for $B\tau_j < 0.1$ but increases rapidly beyond $B\tau_j = 0.1$. A 2-dB penalty occurs for $B\tau_j = 0.16$. Similar to the case of intensity noise, the jitter-induced penalty becomes infinite beyond $B\tau_j = 0.2$. The exact value of $B\tau_j$ at which the penalty becomes infinite depends on the model used to calculate the jitter-induced power penalty. Equation (4.6.20) is obtained by using a specific pulse shape and a specific jitter distribution. It is also based on the use of Eqs. (4.5.9) and (4.6.12), which assumes Gaussian statistics for the receiver current. As evident from Eq. (4.6.16), jitter-induced current fluctuations are not Gaussian in nature. A more accurate calculation shows that Eq. (4.6.20) underestimates the power penalty [97]. The qualitative behavior, however, remains the same. In general, the RMS value of the timing jitter should be below 10% of the bit period for a negligible power penalty. A similar conclusion holds for APD receivers, for which the penalty is generally larger [98].

4.7   RECEIVER PERFORMANCE

The receiver performance is characterized by measuring the BER as a function of the average optical power received. The average optical power corresponding to a BER of 10^{-9} is a measure of receiver sensitivity. Figure 4.23 shows the receiver sensitivity measured in various transmission experiments [99]–[110] by sending a long sequence of pseudo-random bits (typical sequence length $2^{15} - 1$) over a single-mode fiber and then detecting it by using either a p–i–n or an APD receiver. The experiments were performed at the 1.3- or 1.55-μm wavelength, and the bit rate varied from 100 MHz to 10 GHz. The theoretical quantum limit at these two wavelengths is also shown in Fig. 4.23 by using Eq. (4.5.24). A direct comparison shows that the measured receiver sensitivities are worse by 20 dB or more compared with the quantum limit. Most of the degradation is due to the thermal noise that is unavoidable at room temperature and generally dominates the shot noise. Some degradation is due to fiber dispersion, which leads to power penalties; sources of such penalties are discussed in the following chapter. The dispersion-induced sensitivity degradation depends on both the bit rate B and the fiber length L and increases with BL. This is the reason why the sensitivity degradation from the quantum limit is larger (25–30 dB) for systems operating at high bit rates. The receiver sensitivity at 10 Gb/s is typically worse than -20 dBm [110], [111].

It is possible to isolate the extent of sensitivity degradation occurring as a result of signal propagation inside the optical fiber. The common procedure is to perform a separate measurement of the receiver sensitivity by connecting the

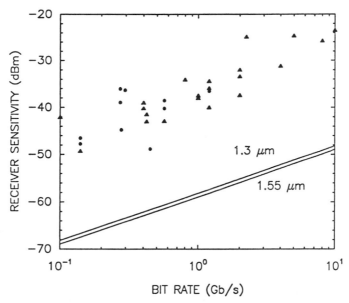

Figure 4.23 Measured receiver sensitivities versus the bit rate for p–i–n (circles) and APD (triangles) receivers in transmission experiments near 1.3- and 1.55-μm wavelengths. The quantum limit of receiver sensitivity is also shown for comparison (solid lines).

transmitter and receiver directly, without the intermediate fiber. Figure 4.24 shows the experimental results of such a measurement for a 1.31-μm system experiment in which the transmitted signal was modulated at 8 Gb/s, propagated over 30-km fiber, and detected by using an InGaAs APD receiver [112]. When the fiber length is reduced to 3 m (the length of pigtails associated with the transmitter and the receiver), the receiver sensitivity improves by 1.5 dB since degradation from fiber dispersion is eliminated. If such an adjustment is made to the data shown in Fig. 4.23, one finds that p–i–n receivers typically operate 25 dB away from the quantum limit. The receiver sensitivity can be improved by about 5–6 dB by using APD receivers, which typically operate 20 dB away from the quantum limit. In terms of the average number of photons/bit, APD receivers require nearly 1000 photons/bit compared with the quantum limit of 10 photons/bit. The receiver performance is generally better for shorter wavelengths in the region 0.8–0.9 μm, where silicon APDs can be used. They perform satisfactorily with about 400 photons/bit; an experiment in 1976 achieved a sensitivity of only 187 photons/bit [113]. It is possible to improve the receiver sensitivity by using coding schemes. A sensitivity of 180 photons/bit was realized in a 1.55-μm system experiment [114] after 305 km of transmission at 140 Mb/s.

The performance of an optical receiver in actual lightwave systems may change with time. Since it is not possible to measure the BER directly for a system in operation, an alternative is needed to monitor system performance. As

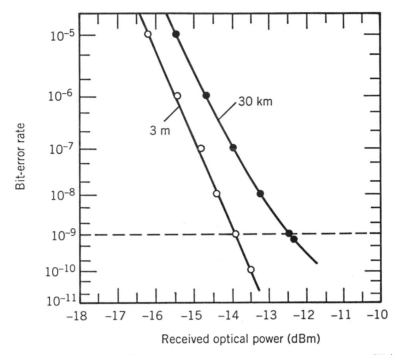

Figure 4.24 BER curves for a 1.3-μm transmission experiment at 8 Gb/s. The horizontal dashed line corresponds to a BER of 10^{-9}. Its intersection with the BER curves determines the receiver sensitivity. (After Ref. [112]. ©1986 IEE. Reprinted with permission.)

discussed in Section 4.3.3, the eye diagram is best suited for this purpose; closing of the eye is a measure of degradation in receiver performance and is associated with a corresponding increase in the BER. Figure 4.25 shows the eye diagrams recorded in a laboratory experiment in which the output of a 1.55-μm optical transmitter was modulated directly at 2.5 Gb/s and detected by using an APD receiver. The eye is wide open in the absence of optical fiber (upper diagram) but is partially closed when the signal is transmitted through a 120-km-long fiber (lower diagram). Closing of the eye is due to fiber dispersion, which leads to considerable distortion of optical pulses as they propagate through the fiber. The eye pattern shown in Fig. 4.25 indicates that the receiver performance is affected by fiber dispersion, although not catastrophically, as the eye is not closed completely. The continuous monitoring of the eye pattern is common in actual systems as a measure of receiver performance.

The performance of optical receivers operating in the wavelength range 1.3–1.6 μm is severely limited by thermal noise, as shown clearly in Fig. 4.23. The use of APD receivers improves the situation, but to a limited extent only, because of the excess noise factor associated with InGaAs APDs. Most receivers operate away from the quantum limit by 20 dB or more. The effect of thermal noise can be considerably reduced by using coherent-detection techniques in which the

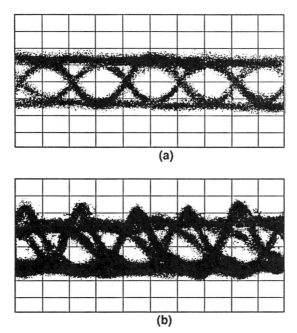

(a)

(b)

Figure 4.25 Eye diagrams at 0 km and at 120 km (lower trace) recorded in a 1.55-μm transmission experiment at 2.5 Gb/s. (After Ref. [115]. ©1990 IEEE. Reprinted with permission.)

received signal is mixed coherently with the output of a narrow-linewidth laser (see Chapter 6). The receiver performance can also be improved by amplifying the optical signal before it is incident on the photodetector. This technique has become practical after the advent of fiber amplifiers and is discussed is Chapter 8.

PROBLEMS

4.1 Calculate the responsivity of a p–i–n photodiode at 1.3 and 1.55 μm if the quantum efficiency is 80%. Why is the photodiode more responsive at 1.55 μm?

4.2 Photons at a rate of 10^{10}/s are incident on an APD with responsivity of 6 A/W. Calculate the quantum efficiency and the photocurrent at the operating wavelength of 1.5 μm for an APD gain of 10.

4.3 Show by solving Eqs. (4.2.3) and (4.2.4) that the multiplication factor M is given by Eq. (4.2.7) for an APD in which electrons initiate the avalanche process. Treat α_e and α_h as constants.

4.4 Draw a block diagram of a digital optical receiver showing its various components. Explain the function of each component. How is the signal used by the decision circuit related to the incident optical power?

4.5 The raised-cosine pulse shape of Eq. (4.3.6) can be generalized to generate a family of such pulses by defining

$$h_{out}(t) = \frac{\sin(\pi Bt)}{\pi Bt} \frac{\cos(\pi \beta Bt)}{1 - (2\beta Bt)^2},$$

where the parameter β varies between 0 and 1. Derive an expression for the transfer function $H_{out}(f)$ given by the Fourier transform of $h_{out}(t)$. Plot $h_{out}(t)$ and $H_{out}(f)$ for $\beta = 0$, 0.5, and 1.

4.6 Consider a 0.8-μm receiver with a silicon p–i–n photodiode. Assume 20 MHz bandwidth, 65% quantum efficiency, 1 nA dark current, 8 pF junction capacitance, and 3 dB amplifier noise figure. The receiver is illuminated with 5 μW of optical power. Determine the RMS noise currents due to shot noise, thermal noise, and amplifier noise. Also calculate the SNR.

4.7 The receiver of Problem 4.6 is used in a digital communication system that requires a SNR of at least 20 dB for satisfactory performance. What is the minimum received power when the detection is limited by (a) shot noise and (b) thermal noise? Also calculate the noise-equivalent power in the two cases.

4.8 The excess noise factor of avalanche photodiodes is often approximated by M^x instead of Eq. (4.4.18). Find the range of M for which Eq. (4.4.18) can be approximated within 10% by $F_A(M) = M^x$ by choosing $x = 0.3$ for Si, 0.7 for InGaAs, and 1.0 for Ge. Use $k_A = 0.02$ for Si, 0.35 for InGaAs, and 1.0 for Ge.

4.9 Derive Eq. (4.4.22). Plot M_{opt} versus k_A by solving the cubic polynomial on a computer by using $R_L = 1$ kW, $F_n = 2$, $R = 1$ A/W, $P_{in} = 1$ μW, and $I_d = 2$ nA. Compare the results with the approximate analytic solution given by Eq. (4.4.23) and comment on its validity.

4.10 Derive an expression for the optimum value of M for which the SNR becomes maximum by using $F_A(M) = M^x$ in Eq. (4.4.19).

4.11 Prove that the bit-error rate (BER) given by Eq. (4.5.6) is minimum when the decision threshold is set close to a value given by Eq. (4.5.8).

4.12 A 1.3-μm digital receiver is operating at 100 Mb/s and has an effective noise bandwidth of 60 MHz. The p–i–n photodiode has negligible dark current and 90% quantum efficiency. The load resistance is 100 Ω and the amplifier noise figure is 3 dB. Calculate the receiver sensitivity corresponding to a BER of 10^{-9}. How much does it change if the receiver is designed to operate reliably up to a BER of 10^{-12}?

4.13 Calculate the receiver sensitivity (at a BER of 10^{-9}) for the receiver in Problem 4.12 in the shot-noise and thermal-noise limits. How many photons are incident during bit 1 in the two limits if the optical pulse can be approximated by a square pulse?

4.14 Derive an expression for the optimum gain M_{opt} of an APD receiver that would maximize the receiver sensitivity by taking the excess-noise factor as M^x. Plot M_{opt} as a function of x for $\sigma_T = 0.2$ mA and $\Delta f = 1$ GHz and estimate its value for InGaAs APDs (see Problem 4.8).

4.15 Derive an expression for the sensitivity of an APD receiver by taking into account a finite extinction ratio for the general case in which both shot noise and thermal noise contribute to the receiver sensitivity. You can neglect the dark current.

4.16 Derive an expression for the intensity-noise-induced power penalty of a p–i–n receiver by taking into account a finite extinction ratio. Shot-noise and intensity-noise contributions can both be neglected compared with the thermal noise in the off-state but not in the on-state.

4.17 Use the result of Problem 4.16 to plot the power penalty as a function of the intensity-noise parameter r_I [see Eq. (4.6.6) for its definition] for several values of the extinction ratio. When does the power penalty become infinite? Explain the meaning of an infinite power penalty.

4.18 Derive an expression for the timing-jitter-induced power penalty by assuming a parabolic pulse shape $I(t) = I_p(1 - B^2 t^2)$ and a Gaussian jitter distribution with a standard deviation τ (RMS value). You can assume that the receiver performance is dominated by thermal noise. Calculate the tolerable value of $B\tau$ that would keep the power penalty below 1 dB.

REFERENCES

[1] S. D. Personick, *Bell Syst. Tech. J.* **52**, 843 (1973); **52**, 875 (1973).

[2] T. P. Lee and T. Li, in *Optical Fiber Telecommunications I*, S. E. Miller and A. G. Chynoweth, Eds., Academic Press, San Diego, CA, 1979, Chap. 18.

[3] R. G. Smith and S. D. Personick, in *Semiconductor Devices for Optical Communications*, H. Kressel, Ed., Springer-Verlag, New York, 1980.

[4] S. R. Forrest, in *Optical Fiber Telecommunications II*, S. E. Miller and I. P. Kaminow, Eds., Academic Press, San Diego, CA, 1988, Chap. 14.

[5] B. L. Kasper, in *Optical Fiber Telecommunications II*, S. E. Miller and I. P. Kaminow, Eds., Academic Press, San Diego, CA, 1988, Chap. 18.

[6] S. B. Alexander, *Optical Communication Receiver Design*, Vol. TT22, SPIE Press, Bellingham, WA, 1995.

[7] M. Razeghi, Ed., *Long Wavelength Infrared Detectors*, Gordon & Breach, Newark, NJ, 1996.

[8] E. L. Dereniak, *Infrared Detectors and Systems*, Wiley, New York, 1996.

[9] J. E. Bowers, C. A. Burrus, and R. J. McCoy, *Electron. Lett.* **21**, 812 (1985).

[10] R. S. Tucker, A. J. Taylor, C. A. Burrus, G. Eisenstein, and J. M. Westfield, *Electron. Lett.* **22**, 917 (1986).

[11] Y.-G. Wey, K. S. Giboney, J. E. Bowers, M. J. Rodwell, P. Silvestre, P. Thiagarajan, and G. Robinson, *J. Lightwave Technol.* **13**, 1490 (1995).

[12] K. Kishino, S. Ünlü, J.-I. Chyi, J. Reed, L. Arsenault, and H. Morkoç, *IEEE J. Quantum Electron.* **27**, 2025 (1991).

[13] C. C. Barron, C. J. Mahon, B. J. Thibeault, G. Wang, W. Jiang, L. A. Coldren, and J. E. Bowers, *Electron. Lett.* **30**, 1796 (1994).

[14] I.-H. Tan, J. Dudley, D. I. Babić, D. A. Cohen, B. D. Young, E. L. Hu, J. E. Bowers, B. I. Miller, U. Koren, and M. G. Young, *IEEE Photon. Technol. Lett.* **6**, 811 (1994).

[15] I.-H. Tan, C.-K. Sun, K. S. Giboney, J. E. Bowers E. L. Hu, B. I. Miller, and R. J. Kapik, *IEEE Photon. Technol. Lett.* **7**, 1477 (1995).

[16] K. Kato, S. Hata, K. Kwano, J. Yoshida, and A. Kozen, *IEEE J. Quantum Electron.* **28**, 2728 (1992).

[17] K. Kato, A. Kozen, Y. Muramoto, Y. Itaya, N. Nagatsuma, and M. Yaita, *IEEE Photon. Technol. Lett.* **6**, 719 (1994).

[18] K. Kato and Y. Akatsu, *Opt. Quantum Electron.* **28**, 557 (1996).

[19] K. S. Giboney, M. J. W. Rodwell, and J. E. Bowers, *IEEE Photon. Technol. Lett.* **4**, 1365 (1992).

[20] K. S. Giboney, R. L. Nagarajan, T. E. Reynolds, S. T. Allen, R. P. Mirin, M. J. W. Rodwell, and J. E. Bowers, *IEEE Photon. Technol. Lett.* **7**, 412 (1995).

[21] G. E. Stillman and C. M. Wolfe, in *Semiconductors and Semimetals*, vol. 12, R. K. Willardson and A. C. Beer, Eds., Academic Press, San Diego, CA, 1977, pp. 291–393.

[22] H. Melchior, in *Laser Handbook*, Vol. 1, F. T. Arecchi and E. O. Schulz-Dubois, Eds., North-Holland, Amsterdam, 1972, pp. 725–835.

[23] J. C. Campbell, A. G. Dentai, W. S. Holden, and B. L. Kasper, *Electron. Lett.* **19**, 818 (1983).

[24] B. L. Kasper and J. C. Campbell, *J. Lightwave Technol.* **5**, 1351 (1987).

[25] L. E. Tarof, *Electron. Lett.* **27**, 34 (1991).

[26] L. E. Tarof, J. Yu, R. Bruce, D. G. Knight, T. Baird, and B. Oosterbrink, *IEEE Photon. Technol. Lett.* **5**, 672 (1993).

[27] J. Yu, L. E. Tarof, R. Bruce, D. G. Knight, K. Visvanatha, and T. Baird, *IEEE Photon. Technol. Lett.* **6**, 632 (1994).

[28] L. E. Tarof, R. Bruce, D. G. Knight, J. Yu, H. B. Kim, and T. Baird, *IEEE Photon. Technol. Lett.* **7**, 1330 (1995).

[29] C. L. F. Ma, M. J. Deen, L. E. Tarof, and J. Yu, *IEEE Trans. Electron. Dev.* **42**, 810 (1995).

[30] C. L. F. Ma, M. J. Deen, and L. E. Tarof, *IEEE J. Quantum Electron.* **31**, 2078 (1995).

[31] F. Capasso, in *Semiconductor and Semimetals*, Vol. 22, Part D, W. T. Tsang, Ed., Academic Press, San Diego, CA, 1985, pp. 1–172.

[32] T. Kagawa, Y. Kawamura, and H. Iwamura, *IEEE J. Quantum Electron.* **28**, 1419 (1992).

[33] I. Watanabe, S. Sugou, H. Ishikawa, T. Anan, K. Makita, M. Tsuji, and K. Taguchi, *IEEE Photon. Technol. Lett.* **5**, 675 (1993).

[34] T. Kagawa, Y. Kawamura, and H. Iwamura, *IEEE J. Quantum Electron.* **29**, 1387 (1993).

[35] S. Hanatani, H. Nakamura, S. Tanaka, T. Ido, and C. Notsu, *Appl. Phys. Lett.* **62**, 1122 (1993).

[36] S. Hanatani, H. Nakamura, S. Tanaka, T. Ido, and C. Notsu, *Microwave Opt. Tech. Lett.* **7**, 103 (1994).

[37] I. Watanabe, M. Tsuji, K. Makita, and K. Taguchi, *IEEE Photon. Technol. Lett.* **8**, 269 (1996).

[38] A. R. Hawkins, W. Wu, P. Abraham, K. Streubel, and J. E. Bowers, *Appl. Phys. Lett.* **70**, 303 (1997).

[39] R. Kuchibhotla, A. Srinivasan, J. C. Campbell, C. Lei, D. Peppe, Y. S. He, and B. G. Streetman, *IEEE Photon. Technol. Lett.* **3**, 354 (1991).

[40] S. S. Murtaza, K. A. Anselm, C. Hu, H. Nie, B. G. Streetman, and J. C. Campbell, *IEEE Photon. Technol. Lett.* **7**, 1486 (1995).

[41] S. Hanatani, H. Kitano, M. Shishikura, and S. Tanaka, *Opt. Quantum Electron.* **28**, 575 (1996).

[42] J. Burm, K. I. Litvin, D. W. Woodard, W. J. Schaff, P. Mandeville, M. A. Jaspan, M. M. Gitin, and L. F. Eastman, *IEEE J. Quantum Electron.* **31**, 1504 (1995).

[43] J. B. D. Soole and H. Schumacher, *IEEE J. Quantum Electron.* **27**, 737 (1991).

[44] J. H. Kim, H. T. Griem, R. A. Friedman, E. Y. Chan, and S. Roy, *IEEE Photon. Technol. Lett.* **4**, 1241 (1992).

[45] R.-H. Yuang, J.-I. Chyi, Y.-J. Chan, W. Lin, and Y.-K. Tu, *IEEE Photon. Technol. Lett.* **7**, 1333 (1995).

[46] O. Vendier, N. M. Jokerst, and R. P. Leavitt, *IEEE Photon. Technol. Lett.* **8**, 266 (1996).

[47] M. C. Hargis, S. E. Ralph, J. Woodall, D. McInturff, A. J. Negri, and P. O. Haugsjaa, *IEEE Photon. Technol. Lett.* **8**, 110 (1996).

[48] W. A. Wohlmuth, P. Fay, C. Caneau, and I. Adesida, *Electron. Lett.* **32**, 249 (1996).

[49] A. Bartels, E. Peiner, G.-P. Tang, R. Klockenbrink, H.-H. Wehmann, and A. Schlachetzki, *IEEE Photon. Technol. Lett.* **8**, 670 (1996).

[50] Y. G. Zhang, A. Z. Li, and J. X. Chen, *IEEE Photon. Technol. Lett.* **8**, 830 (1996).

[51] R. G. Swartz, in *Optical Fiber Telecommunications II*, S. E. Miller and I. P. Kaminow, Eds., Academic Press, San Diego, CA, 1988, Chap. 20.

[52] K. Kobayashi, in *Optical Fiber Telecommunications II*, S. E. Miller and I. P. Kaminow, Eds., Academic Press, San Diego, CA, 1988, Chap. 11.

[53] T. Horimatsu and M. Sasaki, *J. Lightwave Technol.* **7**, 1612 (1989).

[54] O. Wada, H. Hamaguchi, M. Makiuchi, T. Kumai, M. Ito, K. Nakai, T. Horimatsu, and T. Sakurai, *J. Lightwave Technol.* **4**, 1694 (1986).

[55] M. Makiuchi, H. Hamaguchi, T. Kumai, O. Aoki, Y. Oikawa, and O. Wada, *Electron. Lett.* **24**, 995 (1988).

[56] K. Matsuda, M. Kubo, K. Ohnaka, and J. Shibata, *IEEE Trans. Electron. Dev.* **35**, 1284 (1988).

[57] H. Yano, K. Aga, H. Kamei, G. Sasaki, and H. Hayashi, *J. Lightwave Technol.* **8**, 1328 (1990).

[58] H. Hayashi, H. Yano, K. Aga, M. Murata, H. Kamei, and G. Sasaki, *IEE Proc.* **138**, Pt. J, 164 (1991).

[59] H. Yano, G. Sasaki, N. Nishiyama, M. Murata, and H. Hayashi, *IEEE Trans. Electron. Dev.* **39**, 2254 (1992).

[60] Y. Akatsu, M. Miyugawa, Y. Miyamoto, Y. Kobayashi, and Y. Akahori, *IEEE Photon. Technol. Lett.* **5**, 163 (1993).

[61] B.-U. H. Klepser, J. Spicher, C. Bergamaschi, W. Patrick, and W. Bächtold, *Proc. 8th Int. Conf. on InP and Related Materials*, IEEE Press, New York, 1996.

[62] K. Takahata, Y. Muramoto, Y. Akatsu, Y. Akahori, A. Kozen, and Y. Itaya, *IEEE Photon. Technol. Lett.* **8**, 563 (1996).

[63] S. Chandrasekhar, L. M. Lunardi, A. H. Gnauck, R. A. Hamm, and G. J. Qua, *IEEE Photon. Technol. Lett.* **5**, 1316 (1993).

[64] E. Sano, M. Yoneyama, H. Nakajima, and Y. Matsuoka, *J. Lightwave Technol.* **12**, 638 (1994).

[65] J. Cowles, A. L. Gutierrez, P. Bhattacharya, and G. I. Hadad, *IEEE Photon. Technol. Lett.* **6**, 963 (1994).

[66] H. Kamitsuna, *J. Lightwave Technol.* **13**, 2301 (1995).

[67] A. L. Gutierrez-Aitken, K. Yang, X. Zhang, G. I. Haddad, P. Bhattacharya, and L. M. Lunardi, *IEEE Photon. Technol. Lett.* **7**, 1339 (1995).

[68] L. M. Lunardi, S. Chandrasekhar, C. A. Burrus, and R. A. Hamm, *IEEE Photon. Technol. Lett.* **7**, 1201 (1995).

[69] M. Yoneyama, E. Sano, S. Yamahata, and Y. Matsuoka, *IEEE Photon. Technol. Lett.* **8**, 272 (1996).

[70] D. Caffin, M. Bouché, M. Meghelli, A. M. Duchenois, and P. Launay, *Electron. Lett.* **33**, 149 (1997).

[71] E. Sano, K. Kurishima, and S. Yamahata, *Electron. Lett.* **33**, 159 (1997).

[72] W. P. Hong, G. K. Chang, R. Bhat, C. K. Nguyen, and M. Koza, *IEEE Photon. Technol. Lett.* **3**, 156 (1991).

[73] M. Horstmann, K. Schimpf, M. Marso, A. Fox, and P. Kordos, *Electron. Lett.* **32**, 763 (1996).

[74] P. Fay, W. Wohlmuth, C. Caneau, and I. Adesida, *Electron. Lett.* **31**, 755 (1995).

[75] P. Fay, W. Wohlmuth, C. Caneau, and I. Adesida, *IEEE Photon. Technol. Lett.* **8**, 679 (1996).

[76] Y. Oikawa, H. Kuwatsuka, T. Yamamoto, T. Ihara, H. Hamano, and T. Minami, *J. Lightwave Technol.* **12**, 343 (1994).

[77] T. Ohyama, S. Mino, Y. Akahori, M. Yanagisawa, T. Hashimoto, Y. Yamada, Y. Muramoto, and T. Tsunetsugu, *Electron. Lett.* **32**, 845 (1996).

[78] Y. Kobayashi, Y. Akatsu, K. Nakagawa, H. Kikuchi, and Y. Imai, *IEEE Trans. Microwave Theory Tech.* **43**, 1916 (1995).

[79] K. Birch, D. Chown, P. Gibson, T. Goh, I. Gregory, R. Tremaine, and R. Wade, *Proc. Opt. Fiber Commun. Conf.*, Optical Society of America, Washington, DC, 1996.

[80] W. R. Bennett, *Electrical Noise*, McGraw-Hill, New York, 1960.

[81] D. K. C. MacDonald, *Noise and Fluctuations: An Introduction*, Wiley, New York, 1962.

[82] F. N. H. Robinson, *Noise and Fluctuations in Electronic Devices and Circuits*, Oxford University Press, Oxford, 1974.

[83] W. Schottky, *Ann. Phys.* **57**, 541 (1918).

[84] J. B. Johnson, *Phys. Rev.* **32**, 97 (1928).

[85] H. Nyquist, *Phys. Rev.* **32**, 110 (1928).

[86] R. J. McIntyre, *IEEE Trans. Electron. Dev.* **13**, 164 (1966).

[87] P. P. Webb, R. J. McIntyre, and J. Conradi, *RCA Rev.* **35**, 235 (1974).

[88] R. J. McIntyre, *IEEE Trans. Electron. Dev.* **19**, 703 (1972).

[89] P. Balaban, *Bell Syst. Tech. J.* **55**, 745 (1976).

[90] S. D. Personik, P. Balaban, J. H. Bobsin, and P. R. Kumar, *IEEE Trans. Commun.* **25**, 541 (1977).

[91] M. Abramowitz and I. A. Stegun, Eds., *Handbook of Mathematical Functions*, Dover, New York, 1970.

[92] B. E. A. Saleh and M. Teich, *Fundamentals of Photonics*, Wiley, New York, 1991, Chap. 11.

[93] B. E. A. Saleh, *Photoelectron Statistics*, Springer-Verlag, Berlin, 1978.

[94] G. P. Agrawal and T. M. Shen, *Electron. Lett.* **22**, 450 (1986).

[95] J. J. O'Reilly, J. R. F. DaRocha, and K. Schumacher, *IEE Proc.* **132**, Pt. J, 309 (1985).

[96] K. Schumacher and J. J. O'Reilly, *Electron. Lett.* **23**, 718 (1987).

[97] T. M. Shen, *Electron. Lett.* **22**, 1043 (1986).

[98] T. P. Lee, C. A. Burrus, A. G. Dentai, and K. Ogawa, *Electron. Lett.* **16**, 155 (1980).

[99] D. R. Smith, R. C. Hooper, P. P. Smyth, and D. Wake, *Electron. Lett.* **18**, 453 (1982).

[100] J. Yamada, A. Kawana, T. Miya, H. Nagai, and T. Kimura, *IEEE J. Quantum Electron.* **18**, 1537 (1982).

[101] M. C. Brain, P. P. Smyth, D. R. Smith, B. R. White, and P. J. Chidgey, *Electron. Lett.* **20**, 894 (1984).

[102] M. L. Snodgrass and R. Klinman, *J. Lightwave Technol.* **2**, 968 (1984).

[103] S. D. Walker and L. C. Blank, *Electron. Lett.* **20**, 808 (1984).

[104] C. Y. Chen, B. L. Kasper, H. M. Cox, and J. K. Plourde, *Appl. Phys. Lett.* **46**, 379 (1985).

[105] B. L. Kasper, J. C. Campbell, A. H. Gnauck, A. G. Dentai, and J. R. Talman, *Electron. Lett.* **21**, 982 (1985).

[106] B. L. Kasper, J. C. Campbell, J. R. Talman, A. H. Gnauck, J. E. Bowers, and W. S. Holden, *J. Lightwave Technol.* **5**, 344 (1987).

[107] R. Heidemann, U. Scholz, and B. Wedding, *Electron. Lett.* **23**, 1030 (1987).

[108] M. Shikada, S. Fujita, N. Henmi, I. Takano, I. Mito, K. Taguchi, and K. Minemura, *J. Lightwave Technol.* **5**, 1488 (1987).

[109] S. Fujita, M. Kitamura, T. Torikai, N. Henmi, H. Yamada, T. Suzaki, I. Takano, and M. Shikada, *Electron. Lett.* **25**, 702 (1989).

[110] K. Kitamura, K. Ito, H. Matsuda, T. Kaneko, and M. Haneda, *Electron. Lett.* **27**, 1435 (1991).

[111] A. H. Gnauck, J. E. Bowers, and J. C. Campbell, *Electron. Lett.* **22**, 600 (1986).

[112] P. K. Runge, *IEEE Trans. Commun.* **24**, 413 (1976).

[113] L. Pophillat and A. Levasseur, *Electron. Lett.* **27**, 535 (1991).

[114] C. Y. Kuo, M. L. Kao, J. S. French, R. E. Tench, and T. W. Cline, *IEEE Photon. Technol. Lett.* **2**, 911 (1990).

Chapter 5

SYSTEM DESIGN AND PERFORMANCE

In the preceding three chapters we have discussed the three main components of a fiber-optic communication system: optical fibers, optical transmitters, and optical receivers. In this chapter we consider the issues related to system design and performance when the three components are put together to form a practical lightwave system. Section 5.1 provides an overview of various system architectures by considering point-to-point links, broadcast and distribution networks, and local-area networks. The design guidelines for fiber-optic communication systems are discussed in Section 5.2 by considering the effects of fiber loss, dispersion, and nonlinearity. This section also covers various terrestrial and undersea lightwave systems that have been developed since 1977 when the first field trial was completed in Chicago. Section 5.3 covers design issues with a few practical examples by considering both the power budget and rise-time budget. Issues related to system performance are treated in Section 5.4 with emphasis on performance degradation occurring as a result of signal transmission through the optical fiber. The physical mechanisms that can lead to power penalty in actual lightwave systems include modal noise, mode-partition noise, source spectral width, frequency chirp, and reflection feedback; each of them is discussed in separate subsections. In Section 5.5 we emphasize the importance of computer-aided design for lightwave systems.

5.1 SYSTEM ARCHITECTURES

From an architectural standpoint, fiber-optic communication systems [1]–[5] can be classified into three broad categories: (1) point-to-point links, (2) broadcast and distribution networks, and (3) local-area networks. In this section we describe the main characteristics of the three system architectures.

5.1.1 Point-to-Point Links

Point-to-point links constitute the simplest kind of lightwave systems. Their role is to transport information, available in the form of a digital bit stream, from one place to another as accurately as possible. The link length can vary from less than a kilometer (short haul) to thousands of kilometers (long haul), depending on the specific application. For example, optical data links are used to connect computers and terminals within the same building or between two buildings with a relatively short transmission distance (< 10 km). The low loss and the wide bandwidth of optical fibers are not of primary importance for such data links; fibers are used mainly because of their other advantages, such as immunity to electromagnetic interference. In contrast, undersea lightwave systems are used for high-speed transmission across continents with a link length of several thousands of kilometers. Both the low loss and a large bandwidth of optical fibers are important factors in the design of transoceanic systems from the standpoint of reducing the overall operating cost.

When the link length exceeds a certain value, in the range 20–100 km depending on the operating wavelength, it becomes necessary to compensate for the fiber loss, as the signal would otherwise become too weak to be detected reliably. Such a compensation is traditionally carried out by using repeaters, often called *regenerators* because they regenerate the optical signal. A regenerator is nothing but a receiver–transmitter pair that detects the incoming optical signal, recovers the electrical bit stream, and then converts it back into optical form by modulating the optical source. The fiber loss can also be compensated by using optical amplifiers, which amplify the optical bit stream directly without requiring conversion of the signal to the electric domain. In fact, *optical amplifiers* have revolutionized [6]–[8] the development of fiber-optic communication systems (see Chapter 8). Amplifiers are especially valuable for multichannel systems since they can amplify all channels simultaneously (see Chapter 7). However, optical amplifiers add noise and worsen the impact of fiber dispersion and nonlinearity since signal degradation keeps on accumulating over multiple amplification stages. Indeed, periodically amplified systems are generally limited by fiber dispersion unless a dispersion-compensation technique (see Chapter 9) is used. Regenerators do not suffer from this problem, as they regenerate the original bit stream periodically, and thus effectively compensate for both fiber loss and dispersion. It is common to refer to optical amplifiers as optical repeaters to distinguish them from optoelectronic repeaters.

Figure 5.1 shows a lightwave system employing repeaters in the form of either regenerators or optical amplifiers. In both cases the *repeater spacing* L is a major design parameter since the system cost reduces as L increases. However, as discussed in Section 2.4, the transmission distance L depends on the bit rate B because of fiber dispersion. The bit rate–distance product, BL, is generally used as a measure of the system performance for point-to-point links. The BL product depends on the operating wavelength, since both fiber loss and fiber dispersion are wavelength dependent. The first three generations of lightwave systems correspond to three different operating wavelengths, near

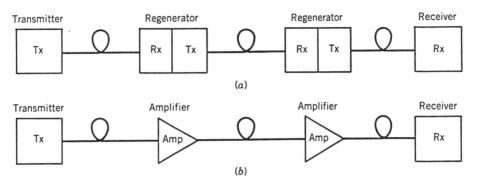

Figure 5.1 Point-to-point fiber links with periodic loss compensation through (a) regenerators and (b) optical amplifiers. A regenerator consists of a receiver (Rx) followed by a transmitter (Tx).

0.85, 1.3, and 1.55 μm [1]. Whereas the BL product is typically $\sim$ 1 (Gb/s)-km for lightwave systems operating near 0.85 μm, it becomes $\sim$ 1 (Tb/s)-km for the third-generation lightwave systems operating near 1.55 μm and can exceed 100 (Tb/s)-km if dispersion-shifted fibers are used.

5.1.2   Broadcast and Distribution Networks

Many applications of optical communication systems require that information not only be transmitted but also distributed to a group of subscribers [2]. Examples include local-loop distribution of telephone services and broadcast of multiple video channels over cable television (CATV) (CATV, short for "common-antenna television"). Considerable effort is directed toward the integration of audio and video services through a broadband *integrated-services digital network* (ISDN). Such a network has the ability to distribute a wide range of services, including telephone, facsimile, computer data, and video broadcasts. Transmission distances are relatively short ($L < 50$ km), but the bit rate can be as high as 10 Gb/s for a super-broadband ISDN.

Figure 5.2 shows two topologies for distribution networks [2]. In the case of *hub topology*, channel distribution takes place at central locations (or hubs) where an automated cross-connect facility switches channels in the electrical domain. The role of fiber is similar to the case of point-to-point links. Since the fiber bandwidth is generally much larger than that required by a single hub office, several offices can share a single fiber headed for the main hub. Telephone networks employ hub topology for distribution of audio channels within a city. A concern for the hub topology is related to its reliability—outage of a single fiber cable can affect the service to a large portion of the network. Additional point-to-point links can be used to guard against such a possibility by connecting important hub locations directly.

In the case of *bus topology*, a single fiber cable carries the multichannel optical signal throughout the area of service. Distribution is done by using optical

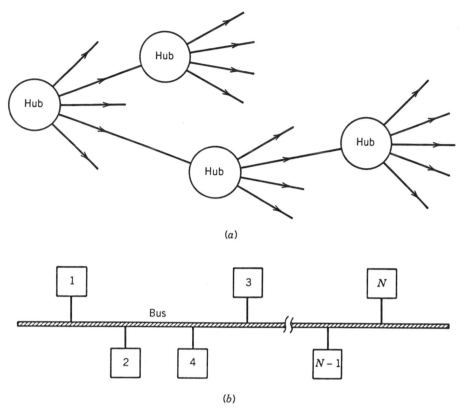

(a)

(b)

Figure 5.2 (a) Hub topology and (b) bus topology for distribution networks.

taps, which divert a small fraction of the optical power to each subscriber. A simple CATV application of bus topology consists of distributing multiple video channels within a city. The use of optical fiber permits distribution of a large number of channels (100 or more) because of its large bandwidth compared with coaxial cables. The advent of *high-definition television* (HDTV) requires lightwave transmission because of a large bandwidth (about 100 Mb/s) of each video channel unless a compression technique (such as MPEG-2) is used.

A problem with the bus topology is that the signal loss increases exponentially with the number of taps and limits the number of subscribers served by a single optical bus. Even when the fiber loss is neglected, the power available at the Nth tap is given by [1]

$$P_N = P_T C[(1-\delta)(1-C)]^{N-1}, \qquad (5.1.1)$$

where P_T is the transmitted power, C is the fraction of power coupled out at each tap, and δ accounts for the insertion loss assumed to be the same at each tap. If we use $\delta = 0.05$, $C = 0.05$, $P_T = 1$ mW, and $P_N = 0.1$ μW as illustrative values, N should not exceed 60. A solution to this problem is offered by optical amplifiers which can boost the optical power of the bus periodically and thus permit distribution to a large number of subscribers as long as the effects of fiber dispersion remain negligible. Multichannel lightwave systems are considered in Chapter 7, where distribution issues are also discussed.

5.1.3 Local-Area Networks

Many applications of fiber-optic communication technology require networks in which a large number of users within a local area (e.g., a university campus) are interconnected in such a way that any user can access the network randomly to transmit data to any other user. Such networks are called *local-area networks* (LANs) and have attracted wide attention [9]–[11]. Since the transmission distances are relatively short (< 10 km), fiber loss is not of much concern for LAN applications. The major motivation behind the use of optical fibers is the large bandwidth offered by fiber-optic communication systems.

The main difference between distribution networks and LANs is related to the random access offered to multiple users of a LAN. The system architecture plays an important role for LANs, since the establishment of predefined protocol rules is a necessity in such an environment. Three commonly used topologies are known as bus, ring, and star configurations [9]–[11]. The bus topology is similar to that shown in Fig. 5.2(b). A well-known example of bus topology is provided by the *Ethernet*, a network protocol used to connect multiple computers and used by the *Internet*. The Ethernet operates at 10 Mb/s (or at 100 Mb/s in the second generation) by using a protocol based on *carrier-sense multiple access* (CSMA) with collision detection. Although the Ethernet LAN architecture has proven to be quite successful when coaxial cables are used for the bus, a number of difficulties arise when optical fibers are used. A major limitation is related to the loss at each tap, which limits the number of users [see Eq. (5.1.1)].

Figure 5.3 shows ring and star topologies for LAN applications. In the ring topology [12], consecutive nodes are connected by point-to-point links to form a closed ring. Each node can transmit and receive the data by using a transmitter–receiver pair, which also acts as a repeater. A token (a predefined bit sequence) is passed around the ring. Each node monitors the bit stream to listen for its own address and to receive the data. It can also transmit by appending the data to an empty token. The use of ring topology for fiber-optic LANs has been commercialized with the advent of a standardized interface known as the *fiber distributed data interface* (FDDI) [12]. The FDDI operates at 100 Mb/s by using multimode fibers and LED-based 1.3-μm transmitters. It is designed to provide backbone services such as the interconnection of lower-speed LANs or mainframe computers.

In the *star topology*, all nodes are connected through point-to-point links to a central node called a hub, or simply a star [2]. Such LANs are further subclassified as *active-star* or *passive-star* networks, depending on whether the central node is an active or passive device. In the active-star configuration, all incoming optical signals are converted to the electrical domain through optical receivers. The electrical signal is then distributed to drive individual node transmitters. Switching operations can also be performed at the central node since distribu-

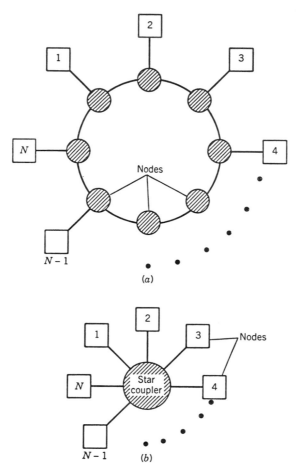

Figure 5.3 (a) Ring topology and (b) star topology for local-area networks.

tion takes place in the electrical domain. In the passive-star configuration [13], distribution takes place in the optical domain through devices such as directional couplers. Since the input from one node is distributed to many output nodes, the power transmitted to each node depends on the number of users. Similar to the case of bus topology, the number of users supported by passive-star LANs is limited by the distribution losses. For an ideal $N \times N$ star coupler, the power reaching each node is simply P_T/N (if we neglect transmission losses) since the transmitted power P_T is divided equally among N users. For a passive star composed of directional couplers (see Section 7.2.4), the power is further reduced because of insertion losses and can be written as [1]

$$P_N = (P_T/N)(1 - \delta)^{\log_2 N}, \qquad (5.1.2)$$

where δ is the insertion loss of each directional coupler. If we use $\delta = 0.05$, $P_T = 1$ mW, and $P_N = 0.1$ μW as illustrative values, N can be as large as 500. This value of N should be compared with $N = 60$ obtained for the case of bus topology by using Eq. (5.1.1). A relatively large value of N makes star topology attractive for LAN applications. LANs are discussed in Chapter 7, which is devoted to multichannel communication systems. The remainder of this chapter focuses on the design and performance of point-to-point links, which constitute a basic element of all communication systems, including LANs and distribution networks.

5.2   DESIGN GUIDELINES

The design of fiber-optic communication systems requires a clear understanding of the limitations imposed by the loss, dispersion, and nonlinearity of the fiber. Since fiber properties are wavelength dependent, the choice of operating wavelength is a major design issue. In this section we discuss how the bit rate and the transmission distance of a fiber link are limited by fiber loss, dispersion, and nonlinearity. The section also outlines the progress realized in the development of terrestrial and undersea lightwave systems since 1977 when the first field trial was completed.

5.2.1   Loss-Limited Lightwave Systems

Except for some short-haul, point-to-point links found in intrabuilding networks, fiber loss plays an important role in the system design. Consider an optical transmitter that is capable of launching an average power $\bar{P}_{tr}$. If the signal is detected by a receiver that requires a minimum average power $\bar{P}_{rec}$ at a bit rate B, the maximum transmission distance is limited by

$$L = \frac{10}{\alpha_f} \log_{10}\left(\frac{\bar{P}_{tr}}{\bar{P}_{rec}}\right), \qquad (5.2.1)$$

where α_f is the net loss (in dB/km) of the fiber cable, including splice and connector losses. The bit-rate dependence of L arises from the linear dependence of $\bar{P}_{rec}$ on the bit rate B, since $\bar{P}_{rec} = \bar{N}_p h\nu B$, where $h\nu$ is the photon energy and $\bar{N}_p$ is the average number of photons/bit required by the receiver [see Eq. (4.5.24)]. Thus L decreases logarithmically as B increases at a given operating wavelength.

The solid lines in Fig. 5.4 show the dependence of L on B for three common operating wavelengths of 0.85, 1.3, and 1.55 μm by using $\alpha_f = 2.5$, 0.4, and 0.25 dB/km, respectively. The transmitted power is taken to be $\bar{P}_{tr} = 1$ mW at the three wavelengths, whereas $\bar{N}_p = 300$ at $\lambda = 0.85$ μm and $\bar{N}_p = 500$ at 1.3 and 1.55 μm. The smallest value of L occurs for lightwave systems operating at 0.85 μm because of a relatively large fiber loss at that wavelength. The repeater spacing of such systems is limited in the range 10–30 km, depending on the bit rate. By contrast, a repeater spacing of more than 100 km is possible for lightwave systems operating near 1.55 μm.

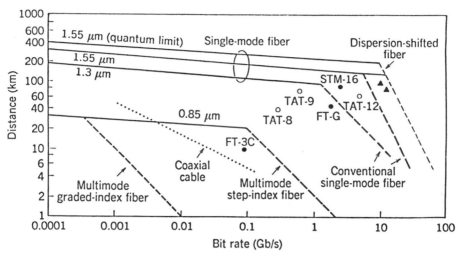

Figure 5.4 Loss limit (solid lines) and dispersion limit (dashed lines) on the transmission distance L as a function of the bit rate B for three commonly used wavelength regions. The dotted line shows the performance of coaxial-cable systems. Filled circles denote terrestrial commercial lightwave systems; empty circles correspond to undersea transmission systems; triangles denote laboratory experiments. (After Ref. [1]. ©1988 Academic Press. Reprinted with permission.)

It is interesting to compare the loss limit of 0.85-μm lightwave systems with that of electrical communication systems based on coaxial cables. The dotted line in Fig. 5.4 shows the bit-rate dependence of L for coaxial cables by assuming that the loss increases as $\sqrt{B}$. The transmission distance is larger for coaxial cables at small bit rates ($B < 5$ Mb/s), but fiber-optic systems take over at bit rates in excess of 5 Mb/s. Since a longer transmission distance translates into a smaller number of repeaters in a long-haul point-to-point link, fiber-optic communication systems offer an economic advantage when the operating bit rate exceeds 10 Mb/s.

5.2.2 Dispersion-Limited Lightwave Systems

In Section 2.4 we discussed how fiber dispersion limits the bit rate–distance product BL because of pulse broadening. When the dispersion-limited transmission distance is shorter than the loss-limited distance of Eq. (5.2.1), the system is said to be dispersion-limited. The dashed lines in Fig. 5.4 show the dispersion-limited transmission distance as a function of the bit rate. Since the physical mechanisms leading to dispersion limitation can be different for different operating wavelengths, let us examine each case separately.

Consider first the case of 0.85-μm lightwave systems, which often use multimode fibers to minimize the system cost. As discussed in Section 2.1, the most limiting factor for multimode fibers is intermodal dispersion. In the case of step-index multimode fibers, Eq. (2.1.6) provides an approximate upper bound on the

BL product. A slightly more restrictive condition $BL = c/(2n_1\Delta)$ is plotted in Fig. 5.4 by using typical values $n_1 = 1.46$ and $\Delta = 0.01$. Even at a low bit rate of 1 Mb/s, such multimode systems are dispersion-limited, and their transmission distance is limited to below 10 km. For this reason, multimode step-index fibers are rarely used in the design of fiber-optic communication systems. Considerable improvement can be realized by using graded-index fibers for which intermodal dispersion limits the BL product to values given by Eq. (2.1.11). The condition $BL = 2c/(n_1\Delta^2)$ is plotted in Fig. 5.4 and shows that 0.85-μm lightwave systems are loss-limited, rather than dispersion-limited, for bit rates up to 100 Mb/s when graded-index fibers are used. The first generation of terrestrial telecommunication systems took advantage of such an improvement and used graded-index fibers. The first commercial system became available in 1980, operating at a bit rate of 45 Mb/s with a repeater spacing below 10 km.

The second generation of lightwave systems used primarily single-mode fibers near the minimum-dispersion wavelength occurring at about 1.31 μm. The most limiting factor for such systems is dispersion-induced pulse broadening dominated by a relatively large source spectral width. As discussed in Section 2.4.3, the BL product is then limited by [see Eq. (2.4.26)]

$$BL \leq (4|D|\sigma_\lambda)^{-1}, \tag{5.2.2}$$

where σ_λ is the root-mean-square (RMS) width of the source spectrum. The actual value of $|D|$ depends on how close the operating wavelength is to the zero-dispersion wavelength of the fiber and is typically ~ 1 ps/(km-nm). Figure 5.4 shows the dispersion limit for 1.3-μm lightwave systems by choosing $|D|\sigma_\lambda = 2$ ps/km so that $BL \leq 125$ (Gb/s)-km. As seen there, such systems are generally loss-limited for bit rates up to 1 Gb/s but become dispersion-limited at higher bit rates.

Third-generation lightwave systems operate near 1.55 μm to take advantage of the smallest fiber loss occurring in this wavelength region. However, fiber dispersion becomes a major problem for such systems since $D \approx 16$ ps/(km-nm) near 1.55 μm for conventional silica fibers. Semiconductor lasers operating in a single longitudinal mode provide a solution to this problem. The ultimate limit is then given by [see Eq. (2.4.31)]

$$B^2L < (16|\beta_2|)^{-1}, \tag{5.2.3}$$

where β_2 is related to D as in Eq. (2.3.5). Figure 5.4 shows this limit by choosing $B^2L = 4000$ (Gb/s)^2-km. As seen there, ideal 1.55-μm systems become dispersion-limited only for $B > 5$ Gb/s. In practice, the frequency chirp imposed on the optical pulse during direct modulation provides a much more severe limitation. The effect of frequency chirp on system performance is discussed in Section 5.4.4. Qualitatively speaking, the frequency chirp manifests through a broadening of the pulse spectrum. If we use Eq. (5.2.2) with $D = 16$ ps/(km-nm) and $\sigma_\lambda = 0.1$ nm, the BL product is limited to 150 (Gb/s)-km. Thus, the frequency chirp limits the transmission distance to 75 km at $B = 2$ Gb/s, even though loss-limited distance exceeds 150 km.

A solution to the chirp problem is offered by *dispersion-shifted fibers* for which dispersion and loss both are minimum near 1.55 μm. Figure 5.4 shows the improvement by using Eq. (5.2.3) with $|\beta_2| = 2$ ps^2/km. Such systems can be operated at 20 Gb/s with a repeater spacing of about 80 km. Further improvement is possible only by operating the lightwave system very close to the zero-dispersion wavelength, a task that requires careful matching of the laser wavelength to the zero-dispersion wavelength and is not always feasible because of variations in the dispersive properties of the fiber along the transmission link. In practice, the frequency chirp makes it difficult to achieve even the limit indicated in Fig. 5.4. By 1989, two laboratory experiments had demonstrated transmission over 81 km at 11 Gb/s [14] and over 100 km at 10 Gb/s [15] by using low-chirp semiconductor lasers together with dispersion-shifted fibers. The triangles in Fig. 5.4 show that such systems are operating quite close to the fundamental limits set by fiber dispersion.

5.2.3   Long-Haul Systems with In-Line Amplifiers

With the advent of optical amplifiers, fiber loss can be compensated by inserting in-line amplifiers periodically in a long-haul fiber link (see Fig. 5.1). At the same time, the effects of group-velocity dispersion (GVD) can be reduced either by operating close to the zero-dispersion wavelength of the fiber or by using a dispersion-compensation scheme (see Chapter 9). Since neither the fiber loss nor the fiber dispersion is then a limiting factor, one may ask how many in-line amplifiers can be cascaded in series, and ultimately, what limits the total link length. This topic is covered in Section 8.6 in the context of erbium-doped fiber amplifiers. In this subsection we focus on the factors that limit the performance of amplified fiber links and provide a few design guidelines.

The most important consideration in the design of fiber links with in-line amplifiers is related to the *nonlinear effects* in optical fibers [16], discussed in Section 2.6. For single-channel lightwave systems, the most dominant nonlinear phenomenon that limits the system performance is *self-phase modulation* (SPM). When optoelectronic regenerators are used, the SPM effects accumulate only over one repeater spacing (typically, < 100 km) and are of little concern if the launch power satisfies Eq. (2.6.8) or the condition $P_{in} \ll 45$ mW. By contrast, the SPM effects accumulate over long lengths ($\sim$ 1000 km) when in-line amplifiers are used periodically for loss compensation. A rough estimate of the limitation imposed by the SPM can be obtained from Eq. (2.6.7) by replacing L_{eff} with the total link length L_T and P_{in} by the average power $\bar{P}$ over one amplification stage. The condition $\phi_{NL} \ll 1$ then limits the total link length to $L_T \ll L_{NL}$, where the nonlinear length is defined as $L_{NL} = (\bar{\gamma}\bar{P})^{-1}$ and $\bar{\gamma}$ is the nonlinear parameter introduced in Eq. (2.6.6). Typically, $\bar{\gamma} \approx 1$ W^{-1}/km, and the link length is limited to below 1000 km even for $\bar{P} = 1$ mW.

The forgoing estimate of the SPM-induced limitation is too simplistic to be accurate since it completely ignores the role of fiber dispersion. In fact, since the dispersive and nonlinear effects act on the optical signal simultaneously, their mutual interplay becomes quite important [16]. The effect of SPM on pulses propagating inside an optical fiber can be included by adding a nonlinear term to the right side of Eq. (2.4.9), which then becomes

$$\frac{\partial A}{\partial z} + \frac{i}{2}\beta_2 \frac{\partial^2 A}{\partial t^2} - \frac{1}{6}\beta_3 \frac{\partial^3 A}{\partial t^3} = i\bar{\gamma}|A|^2 A - \frac{\alpha}{2}A, \qquad (5.2.4)$$

where the fiber loss is α included through the last term. This equation also forms the basis of optical solitons in Chapter 10.

Because of the nonlinear nature of Eq. (5.2.4), it should be solved numerically in general. A numerical approach has indeed been adopted during the 1990s to quantify the impact of SPM on the performance of long-haul lightwave systems making use of in-line amplifiers [17]–[24]. The main conclusion is that the launch power must be optimized to a value that depends on many design parameters, such as the bit rate, total link length, and amplifier spacing. In one study the optimum launch power was found to be about 1 mW for a 5-Gb/s signal transmitted over 9000 km with 40-km amplifier spacing [23].

The combined effects of GVD and SPM also depend on the sign of the dispersion parameter β_2. In the case of anomalous dispersion ($\beta_2 < 0$), the nonlinear phenomenon of *modulation instability* [16] can affect the system performance drastically [24]. This problem can be overcome by using a combination of fibers with normal and anomalous GVD such that the average dispersion over the entire fiber link is "normal." However, a new kind of modulation instability, referred to as *sideband instability* [25], can occur in both the normal and anomalous GVD regions. It has its origin in the periodic variation of the signal power along the fiber link when equally spaced optical amplifiers are used to compensate for the fiber loss. Since the quantity $\bar{\gamma}|A|^2$ in Eq. (5.2.4) is then a periodic function of z, the resulting nonlinear-index grating can initiate a four-wave-mixing process that generates sidebands in the signal spectrum. It can be avoided by making the amplifier spacing nonuniform.

Another factor that plays a crucial role is the noise added by optical amplifiers. Similar to the case of electronic amplifiers (see Section 4.4), the noise of optical amplifiers is quantified through an amplifier noise figure F_n. The noise issue is discussed in Chapter 8. The nonlinear interaction between the amplified spontaneous emission and the signal can lead to a large spectral broadening through the nonlinear phenomena such as cross-phase modulation and four-wave mixing [16]. Since the noise has a much larger bandwidth than the signal, its impact can be reduced by using optical filters. Numerical simulations indeed show a considerable improvement when in-line filters are used [23].

Finally, the polarization effects that are totally negligible in the traditional "nonamplified" lightwave systems become of concern for long-haul systems with in-line amplifiers. The polarization-mode dispersion (PMD) issue was discussed in Section 2.3.5. In addition to PMD, optical amplifiers can also induce polarization-dependent gain and loss [22]. The impact of these effects on lightwave systems is considered in Section 8.6. Although polarization effects must be considered, their impact can be reduced to an acceptable level through proper design. This is also borne out by many system experiments in which in-line optical amplifiers have been used to transmit data over several thousands of kilometers.

The fourth generation of lightwave systems had its beginning in 1995 when amplifier-based lightwave systems became available commercially. Of course, the laboratory demonstrations began as early as 1989 with the advent of the erbium-doped fiber amplifier. Initial long-haul experiments used a recirculating fiber loop to demonstrate system feasibility since it is not practical to use long lengths of fiber in a laboratory setting. Already in 1991, an experiment showed the possibility of data transmission over 21,000 km at 2.5 Gb/s, and over 14,300 km at 5 Gb/s, by using the recirculating-loop configuration [26]. By 1992, a 2.5-Gb/s signal was transmitted over 10,000 km of actual fiber link in a system experiment in which 199 optical amplifiers were inserted with a spacing of about 50 km [27]. The receiver sensitivity degraded by only 5.1 dB despite multiple sources of power penalties acting over such long fiber lengths. In a system trial carried out in 1995 by using actual submarine cables and repeaters [28], a 5.3-Gb/s signal was transmitted over 11,300 km with 60 km of amplifier spacing. This system trial led to the deployment of a commercial transpacific cable (TPC–5) that began operating in 1996. There is considerable interest in extending the bit rate to 10 Gb/s and the amplifier spacing toward 100 km. In one system experiment, the 10-Gb/s signal could be transmitted over 6480 km with 90-km amplifier spacing [29]. With a further increase in the distance, the SNR decreased below the value needed to maintain a BER of 10^{-9} or less. One may think that the performance should improve by operating close to the zero-dispersion wavelength of the fiber. However, an experiment, performed under such conditions, achieved only a distance of 6000 km at 10 Gb/s even with only 40-km amplifier spacing [30], and the situation was worse when the RZ modulation format was used. The combined effects of the higher-order dispersion [the β_3 term in Eq. (5.2.4)] and SPM appear to degrade system performance considerably at a bit rate of 10 Gb/s.

5.2.4 Telecommunication Fiber Links

An important application of point-to-point fiber links is for the worldwide telephone network. Indeed, it is this application that started the field of optical fiber communications in 1977 and has propelled it since then by demanding lightwave systems with higher and higher capacities. This section focuses on the status of commercial telecommunication systems by considering terrestrial and undersea systems separately.

Terrestrial Lightwave Systems

After a successful Chicago field trial in 1977, terrestrial lightwave systems became available commercially beginning in 1980 [31]–[33]. Table 5.1 lists the operating characteristics of several terrestrial systems developed since then. The first generation operated near 0.82 μm and used multimode graded-index fiber as the transmission medium. As seen in Fig. 5.4, the BL product of such

Table 5.1 Terrestrial U.S. lightwave systems

System	Year	λ (μm)	B (Mb/s)	L (km)	Voice Channels
FT–3	1980	0.825	45	< 10	672
FT–3C	1983	0.825	90	< 15	1,344
FT–3X	1984	1.30	180	< 25	2,688
FT–G	1985	1.30	417	< 40	6,048
FT–G-1.7	1987	1.30	1,668	< 46	24,192
STM–16	1991	1.55	2,488	< 85	32,256
STM–64	1996	1.55	9,953	< 90	129,024

systems is limited to 2 (Gb/s)-km. A commercial lightwave system (FT–3C), operating at 90 Mb/s with a repeater spacing of about 12 km realized a BL product of nearly 1 (Gb/s)-km; it is shown by a filled circle in Fig. 5.4. The operating wavelength moved to 1.3 μm in second-generation lightwave systems to take advantage of low fiber loss and low dispersion near this wavelength. Many commercial systems operate near this wavelength. The BL product of 1.3-μm lightwave systems is limited to about 100 (Gb/s)-km when a multimode semiconductor laser is used inside the transmitter. In 1987, a commercial 1.3-μm lightwave system (FT–G) provided data transmission at 1.7 Gb/s with a repeater spacing of about 45 km. A filled circle in Fig. 5.4 shows that this system operates quite close to the dispersion limit.

The third-generation lightwave systems became available commercially in 1991. They operate near 1.55 μm at bit rates in excess of 2 Gb/s, typically at 2.488 Gb/s, corresponding to the OC-48 level of the SONET (or the STS–16 level of the SDH) specifications. The switch to the 1.55-μm wavelength helps to increase the loss-limited transmission distance to more than 150 km because of a fiber loss of only 0.2 dB/km in this wavelength region. However, the repeater spacing is often limited to below 100 km because of the high dispersion of standard telecommunication fibers. In fact, the deployment of third-generation lightwave systems was possible only after the development of DFB semiconductor lasers, which reduce the impact of fiber dispersion by reducing the source spectral width under CW operation to below 100 MHz (see Section 2.4).

Further increase in the bit rate of 1.55-μm lightwave systems requires the use of dispersion-shifted fibers so that both the fiber loss and the GVD are minimum at the operating wavelength. However, more than 50 million kilometers of the standard telecommunication fiber is already installed in the worldwide telephone network. Economic reasons dictate that the fourth generation of lightwave systems make use of this existing base. Two approaches are being used to solve the dispersion problem. First, several dispersion-compensation schemes, discussed in Chapter 9, make it possible to extend the bit rate to 10 Gb/s (STS–64 level) while maintaining a repeater spacing of up to 100 km. Second, several 2.5 Gb/s can be transmitted simultaneously by using the technique of

Table 5.2 Commercial undersea lightwave systems

System	Year	B (Gb/s)	L (km)	Operating Wavelength and Technology
TAT–8	1988	0.28	70	1.3 μm, multimode lasers
TPC–3	1989	0.28	70	1.3 μm, multimode lasers
TAT–9	1991	0.56	80	1.55 μm, DFB lasers
TPC–4	1992	0.56	80	1.55 μm, DFB lasers
TAT–10/11	1993	0.56	80	1.55 μm, DFB lasers
TPC–5	1996	5.30	50	1.55 μm, optical amplifiers
TAT–12/13	1996	5.30	50	1.55 μm, optical amplifiers

wavelength-division multiplexing (WDM), discussed in Chapter 7 in the context of multichannel lightwave systems. The repeater spacing can then be quite large since it is limited by the bit rate of individual channels instead of the total bit rate. Moreover, if the WDM technique is combined with a dispersion-compensation scheme, the limiting transmission distance can be several hundred kilometers provided that the fiber loss is compensated periodically by using the optical amplifiers. Such WDM lightwave systems were deployed commercially by 1997. One WDM system offers a bit rate of 40 Gb/s by multiplexing 16 channels, each operating at 2.5 Gb/s.

Undersea Lightwave Systems

Undersea transmission systems [34]–[38] are used mostly for intercontinental communications (see Fig. 1.3). Reliability is of major concern for such systems, as repairs are expensive. Generally, undersea systems are designed for a 25-year service life, with at most three failures during operation. Table 5.2 lists the main characteristics of transatlantic (TAT) and transpacific (TPC) fiber-optic lightwave systems. The first undersea fiber-optic cable was a second-generation system. It was installed in 1988 in the Atlantic Ocean (TAT–8) for operation at a bit rate of 280 Mb/s with a repeater spacing of up to 70 km. As seen in Fig. 5.4, the system design is on the conservative side, mainly to ensure reliability. The same technology was used for the first transpacific lightwave system (TPC–3), which became operational in 1989.

By 1990 the third-generation lightwave systems had been developed. The TAT–9 and TPC–4 undersea systems used this technology and were designed to operate near 1.55 μm at a bit rate of 560 Mb/s with a repeater spacing of about 80 km. The increasing traffic across the Atlantic Ocean led to the deployment of the TAT–10 and TAT–11 lightwave systems by 1993 with the same technology. The advent of optical amplifiers prompted their use in the next generation of undersea systems, the first two being TPC–5 and TAT–12, both of which became operational by 1996. They belong to the fourth generation of lightwave systems that makes use of optical amplifiers in place of optoelectronic regenerators. They operate at a bit rate of 5.3 Gb/s with an amplifier spacing of about 50 km and with a link length of up to 10,000 km. The bit rate is slightly larger than the STM-32-level bit rate of 5 Gb/s because of the overhead associated with the implementation of a *forward-error-correction* scheme. As discussed earlier, the design of such lightwave systems is much more complex than that of previous undersea systems because of the cumulative effects of fiber dispersion and nonlinearity, which must be controlled over long distances. Moreover, the noise of optical amplifiers also becomes of concern. The transmitter power and the dispersion profile along the link must be optimized to combat such effects. Even then, amplifier spacing is typically limited to 50 km, and an error-correction scheme is needed to ensure a bit-error rate $< 2 \times 10^{-11}$.

Several other undersea systems with link lengths of more than 20,000 km were in the planning or deployment stages in 1996. The 27,300-km *fiber-optic link around the globe* (FLAG, for short) will connect many Asian and European countries at 5.3 Gb/s, with several sections operating at 10.6 Gb/s by using two channels [37]. Another fiber-optic network, known as *Africa One*, will circle the African continent and cover a total transmission distance of about 35,000 km [38]. It is often difficult to use optical amplifiers exclusively over such long distances. The solution is to mix optical amplifiers occasionally with optoelectronic repeaters, which cancel the cumulative degradation occurring over multiple amplification stages.

A second category of undersea lightwave systems requires repeaterless transmission over several hundred kilometers [36]. Such systems are used for inter-island communication or for looping a shoreline such that the signal is regenerated on the shore periodically after a few hundred kilometers of undersea transmission. The dispersive and nonlinear effects are of less concern for such systems than for transoceanic lightwave systems, but the fiber loss becomes a major issue. The reason is easily appreciated by noting that the cable loss exceeds 100 dB over a distance of 500 km even under the best operating conditions. Several laboratory experiments have demonstrated repeaterless transmission at 2.5 Gb/s over more than 500 km by using two in-line amplifiers, pumped remotely from the transmitter and receiver ends by using high-power pump lasers. Another amplifier at the transmitter boosts the launched power to close to 100 mW. Such high input powers exceed the threshold level for stimulated Brillouin scattering (SBS), a nonlinear phenomenon discussed in Section 2.6. The suppression of SBS is realized through phase modulation of the optical carrier that broadens the carrier linewidth to 200 MHz or more [39]). The effect of GVD is reduced by using dispersion-compensating fibers (see Section 9.5).

Directly modulated DFB lasers can also be used for repeaterless transmission. In a 1996 experiment. a 2.5-Gb/s signal was transmitted over 465 km by direct modulation of a DFB laser [40]. Chirping of the modulated signal broadened the spectrum enough that an external phase modulator was not required provided that the launched power was kept below 100 mW. The bit rate of repeaterless undersea systems can be increased to 10 Gb/s by employing the same techniques used at 2.5 Gb/s. In a 1996 experiment [41], the 10-Gb/s signal was transmitted over 442 km by using two remotely pumped in-line amplifiers. Two external modulators were used, one for SBS suppression and another for signal generation. These results indicate that undersea lightwave systems looping a shoreline can operate at 10 Gb/s with only shore-based electronics.

The fourth generation of lightwave systems makes use of optical amplifiers together with the WDM technology (covered in Chapter 7). The undersea transmission systems will also undoubtly make use of the WDM. In fact, the next transpacific link (TPC–6) will operate at a bit rate of 100 Gb/s and is scheduled to go in service by the year 2000. Such a point-to-point link can transmit 1.2 million voice channels simultaneously, a capacity that should be contrasted with the TAT–8 capacity of 8000 voice channels in 1988, which in turn should be compared to the 48-channel capacity of TAT–1 in 1959. Clearly, the use of fiber-optic technology has revolutionized the performance of telecommunication systems.

5.3 SYSTEM DESIGN

In the preceding section we discussed the limitations imposed on the transmission distance and the bit rate of a fiber-optic link because of loss and dispersion associated with the communication channel. The curves shown in Fig. 5.4 provide only a guide to the system design. Many other issues need to be addressed in the design of a realistic fiber-optic communication system. Among them are the operating wavelength, selection of appropriate transmitters, receivers, and fibers, compatibility of various components, cost versus performance, and system reliability and upgradability. In this section we discuss the design process by considering the power and rise-time budgets and illustrate it through specific examples [42]–[44].

The system requirements typically specified in advance are the bit rate B and the transmission distance L. The performance criterion is specified through the bit-error rate (BER), a typical requirement being BER $< 10^{-9}$. The first decision of the system designer concerns the choice of the operating wavelength. As a practical matter, the cost of components is lowest near 0.85 μm and increases as wavelength shifts toward 1.3–1.6 μm. Figure 5.4 can be quite helpful in determining the appropriate operating wavelength. Generally speaking, a fiber-optic link can operate near 0.85 μm if $B \le 100$ Mb/s and $L < 20$ km. This is the case for many LAN applications. On the other hand, the operating wavelength is by necessity in the region 1.3–1.6 μm for long-haul lightwave systems operating at bit rates in excess of 200 Mb/s.

5.3.1 Power Budget

The purpose of the *power budget* is to ensure that enough power will reach the receiver to maintain reliable performance during the entire system lifetime. The minimum average power required by the receiver is the receiver sensitivity $\bar{P}_{\rm rec}$ (see Section 4.4). The average launch power $\bar{P}_{\rm tr}$ is generally specified for each transmitter. The power budget takes an especially simple form in decibel units

Table 5.3 Power budget of a 0.85-μm lightwave system

Quantity	Symbol	Laser	LED
Transmitter power	P_{tr}	0 dBm	−13 dBm
Receiver sensitivity	$\bar{P}_{\text{rec}}$	−42 dBm	−42 dBm
System margin	M_s	6 dB	6 dB
Available channel loss	C_L	36 dB	23 dB
Connector loss	α_{con}	2 dB	2 dB
Fiber cable loss	α_f	3.5 dB/km	3.5 dB/km
Maximum fiber length	L	9.7 km	6 km

with optical powers expressed in dBm (see Appendix B). Specifically,

$$\bar{P}_{\text{tr}} = \bar{P}_{\text{rec}} + C_L + M_s, \tag{5.3.1}$$

where C_L is the total channel loss and M_s is the *system margin*. The purpose of
system margin is to allocate a certain amount of power to additional sources of
power penalty that may develop during system lifetime because of component
degradation or other unforeseen events. A system margin of 6–8 dB is generally
allocated during the design process.

The channel loss C_L should take into account all possible sources of power
loss, including connector and splice losses. If α_f is the fiber loss in dB/km, C_L
can be written as

$$C_L = \alpha_f L + \alpha_{\text{con}} + \alpha_{\text{splice}}, \tag{5.3.2}$$

where α_{con} and α_{splice} account for the connector and splice losses throughout
the fiber link. Sometimes splice loss is included within the specified loss of the
fiber cable. The connector loss α_{con} includes connectors at the transmitter and
receiver ends but must include other connectors if used within the fiber link.

Equations (5.3.1) and (5.3.2) can be used to estimate the maximum transmission distance for a given choice of the components. As an illustration, consider
the design of a fiber link operating at 50 Mb/s and requiring a maximum transmission distance of 8 km. As seen in Fig. 5.4, such a system can be designed
to operate near 0.85 μm provided that graded-index multimode fiber is used
for the fiber cable. The operation near 0.85 μm is desirable from the economic
standpoint. Once the operating wavelength is selected, a decision must be made
about the appropriate transmitters and receivers. The GaAs transmitter can
use a semiconductor laser or a LED as an optical source. Similarly, the receiver
can be designed to use either a p–i–n or an avalanche photodiode. Keeping
the low cost in mind, let us choose a p–i–n receiver. Such a receiver requires
about 5000 photons/bit on average to operate reliably with a BER below 10^{-9}.
By using the relation $\bar{P}_{\text{rec}} = \bar{N}_p h\nu B$ with $\bar{N}_p = 5000$ and $B = 50$ Mb/s, the
receiver sensitivity is given by $\bar{P}_{\text{rec}} = -42$ dBm. The average launch power for
LED and laser-based transmitters is typically 50 μW and 1 mW, respectively.

Table 5.3 shows the power budget for the two transmitters by assuming
that the splice loss is included within the cable loss. The transmission distance

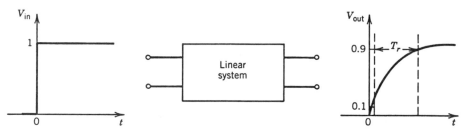

Figure 5.5 Rise time T_r associated with a bandwidth-limited linear system.

L is limited to 6 km for the case of LED-based transmitters. If the system specification is 8 km, a more expensive laser-based transmitter must be used. The alternative is to use an APD receiver. If the receiver sensitivity improves by more than 7 dB when an APD is used in place of a p–i–n photodiode, the transmission distance can be increases to 8 km even for an LED-based transmitter. Economic considerations would then dictate the choice between the laser-based transmitters and APD receivers.

5.3.2   Rise-Time Budget

The purpose of the *rise-time budget* is to ensure that the system is able to operate properly at the intended bit rate. Even if the bandwidth of the individual system components exceeds the bit rate, it is still possible that the total system may not be able to operate at that bit rate. The concept of rise time is used to allocate the bandwidth among various components. The rise time T_r of a linear system is defined as the time during which the response increases from 10% to 90% of its final output value when the input is changed abruptly (a step function). Figure 5.5 illustrates the concept graphically.

There is an inverse relationship between the bandwidth Δf and the rise time T_r of a linear system. This relationship can be understood by considering a simple RC circuit as an example of the linear system. When the input voltage across an RC circuit changes instantaneously from 0 to V_0, the output voltage changes as

$$V_{\text{out}}(t) = V_0[1 - \exp(-t/RC)], \tag{5.3.3}$$

where R is the resistance and C is the capacitance of the RC circuit. The rise time is found to be given by

$$T_r = (\ln 9)RC \approx 2.2RC. \tag{5.3.4}$$

The transfer function $H(f)$ of the RC circuit is obtained by taking the Fourier transform of Eq. (5.3.3) and is

$$H(f) = (1 + i2\pi fRC)^{-1}. \tag{5.3.5}$$

The electrical bandwidth Δf corresponds to the frequency at which $|H(f)|^2 = 1/2$ and is given by the well-known expression $\Delta f = (2\pi RC)^{-1}$. By using Eq.

(5.3.4), Δf and T_r are related as

$$T_r = \frac{2.2}{2\pi\Delta f} = \frac{0.35}{\Delta f}. \tag{5.3.6}$$

One expects an inverse relationship between the rise time and the bandwidth to hold for any linear system. However, the product $T_r\Delta f$ would generally be different than 0.35. It is common to use $T_r\Delta f = 0.35$ in the design of optical communication systems as a conservative guideline. The relationship between the bandwidth Δf and the bit rate B depends on the digital format. In the case of RZ format (see Section 1.2), $\Delta f = B$ and $BT_r = 0.35$. By contrast, $\Delta f \approx B/2$ for the NRZ format and $BT_r = 0.7$. In both cases, the specified bit rate imposes an upper limit on the maximum rise time that can be tolerated: The communication system must be designed to ensure that T_r is below this maximum value, i.e.,

$$T_r \leq \begin{cases} 0.35/B & \text{for RZ format,} \\ 0.70/B & \text{for NRZ format.} \end{cases} \tag{5.3.7}$$

The three components of fiber-optic communication systems have individual rise times. The total rise time of the whole system is related to the individual component rise times approximately as [45]

$$T_r^2 = T_{\text{tr}}^2 + T_{\text{fiber}}^2 + T_{\text{rec}}^2, \tag{5.3.8}$$

where T_{tr}, T_{fiber}, and T_{rec} are the rise times associated with the transmitter, fiber, and receiver, respectively. The rise times of the transmitter and the receiver are generally known to the system designer. The transmitter rise time T_{tr} is determined primarily by the electronic components of the driving circuit and the electrical parasitics associated with the optical source. Typically, T_{tr} is a few nanoseconds for LED-based transmitters but can be as short as 0.1 ns for laser-based transmitters. The receiver rise time T_{rec} is determined primarily by the 3-dB electrical bandwidth of the receiver front end. Equation (5.3.6) can be used to estimate T_{rec} if the front-end bandwidth is specified.

The fiber rise time T_{fiber} should in general include the contributions of both modal and group-velocity dispersions through the relation

$$T_{\text{fiber}}^2 = T_{\text{modal}}^2 + T_{\text{GVD}}^2. \tag{5.3.9}$$

For single-mode fibers, $T_{\text{modal}} = 0$ and $T_{\text{fiber}} = T_{\text{GVD}}$. In principle, one can use the concept of fiber bandwidth discussed in Section 2.4.4 and relate T_{fiber} to the 3-dB fiber bandwidth $f_{3\,\text{dB}}$ through a relation similar to Eq. (5.3.6). In practice it is not easy to calculate $f_{3\,\text{dB}}$, especially in the case of modal dispersion. The reason is that a fiber link consists of many concatenated fiber sections (typical length 5 km), which may have different dispersion characteristics. Furthermore, mode mixing occurring at splices and connectors tends to average out the propagation delay associated with different modes of a multimode fiber. A statistical approach is often necessary to estimate the fiber bandwidth and the corresponding rise time [46]–[49].

In a phenomenological approach, T_{modal} can be approximated by the time delay ΔT given by Eq. (2.1.5) in the absence of mode mixing, i.e.,

$$T_{\text{modal}} \approx (n_1 \Delta / c) L, \tag{5.3.10}$$

where $n_1 \approx n_2$ was used. For graded-index fibers, Eq. (2.1.10) is used in place of Eq. (2.1.5), resulting in $T_{\text{modal}} \approx (n_1 \Delta^2 / 8c) L$. In both cases, the effect of mode mixing is included by changing the linear dependence on L by a sublinear dependence L^q, where q has a value in the range 0.5–1, depending on the extent of mode mixing. A reasonable estimate based on the experimental data is $q = 0.7$. The contribution T_{GVD} can also be approximated by ΔT given by Eq. (2.3.4), so that

$$T_{\text{GVD}} \approx |D| L \Delta \lambda, \tag{5.3.11}$$

where $\Delta \lambda$ is the spectral width of the optical source (taken as a full width at half maximum). The dispersion parameter D may change along the fiber link if different sections have different dispersion characteristics; an average value should be used in Eq. (5.3.11).

As an illustration of the rise-time budget, consider a 1.3-μm lightwave system designed to operate at 1 Gb/s over a single-mode fiber with a repeater spacing of 50 km. The rise times for the transmitter and the receiver have been specified as $T_{\text{tr}} = 0.25$ ns and $T_{\text{rec}} = 0.35$ ns. The source spectral width is specified as $\Delta \lambda = 3$ nm, whereas the average value of D is 2 ps/(km-nm) at the operating wavelength. From Eq. (5.3.11), $T_{\text{GVD}} = 0.3$ ns for a link length $L = 50$ km. Modal dispersion does not occur in single-mode fibers. Hence $T_{\text{modal}} = 0$ and $T_{\text{fiber}} = 0.3$ ns. The system rise time is estimated by using Eq. (5.3.8) and is found to be $T_r = 0.524$ ns. The use of Eq. (5.3.7) indicates that such a system cannot be operated at 1 Gb/s when the RZ format is employed for the optical bit stream. However, it would operate properly if digital format is changed to the NRZ format. If the use of RZ format is a prerequisite, the designer must choose different transmitters and receivers to meet the rise-time budget requirement. The NRZ format is commonly used, as it permits a larger system rise time at the same bit rate.

5.4   SOURCES OF POWER PENALTY

The discussion of Sections 5.2 and 5.3 shows that both fiber loss and fiber dispersion affect the design and the performance of lightwave systems. At low bit rates ($B < 100$ Mb/s), most lightwave systems are limited by fiber loss rather than fiber dispersion as long as the system components are chosen to satisfy the rise-time budget. However, at high bit rates ($B > 500$ Mb/s) fiber dispersion begins to dominate system performance. In particular, the sensitivity of the optical receiver is affected by several physical phenomena which, in combination with fiber dispersion, degrade the signal-to-noise ratio (SNR) at the decision circuit. Among the phenomena that degrade the receiver sensitivity are modal noise, dispersion broadening and intersymbol interference, mode-partition noise, frequency chirp, and reflection feedback. In this section we discuss how the system performance is affected by fiber dispersion by considering the extent of power penalty resulting from these phenomena.

5.4.1 Modal Noise

Modal noise is associated with multimode fibers and has been studied extensively [50]–[64]. Its origin can be understood as follows. Interference among various propagating modes in a multimode fiber creates a *speckle pattern* at the photodetector. The nonuniform intensity distribution associated with the speckle pattern is harmless in itself, as the receiver performance is governed by the total power integrated over the detector area. However, if the speckle pattern fluctuates with time, it will lead to fluctuations in the received power that would degrade the SNR. Such fluctuations are referred to as *modal noise*. They invariably occur in multimode fiber links because of mechanical disturbances such as vibrations and microbends. In addition, splices and connectors act as spatial filters. Any temporal changes in spatial filtering translate into speckle fluctuations and enhancement of the modal noise. Modal noise is strongly affected by the source spectral bandwidth $\Delta\nu$ since mode interference occurs only if the coherence time $(T_c \approx 1/\Delta\nu)$ is longer than the intermodal delay time ΔT given by Eq. (2.1.5). For LED-based transmitters $\Delta\nu$ is large enough $(\Delta\nu \sim 5 \text{ THz})$ that the condition above is not satisfied. Most lightwave systems that use multimode fibers also use LEDs to avoid the modal-noise problem.

Modal noise becomes a serious problem when semiconductor lasers are used in combination with multimode fibers. Attempts have been made to estimate the extent of sensitivity degradation induced by modal noise [54], [55] by calculating the BER after adding modal noise to the other sources of receiver noise. Figure 5.6 shows the power penalty at a BER of 10^{-12} calculated for a 1.3-μm lightwave system operating at 140 Mb/s. The graded-index fiber has a 50-μm core diameter and supports 146 modes. The power penalty depends on the mode-selective coupling loss occurring at splices and connectors. It also depends on the longitudinal-mode spectrum of the semiconductor laser. As expected, power penalty decreases as the number of longitudinal modes increases because of a reduction in the coherence time of the emitted light.

Modal noise can also occur in single-mode systems if short sections of fiber are installed between two connectors or splices during repair or normal maintenance [56]–[59]. A higher-order mode can be excited at the fiber discontinuity occurring at the first splice and then converted back to the fundamental mode at the second connector or splice. Since a higher-order mode cannot propagate far from its excitation point, this problem can be avoided by ensuring that the spacing between two connectors or splices exceeds 2 meters. Generally speaking, modal noise is not a problem for properly designed and maintained single-mode fiber-optic communication systems.

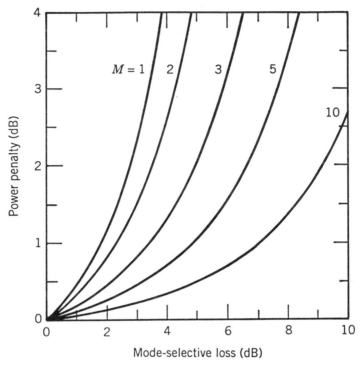

Figure 5.6 Modal-noise power penalty versus mode-selective loss. The parameter M is defined as the total number of longitudinal modes whose power exceeds 10% of the peak power. (After Ref. [54]. ©1986 IEEE. Reprinted with permission.)

With the development of the vertical-cavity surface-emitting laser (VCSEL), the modal-noise issue has resurfaced again in 1990s [60]–[64]. The use of such lasers in short-haul optical data links, making use of multimode fibers (even those made of plastic), is of considerable interest because of the high bandwidth associated with VCSELs. Indeed, bit rates of several Gb/s have been demonstrated in laboratory experiments with plastic-cladded multimode fibers [65]. However, VCSELs have a long coherence length as they oscillate in a single longitudinal mode. The BER measurements show an error floor at a level of 10^{-7} even for a 1-dB mode-selective loss [61]. The problem can be avoided to some extent by using larger-diameter VCSELs which oscillate in several transverse modes and thus have a shorter coherence length. Computer models are generally used to estimate the power penalty for optical data links under realistic operating conditions [63]. Analytic tools such as the saddle-point method can also provide a reasonable estimate of the BER [64].

5.4.2 Dispersion Broadening

The use of single-mode fibers for lightwave systems nearly avoids the problem of intermodal dispersion and the associated modal noise. As discussed in Section

5.2.2, group-velocity dispersion still limits the bit rate–distance product BL by broadening optical pulses beyond their allocated bit slot; Eq. (5.2.2) provides the limiting BL product and shows how it depends on the source spectral width σ_λ. Dispersion-induced pulse broadening can also decrease the receiver sensitivity. In this subsection we discuss the power penalty associated with such a decrease in receiver sensitivity.

Dispersion-induced pulse broadening affects the receiver performance in two ways. First, a part of the pulse energy spreads beyond the allocated bit slot and leads to intersymbol interference (ISI). In practice, the system is designed to minimize the effect of ISI (see Section 4.3.2). Second, the pulse energy within the bit slot is reduced when the optical pulse broadens. Such a decrease in the pulse energy reduces the SNR at the decision circuit. Since the SNR should remain constant to maintain the system performance, the receiver requires more average power. This is the origin of dispersion-induced power penalty δ_d. An exact calculation of δ_d is difficult, as it depends on many details, such as the extent of pulse shaping at the receiver. A rough estimate is obtained by following the analysis of Section 2.4.2, where broadening of Gaussian pulses is discussed. Equation (2.4.16) shows that the optical pulse remains Gaussian, but its peak power is reduced by a pulse-broadening factor given by Eq. (2.4.17). If we define the power penalty δ_d as the increase (in dB) in the received power that would compensate the peak-power reduction, δ_d is given by

$$\delta_d = 10 \log_{10} f_b, \tag{5.4.1}$$

where f_b is the pulse broadening factor. When pulse broadening is due mainly to a wide source spectrum at the transmitter, the broadening factor f_b is given by Eq. (2.4.24), i.e.,

$$f_b = \sigma/\sigma_0 = [1 + (DL\sigma_\lambda/\sigma_0)^2]^{1/2}, \tag{5.4.2}$$

where σ_0 is the RMS width of the optical pulse at the fiber input and σ_λ is the RMS width of the source spectrum assumed to be Gaussian.

Equations (5.4.1) and (5.4.2) can be used to estimate the dispersion penalty for lightwave systems that use single-mode fiber together with a multimode laser or an LED. The ISI is minimized when the bit rate B is such that $4B\sigma \leq 1$, as little pulse energy spreads beyond the bit slot ($T_B = 1/B$). By using $\sigma = (4B)^{-1}$, Eq. (5.4.2) can be written as

$$f_b^2 = 1 + (4BLD\sigma_\lambda f_b)^2. \tag{5.4.3}$$

By solving this equation for f_b and substituting it in Eq. (5.4.1), the power penalty is given by

$$\delta_d = -5 \log_{10}[1 - (4BLD\sigma_\lambda)^2]. \tag{5.4.4}$$

Figure 5.7 shows the power penalty as a function of the dimensionless parameter combination $BLD\sigma_\lambda$. Although the power penalty is negligible ($\delta_d = 0.38$ dB) for $BLD\sigma_\lambda = 0.1$, it increases to 2.2 dB when $BLD\sigma_\lambda = 0.2$ and becomes infinite when $BLD\sigma_\lambda = 0.25$. The BL product, shown in Fig. 5.4, is truly

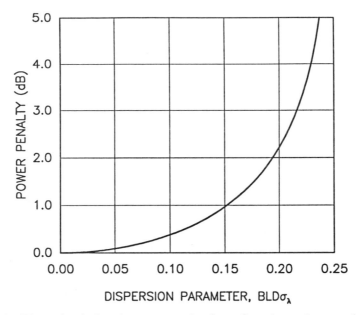

Figure 5.7 Dispersion-induced power penalty for a Gaussian pulse as a function of $BLD\sigma_\lambda$. Source spectrum is also assumed to be Gaussian, with an RMS width σ_λ.

limiting, since receiver sensitivity degrades severely when a system is designed to approach it. Most lightwave systems are designed such that $BLD\sigma_\lambda < 0.2$, so that the dispersion penalty is below 2 dB. It should be stressed that Eq. (5.4.4) provides a rough estimate only as its derivation is based on several simplifying assumptions, such as a Gaussian pulse shape and a Gaussian source spectrum. These assumptions are not always satisfied in practice. Moreover, it is based on the condition $4B\sigma = 1$, so that the pulse remains nearly confined within the bit slot. It is possible to design a system such that the pulse spreads outside the bit slot but ISI is reduced through pulse shaping at the receiver.

5.4.3   Mode-Partition Noise

As discussed in Section 3.3.8, multimode semiconductor lasers exhibit *mode-partition noise* (MPN), a phenomenon occurring because of an anticorrelation among pairs of longitudinal modes. In particular, various longitudinal modes fluctuate in such a way that individual modes exhibit large intensity fluctuations even though the total intensity remains relatively constant. MPN would be harmless in the absence of fiber dispersion, as all modes would remain synchronized during transmission and detection. In practice, different modes become desynchronized, since they travel at slightly different speeds inside the fiber because of group-velocity dispersion. As a result of such desynchronization, the receiver current exhibits additional fluctuations, and the SNR at the decision circuit becomes worse than that expected in the absence of MPN. A power penalty must be paid to improve the SNR to the same value that is necessary to achieve the required BER (see Section 4.5). The effect of MPN on system performance has been studied extensively for both multimode semiconductor lasers [66]–[75] and nearly single-mode lasers [76]–[90].

For multimode semiconductor lasers, the power penalty can be calculated by following an approach similar to that of Section 4.6.2 and is given by [66]

$$\delta_{mpn} = -5\log_{10}(1 - Q^2 r_{mpn}^2), \tag{5.4.5}$$

where r_{mpn} is the relative noise level of the received power in the presence of MPN. A simple model [67] has been used to estimate r_{mpn}. It assumes that laser modes fluctuate in such a way that the total power remains constant under CW operation. It also assumes that the average mode power is distributed according to a Gaussian distribution of RMS width σ_λ and that the pulse shape at the decision circuit of the receiver is described by a cosine function [66]. Different laser modes are assumed to have the same cross-correlation coefficient γ_{cc}, i.e.,

$$\gamma_{cc} = \frac{\langle P_i P_j \rangle}{\langle P_i \rangle \langle P_j \rangle} \tag{5.4.6}$$

for all i and j such that $i \neq j$. The angular brackets denote an average over power fluctuations associated with mode partitioning. A straightforward calculation shows that r_{mpn} is given by [67], [70]

$$r_{mpn} = (k/\sqrt{2})\{1 - \exp[-(\pi BLD\sigma_\lambda)^2]\}, \tag{5.4.7}$$

where the mode-partition coefficient k is related to γ_{cc} as

$$k = \sqrt{1 - \gamma_{cc}}. \tag{5.4.8}$$

The model assumes that mode partition can be quantified in terms of a single parameter k with values in the range 0–1. The numerical value of k is difficult to estimate and is likely to vary from laser to laser. Experimental measurements [67], [72] suggest typical values in the range 0.6–0.8. They also indicate that k is generally different for different mode pairs.

Equations (5.4.5) and (5.4.7) can be used to calculate the MPN-induced power penalty. Figure 5.8 shows the power penalty at a BER of 10^{-9} ($Q = 6$) as a function of the normalized dispersion parameter $BLD\sigma_\lambda$ for several values of the mode-partition coefficient k. For a given value of k, the variation of power penalty is similar to that shown in Fig. 5.7; δ_{mpn} increases rapidly with an increase in $BLD\sigma_\lambda$ and becomes infinite when $BLD\sigma_\lambda$ reaches a critical value. For $k > 0.5$, the MPN-induced power penalty is larger than the penalty occurring due to dispersion-induced pulse broadening (see Fig. 5.7). However, it can be reduced to a negligible level ($\delta_{mpn} < 0.5$ dB) by designing the optical communication system such that $BLD\sigma_\lambda < 0.1$. As an example, consider a 1.3-μm lightwave system. If we assume that the operating wavelength is matched to the zero-dispersion wavelength to within 10 nm, $D \approx 1$ ps/(km-nm). A typical value of σ_λ for multimode semiconductor lasers is 2 nm. The MPN-induced power penalty would be negligible if the BL product were below 50 (Gb/s)-km.

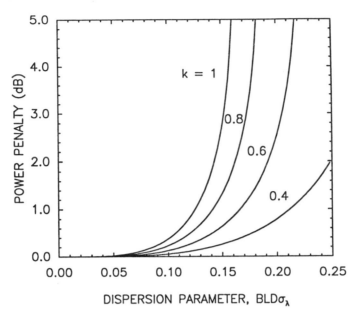

Figure 5.8 MPN-induced Power penalty versus $BLD\sigma_\lambda$ for a multimode semiconductor laser of RMS spectral width σ_λ. Different curves correspond to different values of the mode-partition coefficient k.

At $B = 2$ Gb/s the transmission distance is then limited to 25 km. The situation becomes worse for 1.55-μm lightwave systems for which $D \approx 16$ ps/(km-nm) unless dispersion-shifted fibers are used. In general, the MPN-induced power penalty is quite sensitive to the spectral bandwidth of the multimode laser and can be reduced by reducing the bandwidth. In one study [75], a reduction in the carrier lifetime from 340 ps to 130 ps, realized by p-doping of the active layer, reduced the bandwidth of 1.3-μm semiconductor lasers by only 40% (from 5.6 nm to 3.4 nm), but the power penalty decreased from an infinite value (BER floor above 10^{-9} level) to a mere 0.5 dB.

Most 1.55-μm lightwave systems make use of DFB semiconductor lasers to solve the dispersion problem. One may think that MPN can be avoided completely by using DFB lasers which oscillate in a single longitudinal mode. Unfortunately, this is not the case, as was discovered in many experiments [80]–[83]. The reason is that the main mode of any DFB laser is accompanied by several side modes of much smaller amplitudes. The single-mode nature of DFB lasers is quantified through the *mode-suppression ratio* (MSR), defined as the ratio of the main-mode power P_m to the power P_s of the most dominant side mode. Clearly, the effect of MPN on system performance would depend on the MSR. Attempts have therefore been made to estimate the dependence of the MPN-induced power penalty on the MSR [76]–[90].

A major difference between the multimode and nearly single-mode semiconductor lasers is related to the statistics associated with mode-partition fluctuations [76], [77]. In a multimode laser, both main and side modes are above threshold and their fluctuations are well described by a Gaussian probability density function. By contrast, side modes in a DFB semiconductor laser are typically below threshold, and the optical power associated with them follows an exponential distribution given by [76]

$$p(P_s) = \bar{P}_s^{-1} \exp[-(P_s/\bar{P}_s)], \qquad (5.4.9)$$

where $\bar{P}_s$ is the average value of the random variable P_s.

The effect of side-mode fluctuations on system performance can be appreciated by considering an ideal receiver. Let us assume that the relative delay $\Delta T = DL\Delta\lambda$ between the main and side modes is large enough that the side mode appears outside the bit slot (i.e., $\Delta T > 1/B$ or $BLD\Delta\lambda_L > 1$, where $\Delta\lambda_L$ is the mode spacing). The decision circuit of the receiver would make an error for "0" bits if the side-mode power P_s were to exceed the decision threshold set at $\bar{P}_m/2$, where $\bar{P}_m$ is the average main-mode power. Furthermore, the two modes are anticorrelated in such a way that the main-mode power drops below $\bar{P}_m/2$ whenever side-mode power exceeds $\bar{P}_m/2$, so that the total power remains nearly constant [77]. Thus, an error would occur even for "1" bits whenever $P_s > \bar{P}_m/2$. Since the two terms in Eq. (4.5.2) make equal contributions, the BER is given by [76]

$$\text{BER} = \int_{\bar{P}_m/2}^{\infty} p(P_s)\, dP_s = \exp\left(-\frac{\bar{P}_m}{2\bar{P}_s}\right) = \exp\left(-\frac{R_{ms}}{2}\right). \qquad (5.4.10)$$

The BER depends on the MSR defined as $R_{ms} = \bar{P}_m/\bar{P}_s$ and exceeds 10^{-9} when MSR < 42.

To calculate the MPN-induced power penalty in the presence of receiver noise, one should follow the analysis in Section 4.5.1 and add an additional noise term that accounts for side-mode fluctuations [77], [78]. For a p–i–n receiver the BER is found to be [77]

$$\text{BER} = \frac{1}{2}\,\text{erfc}\left(\frac{Q}{\sqrt{2}}\right) + \exp\left(-\frac{R_{ms}}{2} + \frac{R_{ms}^2}{4Q^2}\right)\left[1 - \frac{1}{2}\,\text{erfc}\left(\frac{Q}{\sqrt{2}} - \frac{R_{ms}}{Q\sqrt{2}}\right)\right],$$
$$(5.4.11)$$

where the parameter Q is defined by Eq. (4.5.10). In the limit of an infinite MSR, Eq. (5.4.11) reduces to Eq. (4.5.9). For a noise-free receiver ($Q = \infty$), Eq. (5.4.11) reduces to Eq. (5.4.10). Figure 5.9 shows the BER versus the power penalty at a BER of 10^{-9} as a function of MSR. As expected, the power penalty becomes infinite for MSR values below 42, since the 10^{-9} BER cannot be realized irrespective of the power received. The penalty can be reduced to a negligible level (< 0.1 dB) for MSR values in excess of 100 (20 dB).

The experimental measurements of the BER in several transmission experiments [80]–[83] show that a BER floor above the 10^{-9} level can occur even for DFB lasers which exhibit a MSR in excess of 30 dB (under CW operation). The reason behind the failure of apparently good lasers is related to the possibility of

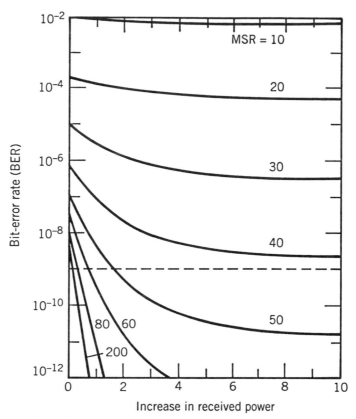

Figure 5.9 Effect of MPN on bit-error rate of DFB lasers for several values of mode-suppression ratio (MSR). Intersection of the dashed line with the solid curves provides MPN-induced power penalty. (After Ref. [77]. ©1985 IEEE. Reprinted with permission.)

side-mode excitation under transient conditions occurring when the laser is repeatedly turned on and off to generate the bit stream. When the laser is biased below threshold and modulated at a high bit rate ($B \geq 1$ Gb/s), the probability of side-mode excitation above $\bar{P}_m/2$ is much higher than that predicted by Eq. (5.4.9). Considerable attention has been paid to calculate, both analytically and numerically, the probability of the transient excitation of side modes and its dependence on various device parameters [79]–[90]. An important device parameter is found to be the *gain margin* between the main and side modes. The gain margin should exceed a critical value which depends on the bit rate. The critical value is about 5–6 cm^{-1} at 500 Mb/s [80] but can exceed 15 cm^{-1} at high bit rates, depending on the bias and modulation currents [85]. The bias current plays a critical role. Numerical simulations show that the the best performance is achieved when the DFB laser is biased close to but slightly below threshold to avoid the bit-pattern effects [90]. Moreover, the effects of MPN are independent of the bit rate as long as the gain margin exceeds a certain value.

The required value of gain margin exceeds 25 cm^{-1} for the 5-GHz modulation frequency. Phase-shifted DFB lasers have a large built-in gain margin and have been developed for this purpose.

5.4.4 Frequency Chirping

Frequency chirping is an important phenomenon that is known to limit the performance of 1.55-μm lightwave systems even when a DFB laser with a large MSR is used to generate the digital bit stream [91]–[104]. As discussed in Section 3.3.7, intensity modulation in semiconductor lasers is invariably accompanied by phase modulation because of the carrier-induced change in the refractive index governed by the linewidth enhancement factor. Optical pulses with a time-dependent phase shift are called chirped. As a result of the frequency chirp imposed on an optical pulse, its spectrum is considerably broadened. Such spectral broadening affects the pulse shape at the fiber output because of fiber dispersion and degrades system performance.

An exact calculation of the chirp-induced power penalty δ_c is difficult because frequency chirp depends on both the shape and the width of the optical pulse [93]–[96]. For nearly rectangular pulses, experimental measurements of time-resolved pulse spectra show that frequency chirp occurs mainly near the leading and trailing edges such that the leading edge shifts toward the blue while the trailing edge shifts toward the red. Because of the spectral shift, the power contained in the chirped portion of the pulse moves out of the bit slot when the pulse propagates inside the optical fiber. Such a power loss decreases the SNR at the receiver and results in power penalty. In a simple model the chirp-induced power penalty is given by [92]

$$\delta_c = -10\log_{10}(1 - 4BLD\Delta\lambda_c), \tag{5.4.12}$$

where $\Delta\lambda_c$ is the spectral shift associated with frequency chirping. This equation applies as long as $LD\Delta\lambda_c < t_c$, where t_c is the chirp duration. Typically, t_c is 100–200 ps, depending on the relaxation-oscillation frequency, since chirping lasts for about one-half of the relaxation-oscillation period. By the time $LD\Delta\lambda_c$ equals t_c, the power penalty stops increasing because all the chirped power has left the bit interval. For $LD\Delta\lambda_c > t_c$, the product $LD\Delta\lambda_c$ in Eq. (5.4.12) should be replaced by t_c.

The model above is overly simplistic, as it does not take into account pulse shaping at the receiver. A more accurate calculation based on raised-cosine filtering (see Section 4.3.2) leads to the following expression [99]:

$$\delta_c = -20\log_{10}\{1 - (4\pi^2/3 - 8)B^2 L_D\Delta\lambda_c t_c[1 + (2B/3)(LD\Delta\lambda_c - t_c)]\}. \tag{5.4.13}$$

The receiver is assumed to contain a p–i–n photodiode. The penalty is larger for an APD, depending on the excess-noise factor of the APD. Figure 5.10 shows the power penalty δ_c as a function of the parameter combination $BLD\Delta\lambda_c$ for several values of the parameter Bt_c, which is a measure of the fraction of the bit period over which chirping occurs. As expected, δ_c increases with both the

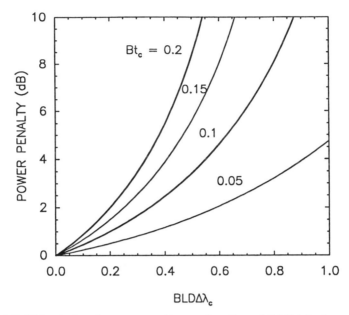

Figure 5.10 Chirp-induced power penalty as a function of $BLD\Delta\lambda_c$ for several values of the parameter Bt_c. $\Delta\lambda_c$ is the wavelength shift occurring because of frequency chirp and t_c is the duration of such a wavelength shift.

chirp $\Delta\lambda_c$ and the chirp duration t_c. The power penalty can be kept below 1 dB if the system is designed such that $BLD\Delta\lambda_c < 0.1$ and $Bt_c < 0.2$. A shortcoming of this model is that $\Delta\lambda_c$ and t_c appear as free parameters and must be determined for each laser through experimental measurements of the frequency chirp. In practice, $\Delta\lambda_c$ itself depends on the bit rate B and increases with it.

For lightwave systems operating at high bit rates ($B > 2$ Gb/s), the bit duration is generally shorter than the total duration $2t_c$ over which chirping is assumed to occur in the foregoing model. The frequency chirp in that case increases almost linearly over the entire pulse width (or bit slot). A similar situation occurs even at low bit rates if the optical pulses do not contain sharp leading and trailing edges but have long rise and fall times (Gaussian-like shape rather than a rectangular shape). If we assume a Gaussian pulse shape and a linear chirp, the analysis of Section 2.4.2 can be used to estimate the chirp-induced power penalty. Equation (2.4.16) shows that the chirped Gaussian pulse remains Gaussian but its peak power decreases because of dispersion-induced pulse broadening. Defining the power penalty as the increase (in dB) in the received power that would compensate the peak-power reduction, δ_c is given by

$$\delta_c = 10\log_{10} f_b, \tag{5.4.14}$$

where f_b is the broadening factor given by Eq. (2.4.22) with $\beta_3 = 0$. The RMS width σ_0 of the input pulse should be such that $4\sigma_0 \leq 1/B$. Choosing the

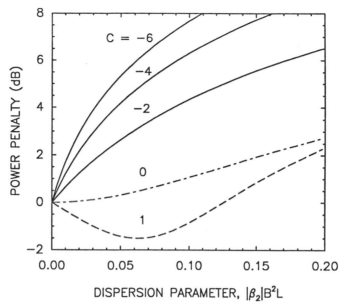

Figure 5.11 Chirp-induced power penalty as a function of $|\beta_2|B^2L$ for several values of the chirp parameter C. The Gaussian optical pulse is assumed to be linearly chirped over its entire width.

worst-case condition $\sigma_0 = 1/4B$, the power penalty is given by

$$\delta_c = 5\log_{10}[(1 + 8C\beta_2B^2L)^2 + (8\beta_2B^2L)^2]. \qquad (5.4.15)$$

Figure 5.11 shows the chirp-induced power penalty as a function of $|\beta_2|B^2L$ for several values of the chirp parameter C. The parameter β_2 is taken to be negative, as is the case for 1.55-μm lightwave systems. The $C = 0$ curve corresponds to the case of a chirp-free pulse. The power penalty is negligible (< 0.1 dB) in this ideal case as long as $|\beta_2|B^2L < 0.05$. However, the penalty can exceed 5 dB if the pulses transmitted are chirped such that $C = -6$. To keep the penalty below 0.1 dB, the system should be designed with $|\beta_2|B^2L < 0.002$. For $|\beta_2| = 20$ ps^2/km, B^2L is limited to 100 (Gb/s)^2-km. Interestingly, system performance is improved for positive values of C since the optical pulse then goes through an initial compression phase (see Section 2.4). Unfortunately, C is negative for semiconductor lasers; it can be approximated by $-\beta_c$, where β_c is the linewidth enhancement factor with positive values of 2–6.

It is important to stress that the analytic results shown in Figs. 5.10 and 5.11 provide only a rough estimate of the power penalty. In practice, the chirp-induced power penalty depends on many system parameters. For instance, several system experiments have shown that the effect of chirp can be reduced by biasing the semiconductor laser above threshold [95]. However, above-threshold biasing increases that extinction ratio r_{ex}, defined in Eq. (4.6.1) as $r_{ex} = P_0/P_1$, where P_0 and P_1 are the powers received for bit 0 and bit 1, respectively. As

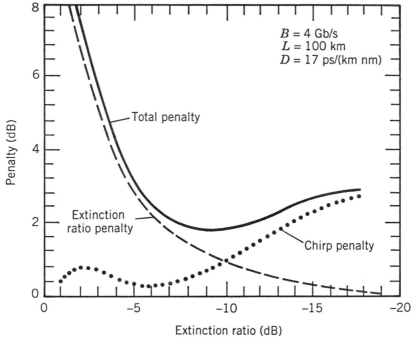

Figure 5.12 Power penalty as a function of the extinction ratio. (After Ref. [97]. ©1987 IEEE. Reprinted with permission.)

discussed in Section 4.6.1, an increase in r_{ex} decreases the receiver sensitivity and leads to its own power penalty. Clearly, r_{ex} cannot be increased indefinitely in an attempt to reduce the chirp penalty. The total system performance can be optimized by designing the system so that it operates with an optimum value of r_{ex} that takes into account the trade-off between the chirp and the extinction ratio. Numerical simulations are often used to understand such trade-offs in actual lightwave systems [102], [103]. Figure 5.12 shows the power penalty as a function of the extinction ratio r_{ex} by simulating numerically the performance of a 1.55-μm lightwave system transmitting at 4 Gb/s over a 100-km-long fiber. The total penalty can be reduced below 2 dB by operating the system with an extinction ratio of about 0.1. The optimum values of r_{ex} and the total penalty are sensitive to many other laser parameters such as the active-region width. A semiconductor laser with a wider active region is found to have a larger chirp penalty [97]. The physical phenomenon behind this width dependence appears to be the nonlinear gain [see Eq. (3.3.40)] and the associated damping of relaxation oscillations. In general, rapid damping of relaxation oscillations decreases the effect of frequency chirp and improves system performance [105].

The origin of chirp in semiconductor lasers is related to carrier-induced index changes governed by the linewidth enhancement factor β_c. The frequency chirp would be absent for a laser with $\beta_c = 0$. Unfortunately, β_c cannot be made zero for semiconductor lasers, although it can be reduced by adopting a multiquantum-well (MQW) design [106]–[110]. The use of MQW active region reduces β_c by about a factor of 2. In one 1.55-μm experiment [112], the 10-Gb/s signal could be transmitted over 60–70 km, despite the high dispersion of standard telecommunication fiber, by biasing the laser above threshold. The MQW DFB laser used in the experiment had $\beta_c \approx 3$. A further reduction in β_c occurs for strained quantum wells [110]. Indeed, $\beta_c \approx 1$ has been measured in modulation-doped strained MQW lasers [111]. Such lasers exhibit low chirp under direct modulation at bit rates as high as 10 Gb/s.

An alternative scheme eliminates the laser-chirp problem completely by operating the laser continuously and using an external modulator to generate the bit stream. This approach has become practical with the development of optical transmitters in which a modulator is integrated monolithically with a DFB laser (see Section 3.4). The chirp parameter C is close to zero in such transmitters. As shown by the $C = 0$ curve in Fig. 5.11, the dispersion penalty is below 2 dB in that case even when $|\beta_2| B^2 L$ is close to 0.2. Moreover, an external modulator can be used to modulate the phase of the optical carrier in such a way that $\beta_2 C < 0$ in Eq. (5.4.15). This feature of external modulators is discussed in Section 9.3.1 in the context of dispersion compensation. As seen in Fig. 5.11, the chirp-induced power penalty becomes negative over a certain range of $|\beta_2| B^2 L$, implying that such frequency chirping is beneficial to combat the effects of dispersion. In a 1996 experiment [113], the 10-Gb/s signal was transmitted penalty free over 100 km of standard telecommunication fiber by using a modulator-integrated transmitter such that C was effectively positive. By using $\beta_2 \approx -20\ \mathrm{ps}^2/\mathrm{km}$, it is easy to verify that $|\beta_2| B^2 L = 0.2$ for this experiment, a value that would have produced a power penalty of more than 8 dB if the DFB laser were modulated directly.

5.4.5   Reflection Feedback and Noise

In fiber-optic communication systems some light invariably gets reflected back because of refractive-index discontinuities occurring at splices, connectors, and fiber ends. The effects of such unintentional feedback have been studied extensively [114]–[132] because they can degrade the performance of lightwave systems considerably. Even a relatively small amount of optical feedback affects the operation of semiconductor lasers [118] and can lead to excess noise in the transmitter output. Even when an isolator is used between the transmitter and the fiber, multiple reflections between splices and connectors can generate additional intensity noise and degrade receiver performance [120]. This subsection is devoted to the effect of reflection-induced noise on receiver sensitivity.

Most reflections in a fiber link originate at glass–air interfaces whose reflectivity can be estimated by using $R_f = (n_f - 1)^2/(n_f + 1)^2$, where n_f is the refractive index of the fiber material. For silica fibers $R_f = 3.6\%$ (-14.4 dB) if we use $n_f = 1.47$. This value increases to 5.3% for polished fiber ends since polishing can create a thin surface layer with a refractive index of about 1.6. In the case of multiple reflections occurring between two splices or connectors, the reflection feedback can increase considerably because the two reflecting surfaces act as mirrors of a Fabry–Perot interferometer. When the resonance condition is satisfied, the reflectivity increases to 14% for unpolished surfaces and to over 22% for polished surfaces. Clearly, a considerable fraction of the signal transmitted can be reflected back unless precautions are taken to reduce the optical feedback. A common technique for reducing reflection feedback is to use index-matching oil or gel near glass–air interfaces. Sometimes the tip of the fiber is curved or cut at an angle so that the reflected light deviates from the fiber axis. Reflection feedback can be reduced to below 0.1% by such techniques.

Semiconductor lasers are extremely sensitive to optical feedback [125]; their operating characteristics can be affected by feedback as small as −80 dB [118]. The most dramatic effect of feedback is on the laser linewidth, which can narrow or broaden by several order of magnitude, depending on the exact location of the surface where feedback originates [114]. The reason behind such a sensitivity is related to the fact that the phase of the reflected light can perturb the laser phase significantly even for relatively weak feedback levels. Such feedback-induced phase changes are detrimental mainly for coherent communication systems. The performance of direct-detection lightwave systems is affected by intensity noise rather than phase noise.

Optical feedback can increase the intensity noise significantly. Several experiments have shown a feedback-induced enhancement of the intensity noise occurring at frequencies corresponding to multiples of the external-cavity mode spacing [115], [116]. In fact, there are several mechanisms through which the relative intensity noise (RIN) of a semiconductor laser can be enhanced by the external optical feedback. In a simple model [119], the feedback-induced enhancement of the intensity noise is attributed to the onset of multiple, closely spaced, external-cavity longitudinal modes whose spacing is determined by the distance between the laser output facet and the glass–air interface where feedback originates. The number and the amplitudes of the external-cavity modes depend on the amount of feedback. In this model, the RIN enhancement is due to intensity fluctuations of the feedback-generated side modes. Another source of RIN enhancement has its origin in the feedback-induced chaos in semiconductor lasers. Numerical simulations of the rate equations show that the RIN can be enhanced by 20 dB or more when the feedback level exceeds a certain value [126]. Even though the feedback-induced chaos is deterministic in nature, it manifests as an apparent RIN increase.

Experimental measurement of the RIN and the BER in the presence of optical feedback confirm that the feedback-induced RIN enhancement leads to a power penalty in lightwave systems [129]–[132]. Figure 5.13 shows the results of the BER measurements for a VCSEL operating at 958 nm. Such a laser operates in a single longitudinal mode because of an ultrashort cavity length ($\sim 1~\mu$m) and exhibits a RIN near −130 dB/Hz in the absence of reflection feedback. However, the RIN increases by as much as 20 dB when the feedback exceeds the −30-dB level. The BER measurements at a bit rate of 500 Mb/s show a power penalty of 0.8 dB at a BER of 10^{-9} for −30-dB feedback, and the penalty increases rapidly at higher feedback levels [131].

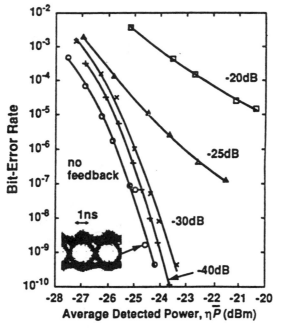

Figure 5.13 Experimentally measured BER at 500 Mb/s for a VCSEL under optical feedback. The BER is measured at several feedback levels. (After Ref. [131]. ©1993 IEEE. Reprinted with permission.)

The power penalty can be calculated by following the analysis of Section 4.6.2 and is given by

$$\delta_{\text{ref}} = -10 \log_{10}(1 - r_{\text{eff}}^2 Q^2), \qquad (5.4.16)$$

where r_{eff} is the effective intensity noise over the receiver bandwidth Δf and is obtained from

$$r_{\text{eff}}^2 = \frac{1}{2\pi} \int_{-\infty}^{\infty} \text{RIN}(\omega)\, d\omega = 2(\text{RIN})\Delta f. \qquad (5.4.17)$$

In the case of feedback-induced external-cavity modes, r_{eff} can be calculated by using a simple model [119] and is found to be

$$r_{\text{eff}}^2 \approx r_I^2 + N/(\text{MSR})^2, \qquad (5.4.18)$$

where r_I is the relative noise level in the absence of reflection feedback, N is the number of external-cavity modes, and MSR is the factor by which the external-cavity modes remain suppressed. Figure 5.14 shows the reflection-noise power penalty as a function of MSR for several values of N by choosing $r_I = 0.01$. The penalty is negligible in the absence of feedback ($N = 0$). However, it increases with an increase in N and a decrease in MSR. In fact, the penalty becomes infinite when MSR is reduced below a critical value. Thus, reflection feedback can degrade system performance to the extent that the system cannot achieve the desired BER despite an indefinite increase in the power received.

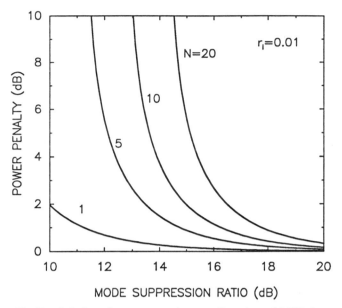

Figure 5.14 Feedback-induced power penalty as a function of MSR for several values of N and $r_I = 0.01$. Reflection feedback into the laser is assumed to generate N side modes of the same amplitude.

Such reflection-induced BER floors have been observed experimentally [117] and indicate the severe impact of reflection noise on the performance of lightwave systems. An example of the reflection-induced BER floor is seen in Fig. 5.13, where the BER remains above 10^{-9} for feedback levels in excess of -25 dB. Generally speaking, most lightwave systems operate satisfactorily when the reflection feedback is below -30 dB. In practice, the problem can be nearly eliminated by using an optical isolator within the transmitter module.

Even when an isolator is used, reflection noise can be a problem for lightwave systems. In long-haul fiber links making use of optical amplifiers, fiber dispersion can convert the phase noise to intensity noise, leading to performance degradation [122], [128]. Similarly, two reflecting surfaces anywhere along the fiber link act as a Fabry-Perot interferometer which can convert phase noise into intensity noise [120]. Such a conversion can be understood by noting that multiple reflections inside a Fabry–Perot interferometer lead to a phase-dependent term in the transmitted intensity which fluctuates in response to phase fluctuations. As a result, the RIN of the signal incident on the receiver is higher than that occurring in the absence of reflection feedback. Most of the RIN enhancement occurs over a narrow frequency band whose spectral width is governed by the laser linewidth (~ 100 MHz). Since the total noise is obtained by integrating over the receiver bandwidth, it can affect system performance considerably at bit rates larger than the laser linewidth. The power penalty can still be calculated by using Eq. (5.4.16). A simple model that includes only two reflections

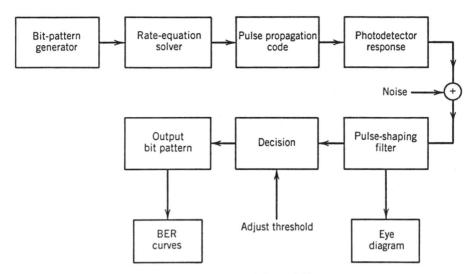

Figure 5.15 Steps involved in computer modeling of fiber-optic communication systems.

between the reflecting interfaces shows that r_{eff} is proportional to $(R_1 R_2)^{1/2}$, where R_1 and R_2 are the reflectivities of the two interfaces [120]. Figure 4.19 can be used to estimate the power penalty. It shows that power penalty can become infinite and lead to BER floors when r_{eff} exceeds 0.2. Such BER floors have been observed experimentally [120]. They can be avoided only by eliminating or reducing parasitic reflections along the entire fiber link. It is therefore necessary to employ connectors and splices that reduce reflections through the use of index matching or other techniques.

5.5   COMPUTER-AIDED DESIGN

The design of a fiber-optic communication system involves optimization of a large number of parameters associated with the transmitter, optical fiber, and receiver. The design aspects discussed in Section 5.3 are too simple to provide optimized values of system parameters. The power budget and the rise-time budget are used to obtain a conservative estimate of the transmission distance (repeater spacing) and the bit rate. The system margin in Eq. (5.3.1) is used as a vehicle to include various sources of power penalties discussed in Section 5.4. Often, the tendency is to overdesign the system, to be on the safe side.

An alternative approach uses computer simulations based on realistic modeling of fiber-optic communication systems [133]–[146]. Such models are capable of optimizing the whole system and provide the optimum values of the system parameters such that the design objectives are met at a minimum cost. A general approach consists of simulating the generation of an optical bit stream, its transmission through the fiber, and its detection at the receiver. Figure 5.15 illustrates the various steps involved in the simulation process.

Each step in the block diagram can be carried out numerically by using the theory discussed in Chapters 2–4. The input to the optical transmitter is a pseudo-random sequence of electrical pulses which represent "1" and "0" bits. The length N of the pseudo-random bit sequence determines the computing time and should be chosen judiciously. Typically, $N = 2^M$, where M is in the range 6–10. The optical bit stream can be obtained by solving the rate equations that govern the modulation response of semiconductor lasers (see Section 3.3.7). The phase information is automatically included to account for the effects of frequency chirping [97]. Deformation of the optical bit stream during its transmission inside the optical fiber is calculated by using Eq. (5.2.4). If nonlinear effects are not important, the analysis of Section 2.4 can be used. The optical signal transmitted is converted into the electrical domain at the receiver. The receiver noise is stimulated by adding a fluctuating term with Gaussian statistics. The bit stream is shaped by passing it through a filter whose bandwidth is also a design parameter. An eye diagram is constructed numerically. The effect of varying system parameters is studied by monitoring eye degradation. Such an approach can be used to obtain the power penalty associated with various mechanisms discussed in Section 5.4. It can also be used to investigate trade-offs that would optimize the overall system performance. An example is shown in Fig. 5.12, where the dependence of the calculated system penalty on the frequency chirp and extinction ratio is demonstrated. Numerical simulations reveal the existence of an optimum extinction ratio for which the system penalty is minimum.

Computer-aided design has another important role to play. A long-haul lightwave system may contain many repeaters, both optical and electrical. Transmitters, receivers, and amplifiers used at repeaters, although chosen to satisfy nominal specifications, are never identical. Similarly, fiber cables are constructed by splicing many different pieces (typical length 4–8 km) which have slightly different loss and dispersion characteristics. The net result is that many system parameters vary around their nominal values. For example, the dispersion parameter D, responsible not only for pulse broadening but also for other sources of power penalty, such as mode-partition noise, can vary significantly in different sections of the fiber link because of variations in the zero-dispersion wavelength and the transmitter wavelength. A statistical approach is often used to estimate the effect of such inherent variations in a realistic lightwave system [138]–[142]. The idea behind such an approach is that it is extremely unlikely that all system parameters would take their worst-case values at the same time. Thus, repeater spacing can be increased well above its worst-case value if the system is designed to operate reliably at the specific bit rate with a high probability (say 99%).

The importance of computer-aided design for fiber-optic communication systems became apparent during the decade of the 1990s when the dispersive and nonlinear effects in optical fibers became of paramount concern with increasing bit rates and transmission distances. By 1997, several software packages were available commercially [146]. A general-purpose simulation tool capable of handling almost all design aspects of lightwave systems still remains to be developed. As the bit rate of commercial lightwave systems approaches 100 Gb/s and the use of optical amplifiers becomes widespread, the long-haul fiber links are likely to rely heavily on computer-aided design.

PROBLEMS

5.1 A distribution network uses an optical bus to distribute the signal to 10 users. Each optical tap couples 10% of the power to the user and has 1-dB insertion loss. Assuming that the station 1 transmits 1 mW of power over the optical bus, calculate the power received by the stations 8, 9, and 10.

5.2 A CATV operator uses an optical bus to distribute the video signal to its subscribers. Each receiver needs a minimum of 100 nW to operate satisfactorily. Optical taps couple 5% of the power to each subscriber. Assume 0.5 dB insertion loss for each tap and 1-mW transmitter power. How many subscribers can be added to the optical bus?

5.3 A star network uses directional couplers with 0.5-dB insertion loss to distribute data to its subscribers. If each receiver requires a minimum of 100 nW and each transmitter is capable of emitting 0.5 mW, calculate the maximum number of subscribers served by the network.

5.4 Make the power budget and calculate the maximum transmission distance for a 1.3-μm lightwave system operating at 100 Mb/s by using an InGaAsP LED capable of coupling 0.1 mW of average power into a single-mode fiber. Assume 1-dB/km attenuation, 0.2-dB splice loss at 2-km intervals, 1-dB connector loss at each end of fiber link, and 100-nW sensitivity for the p–i–n receiver. Allow a 6-dB system margin.

5.5 A 1.3-μm long-haul lightwave system is designed to operate at 1.5 Gb/s. It uses semiconductor lasers capable of coupling 1 mW of average power into the single-mode fiber. The fiber-cable loss is specified as 0.5 dB/km and includes splice losses. The connectors at each end have 1-dB loss. The InGaAs p–i–n receiver has a sensitivity of 250 nW. Make the power budget and estimate the repeater spacing.

5.6 Prove that the rise time T_r and the 3-dB bandwidth Δf of a RC circuit are related by $T_r \Delta f = 0.35$.

5.7 Consider a super-Gaussian optical pulse with the power distribution

$$P(t) = P_0 \exp[-(t/T_0)^{2m}],$$

where the parameter m controls the pulse shape. Derive an expression for the rise time T_r of such a pulse. Calculate the ratio T_r/T_{FWHM}, where T_{FWHM} is the full width at half maximum, and show that for a Gaussian pulse ($m = 1$) this ratio equals 0.716.

5.8 Prove that for a Gaussian optical pulse, the rise time T_r and the 3-dB optical bandwidth Δf are related by $T_r \Delta f = 0.316$.

5.9 Make the rise-time budget for a 0.85-μm, 10-km fiber link designed to operate at 50 Mb/s. The LED transmitter and the Si p–i–n receiver have rise times of 10 and 15 ns, respectively. The graded-index fiber has a core index of 1.46, $\Delta = 0.01$, and $D = 80$ ps/(km-nm). The LED spectral width is 50 nm. Can the system be designed to operate with the NRZ format?

5.10 A 1.3-μm lightwave system is designed to operate at 1.7 Gb/s with a repeater spacing of 45 km. The single-mode fiber has a dispersion slope of 0.1 ps/(km-nm^2) in the vicinity of the zero-dispersion wavelength occurring at 1.308 μm. Calculate the wavelength range of multimode semiconductor lasers for which the mode-partition-noise power penalty remains below 1 dB. Assume that the RMS spectral width of the laser is 2 nm and the mode-partition coefficient $k = 0.7$.

5.11 Generalize Eq. (5.4.5) for the case of APD receivers by including the excess-noise factor in the form $F(M) = M^x$.

5.12 Consider a 1.55-μm lightwave system operating at 1 Gb/s by using multimode semiconductor lasers of 2-nm (RMS) spectral width. Calculate the maximum transmission distance that would keep the mode-partition-noise power penalty below 2 dB. Use $k = 0.8$ for the mode-partition coefficient.

5.13 Follow the rate-equation analysis of Section 3.3.8 (see also Ref. [76]) to prove that the side-mode power P_s follows an exponential probability density function given by Eq. (5.4.9).

5.14 Use Eq. (5.4.15) to determine the maximum transmission distance for a 1.55-μm lightwave system operating at 4 Gb/s such that the chirp-induced power penalty is below 1 dB. Assume that $C = -6$ for the single-mode semiconductor laser and $\beta_2 = -20$ ps^2/km for the single-mode fiber.

5.15 Repeat Problem 5.14 for the case of 8-Gb/s bit rate.

5.16 Use the results of Problem 4.16 to obtain an expression of the reflection-induced power penalty in the case of a finite extinction ratio r_{ex}. Reproduce the penalty curves shown in Fig. 5.13 for the case $r_{ex} = 0.1$.

5.17 Consider a Fabry-Perot interferometer with two surfaces of reflectivity R_1 and R_2. Follow the analysis of Ref. [120] to derive an expression of the relative intensity noise RIN(ω) of the transmitted light as a function of the linewidth of the incident light. Assume that R_1 and R_2 are small enough that it is enough to consider only a single reflection at each surface.

5.18 Follow the analysis of Ref. [134] to obtain an expression for the total receiver noise by including thermal noise, shot noise, intensity noise, mode-partition noise, chirp noise, and reflection noise.

REFERENCES

[1] P. S. Henry, R. A. Linke, and A. H. Gnauck, in *Optical Fiber Telecommunications II*, S. E. Miller and I. P. Kaminow, Eds., Academic Press, San Diego, CA, 1988, Chap. 21.

[2] S. E. Miller and I. P. Kaminow, Eds., *Optical Fiber Telecommunications II*, Academic Press, San Diego, CA, 1988, Chaps. 22–25.

[3] T. Li, Ed., *Topics in Lightwave Transmission Systems*, Academic Press, San Diego, CA, 1991.

[4] P. E. Green, Jr., *Fiber-Optic Networks*, Prentice Hall, Upper Saddle River, NJ, 1993.

[5] M. M.-K. Liu, *Principles and Applications of Optical Communications*, Irwin, Chicago, 1996.

[6] T. Li, *Proc. IEEE* **81**, 1568 (1993).

[7] E. Desurvire, *Erbium-Doped Fiber Amplifiers*, Wiley, New York, 1994.

[8] R. Giles and T. Li, *Proc. IEEE* **84**, 870 (1996).

[9] E. G. Rawson, Ed., *Selected Papers on Fiber-Optic Local-Area Networks*, SPIE, Bellingham, WA, 1994.

[10] D. W. Smith, *Optical Network Technology*, Chapman & Hall, New York, 1995.

[11] P. E. Green, Jr., *IEEE J. Sel. Areas Commun.* **14**, 764 (1996).

[12] F. E. Ross, *IEEE J. Sel. Areas Commun.* **7**, 1043 (1989).

[13] F. W. Scholl and M. H. Coden, *IEEE J. Sel. Areas Commun.* **6**, 913 (1988).

[14] J. L. Gimlett, M. Z. Iqbal, J. Young, L. Curtis, R. Spicer, and N. K. Cheung, *Electron. Lett.* **25**, 596 (1989).

[15] S. Fujita, M. Kitamura, T. Torikai, N. Henmi, H. Yamada, T. Suzaki, I. Takano, and M. Shikada, *Electron. Lett.* **25**, 702 (1989).

[16] G. P. Agrawal, *Nonlinear Fiber Optics*, 2nd ed., Academic Press, San Diego, CA, 1995.

[17] J. P. Hamaide, P. Emplit, and J. M. Gabriagues, *Electron. Lett.* **26**, 1451 (1990).

[18] D. Marcuse, *J. Lightwave Technol.* **9**, 1330 (1991).

[19] A. Naka and S. Saito, *Electron. Lett.* **28**, 2221 (1992).

[20] A. Mecozzi, *J. Opt. Soc. Am. B* **11**, 462 (1994).

[21] A. Naka and S. Saito, *J. Lightwave Technol.* **12**, 280 (1994).

[22] E. Lichtman, *J. Lightwave Technol.* **13**, 898 (1995).

[23] F. Matera and M. Settembre, *J. Lightwave Technol.* **14**, 1 (1996).

[24] N. Kikuchi and S. Sasaki, *Electron. Lett.* **32**, 570 (1996).

[25] F. Matera, A. Mecozzi, M. Romagnoli, and M. Settembre, *Opt. Lett.* **18**, 1499 (1993).

[26] N. S. Bergano, J. Aspell, C. R. Davidson, P. R. Trischitta, B. M. Nyman, and F. W. Kerfoot, *Electron. Lett.* **27**, 1889 (1991).

[27] T. Imai, M. Murakami, Y. Fukuda, M. Aiki, and T. Ito, *Electron. Lett.* **28**, 1484 (1992).

[28] T. Otani, K. Goto, H. Abe, M. Tanaka, H. Yamamoto, and H. Wakabayashi, *Electron. Lett.* **31**, 380 (1995).

[29] M. Murakami, T. Takahashi, M. Aoyama, M. Amemiya, M. Sumida, N. Ohkawa, Y. Fukuda, T. Imai, and M. Aiki, *Electron. Lett.* **31**, 814 (1995).

[30] T. Matsuda, A. Naka, and S. Saito, *Electron. Lett.* **32**, 229 (1996).

[31] R. J. Sanferrare, *AT&T Tech. J.* **66** (1), 95 (1987).

[32] C. Fan and L. Clark, *Opt. Photon. News* **6** (2), 26 (1995).

[33] I. Jacobs, *Opt. Photon. News* **6** (2), 19 (1995).

[34] P. K. Runge, *AT&T Tech. J.* **71** (1), 5 (1992).

[35] J. M. Sipress, Special issue, *AT&T Tech. J.* **73** (1), 4 (1995).

[36] E. K. Stafford, J. Mariano, and M. M. Sanders, *AT&T Tech. J.* **73** (1), 47 (1995).

[37] T. Welsh, R. Smith, H. Azami, and R. Chrisner, *IEEE Commun. Mag.* **34** (2), 30 (1996).

[38] W. C. Marra and J. Schesser, *IEEE Commun. Mag.* **34** (2), 50 (1996).

[39] P. B. Hansen, L. Eskildsen, S. G. Grubb, A. M. Vengsarkar, S. K. Korotky, T. A. Strasser, J. E. J. Alphonsus, J. J. Veselka, D. J. DiGiovanni, D. W. Peckham, E. C. Beck, D. Truxal, W. Y. Cheung, S. G. Kosinski, D. Gasper, P. F. Wysocki, V. L. da Silva, and J. R. Simpson, *Electron. Lett.* **31**, 1460 (1995).

[40] L. Eskildsen, P. B. Hansen, S. G. Grubb, A. M. Vengsarkar, T. A. Strasser, J. E. J. Alphonsus, D. J. DiGiovanni, D. W. Peckham, D. Truxal, and W. Y. Cheung, *IEEE Photon. Technol. Lett.* **8**, 724 (1996).

[41] P. B. Hansen, L. Eskildsen, S. G. Grubb, A. M. Vengsarkar, S. K. Korotky, T. A. Strasser, J. E. J. Alphonsus, J. J. Veselka, D. J. DiGiovanni, D. W. Peckham, and D. Truxal, *Electron. Lett.* **32**, 1018 (1996).

[42] G. Keiser, *Optical Fiber Communications*, 2nd ed., McGraw-Hill, New York, 1991, Chap. 8.

[43] J. M. Senior, *Optical Fiber Communications: Principles and Practice*, 2nd ed., Prentice Hall, Upper Saddle River, NJ, 1992, Chap. 11.

[44] R. L. Freeman, *Telecommunication System Engineering*, 3rd ed., Wiley, New York, 1996.

[45] C. Kleekamp and B. Metcalf, *Designer's Guide to Fiber Optics*, Cahners, Boston, 1978.

[46] M. Eve, *Opt. Quantum Electron.* **10**, 45 (1978).

[47] P. M. Rodhe, *J. Lightwave Technol.* **3**, 145 (1985).

[48] D. A. Nolan, R. M. Hawk, and D. B. Keck, *J. Lightwave Technol.* **5**, 1727 (1987).

[49] R. D. de la Iglesia and E. T. Azpitarte, *J. Lightwave Technol.* **5**, 1768 (1987).

[50] R. E. Epworth, *IEEE J. Quantum Electron.* **18**, 543 (1982).

[51] D. R. Hjelme and A. R. Michelson, *Appl. Opt.* **22**, 3874 (1983).

[52] P. E. Couch and R. E. Epworth, *J. Lightwave Technol.* **1**, 591 (1983).

[53] T. Kanada, *J. Lightwave Technol.* **2**, 11 (1984).

[54] A. M. J. Koonen, *IEEE J. Sel. Areas Commun.* **4**, 1515 (1986).

[55] P. Chan and T. T. Tjhung, *J. Lightwave Technol.* **7**, 1285 (1989).

[56] P. M. Shankar, *J. Opt. Commun.* **10**, 19 (1989).

[57] G. A. Olson and R. M. Fortenberry, *Fiber Integ. Opt.* **9**, 237 (1990).

[58] J. C. Goodwin and P. J. Vella, *J. Lightwave Technol.* **9**, 954 (1991).

[59] C. M. Olsen, *J. Lightwave Technol.* **9**, 1742 (1991).

[60] K. Abe, Y. Lacroix, L. Bonnell, and Z. Jakubczyk, *J. Lightwave Technol.* **10**, 401 (1992).

[61] D. M. Kuchta and C. J. Mahon, *IEEE Photon. Technol. Lett.* **6**, 288 (1994).

[62] C. M. Olsen and D. M. Kuchta, *Fiber Integ. Opt.* **14**, 121 (1995).

[63] R. J. S. Bates, D. M. Kuchta, and K. P. Jackson, *Opt. Quantum Electron.* **27**, 203 (1995).

[64] C.-L. Ho, *J. Lightwave Technol.* **13**, 1820 (1995).

[65] H. Kosaka, A. K. Dutta, K. Kurihara, Y. Sugimoto, and K. Kasahara, *IEEE Photon. Technol. Lett.* **7**, 926 (1995).

[66] K. Ogawa, *IEEE J. Quantum Electron.* **18**, 849 (1982).

[67] K. Ogawa, in *Semiconductors and Semimetals*, vol. 22C, W. T. Tsang, Ed., Academic Press, San Diego, CA, 1985, pp. 299-330.

[68] W. R. Throssell, *J. Lightwave Technol.* **4**, 948 (1986).

[69] J. C. Campbell, *J. Lightwave Technol.* **6**, 564 (1988).

[70] G. P. Agrawal, P. J. Anthony, and T. M. Shen, *J. Lightwave Technol.* **6**, 620 (1988).

[71] C. M. Olsen, K. E. Stubkjaer, and H. Olesen, *J. Lightwave Technol.* **7**, 657 (1989).

[72] M. Mori, Y. Ohkuma, and N. Yamaguchi, *J. Lightwave Technol.* **7**, 1125 (1989).

[73] W. Jiang, R. Feng, and P. Ye, *Opt. Quantum Electron.* **22**, 23 (1990).

[74] R. S. Fyath and J. J. O'Reilly, *IEE Proc.* **137**, Pt. J, 230 (1990).

[75] W.-H. Cheng and A.-K. Chu, *IEEE Photon. Technol. Lett.* **8**, 611 (1996).

[76] C. H. Henry, P. S. Henry, and M. Lax, *J. Lightwave Technol.* **2**, 209 (1984).

[77] R. A. Linke, B. L. Kasper, C. A. Burrus, I. P. Kaminow, J. S. Ko, and T. P. Lee, *J. Lightwave Technol.* **3**, 706 (1985).

[78] E. E. Basch, R. F. Kearns, and T. G. Brown, *J. Lightwave Technol.* **4**, 516 (1986).

[79] J. C. Cartledge, *J. Lightwave Technol.* **6**, 626 (1988).

[80] N. Henmi, Y. Koizumi, M. Yamaguchi, M. Shikada, and I. Mito, *J. Lightwave Technol.* **6**, 636 (1988).

[81] M. M. Choy, P. L. Liu, and S. Sasaki, *Appl. Phys. Lett.* **52**, 1762 (1988).

[82] P. L. Liu and M. M. Choy, *IEEE J. Quantum Electron.* **25**, 854 (1989); *IEEE J. Quantum Electron.* **25**, 1767 (1989).

[83] D. A. Fishmen, *J. Lightwave Technol.* **8**, 634 (1990).

[84] S. E. Miller, *IEEE J. Quantum Electron.* **25**, 1771 (1989); **26**, 242 (1990).

[85] A. Mecozzi, A. Sapia, P. Spano, and G. P. Agrawal, *IEEE J. Quantum Electron.* **27**, 332 (1991).

[86] K. Kiasaleh and L. S. Tamil, *Opt. Commun.* **82**, 41 (1991).

[87] T. B. Anderson and B. R. Clarke, *IEEE J. Quantum Electron.* **29**, 3 (1993).

[88] A. Valle, P. Colet, L. Pesquera, and M. San Miguel, *IEE Proc.* **140**, Pt. J, 237 (1993).

[89] H. Wu and H. Chang, *IEEE J. Quantum Electron.* **29**, 2154 (1993).

[90] A. Valle, C. R. Mirasso, and L. Pesquera, *IEEE J. Quantum Electron.* **31**, 876 (1995).

[91] D. A. Frisch and I. D. Henning, *Electron. Lett.* **20**, 631 (1984).

[92] R. A. Linke, *Electron. Lett.* **20**, 472 (1984); *IEEE J. Quantum Electron.* **21**, 593 (1985).

[93] T. L. Koch and J. E. Bowers, *Electron. Lett.* **20**, 1038 (1984).

[94] F. Koyama and Y. Suematsu, *IEEE J. Quantum Electron.* **21**, 292 (1985).

[95] A. H. Gnauck, B. L. Kasper, R. A. Linke, R. W. Dawson, T. L. Koch, T. J. Bridges, E. G. Burkhardt, R. T. Yen, D. P. Wilt, J. C. Campbell, K. C. Nelson, and L. G. Cohen, *J. Lightwave Technol.* **3**, 1032 (1985).

[96] G. P. Agrawal and M. J. Potasek, *Opt. Lett.* **11**, 318 (1986).

[97] P. J. Corvini and T. L. Koch, *J. Lightwave Technol.* **5**, 1591 (1987).

[98] J. J. O'Reilly and H. J. A. da Silva, *Electron. Lett.* **23**, 992 (1987).

[99] S. Yamamoto, M. Kuwazuru, H. Wakabayashi, and Y. Iwamoto, *J. Lightwave Technol.* **5**, 1518 (1987).

[100] D. A. Atlas, A. F. Elrefaie, M. B. Romeiser, and D. G. Daut, *Opt. Lett.* **13**, 1035 (1988).

[101] K. Hagimoto and K. Aida, *J. Lightwave Technol.* **6**, 1678 (1988).

[102] H. J. A. da Silva, R. S. Fyath, and J. J. O'Reilly, *IEE Proc.* **136**, Pt. J, 209 (1989).

[103] J. C. Cartledge and G. S. Burley, *J. Lightwave Technol.* **7**, 568 (1989).

[104] J. C. Cartledge and M. Z. Iqbal, *IEEE Photon. Technol. Lett.* **1**, 346 (1989).

[105] G. P. Agrawal and T. M. Shen, *Electron. Lett.* **22**, 1087 (1986).

[106] K. Uomi, S. Sasaki, T. Tsuchiya, H. Nakano, and N. Chinone, *IEEE Photon. Technol. Lett.* **2**, 229 (1990).

[107] S. Kakimoto, Y. Nakajima, Y. Sakakibara, H. Watanabe, A. Takemoto, and N. Yoshida, *IEEE J. Quantum Electron.* **26**, 1460 (1990).

[108] K. Uomi, T. Tsuchiya, H. Nakano, M. Suzuki, and N. Chinone, *IEEE J. Quantum Electron.* **27**, 1705 (1991).

[109] M. Blez, D. Mathoorasing, C. Kazmierski, M. Quillec, M. Gilleron, J. Landreau, and H. Nakajima, *IEEE J. Quantum Electron.* **29**, 1676 (1993).

[110] H. D. Summers and I. H. White, *Electron. Lett.* **30**, 1140 (1994).

[111] F. Kano, T. Yamanaka, N. Yamamoto, H. Mawatan, Y. Tohmori, and Y. Yoshikuni, *IEEE J. Quantum Electron.* **30**, 533 (1994).

[112] S. Mohredik, H. Burkhard, F. Steinhagen, H. Hilmer, R. Lösch, W. Scalapp, and R. Göbel, *IEEE Photon. Technol. Lett.* **7**, 1357 (1996).

[113] K. Morito, R. Sahara, K. Sato, and Y. Kotaki, *IEEE Photon. Technol. Lett.* **8**, 431 (1996).

[114] G. P. Agrawal, *IEEE J. Quantum Electron.* **20**, 468 (1984).

[115] G. P. Agrawal, N. A. Olsson, and N. K. Dutta, *Appl. Phys. Lett.* **45**, 597 (1984).

[116] T. Fujita, S. Ishizuka, K. Fujito, H. Serizawa, and H. Sato, *IEEE J. Quantum Electron.* **20**, 492 (1984).

[117] N. A. Olsson, W. T. Tsang, H. Temkin, N. K. Dutta, and R. A. Logan, *J. Lightwave Technol.* **3**, 215 (1985).

[118] R. W. Tkach and A. R. Chraplyvy, *J. Lightwave Technol.* **4**, 1655 (1986).

[119] G. P. Agrawal and T. M. Shen, *J. Lightwave Technol.* **4**, 58 (1986).

[120] J. L. Gimlett and N. K. Cheung, *J. Lightwave Technol.* **7**, 888 (1989).

[121] M. Tur and E. L. Goldstein, *J. Lightwave Technol.* **7**, 2055 (1989).

[122] S. Yamamoto, N. Edagawa, H. Taga, Y. Yoshida, and H. Wakabayashi, *J. Lightwave Technol.* **8**, 1716 (1990).

[123] G. P. Agrawal, *Proc. SPIE* **1376**, 224 (1991).

[124] K. Petermann, *Laser Diode Modulation and Noise*, 2nd ed., Kluwer Academic, Dordrecht, The Netherlands, 1991.

[125] G. P. Agrawal and N. K. Dutta, *Semiconductor Lasers*, 2nd ed., Van Nostrand Reinhold, New York, 1993.

[126] A. T. Ryan, G. P. Agrawal, G. R. Gray, and E. C. Gage, *IEEE J. Quantum Electron.* **30**, 668 (1994).

[127] K. Petermann, *IEEE J. Sel. Topics Quantum Electron.* **1**, 480 (1995).

[128] R. S. Fyath and R. S. A. Waily, *Int. J. Optoelectron.* **10**, 195 (1995).

[129] M. Shikada, S. Takano, S. Fujita, I. Mito, and K. Minemura, *J. Lightwave Technol.* **6**, 655 (1988).

[130] R. Heidemann, *J. Lightwave Technol.* **6**, 1693 (1988).

[131] K.-P. Ho, J. D. Walker, and J. M. Kahn, *IEEE Photon. Technol. Lett.* **5**, 892 (1993).

[132] U. Fiedler, E. Zeeb, G. Reiner, G. Jost, and K. J. Ebeling, *Electron. Lett.* **30**, 1898 (1994).

[133] D. G. Duff, *IEEE J. Sel. Areas Commun.* **2**, 171 (1984).

[134] T. M. Shen and G. P. Agrawal, *J. Lightwave Technol.* **5**, 653 (1987).

[135] A. F. Elrefaie, J. K. Townsend, M. B. Romeiser, and K. S. Shanmugan, *IEEE J. Sel. Areas Commun.* **6**, 94 (1988).

[136] K. S. Shanmugan, *IEEE J. Sel. Areas Commun.* **6**, 5 (1988).

[137] W. H. Tranter and C. R. Ryan, *IEEE J. Sel. Areas Commun.* **6**, 13 (1988).

[138] R. D. de la Iglesia, *J. Lightwave Technol.* **4**, 767 (1986).

[139] T. J. Batten, A. J. Gibbs, and G. Nicholson, *J. Lightwave Technol.* **7**, 209 (1989).

[140] K. Kikushima and K. Hogari, *J. Lightwave Technol.* **8**, 11 (1990).

[141] M. K. Moaveni and M. Shafi, *J. Lightwave Technol.* **8**, 1064 (1990).

[142] M. C. Jeruchim, P. Balaban, and K. S. Shamugan, *Simulation of Communication Systems*, Plenum Press, New York, 1992.

[143] K. Hinton and T. Stephens, *IEEE J. Sel. Areas Commun.* **11**, 380 (1993).

[144] K. B. Letaief, *IEEE Trans. Commun.* **43**, Pt. 1, 240 (1995).

[145] A. J. Lowery, P. C. R. Gurney, X.-H. Wang, L. V. T. Nguyen, Y.-C. Chan, and M. Premaratne, *Proc. SPIE* **2693**, 624 (1996).

[146] A. J. Lowery, *IEEE Spectrum* **34** (4), 26 (1997).

Chapter 6

COHERENT LIGHTWAVE SYSTEMS

The lightwave systems discussed in Chapter 5 are based on a simple digital transmission scheme in which an electrical bit stream is used to modulate the intensity of the optical carrier and the optical signal is detected directly at a photodiode to convert it to the original digital signal in the electrical domain. Such a scheme is referred to as intensity modulation with direct detection (IM/DD). Many alternative schemes, well known in the context of radio and microwave communication systems [1]–[6], transmit information by modulating the frequency or the phase of the optical carrier and detect the transmitted signal by using homodyne or heterodyne detection techniques. Since phase coherence of the optical carrier plays an important role in the implementation of such schemes, they are referred to as coherent communication techniques, and the fiber-optic communication systems making use of them are called coherent lightwave systems. Coherent transmission techniques were explored during the 1980s, and many field trials established their feasibility by 1990 [7]–[16]. Their deployment has been delayed with the advent of optical amplifiers, although the research and development phase has continued worldwide.

The motivation behind using the coherent communication techniques is twofold. First, the receiver sensitivity can be improved by up to 20 dB compared with IM/DD systems. Such an improvement permits much longer transmission distance (up to an additional 100 km near 1.55 μm) for the same amount of transmitter power. Second, the use of coherent detection allows an efficient use of fiber bandwidth. Many channels can be transmitted simultaneously over the same fiber by using frequency-division multiplexing (FDM) with a channel spacing as small as 1–10 GHz. Although FDM techniques can also be used with the IM/DD scheme, channel spacing typically exceeds 100 GHz in that case. This aspect of coherent lightwave systems is discussed in Chapter 7. In this chapter the focus is on the design and the performance of single-channel coherent systems. The basic concepts behind coherent detection are discussed

239

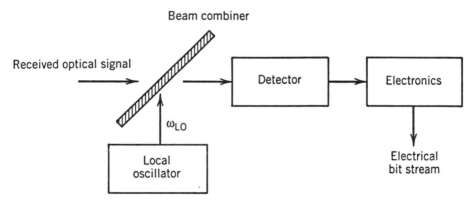

Figure 6.1 Coherent detection scheme. The optical signal is mixed coherently with the output of a local oscillator before it falls on the detector.

in Section 6.1. In Section 6.2 we present various modulation formats and their implementation for coherent optical communications. Section 6.3 is devoted to a discussion of synchronous and asynchronous demodulation schemes used by the coherent receivers. The bit-error rate (BER) for various modulation and demodulation schemes is considered in Section 6.4, where the improvement in receiver sensitivity realized by the use of coherent communication techniques is discussed. In Section 6.5 we discuss the degradation of receiver sensitivity through various mechanisms such as phase noise, intensity noise, polarization mismatch, fiber dispersion, and reflection feedback, with emphasis on the design issues. In the final section we review the performance of coherent lightwave systems by considering system experiments and field trials.

6.1   BASIC CONCEPTS

6.1.1   Local Oscillator

The basic idea behind coherent lightwave systems is to mix the received signal coherently with another optical wave before it is incident on the photodetector (see Fig. 6.1). The optical wave is generated locally at the receiver by using a narrow-linewidth laser, called a *local oscillator* (LO)—a term borrowed from radio communication literature. To see how mixing the received signal with the LO output can improve receiver performance, consider the optical signal and write it as (in complex notation)

$$E_s = A_s \exp[-i(\omega_0 t + \phi_s)], \qquad (6.1.1)$$

where ω_0 is the carrier frequency, A_s is the amplitude, and ϕ_s is the phase. The optical field associated with the local oscillator is given by a similar expression,

$$E_{LO} = A_{LO} \exp[-i(\omega_{LO} t + \phi_{LO})], \qquad (6.1.2)$$

where A_{LO}, ω_{LO}, and ϕ_{LO} are the amplitude, frequency, and phase of the local oscillator, respectively. The scalar notation is used for both E_s and E_{LO} by assuming that the two fields are identically polarized; polarization-mismatch issues are discussed in Section 6.5.3. The photodetector in Fig. 6.1 responds to the intensity $|E_s+E_{LO}|^2$. Since the optical power is proportional to the intensity, the power received at the photodetector is given by $P = K|E_s + E_{LO}|^2$, where K is a constant of proportionality. Thus,

$$P(t) = P_s + P_{LO} + 2\sqrt{P_sP_{LO}}\,\cos(\omega_{IF}t + \phi_s - \phi_{LO}), \qquad (6.1.3)$$

where

$$P_s = KA_s^2, \quad P_{LO} = KA_{LO}^2, \quad \omega_{IF} = \omega_0 - \omega_{LO}. \qquad (6.1.4)$$

The frequency ν_{IF}, defined as $\nu_{IF} = \omega_{IF}/2\pi$, is known as the *intermediate frequency* (IF). When $\omega_0 \neq \omega_{LO}$, the optical signal is demodulated in two stages; its carrier frequency is first converted to an intermediate frequency ν_{IF} (typically 0.1–5 GHz) before the signal is demodulated to the baseband. It is not always necessary to use an intermediate frequency. In fact, there are two different coherent detection techniques to choose from, depending on whether or not ω_{IF} equals zero. They are known as *homodyne* and *heterodyne* detection techniques and are discussed in the following two subsections.

6.1.2   Homodyne Detection

In this coherent-detection technique, the local-oscillator frequency ω_{LO} is selected to coincide with the signal-carrier frequency ω_0 so that $\omega_{IF} = 0$. By using Eq. (6.1.3), the photocurrent ($I = RP$, where R is the detector responsivity) is given by

$$I(t) = R(P_s + P_{LO}) + 2R\sqrt{P_sP_{LO}}\,\cos(\phi_s - \phi_{LO}). \qquad (6.1.5)$$

Typically, $P_{LO} \gg P_s$, and $P_s+P_{LO} \approx P_{LO}$. The last term in Eq. (6.1.5) contains the information transmitted and is used by the decision circuit. Consider the case in which the local-oscillator phase is locked to the signal phase so that $\phi_s = \phi_{LO}$. The homodyne signal is then given by

$$I_p(t) = 2R\sqrt{P_sP_{LO}}. \qquad (6.1.6)$$

The advantage of homodyne detection is evident from Eq. (6.1.6) if we note that the signal current for the direct-detection case is given by $I_{dd}(t) = RP_s(t)$. The average electrical signal power is increased by a factor of $4P_{LO}/\bar{P}_s$ by the use of homodyne detection. Since $4P_{LO}/\bar{P}_s$ can be made $\gg 1$, the enhancement can be by orders of magnitude. Although shot noise is also enhanced, it is shown in Section 6.1.4 that homodyne detection improves the SNR by a large factor.

Another advantage of coherent detection is evident from Eq. (6.1.5). Since the last term contains the signal phase explicitly, it is possible to transmit information by modulating the phase of the optical carrier. Direct detection does not allow phase or frequency modulation, as all phase information about the signal phase is lost. The modulation formats for coherent systems are discussed separately in Section 6.2.

A disadvantage of homodyne detection also results from its phase sensitivity. Since the last term in Eq. (6.1.5) contains the local-oscillator phase ϕ_{LO} explicitly, clearly ϕ_{LO} should be controlled. Ideally, ϕ_s and ϕ_{LO} should stay constant except for the intentional modulation of ϕ_s. In practice, both ϕ_s and ϕ_{LO} fluctuate with time in a random manner. However, their difference $\phi_s - \phi_{LO}$ can be made to remain nearly constant through an optical phase-locked loop. The implementation of such a loop is not simple and makes the design of optical homodyne receivers quite complicated. In addition, matching the transmitter and local-oscillator frequencies puts stringent requirements on the two optical sources. These problems can be overcome by the use of heterodyne detection, discussed next.

6.1.3 Heterodyne Detection

In the case of heterodyne detection the local-oscillator frequency ω_{LO} is chosen to differ form the signal-carrier frequency ω_0 such that the intermediate frequency ω_{IF} is in the microwave region ($\nu_{IF} \sim 1$ GHz). By using Eq. (6.1.3) together with $I = RP$, the detector current is given by

$$I(t) = R(P_s + P_{LO}) + 2R\sqrt{P_s P_{LO}} \cos(\omega_{IF} t + \phi_s - \phi_{LO}). \qquad (6.1.7)$$

Since $P_{LO} \gg P_s$ in practice, the nearly constant direct-current (dc) term can be filtered out easily. The heterodyne signal is thus given by the alternating-current (ac) term in Eq. (6.1.7) or by

$$I_{ac}(t) = 2R\sqrt{P_s P_{LO}} \cos(\omega_{IF} t + \phi_s - \phi_{LO}). \qquad (6.1.8)$$

As for homodyne detection, information can be transmitted through amplitude, phase, or frequency modulation of the optical carrier. Furthermore, similar to the homodyne case, the local oscillator amplifies the received signal, thereby improving the SNR. However, the SNR improvement is lower by a factor of 2 (or by 3 dB) than in the homodyne case. This reduction is referred to as the 3-dB heterodyne-detection penalty. The origin of the 3-dB penalty can be seen by considering the signal power (proportional to the square of the current). Because of the ac nature of I_{ac}, the average signal power is lower by a factor of 2 when I_{ac}^2 is averaged over a cycle at the intermediate frequency (recall that the average of $\cos^2 \theta$ over θ is $1/2$).

The advantage gained at the expense of the 3-dB penalty is that the receiver design is considerably simplified since an optical phase-locked loop is no longer needed. Fluctuations in both ϕ_s and ϕ_{LO} still need to be controlled by using narrow-linewidth semiconductor lasers for both optical sources. However, as discussed in Section 6.5.1, the linewidth requirements are quite moderate for the case of an asynchronous demodulation scheme. This feature makes heterodyne-detection schemes suitable for practical implementation of coherent lightwave systems.

6.1.4 Signal-to-Noise Ratio

The advantage of coherent detection for lightwave systems can be made more quantitative by considering the SNR of the receiver current. For this purpose, it is necessary to extend the analysis of Section 4.4 to heterodyne detection. The receiver current fluctuates because of shot noise and thermal noise. The variance σ^2 of current fluctuations is obtained by adding the two contributions so that

$$\sigma^2 = \sigma_s^2 + \sigma_T^2, \tag{6.1.9}$$

where

$$\sigma_s^2 = 2q(I + I_d)\Delta f, \quad \sigma_T^2 = (4k_BT/R_L)F_n\Delta f. \tag{6.1.10}$$

The notation used here is the same as in Section 4.4. The main difference from the analysis of Section 4.4 occurs in the shot-noise contribution. The current I in Eq. (6.1.10) is the total photocurrent generated at the detector and is given by Eq. (6.1.5) or Eq. (6.1.7), depending on whether homodyne or heterodyne detection is employed. In practice, $P_{LO} \gg P_s$, and I in Eq. (6.1.10) can be replaced by the dominant term RP_{LO} for both cases.

The SNR is obtained by dividing the average signal power by the average noise power. In the heterodyne case, SNR is given by

$$\text{SNR} = \frac{\langle I_{ac}^2 \rangle}{\sigma^2} = \frac{2R^2\bar{P}_s P_{LO}}{2q(RP_{LO} + I_d)\Delta f + \sigma_T^2}. \tag{6.1.11}$$

In the homodyne case the SNR is larger by a factor of 2 if we assume that $\phi_s = \phi_{LO}$ in Eq. (6.1.5). The main advantage of coherent detection can be seen from Eq. (6.1.11). Since the local-oscillator power P_{LO} can be controlled at the receiver, it can be made large enough that the receiver noise is dominated by shot noise. More specifically, $\sigma_s^2 \gg \sigma_T^2$ when

$$P_{LO} \gg \sigma_T^2/(2qR\Delta f). \tag{6.1.12}$$

Under the same conditions, the dark-current contribution to the shot noise is negligible ($I_d \ll RP_{LO}$). The SNR is then given by

$$\text{SNR} \approx \frac{R\bar{P}_s}{q\Delta f} = \frac{\eta \bar{P}_s}{h\nu\Delta f}, \tag{6.1.13}$$

where $R = \eta q/h\nu$ was used from Eq. (4.1.3). The use of coherent detection allows one to achieve the shot-noise limit even for p–i–n receivers whose performance is generally dominated by thermal noise. Moreover, in contrast with the case of APD receivers, this limit is realized without adding excess shot noise.

It is useful to express SNR in terms of the number of photons, N_p, received within a single bit. At the bit rate B, the average signal power $\bar{P}_s$ is related to N_p as $\bar{P}_s = N_p h\nu B$. Typically, $\Delta f \approx B/2$. By using these values of $\bar{P}_s$ and Δf in Eq. (6.1.13), the SNR is given by a simple expression

$$\text{SNR} = 2\eta N_p. \tag{6.1.14}$$

For the case of homodyne detection, SNR is larger by a factor of 2 and is given by SNR $= 4\eta N_p$. Section 6.4 discusses the dependence of the BER on SNR and shows how receiver sensitivity is improved by the use of coherent detection.

6.2 MODULATION FORMATS

As discussed in Section 6.1, an important advantage of using the coherent detection techniques is that both the amplitude and the phase of the received optical signal can be detected and measured. This feature opens up the possibility of sending information by modulating either the amplitude, or the phase, or the frequency of the optical carrier. In the case of digital communication systems, the three possibilities give rise to three modulation formats: *amplitude-shift keying* (ASK), *phase-shift keying* (PSK), and *frequency-shift keying* (FSK) [1]–[6]. Figure 6.2 shows schematically the three modulation formats for a specific bit pattern. In the following subsections we consider each format separately and discuss its implementation in practical lightwave systems.

6.2.1 ASK Format

The electric field associated with the optical signal can be written as [by taking the real part of Eq. (6.1.1)]

$$E_s(t) = A_s(t) \cos[\omega_0 t + \phi_s(t)]. \qquad (6.2.1)$$

In the case of ASK format, the amplitude A_s is modulated while keeping ω_0 and ϕ_s constant. For binary digital modulation, A_s takes one of the two fixed values during each bit period, depending on whether "1" or "0" bit is being transmitted. In most practical situations, A_s is set to zero during transmission of "0" bits. The ASK format is then called *on–off keying* (OOK) and is identical with the modulation scheme commonly used for noncoherent (IM/DD) digital lightwave systems.

The implementation of ASK for coherent systems differs considerably from the case of the direct-detection systems discussed in Chapter 5. Whereas the optical bit stream for direct-detection systems is often generated by modulating a LED or a semiconductor laser directly, an external modulator becomes a necessity for coherent communication systems. The reason behind this necessity is related to phase changes that invariably occur when the amplitude A_s (or the power) is changed by modulating the current applied to a semiconductor laser (see Section 3.3.7). For IM/DD systems, such unintentional phase changes are not seen by the detector (as the detector responds only to the optical power) and are not of major concern except for the chirp-induced power penalty discussed in Section 5.4.4. The situation is entirely different in the case of coherent systems, where the detector response depends on the phase of the received signal. The implementation of ASK format for coherent systems requires the phase ϕ_s to remain nearly constant. This is achieved by operating the semiconductor laser continuously at a constant current and modulating its output by using an

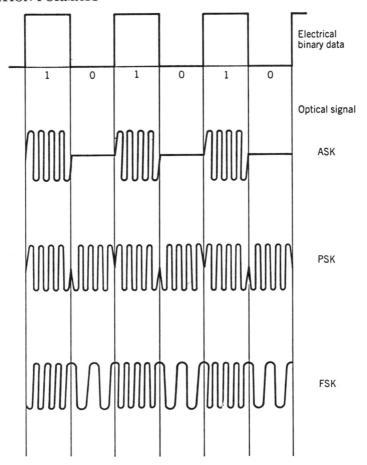

Figure 6.2 ASK, PSK, and FSK modulation formats for a specific bit pattern shown on the top. In each case rapid oscillations correspond to variations of the electromagnetic field at the optical carrier frequency.

external modulator. Since all external modulators have some insertion loss, a power penalty is generally paid whenever an external modulator is used; it can be reduced to below 1 dB for monolithically integrated modulators.

Commonly used external modulators make use of titanium-diffused $LiNbO_3$ waveguides in a Mach–Zehnder or a directional-coupler configuration [17]–[20]. The Mach–Zehnder interferometer design is shown in Fig. 6.3(a). The refractive index of electro-optic materials such as $LiNbO_3$ can be changed by applying an external voltage. In the absence of external voltage, the optical fields in the two arms of the Mach–Zehnder interferometer experience identical phase shifts and interfere constructively. The additional phase shift introduced in one of the arms through voltage-induced index changes destroys the constructive nature of the interference and reduces the transmitted intensity. In particular, no light is transmitted when the phase difference between the two arms equals

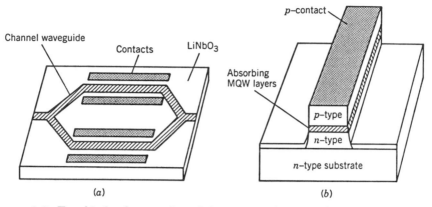

Figure 6.3 Two kinds of external modulators used for ASK: (a) LiNbO$_3$ waveguide modulator in the Mach–Zehnder configuration; (b) semiconductor waveguide modulator based on electroabsorption.

π, because of destructive interference occurring in that case. As a result, the electrical bit stream applied to the modulator produces an optical replica of the bit stream. The performance of external modulators is quantified through the on–off ratio (also called extinction ratio) and the modulation bandwidth. LiNbO$_3$ modulators provide an on–off ratio in excess of 20 and can be modulated at speeds up to 75 GHz [18], [19]. The driving voltage is typically 5 V but can be reduced to near 3 V with a suitable design [20]. Other materials can also be used to make external modulators. For example, modulators have been fabricated using electro-optic polymers, and modulation bandwidths of up to 60 GHz have been demonstrated [21]. Such modulators can be integrated monolithically with the driving circuitry [22].

Modulators can also be made using semiconductors. The basic mechanism behind the operation of semiconductor modulators is electroabsorption discussed in Section 3.4.4. All semiconductors begin to absorb incident light when photon energy exceeds the bandgap energy. Since the electroabsorption effect is stronger in multiquantum-well (MQW) structures, such a structure [see Fig. 6.3(b)] has been studied extensively [23]–[26]. Modulation bandwidths in excess of 20 GHz were demonstrated [23] as early as 1989 and later extended to 50 GHz. By 1996, polarization-insensitive 1.55-μm modulators with a large bandwidth (42 GHz) and a low operating voltage (< 2 V) have been made using a strained-MQW design [26]. An advantage of semiconductor modulators is that they can be integrated with the semiconductor laser on the same chip (see Section 3.4.4). Such monolithically integrated transmitters can operate at bit rates as high as 20 Gb/s and are available commercially.

6.2.2 PSK Format

In the case of PSK format, the optical bit stream is generated by modulating the phase ϕ_s in Eq. (6.2.1) while the amplitude A_s and the frequency ω_0 of the optical carrier are kept constant. For binary PSK, the phase ϕ_s takes two values, commonly chosen to be 0 and π. Figure 6.2 shows the binary PSK format schematically for a specific bit pattern. An interesting aspect of the PSK format is that the optical intensity remains constant during all bits and the signal appears to have a continuous-wave (CW) form. Coherent detection is a necessity for PSK, as all information would be lost if the optical signal were detected directly without mixing it with the output of a local oscillator.

The implementation of PSK requires an external modulator capable of changing the optical phase in response to an applied voltage. The physical mechanism used by such modulators is called electrorefraction. Any electro-optic crystal with proper orientation can be used for phase modulation. A LiNbO$_3$ crystal is commonly used in practice. The design of LiNbO$_3$-based phase modulators is much simpler than that shown in Fig. 6.3(a), as a Mach–Zehnder interferometer is no longer needed; a single waveguide is used in its place. The phase shift $\delta\phi$ is related to the index change δn by the simple relation

$$\delta\phi = (2\pi/\lambda)(\delta n)l_m, \tag{6.2.2}$$

where l_m is the length over which index change is induced by the applied voltage. The index change δn is proportional to the applied voltage, which is chosen such that $\delta\phi = \pi$. A π phase shift can be imposed on the optical carrier by applying the required voltage for the duration of each "1" bit.

Semiconductors can also be used to make phase modulators, especially if a MQW structure is used, since the electrorefraction effect originating from the *quantum-confinement Stark effect* is enhanced for a quantum-well design. Such MQW phase modulators have been developed [27]–[32] and are able to operate at a bit rate of up to 20 Gb/s in the wavelength range 1.3–1.6 μm. By 1992, MQW devices had a modulation bandwidth of 20 GHz and required only 3.85 V for introducing a π phase shift when operated near 1.55 μm [27]. The operating voltage was further reduced to 2.8 V in a phase modulator based on the electro-absorption effect in a MQW waveguide [28]. However, the module exhibited an insertion loss of 8 dB because of high coupling losses. A spot-size converter is sometimes integrated with the phase modulator to reduce coupling losses [29]. The best performance is achieved when a semiconductor phase modulator is monolithically integrated within the transmitter [30]. Such transmitters are quite useful for coherent lightwave systems. Several other techniques can be used to make phase modulators [31], [32].

The use of PSK format requires that the phase of the optical carrier remain stable so that phase information can be extracted at the receiver without ambiguity. This requirement puts a stringent condition on the tolerable linewidths of the transmitter laser and the local oscillator. As discussed later in Section 6.5.1, the linewidth requirement can be somewhat relaxed by using a variant of the PSK format, known as *differential phase-shift keying* (DPSK). In the case of DPSK, information is coded by using the phase difference between two neighboring bits. For instance, if ϕ_k represents the phase of the kth bit, the phase difference $\Delta\phi = \phi_k - \phi_{k-1}$ is changed by π or 0, depending on whether kth bit is a 1 or 0 bit. The advantage of DPSK is that the transmittal signal can be demodulated successfully as long as the carrier phase remains relatively stable over a duration of two bits.

6.2.3   FSK Format

In the case of FSK modulation, information is coded on the optical carrier by shifting the carrier frequency ω_0 itself [see Eq. (6.2.1)]. For a binary digital signal, ω_0 takes two values, $\omega_0 + \Delta\omega$ and $\omega_0 - \Delta\omega$, depending on whether a 1 or 0 bit is being transmitted. The shift $\Delta f = \Delta\omega/2\pi$ is called the *frequency deviation*. The quantity $2\Delta f$ is sometimes called *tone spacing*, as it represents the frequency spacing between 1 and 0 bits. The optical field for FSK format can be written as

$$E_s(t) = A_s \cos[(\omega_0 \pm \Delta\omega)t + \phi_s], \tag{6.2.3}$$

where $+$ and $-$ signs correspond to 1 and 0 bits. By noting that the argument of cosine can be written as $\omega_0 t + (\phi_s \pm \Delta\omega t)$, the FSK format can also be viewed as a kind of PSK modulation such that the carrier phase increases or decreases linearly over the bit duration.

The choice of the frequency deviation Δf depends on the available bandwidth. The total bandwidth of a FSK signal is given approximately by $2\Delta f + 2B$, where B is the bit rate [1]. When $\Delta f \gg B$, the bandwidth approaches $2\Delta f$ and is nearly independent of the bit rate. This case is often referred to as *wide-deviation* or wideband FSK. In the opposite case of $\Delta f \ll B$, called *narrow-deviation* or narrowband FSK, the bandwidth approaches $2B$. The ratio $\beta_{FM} = \Delta f/B$, called the FM index, serves to distinguish the two cases, depending on whether $\beta_{FM} \gg 1$ or $\beta_{FM} \ll 1$.

The implementation of FSK requires modulators capable of shifting the frequency of the incident optical signal. Electro-optic materials such as $LiNbO_3$ normally produce a phase shift proportional to the applied voltage. They can be used for FSK by applying a triangular voltage pulse (sawtooth-like), since a linear phase change corresponds to a frequency shift. An alternative technique makes use of Bragg scattering from acoustic waves. Such modulators are called acousto-optic modulators. Their use is somewhat cumbersome in the bulk form. However, they can be fabricated in compact form by using surface acoustic waves on a slab waveguide. A problem with the acousto-optic modulators is that a frequency shift is accompanied by a shift in the diffraction angle, making alignment difficult for large frequency shifts. Generally, the frequency shifts are limited to below 1 GHz for such techniques.

The simplest and most commonly used method for the FSK format makes use of the direct-modulation capability of semiconductor lasers. As discussed in Section 3.4, a change in the operating current of a semiconductor laser leads to changes in both the amplitude and frequency of emitted light. In the case of ASK or OOK, the frequency shift chirps the emitted optical pulse and is undesirable. But the same frequency shift can be used to advantage for the purpose of FSK. Typical values of frequency shifts are 0.1–1 GHz/mA. Therefore, only a small

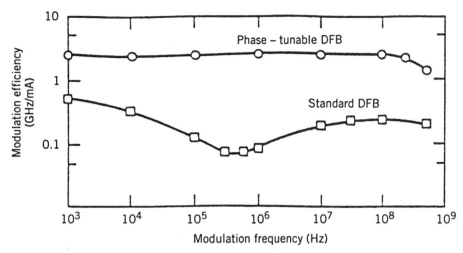

Figure 6.4 FM response of a typical DFB semiconductor laser exhibiting a dip in the frequency range 0.1–10 MHz. (After Ref. [12]. ©1988 IEEE. Reprinted with permission.)

change in the operating current (~ 1 mA) is required to produce frequency shifts of 1 GHz. Such current changes are small enough that the amplitude changes little from bit to bit.

DFB semiconductor lasers are often used for coherent transmission because of their ability to operate in a single longitudinal mode with a narrow linewidth. For the purpose of FSK, the FM response of such lasers should be flat over the entire bandwidth. Unfortunately, this is not the case in practice. Semiconductor lasers typically exhibit a dip in their FM response over the frequency range 0.1–10 MHz [33]. The reason is that two different physical phenomena contribute to the frequency shift when the device current is changed. Changes in the refractive index, responsible for the frequency shift, can occur either because of a temperature shift or because of a change in the carrier density. The thermal effects contribute only up to modulation frequencies of about 1 MHz because of a slow thermal response. The FM response decreases in the frequency range 0.1–10 MHz because the thermal contribution and the carrier-density contribution occur with opposite phases. Figure 6.4 shows the FM response of a typical DFB laser.

The nonuniformity of FM response is undesirable for FSK coherent lightwave systems, and many techniques are used to make it uniform. An equalization circuit can be used, although it generally reduces the modulation efficiency. Another technique makes use of transmission codes which reduce the low-frequency components of the data where distortion is highest. New types of multisection DFB lasers have been developed to achieve uniform FM response [34]–[40]. Figure 6.4 shows the FM response of a two-section DFB laser. It is not only uniform in the range up to about 1 GHz, but its modulation efficiency is also high. Even better performance has been realized by using three-section DFB lasers, which are described in Section 3.3.5. Such lasers consist of an active section, a phase-shift section, and a Bragg-mirror section, each of which can be independently biased (see Fig. 3.20). Flat FM response from 100 kHz to 15 GHz was demonstrated [34] in 1990 in such a three-section DFB laser, which continued to operate in a single longitudinal mode with a narrow linewidth (< 1 MHz). By 1995, the use of gain-coupled, phase-shifted, DFB lasers has extended the range of uniform FM response from 10 kHz to 20 GHz [38]. When FSK is performed through direct modulation, the carrier phase varies continuously from bit to bit. The FSK format in that case is often referred to as continuous-phase FSK (CPFSK). When the tone spacing $2\Delta f$ is chosen to be $B/2$ ($\beta_{FM} = 1/2$), CPFSK is also called minimum-shift keying (MSK).

6.3   DEMODULATION SCHEMES

As discussed in Section 6.1, either homodyne or heterodyne detection can be used to convert the received optical signal into electrical form. In the case of homodyne detection, the optical signal is demodulated directly to the baseband. Although simple in concept, homodyne detection is difficult to implement in practice, as it requires a local oscillator whose frequency matches the carrier frequency exactly and whose phase is locked to the incoming signal. Such a demodulation scheme is called synchronous and is essential for homodyne detection. Although optical phase-locked loop have been developed for this purpose, their use is complicated in practice. Heterodyne detection simplifies the receiver design, as neither optical phase locking nor frequency matching of the local oscillator is required. However, the electrical signal is in the form of microwaves and must be demodulated from the intermediate frequency to the baseband by using techniques developed for microwave communication systems [1]–[6]. Demodulation can be carried out either synchronously or asynchronously. Asynchronous demodulation is also called incoherent in radio communication literature. In optical communication literature, the term *coherent detection* is used in a wider sense. A lightwave system is called coherent as long as it uses a local oscillator for homodyne or heterodyne mixing of the optical signal irrespective of the demodulation technique used to convert the intermediate-frequency signal to baseband frequencies. In the following subsections we describe synchronous and asynchronous demodulation schemes for the heterodyne signal.

6.3.1   Heterodyne Synchronous Demodulation

Figure 6.5 shows a block diagram of a synchronous heterodyne receiver. The current generated at the photodiode is passed through a bandpass filter (BPF) centered at the intermediate frequency ω_{IF}. The filtered current in the absence of noise can be written as [see Eq. (6.1.8)]

$$I_f(t) = I_p \cos(\omega_{IF} t - \phi), \qquad (6.3.1)$$

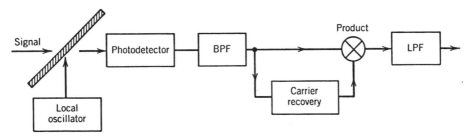

Figure 6.5 Block diagram of a synchronous heterodyne receiver.

where I_p is given by Eq. (6.1.6) and ϕ is the phase difference between the local oscillator and signal phases. The receiver noise is also filtered by the BPF. If we use the in-phase and out-of-phase quadrature components of the filtered Gaussian noise [1], the receiver noise is included through

$$I_f(t) = (I_p \cos \phi + i_c) \cos(\omega_{IF} t) + (I_p \sin \phi + i_s) \sin(\omega_{IF} t), \qquad (6.3.2)$$

where i_c and i_s are Gaussian random variables of zero mean whose variance σ^2 is given by Eq. (6.1.9). For synchronous demodulation, $I_f(t)$ is multiplied by $\cos(\omega_{IF} t)$ and filtered by a low-pass filter. The resulting baseband signal is

$$I_d = \langle I_f \cos(\omega_{IF} t) \rangle = \tfrac{1}{2}(I_p \cos \phi + i_c), \qquad (6.3.3)$$

where angle brackets denote low-pass filtering that leads to rejection of ac components oscillating at $2\omega_{IF}$. Equation (6.3.3) shows that only the in-phase noise component affects the performance of synchronous heterodyne receivers.

Synchronous demodulation requires recovery of the microwave carrier at the intermediate frequency ω_{IF}. Several electronic schemes can be used for this purpose, all requiring a kind of electrical phase-locked loop [41]. Two commonly used loops are the *squaring loop* and the *Costas loop*. A squaring loop uses a square-law device to obtain a signal of the form $\cos^2(\omega_{IF} t)$ that has a frequency component at $2\omega_{IF}$. This component can be used to generate a microwave signal at ω_{IF}.

6.3.2 Heterodyne Asynchronous Demodulation

Figure 6.6 shows a block diagram of an asynchronous heterodyne receiver. It does not require recovery of the microwave carrier at the intermediate frequency, resulting in a much simpler receiver design. The filtered signal $I_f(t)$ is converted to the baseband by using an *envelope detector*, followed by a low-pass filter. The signal received by the decision circuit is just $I_d = |I_f|$, where I_f is given by Eq. (6.3.2). It can be written as

$$I_d = |I_f| = [(I_p \cos \phi + i_c)^2 + (I_p \sin \phi + i_s)^2]^{1/2}. \qquad (6.3.4)$$

The main difference is that both the in-phase and out-of-phase quadrature components of the receiver noise affect the signal. The SNR is thus degraded compared with the case of synchronous demodulation. As discussed in Section 6.4,

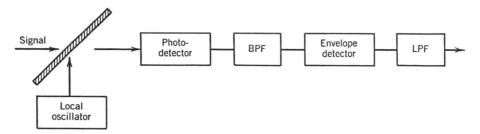

Figure 6.6 Block diagram of an asynchronous heterodyne receiver.

asynchronous receivers are less sensitive because of the SNR degradation. However, sensitivity degradation is quite small (about 0.5 dB). At the same time, the linewidth requirements for the transmitter and the local oscillator are quite modest in the case of asynchronous demodulation (see Section 6.5). For this reason, asynchronous heterodyne receivers play an important role in the design of coherent lightwave systems.

The asynchronous heterodyne receiver shown in Fig. 6.6 requires modifications when the FSK and PSK modulation formats are used. Figure 6.7 shows two demodulation schemes. The FSK *dual-filter receiver* uses two separate branches to process the 1 and 0 bits whose carrier frequencies, and hence the intermediate frequencies, are different. The scheme can be used whenever the tone spacing is much larger than the bit rates, so that the spectra of 1 and 0 bits have negligible overlap (wide-deviation FSK). The two BPFs have their center frequencies separated exactly by the tone spacing so that each BPF passes either 1 or 0 bits only. The FSK dual-filter receiver can be thought of as two ASK single-filter receivers in parallel whose outputs are combined before reaching the decision circuit. A single-filter receiver of Fig. 6.6 can be used for FSK demodulation if its bandwidth is chosen to be wide enough to pass the entire bit stream. The signal is then processed by a frequency discriminator to identify 1 and 0 bits. This scheme works well only for narrow-deviation FSK, for which tone spacing is less than or comparable to the bit rate ($\beta_{FM} \leq 1$).

Asynchronous demodulation cannot be used for the PSK format, as the phase of the transmitter laser and the local oscillator are not locked and can drift with time. However, the use of DPSK format permits asynchronous demodulation by using the delay scheme shown in Fig. 6.7(b). The idea is to multiply the received bit stream by a replica of it that has been delayed by one bit period. The resulting signal has a component of the form $\cos(\phi_k - \phi_{k-1})$, where ϕ_k is the phase of kth bit, which can be used to recover the bit pattern since information is encoded in the phase difference $\phi_k - \phi_{k-1}$. Such a scheme requires phase stability only over a few bits and can be implemented by using DFB semiconductor lasers. The delay-demodulation scheme can also be used for CPFSK. The amount of delay in that case depends on the tone spacing and is chosen such that the phase is shifted by π for the delayed signal.

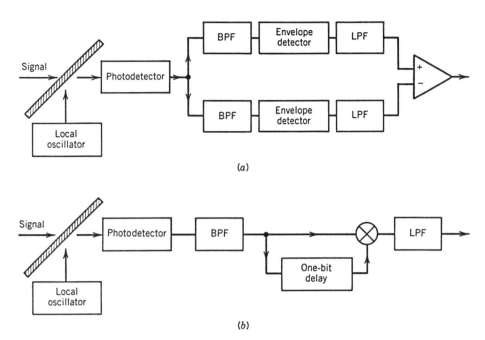

(a)

(b)

Figure 6.7 (a) Dual-filter FSK and (b) delay-demodulation DPSK asynchronous heterodyne receivers.

6.4   BIT-ERROR RATE

The preceding three sections have provided enough background material for calculating the bit-error rate (BER) of coherent lightwave systems. However, the BER, and hence the receiver sensitivity, depend on the modulation format as well as on the demodulation scheme used by the coherent receiver. The section considers each case separately.

6.4.1   Synchronous ASK Receivers

Consider first the case of heterodyne detection. The signal used by the decision circuit is given by Eq. (6.3.3). The phase ϕ generally varies randomly because of phase fluctuations associated with the transmitter laser and the local oscillator. As discussed in Section 6.5, the effect of phase fluctuations can be made negligible by using semiconductor lasers whose linewidth is a small fraction of the bit rate. Assuming this to be the case and setting $\phi = 0$ in Eq. (6.3.2), the decision signal is given by

$$I_d = \tfrac{1}{2}(I_p + i_c), \qquad (6.4.1)$$

where $I_p = 2R(P_s P_{LO})^{1/2}$ from Eq. (6.1.6). I_p takes values I_1 or I_0 depending on whether 1 or 0 bit is being detected. For simplicity, consider the case $I_0 = 0$, in which no power is transmitted during the 0 bits. Except for the factor of $1/2$ in Eq. (6.4.1), the situation is analogous to the case of direct detection discussed in Section 4.5. The factor of 1/2 does not affect the BER since both the signal and the noise are reduced by the same factor, leaving the SNR unchanged. In fact, one can use the same result [Eq. (4.5.9)],

$$\text{BER} = \frac{1}{2}\,\text{erfc}\left(\frac{Q}{\sqrt{2}}\right), \qquad (6.4.2)$$

where Q is given by Eq. (4.5.10) and can be written as

$$Q = \frac{I_1 - I_0}{\sigma_1 + \sigma_0} \approx \frac{I_1}{2\sigma_1} = \frac{1}{2}(\text{SNR})^{1/2}. \qquad (6.4.3)$$

In relating Q to SNR, we used $I_0 = 0$ and set $\sigma_0 \approx \sigma_1$. The latter approximation is justified for most coherent receivers whose noise is dominated by the shot noise induced by local-oscillator power and remains the same irrespective of the received signal power. Indeed, as shown in Section 6.1.4, the SNR of such receivers can be related to the number of photons received during each 1 bit by the simple relation SNR $= 2\eta N_p$ [see Eq. (6.1.15)], where η is the quantum efficiency of the photodetector. Equations (6.4.2) and (6.4.3) together with the relation above provide the following expression for the BER:

$$\text{BER} = \tfrac{1}{2}\,\text{erfc}(\sqrt{\eta N_p/4}\,). \qquad \text{[ASK heterodyne]} \qquad (6.4.4)$$

One can use the same method to calculate the BER for the case of ASK homodyne receivers. Equations (6.4.2) and (6.4.3) still remain applicable. However, the SNR is improved by 3 dB for the homodyne case, so that SNR $= 4\eta N_p$. The BER is thus given by

$$\text{BER} = \tfrac{1}{2}\,\text{erfc}(\sqrt{\eta N_p/2}\,). \qquad \text{[ASK homodyne]} \qquad (6.4.5)$$

Equations (6.4.4) and (6.4.5) can be used to calculate the receiver sensitivity at a specific BER. Similar to the direct-detection case discussed in Section 4.4, one can define the receiver sensitivity $\bar{P}_{\text{rec}}$ as the average received power corresponding to a BER of 10^{-9}. From Eqs. (6.4.2) and (6.4.3), BER $= 10^{-9}$ when $Q \approx 6$ or when SNR $= 144$ (21.6 dB). For the ASK heterodyne case we can use Eq. (6.1.14) to relate SNR to $\bar{P}_{\text{rec}}$ if we note that $\bar{P}_{\text{rec}} = P_s/2$ simply because signal power is zero during the 0 bits. The result is

$$\bar{P}_{\text{rec}} = 2Q^2 h\nu\Delta f/\eta = 72h\nu\Delta f/\eta. \qquad (6.4.6)$$

For the ASK homodyne case, $\bar{P}_{\text{rec}}$ is smaller by a factor of 2 because of the 3-dB homodyne-detection advantage discussed in Section 6.1.3. As an example, for a 1.55-μm ASK heterodyne receiver with $\eta = 0.8$ and $\Delta f = 1$ GHz, the receiver sensitivity is about 12 nW and reduces to 6 nW if homodyne detection is used.

The receiver sensitivity is often quoted in terms of the number of photons N_p by using Eqs. (6.4.4) and (6.4.5), as such a definition makes it independent of the receiver bandwidth and the operating wavelength. Furthermore, η is also set to 1 so that the sensitivity corresponds to an ideal photodetector. It is easy to verify that for BER $= 10^{-9}$, $N_p = 72$ and 36 for the heterodyne and homodyne cases, respectively. It is important to remember that N_p corresponds to the number of photons within a single 1 bit. The average number of photons per bit, $\bar{N}_p$, is reduced by a factor of 2 if we assume that 0 and 1 bits are equally likely to occur in a long bit sequence, so that $\bar{N}_p = N_p/2$.

6.4.2   Synchronous PSK Receivers

Consider first the case of heterodyne detection. The signal at the decision circuit is given by Eq. (6.3.3) or by

$$I_d = \tfrac{1}{2}\left(I_p \cos\phi + i_c\right). \tag{6.4.7}$$

The main difference from the ASK case is that I_p is constant, but the phase ϕ takes values 0 or π depending on whether a 1 or 0 is transmitted. In both cases, I_d is a Gaussian random variable but its average value is either $I_p/2$ or $-I_p/2$, depending on the received bit. The situation is analogous to the ASK case with the difference that $I_0 = -I_1$ in place of being zero. In fact, one can use Eq. (6.4.2) for the BER, but Q is now given by

$$Q = \frac{I_1 - I_0}{\sigma_1 + \sigma_0} \approx \frac{2I_1}{2\sigma_1} = (\text{SNR})^{1/2}, \tag{6.4.8}$$

where $I_0 = -I_1$ and $\sigma_0 = \sigma_1$ was used. By using SNR $= 2\eta N_p$ from Eq. (6.1.15), the BER is given by

$$\text{BER} = \tfrac{1}{2}\,\text{erfc}(\sqrt{\eta N_p}\,). \qquad \text{[PSK heterodyne]} \tag{6.4.9}$$

As before, the SNR is improved by 3 dB, or by a factor of 2, in the case of PSK homodyne detection, so that

$$\text{BER} = \tfrac{1}{2}\,\text{erfc}(\sqrt{2\eta N_p}\,). \qquad \text{[PSK homodyne]} \tag{6.4.10}$$

The receiver sensitivity at a BER of 10^{-9} can be obtained by using $Q = 6$ and Eq. (6.1.14) for SNR. For the purpose of comparison, it is useful to express the receiver sensitivity in terms of the number of photons N_p. It is easy to verify that $N_p = 18$ and 9 for the cases of heterodyne and homodyne PSK detection, respectively. The average number of photons/bit, $\bar{N}_p$, equals N_p for the PSK format because the same power is transmitted during 1 and 0 bits. A PSK homodyne receiver is the most sensitive receiver, requiring only 9 photons/bit. It should be emphasized that this conclusion is based on the Gaussian approximation for the receiver noise [42].

It is interesting to compare the sensitivity of coherent receivers with that of a direct-detection receiver. Table 6.1 shows such a comparison. As discussed in Section 4.5.3, an ideal direct-detection receiver requires 10 photons/bit to operate at a BER of $\leq 10^{-9}$. This value is only slightly inferior to the best case of a PSK homodyne receiver and considerably superior to that of heterodyne

Table 6.1 Sensitivity of synchronous receivers

Modulation Format	Bit-Error Rate	N_p	$\bar{N}_p$
ASK heterodyne	$\frac{1}{2}\mathrm{erfc}(\sqrt{\eta N_p/4})$	72	36
ASK homodyne	$\frac{1}{2}\mathrm{erfc}(\sqrt{\eta N_p/2})$	36	18
PSK heterodyne	$\frac{1}{2}\mathrm{erfc}(\sqrt{\eta N_p})$	18	18
PSK homodyne	$\frac{1}{2}\mathrm{erfc}(\sqrt{2\eta N_p})$	9	9
FSK heterodyne	$\frac{1}{2}\mathrm{erfc}(\sqrt{\eta N_p/2})$	36	36
Direct detection	$\frac{1}{2}\exp(-\eta N_p)$	20	10

schemes. However, it is never achieved in practice because of thermal noise, dark current, and many other factors, which degrade the sensitivity to the extent that $\bar{N}_p \approx 1000$ is usually required. In the case of coherent receivers, $\bar{N}_p$ below 100 can be realized simply because shot noise can be made dominant by increasing the local-oscillator power. The performance of coherent receivers is discussed in Section 6.6.

6.4.3 Synchronous FSK Receivers

Synchronous FSK receivers generally use a dual-filter scheme similar to that shown in Fig. 6.7(a) for the asynchronous case. Each filter passes only 1 or 0 bits. The scheme is equivalent to two complementary ASK heterodyne receivers operating in parallel. This feature can be used to calculate the BER of dual-filter synchronous FSK receivers. Indeed, one can use Eqs. (6.4.2) and (6.4.3) for the FSK case also. However, the SNR is improved by a factor of 2 compared with the ASK case. The improvement is due to the fact that whereas no power is received, on average, half the time for ASK receivers, the same amount of power is received all the time for FSK receivers. Hence the signal power is enhanced by a factor of 2, whereas the noise power remains the same if we assume the same receiver bandwidth in the two cases. By using SNR $= 4\eta N_p$ in Eq. (6.4.3), the BER is given by

$$\mathrm{BER} = \tfrac{1}{2}\,\mathrm{erfc}(\sqrt{\eta N_p/2}). \qquad \text{[FSK heterodyne]} \qquad (6.4.11)$$

The receiver sensitivity is obtained from Eq. (6.4.6) by replacing the factor of 72 by 36. In terms of the number of photons, the sensitivity is given by $N_p = 36$. The average number of photons/bit, $\bar{N}_p$, also equals 36, since each bit carries the same energy. A comparison of ASK and FSK heterodyne schemes in Table 6.1 shows that $\bar{N}_p = 36$ for both schemes. Therefore even though the ASK heterodyne receiver requires 72 photons within the 1 bit, the receiver sensitivity (average received power) is the same for both the ASK and FSK

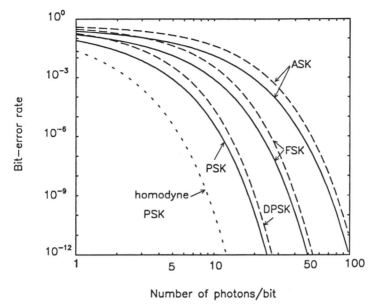

Number of photons/bit

Figure 6.8 Bit-error-rate curves for various modulation formats. The solid and dashed lines correspond to the cases of synchronous and asynchronous demodulation, respectively.

schemes. Figure 6.8 plots the BER as a function of N_p for the ASK, PSK, and FSK formats by using Eqs. (6.4.4), (6.4.9), and (6.4.11). The dotted curve shows the BER for the case of synchronous PSK homodyne receiver discussed in Section 6.4.2. The dashed curves correspond to the case of asynchronous receivers discussed in the following subsections.

6.4.4   Asynchronous ASK Receivers

The BER calculation for asynchronous receivers is slightly more complicated than for synchronous receivers because the noise does not remain Gaussian when an envelope detector is used (see Fig. 6.6). The reason can be understood from Eq. (6.3.4), which shows the signal received by the decision circuit. For the case of an ideal ASK heterodyne receiver without phase fluctuations, ϕ can be set to zero so that (subscript d is dropped for simplicity of notation)

$$I = [(I_p + i_c)^2 + i_s^2]^{1/2}. \qquad (6.4.12)$$

Even though both $I_p + i_c$ and i_s are Gaussian random variables, the probability density function (PDF) of I is not Gaussian. It can be calculated by using the standard techniques [43] and is found to be given by [44]

$$p(I, I_p) = \frac{I}{\sigma^2} \exp\left(-\frac{I^2 + I_p^2}{2\sigma^2}\right) I_0\left(\frac{I_p I}{\sigma^2}\right), \qquad (6.4.13)$$

where I_0 represents the modified Bessel function of the first kind. Both i_c and i_s are assumed to have a Gaussian PDF with zero mean and the same standard deviation σ, where σ is the RMS noise current. The PDF given by Eq. (6.4.13) is known as the *Rice distribution* [44]. Note that I varies in the range 0 to ∞, since the output of an envelope detector can have only positive values. When $I_p = 0$, the Rice distribution reduces to the *Rayleigh distribution*, well known in statistical optics [43].

The BER calculation follows the analysis of Section 4.5.1 with the only difference that the Rice distribution be used in place of the Gaussian distribution. The BER is given by Eq. (4.5.2) or by

$$\text{BER} = \tfrac{1}{2}[P(0/1) + P(1/0)], \tag{6.4.14}$$

where

$$P(0/1) = \int_0^{I_D} p(I, I_1)\, dI, \qquad P(1/0) = \int_{I_D}^{\infty} P(I, I_0)\, dI. \tag{6.4.15}$$

The notation is the same as that of Section 4.5.1. In particular, I_D is the decision level and I_1 and I_0 are values of I_p for 1 and 0 bits. The noise is taken to be the same for all bits ($\sigma_0 = \sigma_1 = \sigma$) by assuming that it is dominated by the local-oscillator shot noise. The integrals in Eq. (6.4.15) can be expressed in terms of Marcum's Q function, defined as [45]

$$Q(\alpha, \beta) = \int_{\beta}^{\infty} x \exp\left(-\frac{x^2 + \alpha^2}{2}\right) I_0(\alpha x)\, dx. \tag{6.4.16}$$

The result for the BER is

$$\text{BER} = \frac{1}{2}\left[1 - Q\left(\frac{I_1}{\sigma}, \frac{I_D}{\sigma}\right) + Q\left(\frac{I_0}{\sigma}, \frac{I_D}{\sigma}\right)\right]. \tag{6.4.17}$$

The decision level I_D is chosen such that the BER is minimum for given values of I_1, I_0, and σ. It is difficult to obtain an analytic expression of I_D under general conditions. However, under typical operating conditions, $I_0 \approx 0$, $I_1/\sigma \gg 1$, and I_D is well approximated by $I_1/2$. The BER then becomes

$$\text{BER} \approx \tfrac{1}{2}\exp(-I_1^2/8\sigma^2) = \tfrac{1}{2}\exp(-\text{SNR}/8). \tag{6.4.18}$$

When the receiver noise σ is dominated by the shot noise, the SNR is given by Eq. (6.1.15). By using $\text{SNR} = 2\eta N_p$, we obtain the final result,

$$\text{BER} = \tfrac{1}{2}\exp(-\eta N_p/4), \tag{6.4.19}$$

which should be compared with Eq. (6.4.4) obtained for the case of synchronous ASK heterodyne receivers. Equation (6.4.19) is plotted in Fig. 6.8 with a dashed line. It shows that the BER is larger for the asynchronous case for the same value of ηN_p. However, the difference is so small that the receiver sensitivity at a BER of 10^{-9} is degraded by only about 0.5 dB. If we assume that $\eta = 1$, Eq. (6.4.19) shows that BER $= 10^{-9}$ for $N_p = 80$ ($N_p = 72$ for the synchronous case). Asynchronous receivers hence provide performance comparable to that of synchronous receivers and are often used in practice because of their simpler design.

6.4.5   Asynchronous FSK Receivers

Although a single-filter heterodyne receiver can be used for FSK, it has the disadvantage that one-half of the received power is rejected, resulting in an obvious 3-dB penalty. For this reason, a dual-filter FSK receiver [see Fig. 6.7(a)] is commonly employed in which 1 and 0 bits pass through separate filters. The output of two envelope detectors are subtracted, and the resulting signal is used by the decision circuit. Since the average current takes values I_p and $-I_p$ for 1 and 0 bits, the decision threshold is set in the middle ($I_D = 0$). Let I and I' be the currents generated in the upper and lower branches of the dual filter receiver, where both of them include noise currents through Eq. (6.4.12). Consider the case in which 1 bits are received in the upper branch. The current I is then given by Eq. (6.4.12) and follows a Rice distribution with $I_p = I_1$ in Eq. (6.4.13). On the other hand, I' consists only of noise and its distribution is obtained by setting $I_p = 0$ in Eq. (6.4.13). An error is made when $I' > I$, as the signal is then below the decision level, resulting in

$$P(0/1) = \int_0^\infty p(I, I_1) \left[\int_I^\infty p(I', 0) \, dI' \right] dI, \qquad (6.4.20)$$

where the inner integral provides the error probability for a fixed value of I and the outer integral sums it over all possible values of I. The probability $P(1/0)$ can be obtained similarly. In fact, $P(1/0) = P(0/1)$ because of the symmetric nature of a dual-filter receiver.

The integral in Eq. (6.4.20) can be evaluated analytically. By using Eq. (6.4.13) in the inner integral with $I_p = 0$, it is easy to verify that

$$\int_I^\infty p(I', 0) \, dI' = \exp\left(-\frac{I^2}{2\sigma^2} \right). \qquad (6.4.21)$$

By using Eqs. (6.4.14), (6.4.20), and (6.4.21) with $P(1/0) = P(0/1)$, the BER is given by

$$\text{BER} = \int_0^\infty \frac{I}{\sigma^2} \exp\left(-\frac{I^2 + I_1^2}{2\sigma^2} \right) I_0 \left(\frac{I_1 I}{\sigma^2} \right) \exp\left(-\frac{I^2}{2\sigma^2} \right) dI, \qquad (6.4.22)$$

where $p(I, I_p)$ was substituted from Eq. (6.4.13). By introducing the variable $x = \sqrt{2}\, I$, Eq. (6.4.22) can be written as

$$\text{BER} = \frac{1}{2} \exp\left(-\frac{I^2}{4\sigma^2} \right) \int_0^\infty \frac{x}{\sigma^2} \exp\left(-\frac{x^2 + I_1^2/2}{2\sigma^2} \right) I_0 \left(\frac{I_1 x}{\sigma^2 \sqrt{2}} \right) dx. \qquad (6.4.23)$$

The integrand in Eq. (6.4.23) is just $p(x, I_1/\sqrt{2})$ and the integral must be 1. The BER is thus simply given by

$$\text{BER} = \tfrac{1}{2} \exp(-I_1^2/4\sigma^2) = \tfrac{1}{2} \exp(-\text{SNR}/4). \qquad (6.4.24)$$

By using SNR $= 2\eta N_p$ from Eq. (6.1.15), we obtain the final result

$$\text{BER} = \tfrac{1}{2} \exp(-\eta N_p/2), \qquad (6.4.25)$$

Table 6.2 Sensitivity of asynchronous receivers

Modulation Format	Bit-Error Rate	N_p	$\bar{N}_p$
ASK heterodyne	$\frac{1}{2}\exp(-\eta N_p/4)$	80	40
FSK heterodyne	$\frac{1}{2}\exp(-\eta N_p/2)$	40	40
DPSK heterodyne	$\frac{1}{2}\exp(-\eta N_p)$	20	20
Direct detection	$\frac{1}{2}\exp(-\eta N_p)$	20	10

which should be compared with Eq. (6.4.11) obtained for the case of synchronous FSK heterodyne receivers. Figure 6.8 compares the BER in the two cases. Just as in the ASK case, the BER is larger for asynchronous demodulation. However, the difference is small, and the receiver sensitivity is degraded by only about 0.5 dB compared with the synchronous case. If we assume that $\eta = 1$, $N_p = 40$ at a BER of 10^{-9} ($N_p = 36$ in the synchronous case). $\bar{N}_p$ also equals 40, since the same number of photons are received during 1 and 0 bits. Similar to the synchronous case, $\bar{N}_p$ is the same for both the ASK and FSK formats.

6.4.6 Asynchronous DPSK Receivers

As mentioned in Section 6.2.2, asynchronous demodulation cannot be used for PSK signals. A variant of PSK, known as DPSK, can be demodulated by using an asynchronous DPSK receiver [see Fig. 6.7(b)]. The filtered current is divided into two parts, and one part is delayed by exactly one bit period. The product of two currents contains information about the phase difference between the two neighboring bits and is used by the decision current to determine the bit pattern.

The BER calculation is slightly more complicated for the DPSK case, since the signal is formed by the product of two currents. The final result is, however, quite simple and is given by [11]

$$\text{BER} = \tfrac{1}{2}\exp(-\eta N_p). \qquad (6.4.26)$$

It can be obtained from the FSK result, Eq. (6.4.24), by using a simple argument [13] which shows that the demodulated DPSK signal corresponds to the FSK case if we replace I_1 by $2I_1$ and σ^2 by $2\sigma^2$. Figure 6.8 shows the BER by a dashed line (the curve marked DPSK). For $\eta = 1$, a BER of 10^{-9} is obtained for $N_p = 20$. Thus DPSK is more sensitive by 3 dB compared with both ASK and FSK. Table 6.2 lists the BER and the receiver sensitivity for the three modulation schemes used with asynchronous demodulation. The direct-detection case is also listed for comparison.

6.5   SENSITIVITY DEGRADATION

The sensitivity analysis of the preceding section assumes ideal operating conditions for a coherent lightwave system with perfect components. The receiver sensitivity obtained there corresponds to the *quantum limit* that is difficult to achieve in practice. Many physical mechanisms degrade the receiver sensitivity in practical coherent systems; among them are phase noise, intensity noise, polarization mismatch, and fiber dispersion. In this section we discuss sensitivity-degradation mechanisms and the ways to improve system performance by a proper design.

6.5.1   Phase Noise

An important source of sensitivity degradation in coherent lightwave systems is the phase noise associated with the transmitter laser and the local oscillator. The reason can be understood from Eqs. (6.1.5) and (6.1.7), which show the current generated by the photodetector for homodyne and heterodyne receivers, respectively. In both cases, phase fluctuations lead to current fluctuations because of the coherent nature of the photodetection process, and degrade the SNR. Both the signal phase ϕ_s and the local-oscillator phase ϕ_{LO} should remain relatively stable to avoid the sensitivity degradation. A measure of the duration over which the laser phase remains relatively stable is provided by the *coherence time.* As the coherence time is inversely related to the laser linewidth $\Delta\nu$, it is common to use the linewidth-to-bit rate ratio, $\Delta\nu/B$, to characterize the effect of phase noise on the performance of coherent lightwave systems. Since both ϕ_s and ϕ_{LO} fluctuate independently, $\Delta\nu$ is actually the sum of the linewidths $\Delta\nu_T$ and $\Delta\nu_{LO}$ associated with the transmitter and the local oscillator, respectively. The quantity $\Delta\nu = \Delta\nu_T + \Delta\nu_{LO}$ is often called the IF linewidth.

Considerable attention has been paid to evaluate the BER in the presence of phase noise and to estimate the dependence of the power penalty on the ratio $\Delta\nu/B$ [46]–[60]. The tolerable value of $\Delta\nu/B$, often defined such that the power penalty is below 1 dB, depends on the modulation format as well as on the demodulation technique. In general, the linewidth requirements are most stringent for homodyne receivers. Although the tolerable linewidth depends to some extent on the design of phase-locked loop, typically $\Delta\nu/B$ should be $< 5 \times 10^{-4}$ to realize a power penalty of less than 1 dB [48]. The requirement becomes $\Delta\nu/B < 1 \times 10^{-4}$ if the penalty is to be kept below 0.5 dB [49].

The linewidth requirements are relaxed considerably for heterodyne receivers, especially in the case of asynchronous demodulation with ASK or FSK modulation format. For synchronous heterodyne receivers $\Delta\nu/B < 5 \times 10^{-3}$ is required [48], [51]. By contrast, $\Delta\nu/B$ can exceed 0.1 in the case of asynchronous ASK and FSK receivers [54]–[57]. The reason is related to the fact that such receivers use an envelope detector (see Fig. 6.6), which throws away the phase information. The effect of phase fluctuations is mainly to broaden the signal bandwidth. The signal can be recovered by increasing the bandwidth of the

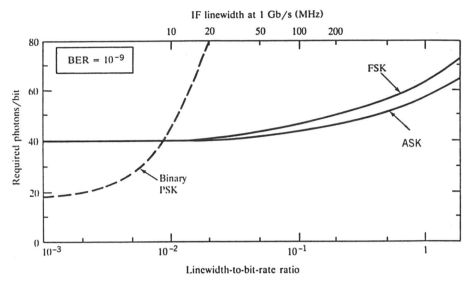

Figure 6.9 Receiver sensitivity $\bar{N}_p$ versus $\Delta\nu/B$ for asynchronous ASK and FSK heterodyne receivers. The dashed line shows the sensitivity degradation for a synchronous PSK heterodyne receiver. (After Ref. [54]. ©1988 IEEE. Reprinted with permission.)

bandpass filter (BPF). In principle, any linewidth can be tolerated if the BPF bandwidth is suitably increased. However, a penalty must be paid since receiver noise increases with an increase in the BPF bandwidth. Figure 6.9 shows how receiver sensitivity (expressed in average number of photons/bit, $\bar{N}_p$) degrades with $\Delta\nu/B$ for ASK and FSK formats. The BER calculation is rather cumbersome and requires numerical simulations for an exact analysis [54], [56]. Approximate methods have been developed to provide the analytic results accurate to within 1 dB [57].

The DPSK format requires narrower linewidths compared with the ASK and FSK formats when asynchronous demodulation based on the delay scheme [see Fig. 6.7(b)] is used. The reason is that information is contained in the phase difference between the two neighboring bits, and the phase should remain stable at least over the duration of two bits. Several calculations show [47], [48] that generally $\Delta\nu/B$ should be less than 1% to operate with a < 1 dB power penalty. For a 1-Gb/s bit rate, the required linewidth is ∼ 1 MHz but becomes < 1 MHz at low bit rates.

The design of coherent lightwave systems requires semiconductor lasers that operate in a single longitudinal mode with a narrow linewidth and whose wavelength can be tuned (at least over a few nanometers) to match the carrier frequency ω_0 and the local-oscillator frequency ω_{LO} either exactly (homodyne detection) or to the required intermediate frequency. Multisection DFB lasers, discussed in Section 3.3.5, have been developed to meet these requirements. Narrow linewidth can also be obtained by using a MQW design for the active

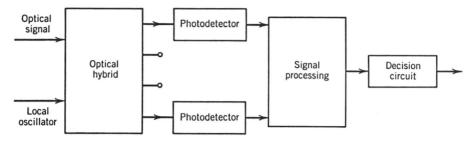

Figure 6.10 Multiport phase-diversity receiver.

region of a single-section DFB laser. Values as small as 0.1 MHz have been realized using strained MQW lasers [61].

An alternative approach solves the phase-noise problem by designing special receivers known as *phase-diversity receivers* [62]–[66]. Such receivers use two or more photodetectors whose outputs are combined to produce a signal that is independent of the phase difference $\phi_{IF} = \phi_s - \phi_{LO}$. The technique works quite well for ASK, FSK, and DPSK formats. Figure 6.10 shows schematically a multiport phase-diversity receiver. An optical component known as an *optical hybrid* combines the signal and local-oscillator inputs and provides its output through several ports with appropriate phase shifts introduced into different branches. The output from each port is processed electronically and combined to provide a current that is independent of ϕ_{IF}. For example, in the case of a two-port homodyne receiver, the two output branches have a relative phase shift of 90°, so that the currents in the two branches vary as $I_p \cos \phi_{IF}$ and $I_p \sin \phi_{IF}$. When the two currents are squared and added, the signal becomes independent of ϕ_{IF}. In the case of three-port receivers, the three branches have relative phase shifts of 0, 120°, and 240°. Again, when the currents are added and squared, the signal becomes independent of ϕ_{IF}. The concept can be extended to design receivers with four or more branches. However, the receiver design becomes more complex as more branches are added. Moreover, high-power local oscillators are needed to supply enough power to each branch. For these reasons, most phase-diversity receivers use two or three ports. Several system experiments have shown that the linewidth can approach the bit rate without introducing a significant power penalty even for homodyne receivers [63]–[66]. Numerical simulations of phase-diversity receivers show that the noise is far from being Gaussian [67]. In general, the BER is affected not only by the laser linewidth but also by other factors, such as the the the BPF bandwidth.

6.5.2 Intensity Noise

The effect of intensity noise on the performance of direct-detection receivers was discussed in Section 4.6.2 and found to be negligible in most cases of practical interest. However, this is not the case for coherent receivers [68]–[72]. To understand why intensity noise plays an important role in coherent receivers,

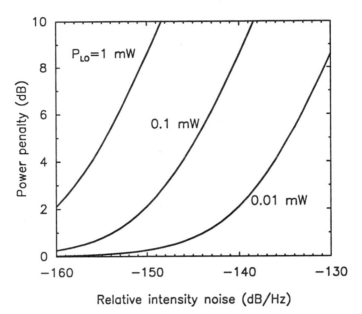

Figure 6.11 Power penalty versus RIN for several values of the local-oscillator power.

one can follow the analysis of Section 4.6.2 and add the contribution of the intensity noise to the current variance in Eq. (6.1.9), to obtain

$$\sigma^2 = \sigma_s^2 + \sigma_T^2 + \sigma_I^2, \qquad (6.5.1)$$

where

$$\sigma_I = RP_{LO}r_I \qquad (6.5.2)$$

and r_I is related to the *relative intensity noise* (RIN) of the local oscillator as defined in Eq. (4.6.7). If the RIN spectrum is flat up to the receiver bandwidth Δf, r_I^2 can be approximated by $2(\text{RIN})\Delta f$. The SNR is obtained by using Eqs. (6.5.1) and (6.5.2) in Eq. (6.1.11) and is given by

$$\text{SNR} = \frac{2R^2 \bar{P}_s P_{LO}}{2q(RP_{LO} + I_d)\Delta f + \sigma_T^2 + 2R^2 P_{LO}^2(\text{RIN})\Delta f}. \qquad (6.5.3)$$

The local-oscillator power P_{LO} should be large enough to satisfy Eq. (6.1.12) if the receiver were to operate in the shot-noise limit. However, an increase in P_{LO} increases the contribution of intensity noise quadratically as seen from Eq. (6.5.3). If the intensity-noise contribution becomes comparable to shot noise, the SNR would decrease unless the signal power $\bar{P}_s$ is increased to offset the increase in receiver noise. This increase in $\bar{P}_s$ is just the power penalty δ_I resulting from the local-oscillator intensity noise. If we neglect I_d and σ_T^2 in Eq. (6.5.3) by assuming that the receiver is designed to operate in the shot-noise limit, the power penalty (in dB) is given by the simple expression

$$\delta_I = 10 \log_{10}[1 + (\eta/h\nu)P_{LO}(\text{RIN})]. \qquad (6.5.4)$$

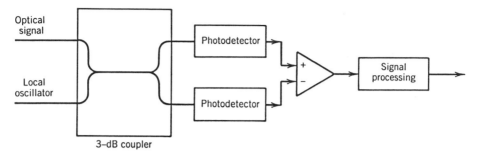

Figure 6.12 Two-port balanced coherent receiver.

Figure 6.11 shows δ_I as a function of RIN for several values of P_{LO} by using $\eta = 0.8$ and $h\nu = 0.8$ eV for 1.55-μm coherent receivers. The power penalty exceeds 2 dB when $P_{LO} = 1$ mW even for a local oscillator with a RIN of -160 dB/Hz, a value difficult to achieve for DFB semiconductor lasers. For a local oscillator with a RIN of -150 dB/Hz, P_{LO} should be less than 0.1 mW to keep the power penalty below 2 dB. The power penalty can be made negligible for a RIN of -150 dB/Hz if only 10 μW of local-oscillator power is used. However, Eq. (6.1.13) is unlikely to be satisfied for such small values of P_{LO}, and receiver performance would be limited by thermal noise. Sensitivity degradation as a result of local-oscillator intensity noise has been observed in a two-port ASK homodyne receiver [68]. The power penalty is reduced for three-port receivers [66], but intensity noise remains a limiting factor for $P_{LO} > 0.1$ mW. It should be stressed that the derivation of Eq. (6.5.4) is based on the assumption that the receiver noise is Gaussian. A numerical approach is necessary for a more accurate analysis of the intensity noise [70]–[72].

A solution to the intensity-noise problem is offered by the *balanced receiver* [73], which uses a two-port design with two photodetectors [74]–[76]. Figure 6.12 shows such a receiver schematically. It uses a 3-dB fiber coupler as an optical hybrid. The fiber coupler mixes the received optical signal with the local oscillator and splits the combined optical signal into two equal parts with an appropriate relative phase shift. The operation of a balanced receiver can be understood by considering the photocurrents I_+ and I_- generated in each branch. If the interference term in Eq. (6.1.5) or Eq. (6.1.7) has opposite signs for the two branches, the currents I_+ and I_- are given by

$$I_+ = \tfrac{1}{2}R(P_s + P_{LO}) + R\sqrt{P_s P_{LO}}\cos(\omega_{IF}t + \phi_{IF}), \qquad (6.5.5)$$

$$I_- = \tfrac{1}{2}R(P_s + P_{LO}) - R\sqrt{P_s P_{LO}}\cos(\omega_{IF}t + \phi_{IF}), \qquad (6.5.6)$$

where ϕ_{IF} is related to the phase difference $\phi_s - \phi_{LO}$. The subtraction of the two currents provides the heterodyne signal. The dc term is eliminated completely during the subtraction process when the two branches are balanced in such a way that each branch receives equal signal and local-oscillator powers. This occurs for a perfect 3-dB coupler with a 50% splitting ratio for each branch. The important point is that the intensity noise associated with the dc term is also eliminated during the subtraction process. The reason is that the same local oscillator provides power to each branch so that intensity fluctuations in the two branches are perfectly correlated and cancel out during subtraction of the photocurrents I_+ and I_-. It should be noted that intensity fluctuations associated with the ac term are not canceled even in a balanced receiver. However, their impact is less severe on the system performance because of the square-root dependence of the ac term on the local-oscillator power.

Balanced receivers are commonly used in the design of coherent lightwave systems because of the two advantages offered by them. First, the intensity-noise problem is nearly eliminated. Second, all of the signal and local-oscillator power is used effectively. A single-port receiver (see Fig. 6.1) rejects parts of both P_s and P_{LO} during the mixing process. Any loss in P_s is equivalent to a power penalty. Balanced receivers use all of the signal power and avoid this power penalty. At the same time, all of the local-oscillator power is used by the balanced receiver, making it easier to operate in the shot-noise limit.

6.5.3 Polarization Mismatch

The polarization state of the received optical signal plays no role in the case of direct-detection receivers simply because the photocurrent generated in such receivers depends only on the number of incident photons, irrespective of their state of polarization. This is not the case for coherent receivers, whose operation requires matching the state of polarization of the local oscillator to that of the signal received. The polarization-matching requirement can be understood from the analysis of Section 6.1, where the use of scalar fields E_s and E_{LO} implicitly assumed the same polarization state for the two optical fields. If $\hat{e}_s$ and $\hat{e}_{LO}$ represent the unit vectors along the direction of polarization of E_s and E_{LO}, respectively, the interference term in Eq. (6.1.3) contains an additional factor $\cos\theta$, where θ is the angle between $\hat{e}_s$ and $\hat{e}_{LO}$. Since the interference term is used by the decision circuit to reconstruct the transmitted bit stream, any change in θ from its ideal value of $\theta = 0$ reduces the signal (fading) and affects the receiver performance. In particular, if the polarization states of E_s and E_{LO} are orthogonal to each other ($\theta = 90°$), the signal disappears (complete fading). Any change in θ affects the BER through changes in the receiver current and SNR.

The polarization state $\hat{e}_{LO}$ of the local oscillator is determined by the laser and remains fixed. This is also the case for the transmitted signal before it is launched into the fiber. However, at the fiber output, the polarization state $\hat{e}_s$ of the signal received differs from that of the signal transmitted because of fiber birefringence, as discussed in Section 2.2.3 in the context of single-mode fibers. Such a change would not be a problem if $\hat{e}_s$ remained constant with time because one could match it with $\hat{e}_{LO}$ by simple optical techniques. The source of the problem lies in the fact that for most fibers $\hat{e}_s$ changes randomly because of birefringence fluctuations related to environmental changes (nonuniform stress, temperature variations, etc.). Such changes occur on a time scale ranging from seconds to microseconds. They lead to random changes in the

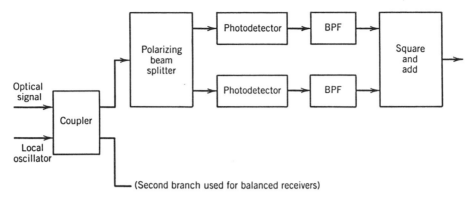

Figure 6.13 Polarization-diversity coherent receiver. Two branches are identical for balanced receivers.

BER and render coherent receivers unusable unless some scheme is devised to make the BER independent of polarization fluctuations. Although polarization fluctuations do not occur in polarization-preserving fibers, such fibers are not used in practice, as they are difficult to work with and have higher losses than those of conventional fibers. Moreover, the fiber already installed is not of polarization-preserving type. Thus, a different solution to the polarization-mismatch problem is required.

Several schemes have been proposed to solve the polarization-mismatch problem [77]–[82]. In one scheme [77], the polarization state of the optical signal received is tracked electronically and a feedback-control technique is used to match $\hat{e}_{LO}$ with $\hat{e}_s$. In another, polarization scrambling or spreading is used to force $\hat{e}_s$ to change randomly during a bit period [78]–[81]. Rapid changes of $\hat{e}_s$ are less of a problem than slow changes since, on average, the same power is received during each bit. A third scheme makes use of optical phase conjugation to solve the polarization problem [82]. The phase-conjugated signal can be generated inside a dispersion-shifted fiber by using four-wave mixing such that its polarization matches that of the incident signal (see Section 9.7.3). The pump laser used for four-wave mixing also plays the role of the local oscillator. The resulting photocurrent has a frequency component at twice the pump-signal detuning that can be used for recovering the bit stream.

The most commonly used approach solves the polarization problem by using a two-port receiver similar to that shown in Fig. 6.12, with the difference that the two branches process orthogonal polarization components. Such receivers are called *polarization-diversity receivers* [83]–[87], as their operation is independent of the polarization state of the signal received. The polarization-control problem has been studied extensively [88]–[95] because of its importance in the design of coherent lightwave systems.

Figure 6.13 shows a block digram of a polarization-diversity receiver. A polarization beam splitter is used to obtain the orthogonally polarized components, which are then processed by separate branches of the two-port receiver.

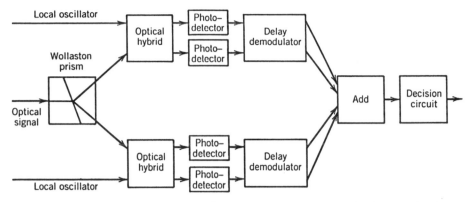

Figure 6.14 Four-port coherent DPSK receiver employing both phase and polarization diversity. (After Ref. [85]. ©1987 IEE. Reprinted with permission.)

When the photocurrents generated in the two branches are squared and added, the signal becomes polarization independent. The penalty paid to realize polarization diversity depends on the modulation and demodulation techniques used by the receiver. For synchronous demodulation, the power penalty can be as large as 3 dB [90]. However, the penalty is only 0.4–0.6 dB for optimized asynchronous receivers [83]. The technique of polarization diversity can be combined with phase diversity to realize a receiver that is independent of both phase and polarization fluctuations of the signal received [85], [96]. Figure 6.14 shows such a four-port receiver having four branches, each with its own photodetector. The performance of such receivers would be limited by the intensity noise of the local oscillator, as discussed in Section 6.5.2. The next step consists of designing a balanced phase- and polarization-diversity receiver by using eight branches with their own photodetectors. Such a receiver has been demonstrated by using a compact bulk optical hybrid [97]. In practical coherent systems, a balanced, polarization-diversity receiver is used in combination with narrow-linewidth lasers to simplify the receiver design, yet avoid the limitations imposed by intensity noise and polarization fluctuations.

6.5.4 Fiber Dispersion

Section 5.4 discussed how fiber dispersion limits the bit-rate–distance product (BL) of direct-detection (IM/DD) lightwave systems. Fiber dispersion also affects the performance of coherent systems [98]–[101], although its impact is less severe than for IM/DD systems. The reason is that coherent systems, by necessity, use a semiconductor laser operating in a single longitudinal mode with a narrow linewidth. Frequency chirping is avoided by using external modulators. Moreover, it is possible to compensate for fiber dispersion (see Section 9.2) through electronic equalization techniques in the IF domain [102], [103].

The effect of fiber dispersion on the transmitted signal can be calculated by using the analysis of Section 2.4. In particular, Eq. (2.4.15) can be used to

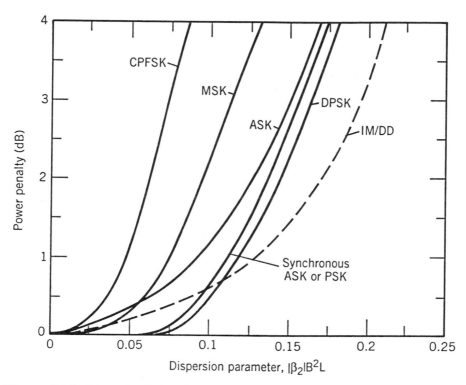

Figure 6.15 Dispersion-induced power penalty as a function of the dimensionless parameter $|\beta_2|B^2L$ for several modulation formats. The dashed line shows power penalty for a direct-detection system. (After Ref. [100]. ©1988 IEEE. Reprinted with permission.)

calculate the optical field at the fiber output for any modulation technique. The power penalty has been calculated for various modulation formats [100] through numerical simulations of the eye degradation occurring when a pseudo-random bit sequence is propagated through a single-mode fiber and demodulated by using a synchronous or asynchronous receiver. Figure 6.15 shows the power penalty as a function of the dimensionless parameter combination $|\beta_2|B^2L$ for several kinds of modulation formats. The dashed line shows, for comparison, the case of an IM/DD system. In all cases, the low-pass filter (before the decision circuit) is taken to be a second-order *Butterworth filter* [104], with the 3-dB bandwidth equal to 65% of the bit rate.

As seen in Fig. 6.15, fiber dispersion affects the performance of a coherent lightwave system qualitatively in the same way for all modulation formats, although quantitative differences do occur. The power penalty increases most rapidly for CPFSK and MSK formats, for which tone spacing is smaller than the bit rate. In all cases system performance depends on the product B^2L rather than BL. One can estimate the limiting value of B^2L by noting that the power penalty can be reduced to below 1 dB in most cases if the system is designed such that $|\beta_2| B^2 L < 0.1$. For conventional fibers, $\beta_2 = -20$ ps^2/km near 1.55 μm. The product $B^2 L$ is thus limited to 5000 (Gb/s)^2-km. For bit rates up to about 6 Gb/s, the transmission distance is limited by fiber loss rather than by fiber dispersion. The system becomes severely dispersion limited at bit rates in excess of 8 Gb/s; L is limited to 50 km for $B = 10$ Gb/s. An order-of-magnitude improvement in $B^2 L$ can be realized by operating the system near the zero-dispersion wavelength so that $|\beta_2| \leq 2$ ps^2/km. Dispersion becomes a major limiting factor when transmission distance is increased by using in-line optical amplifiers ($L \sim 1000$ km). As mentioned earlier, electronic equalization can be used to compensate for dispersion in such fiber links [102], [103]. The basic idea is to pass the intermediate-frequency signal through a filter whose transfer function is the inverse of the transfer function associated with the fiber (see Section 9.2). It is also possible to compensate fiber dispersion through optical equalization by using techniques described in Chapter 9 [105], [106]. Polarization-mode dispersion then becomes a limiting factor for long-haul coherent systems [107], [108].

6.5.5   Other Limiting Factors

Many other mechanisms can degrade the performance of coherent lightwave systems and should be considered during system design. Reflection feedback is one such limiting factor. The effect of reflection feedback on IM/DD systems has been discussed in Section 5.4.5. Essentially the same discussion applies to coherent lightwave systems. Any feedback into the laser transmitter or the local oscillator must be avoided, as it can lead to linewidth broadening or multimode operation of the semiconductor laser, both of which cannot be tolerated for coherent systems. Indeed, optical isolators are commonly used to reduce the optical feedback into semiconductor lasers.

Multiple reflections between two reflecting surfaces along the fiber cable can convert phase noise into intensity noise and affect system performance as discussed in Section 5.4.5. For coherent systems such conversion can occur even inside the receiver, where short fiber segments are used to connect the local oscillator to other receiver components, such as an optical hybrid (see Fig. 6.10). Calculations for phase-diversity receivers show that the reflectivity of splices and connectors should be below -35 dB under typical operating conditions [109]. Such reflection effects become less important for balanced receivers, where the impact of intensity noise on receiver performance is considerably reduced. Conversion of phase noise into intensity noise can occur even without parasitic reflections. However, the power penalty can be reduced to below 0.5 dB by reducing the ratio $\Delta\nu/B$ to below 20% in two-branch phase diversity ASK receivers [110].

Nonlinear effects in optical fibers, discussed in Section 2.6, may also limit coherent systems, depending on the transmitter power launched into the fiber [111]. Stimulated Raman scattering is not likely to be a limiting factor for single-channel coherent systems because of its high threshold (~ 500 mW) but becomes important for multichannel coherent systems (see Section 7.3.2). On the other hand, stimulated Brillouin scattering (SBS) has a low threshold and can affect single-channel coherent systems. The SBS threshold depends on both the modulation format and the bit rate [112]. Its effect on coherent systems has been studied extensively [113]. Nonlinear refraction converts intensity fluctuations into phase fluctuation through self-phase modulation [111]. It becomes important when the transmitted power is relatively large (> 30 mW) or when in-line optical amplifiers are used. Four-wave mixing is a limiting factor only for multichannel coherent systems.

6.6   SYSTEM PERFORMANCE

A large number of laboratory transmission experiments were performed during the 1980s to demonstrate the potential of coherent lightwave systems. A major objective was to demonstrate that coherent receivers are more sensitive than IM/DD receivers and allow operation closer to the quantum limit. The system performance is generally quantified through the receiver sensitivity, expressed as the average number of photons/bit, $\bar{N}_p$, required to realize a BER of 10^{-9}. In this section we discuss the system performance demonstrated in laboratory experiments and field trials conducted for commercial applications.

6.6.1   Asynchronous Heterodyne Systems

Asynchronous heterodyne systems have attracted the most attention in practice simply because the linewidth requirements for the transmitter laser and the local oscillator are so relaxed that standard DFB lasers can be used. Experiments have been performed with the ASK, FSK, and DPSK modulation formats [114]–[116]. An ASK experiment in 1990 showed that a baseline receiver sensitivity (without the fiber) of 175 photons/bit can be realized at 4 Gb/s [116]. This value is only 6.4 dB away from the quantum limit of 40 photons/bit obtained in Section 6.4.4. The sensitivity degraded by just 1 dB when the signal was transmitted through 160 km of conventional fiber with $D \approx 17$ ps/(nm-km). The system performance was similar when the FSK format was used in place of ASK. The baseline receiver sensitivity at 4 Gb/s was 191 photons/bit and increased by only 0.7 dB with transmission over 160 km of optical fiber. The frequency separation (tone spacing) was equal to the bit rate in this experiment.

The same experiment was repeated with the DPSK format using a LiNbO$_3$ phase modulator [116]. The baseline receiver sensitivity at 4 Gb/s was 209 photons/bit and degraded by 1.8 dB when the signal was transmitted over 160 km of conventional fiber. These experiments show the potential of asynchronous heterodyne lightwave systems for long-haul transmission. For comparison, the receiver sensitivity of 1.55-μm IM/DD receivers is such that $\bar{N}_p$ typically exceeds 1000 photons/bit even when APDs are used. The more than 6-dB improvement realized by the use of coherent-detection techniques translates into an additional 30 km of repeater spacing if we use 0.2 dB/km as the fiber loss. Even better performance is possible for DPSK systems operating at lower bit rates.

A record sensitivity of only 45 photons/bit was realized in 1986 by using DPSK at 400 Mb/s [114]. This value is only 3.5 dB away from the quantum limit of 20 photons/bit (see Section 6.4.6).

DPSK receivers continue to attract attention because of their high sensitivity and relative ease of implementation [117]–[125]. Several schemes make their use more practical. The DPSK signal at transmitter can be generated through direct modulation of a DFB laser [117]. Demodulation of the DPSK signal can be done optically by using a Mach-Zehnder interferometer with a one-bit delay in one arm, followed by two photodetectors at each output port of the interferometer. Such receivers are called direct-detection DPSK receivers since they do not use a local oscillator and exhibit performance comparable to their heterodyne counterparts [118]. In a 3-Gb/s experiment making use of this scheme, only 62 photons/bit were needed by an optically demodulated DPSK receiver with an optical preamplifier [119]. In another variant, the transmitter sends a PSK signal but the receiver is designed to detect the phase difference such that a local oscillator is not needed [120]. Considerable work has been done to quantify the performance of various DPSK and FSK schemes through numerical models that include the effects of phase noise and the preamplifier noise [121]–[125].

Asynchronous heterodyne schemes have also been used for long-haul coherent systems that use in-line optical amplifiers to increase the transmission distance. A 1991 experiment realized a transmission distance of 2223 km at 2.5 Gb/s by using 25 erbium-doped fiber amplifiers at approximately 80-km intervals [126]. The performance of long-haul coherent systems is affected by the amplifier noise as well as by the nonlinear effects in optical fibers. Their design requires optimization of many operating parameters, such as amplifier spacing, launch power, laser linewidth, IF bandwidth, and decision threshold [127]–[129].

6.6.2   Synchronous Heterodyne Systems

As discussed in Section 6.4, synchronous heterodyne receivers are more sensitive than asynchronous receivers. They are also more difficult to implement, as the microwave carrier must be recovered from the received data for synchronous demodulation. Since the sensitivity advantage is minimal (less than 0.5 dB) for ASK and FSK formats (compare Tables 6.1 and 6.2), most of the laboratory demonstrations have focused on the PSK format [130]–[133], for which the quantum-limited sensitivity is only 18 photons/bit. A problem with the PSK format is that the carrier is suppressed when the phase shift between 1 and 0 bits is exactly 180°, since the transmitted power is then entirely contained in the modulation sidebands. This feature poses a problem for carrier recovery. A solution is offered by the pilot-carrier scheme in which the phase shift is reduced below 180° (typically 150–160°) so that a few percent of the power remains in the carrier and can be used for synchronous demodulation at the receiver.

Phase noise is a serious problem for synchronous heterodyne receivers. As discussed in Section 6.5.1, the ratio $\Delta\nu/B$ must be less than 5×10^{-3}, where $\Delta\nu = \Delta\nu_T + \Delta\nu_{LO}$ is the IF linewidth. For bit rates below 1 Gb/s, the laser linewidth should be less than 2 MHz. External-cavity semiconductor lasers are often used in the synchronous experiments, as they can provide linewidths below 0.1 MHz. Several experiments have been performed using diode-pumped Nd:YAG lasers [130]–[132], which operate at a fixed wavelength near 1.32 μm but provide linewidths as small as 1 kHz. In one experiment [132], the bit rate was 4 Gb/s, but the receiver sensitivity of 631 photons/bit was 15.4 dB away from the quantum limit of 18 photons/bit, mainly because of the residual thermal noise and the intensity noise, since the balanced configuration was not used. The receiver sensitivity could be improved to 235 photons/bit at a lower bit rate of 2 Gb/s [132]. This sensitivity is still not as good as that obtained for asynchronous heterodyne receivers.

6.6.3   Homodyne Systems

As seen in Table 6.1, homodyne systems with the PSK format offer the best receiver sensitivity; the quantum-limited value is only 9 photons/bit. Implementation of such systems requires an optical phase-locked loop [134]–[139]. Many transmission experiments have been carried out to demonstrate the potential of PSK homodyne systems [140]–[150] using He–Ne lasers, Nd:YAG lasers, and semiconductor lasers. The receiver sensitivity achieved in these experiments depends on the bit rate. At a relatively low bit rate of 140 Mb/s, receiver sensitivities of 26 photons/bit at 1.52 μm [141] and 25 photons/bit at 1.32 μm [142] have been obtained by using He–Ne and Nd:YAG lasers, respectively. In a 1992 experiment, a sensitivity of 20 photons/bit at 565 Mb/s was obtained using synchronization bits for phase locking [145]. These values, although about 4 dB away from the quantum limit of 9 photons/bit, illustrate the potential of homodyne systems. In terms of the bit energy, 20 photons at 1.52 μm correspond to an energy of only 3 attojoule!

Receiver sensitivities realized in practice decrease as the bit rate increases. A sensitivity of 46 photons/bit was demonstrated in a 1-Gb/s experiment [140] that used external-cavity semiconductor lasers operating near 1.5 μm and transmitted the signal over 209 km of a standard single-mode fiber. The dispersion penalty was negligible (about 0.1 dB) in this experiment, as expected from Fig. 6.15. In another experiment [143] the bit rate was extended to 4 Gb/s. The baseline receiver sensitivity (without the fiber) was 72 photons/bit. When the signal was transmitted over 167 km of standard single-mode fiber, the receiver sensitivity degraded by only 0.6 dB (83 photons/bit), indicating that dispersion was not a problem even at 4 Gb/s. Figure 6.16 shows the BER curves obtained in this experiment with and without fiber together with the eye diagram (inset) obtained after 167 km of fiber. Another experiment [144] increased the bit rate to 10 Gb/s by using a 1.55-μm external-cavity DFB laser whose output was phase-modulated through a LiNbO$_3$ external modulator. The receiver sensitivity at 10 Gb/s was 297 photons/bit. The signal was transmitted through 151 km of dispersion-shifted fiber without a dispersion-induced power penalty. The use of dispersion-shifted fiber was necessary in this experiment, since B^2L exceeded 15,000 (Gb/s)^2-km (see Section 6.5.4).

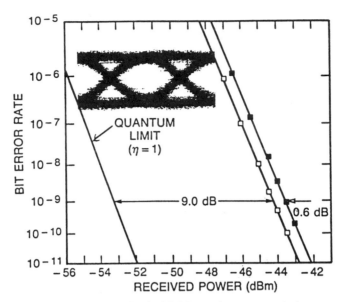

Figure 6.16 BER curves for a 4-Gb/s PSK homodyne transmission experiment with 5 m (empty rectangles) and 167 km (filled rectangles) of fiber. The quantum limit is shown for comparison. Inset shows the eye pattern after 167 km transmission at −44 dBm received power. (After Ref. [143]. ©1990 IEEE. Reprinted with permission.)

Long-haul homodyne systems use in-line amplifiers to compensate for fiber loss together with a dispersion-compensation scheme. In a 1993 experiment [147], a 6-Gb/s PSK signal was transmitted over 270 km using multiple in-line amplifiers. A microstrip line was used as a delay equalizer (see Section 9.2) to compensate for the fiber dispersion. Its use was feasible because of the implementation of the single-sideband technique. In a later experiment [148], the bit rate was extended to 10 Gb/s by using the vestigial-sideband technique. The 1.55-μm PSK signal could be transmitted over 126 km of standard telecommunication fiber with dispersion compensation provided by a 10-cm microstrip line. The design of long-haul homodyne systems with in-line amplifiers requires consideration of many factors such as phase noise, shot noise, imperfect phase recovery, and amplifier noise. Numerical simulations are often used to optimize the system performance [149], [150].

6.6.4 Field Trials

Any new technology must be tested through field trials before it can be commercialized. Several field trials for coherent lightwave systems [151]–[156] have been carried out since 1988. In all cases, an asynchronous heterodyne receiver was used because of its simplicity and not-so-stringent linewidth requirements. The modulation format of choice was the CPFSK format. This choice avoids the use of an external modulator, thereby simplifying the transmitter design. Furthermore, the laboratory experiments have shown that high-sensitivity receivers can be designed at bit rates as high as 10 Gb/s. A balanced polarization-diversity heterodyne receiver is used to demodulate the transmitted signal.

Field trials have included testing of both land- and sea-based telecommunication systems. In the case of one submarine trial [155], the system was operated at 560 Mb/s with the CPFSK format over 90 km of fiber cable. In another submarine trial [156], the system was operated at 2.5 Gb/s with the CPFSK format over fiber lengths of up to 431 km by using regenerators. Both trials showed that the use of polarization-diversity receivers is essential for practical coherent systems. In addition, the receiver incorporated electronic circuitry for automatic gain and frequency controls.

In spite of the successful field trials, coherent lightwave systems had not reached the commercial stage in 1996. Part of the reason is related to the complexity and reliability of coherent transmitters and receivers. The integration of these components on a single chip should improve reliability. Considerable development effort was directed in the 1990s toward designing optoelectronic integrated circuits (OEICs) for coherent lightwave systems [157]–[163]. By 1994, a balanced, polarization-diversity heterodyne receiver containing four photodiodes and made by using the InP/InGaAsP material system, exhibited a bandwidth in excess of 10 GHz [159]. A tunable local oscillator can also be integrated on the same chip. Such a tunable polarization diversity heterodyne OEIC receiver was used in a 140-Mb/s system experiment, intended mainly for video distribution [160]. A balanced heterodyne OEIC receiver, integrating a local oscillator and two photodiodes with a 3-dB coupler, has also been fabricated; it exhibited a 9-GHz bandwidth [163].

The main reason for the delay in the commercialization of coherent systems is the advent of the erbium-doped fiber amplifier. As discussed in Chapter 8, such amplifiers can be used to improve receiver sensitivity in IM/DD systems through preamplification of the optical signal received, and provide an improvement comparable to that of coherent detection. Coherent techniques are much more likely to be deployed for multichannel distribution systems requiring a small channel spacing (< 10 GHz). Indeed, the development effort in 1996 is directed toward that goal, as indicated by several recent field trials [164]. This aspect is covered in the next chapter.

PROBLEMS

6.1 Prove the 3-dB advantage of homodyne detection by showing that the average electrical power generated by a coherent receiver is twice as large for homodyne detection as for heterodyne detection under identical operating conditions.

6.2 Derive an expression for the SNR of a homodyne receiver by taking into account both the shot noise and the thermal noise.

6.3 Consider a 1.55-μm heterodyne receiver with a p–i–n photodiode of 90% quantum efficiency connected to a 50-Ω load resistance. How much local-oscillator power is needed to operate in the shot-noise limit? Assume that shot-noise limit is achieved when the thermal-noise contribution at room temperature to the noise power is below 1%.

6.4 Prove that the SNR of an ideal PSK homodyne receiver (perfect phase locking and 100% quantum efficiency) approaches $4\bar{N}_p$, where $\bar{N}_p$ is the average number of photons/bit. Assume that the receiver bandwidth equals half the bit rate and that the receiver operates in the shot-noise limit.

6.5 Show how an electro-optic material such as $LiNbO_3$ can be used for generating optical bit streams with ASK, PSK, and FSK modulation formats. Use diagrams as necessary.

6.6 A 1.55-μm DFB laser is used for the FSK modulation at 100 Mb/s with a tone spacing of 300 MHz. The modulation efficiency is 500 MHz/mA and the differential quantum efficiency equals 50% at the bias level of 3 mW. Estimate the power change associated with FSK by assuming that the two facets emit equal powers.

6.7 Derive an expression for the BER of a synchronous heterodyne ASK receiver by assuming that the in-phase noise component i_c has a probability density function

$$p(i_c) = \frac{1}{\sigma\sqrt{2}} \exp\left(-\frac{\sqrt{2}}{\sigma}|i_c|\right).$$

Determine the SNR required to achieve a BER of 10^{-9}.

6.8 Calculate the sensitivity (in dBm units) of a homodyne ASK receiver operating at 1.55 μm in the shot-noise limit by using the SNR expression obtained in Problem 6.2. Assume that $\eta = 0.8$ and $\Delta f = 1$ GHz. What is the receiver sensitivity when the PSK format is used in place of ASK?

6.9 Derive the Rice distribution [Eq. (6.4.13)] for the signal current I given Eq. (6.4.12) for an asynchronous heterodyne ASK receiver. Assume that both quadrature components of noise obey Gaussian statistics with standard deviation σ.

6.10 Show that the BER of an asynchronous heterodyne ASK receiver [Eq. (6.4.17)] can be approximated as

$$BER = \tfrac{1}{2} \exp[-I_1^2/(8\sigma^2)]$$

when $I_1/\sigma \gg 1$ and $I_0 = 0$. Assume that $I_D = I_1/2$.

6.11 Asynchronous heterodyne FSK receivers are commonly used for coherent lightwave systems. What is the SNR required by such receivers to operate at a BER of 10^{-9}? Calculate the receiver sensitivity (in dBm units) at 2 Gb/s in the shot-noise limit by assuming 1.2-GHz receiver bandwidth, 80% quantum efficiency, and a 1.55-μm operating wavelength.

6.12 Derive an expression for the SNR in terms of the intensity noise parameter r_I by including intensity noise through Eq. (6.5.2). Prove that the optimum value of P_{LO} at which the SNR is maximum is given by $P_{LO} = \sigma_T/(Rr_I)$ when the dark-current contribution to the shot noise is neglected.

6.13 Derive an expression for the power penalty as a function of r_I by using the SNR obtained in Problem 6.12.

6.14 Consider an optical carrier whose amplitude and frequency are constant but whose phase is modulated sinusoidally as $\phi(t) = \phi_0 \sin(\omega_m t)$. Show that the amplitude becomes modulated during propagation inside the fiber because of fiber dispersion.

6.15 Discuss the effect of laser linewidth on coherent communication systems. Why is the homodyne PSK receiver most sensitive to phase fluctuations? How is this sensitivity reduced for asynchronous heterodyne receivers?

REFERENCES

[1] M. Schwartz, *Information Transmission, Modulation, and Noise*, 4th ed., McGraw-Hill, New York, 1990.

[2] R. E. Ziemer, *Principles of Communications; Systems, Modulation and Noise*, Wiley, New York, 1994.

[3] L. W. Couch II, *Digital and Analog Communication Systems*, 5th ed., Prentice Hall, Upper Saddle River, NJ, 1995.

[4] M. S. Roden, *Analog and Digital Communication Systems*, Prentice Hall, Upper Saddle River, NJ, 1995.

[5] B. P. Lathi, *Modern Digital and Analog Communication Systems*, Oxford University Press, New York, 1995.

[6] W. R. Bennett, *Communication Systems and Techniques*, IEEE Press, Piscataway, NJ, 1995.

[7] J. Salz, *AT&T Tech. J.* **64**, 2153 (1985); *IEEE Commun. Mag.* **24** (6), 38 (1986).

[8] E. Basch and T. Brown, in *Optical Fiber Transmission*, E. E. Basch, Ed., SAMS, Indianapolis, IN, 1986, Chap. 16.

[9] T. Okoshi, *J. Lightwave Technol.* **5**, 44 (1987).

[10] T. Kimura, *J. Lightwave Technol.* **5**, 414 (1987)

[11] T. Okoshi and K. Kikuchi, *Coherent Optical Fiber Communications*, Kluwer Academic, Boston, 1988.

[12] R. A. Linke and A. H. Gnauck, *J. Lightwave Technol.* **6**, 1750 (1988).

[13] J. R. Barry and E. A. Lee, *Proc. IEEE* **78**, 1369 (1990).

[14] P. S. Henry and S. D. Persoinick, Eds., *Coherent Lightwave Communications*, IEEE Press, Piscataway, NJ, 1990.

[15] S. Betti, G. de Marchis, and E. Iannone, *Coherent Optical Communication Systems*, Wiley, New York, 1995.

[16] S. Ryu, *Coherent Lightwave Communication Systems*, Artec House, Boston, 1995.

[17] L. Thylen, *J. Lightwave Technol.* **6**, 847 (1988).

[18] D. W. Dolfi and T. R. Ranganath, *Electron. Lett.* **28**, 1197 (1992).

[19] K. Noguchi, H. Miyazawa, and O. Mitomi, *Electron. Lett.* **30**, 949 (1994).

[20] K. Noguchi, O. Mitomi, and H. Miyazawa, *IEICE Trans. Electron. Jpn.* **E79**, 27 (1996).

[21] W. Wang, D. Chen, H. R. Fetterman, Y. Shi, W. H. Steier, L. R. Dalton, and P.-M. Chow, *Appl. Phys. Lett.* **67**, 1806 (1995).

[22] S. Kalluri, M. Ziari, A. Chen, V. Chuyanov, W. H. Steier, D. Chen, B. Jalali, H. Fetterman, and L. R. Dalton, *IEEE Photon. Technol. Lett.* **8**, 644 (1996).

[23] I. Kotaka, K. Wakita, O. Mitomi, H. Isai, and Y. Kawamura, *IEEE Photon. Technol. Lett.* **1**, 100 (1989).

[24] T. Kataoka, Y. Miyamoto, K. Wakita, and I. Kotaka, *Electron. Lett.* **28**, 897 (1992).

[25] F. Devaux, S. Chelles, A. Ougazzaden, A. Mircea, M. Carré, F. Huet, A. C. Y. Sorel. J. F. Kerlides, and M. Henry, *IEEE Photon. Technol. Lett.* **6**, 1203 (1994).

[26] R. Weinmann, D. Baums, U. Cabulla, H. Haisch, D. Kaiser, E. Kühn, E. Lach, K. Satzke, J. Weber, P. Wiedemann, and E. Zielinski, *IEEE Photon. Technol. Lett.* **8**, 891 (1996).

[27] K. Wakita, I. Kotaka, and H. Asai, *IEEE Photon. Technol. Lett.* **4**, 29 (1992).

[28] S. Yoshida, Y. Tada, I. Kotaka, and K. Wakita, *Electron. Lett.* **30**, 1795 (1994).

[29] N. Yoshimoto, K. Kawano, Y. Hasumi, H. Takeuchi, S. Kondo, and Y. Noguchi, *IEEE Photon. Technol. Lett.* **6**, 208 (1994).

[30] K. Wakita and I. Kotaka, *Microwave Opt. Tech. Lett.* **7**, 120 (1994).

[31] E. M. Goldys and T. L. Tansley, *Microelectron. J.* **25**, 697 (1994).

[32] A. Segev, A. Saar, J. Oiknine-Schlesinger, and E. Ehrenfreund, *Superlattices Microstruct.* **19**, 47 (1996).

[33] S. Kobayashi, Y. Yamamoto, M. Ito, and T. Kimura, *IEEE J. Quantum Electron.* **18**, 582 (1982).

[34] S. Ogita, Y. Kotaki, M. Matsuda, Y. Kuwahara, H. Onaka, H. Miyata, and H. Ishikawa, *IEEE Photon. Technol. Lett.* **2**, 165 (1990).

[35] M. Kitamura, H. Yamazaki, H. Yamada, S. Takano, K. Kosuge, Y. Sugiyama, M. Yamaguchi, and I. Mito, *IEEE J. Quantum Electron.* **29**, 1728 (1993).

[36] M. Okai, M. Suzuki, and T. Taniwatari, *Electron. Lett.* **30**, 1135 (1994).

[37] B. Tromborg, H. E. Lassen, and H. Olesen, *IEEE J. Quantum Electron.* **30**, 939 (1994).

[38] M. Okai, M. Suzuki, and M. Aoki, *IEEE J. Sel. Topics Quantum Electron.* **1**, 461 (1995).

[39] J.-I. Shim, H. Olesen, H. Yamazaki, M. Yamaguchi, and M. Kitamura, *IEEE J. Sel. Topics Quantum Electron.* **1**, 516 (1995).

[40] M. Ferreira, *IEEE J. Quantum Electron.* **32**, 851 (1996).

[41] F. M. Gardner, *Phaselock Techniques*, Wiley, New York, 1985.

[42] M. I. Irshid and S. Y. Helo, *J. Opt. Commun.* **15**, 133 (1994).

[43] J. W. Goodman, *Statistical Optics*, Wiley, New York, 1985.

[44] S. O. Rice, *Bell Syst. Tech. J.* **23**, 282 (1944); **24**, 96 (1945).

[45] J. I. Marcum, *IRE Trans. Inform. Theory* **6**, 259 (1960).

[46] K. Kikuchi, T. Okoshi, M. Nagamatsu, and H. Henmi, *J. Lightwave Technol.* **2**, 1024 (1984).

[47] G. Nicholson, *Electron. Lett.* **20**, 1005 (1984).

[48] L. G. Kazovsky, *J. Lightwave Technol.* **3**, 1238 (1985); *J. Opt. Commun.* **7**, 66 (1986); *J. Lightwave Technol.* **4**, 415 (1986).

[49] B. Glance, *J. Lightwave Technol.* **4**, 228 (1986).

[50] I. Garrett and G. Jacobsen, *Electron. Lett.* **21**, 280 (1985); *J. Lightwave Technol.* **4**, 323 (1986); **5**, 551 (1987).

[51] T. G. Hodgkinson, *J. Lightwave Technol.* **5**, 573 (1987).

[52] G. Jacobsen and I. Garrett, *IEE Proc.* **134**, Pt. J, 303 (1987); *J. Lightwave Technol.* **5**, 478 (1987).

[53] L. G. Kazovsky, P. Meissner, and E. Patzak, *J. Lightwave Technol.* **5**, 770 (1987).

[54] G. J. Foschini, L. J. Greenstein, and G. Vannuchi, *IEEE Trans. Commun.* **36**, 306 (1988).

[55] L. J. Greenstein, G. Vannuchi, and G. J. Foschini, *IEEE Trans. Commun.* **37**, 405 (1989).

[56] I. Garrett, D. J. Bond, J. B. Waite, D. S. L. Lettis, and G. Jacobsen, *J. Lightwave Technol.* **8**, 329 (1990).

[57] L. G. Kazovsky and O. K. Tonguz, *J. Lightwave Technol.* **8**, 338 (1990).

[58] R. Corvaja and G. L. Pierobon, *J. Lightwave Technol.* **12**, 519 (1994).

[59] R. Corvaja, G. L. Pierobon, and L. Tomba, *J. Lightwave Technol.* **12**, 1665 (1994).

[60] H. Ghafouri-Shiraz, Y. H. Heng, and T. Aruga, *Microwave Opt. Tech. Lett.* **11**, 14 (1996).

[61] F. Kano, T. Yamanaka, N. Yamamoto, H. Mawatan, Y. Tohmori, and Y. Yoshikuni, *IEEE J. Quantum Electron.* **30**, 533 (1994).

[62] T. G. Hodgkinson, R. A. Harmon, and D.W. Smith, *Electron. Lett.* **21**, 867 (1985).

[63] A. W. Davis and S. Wright, *Electron. Lett.* **22**, 9 (1986).

[64] A. W. Davis, M. J. Pettitt, J. P. King, and S. Wright, *J. Lightwave Technol.* **5**, 561 (1987).

[65] L. G. Kazovsky, R. Welter, A. F. Elrefaie, and W. Sessa, *J. Lightwave Technol.* **6**, 1527 (1988).

[66] L. G. Kazovsky, *J. Lightwave Technol.* **7**, 279 (1989).

[67] C.-L. Ho and H.-N. Wang, *J. Lightwave Technol.* **13**, 971 (1995).

[68] L. G. Kazovsky, A. F. Elrefaie, R. Welter, P. Crepso, J. Gimlett, and R. W. Smith, *Electron. Lett.* **23**, 871 (1987).

[69] A. F. Elrefaie, D. A. Atlas, L. G. Kazovsky, and R. E. Wagner, *Electron. Lett.* **24**, 158 (1988).

[70] R. Gross, P. Meissner, and E. Patzak, *J. Lightwave Technol.* **6**, 521 (1988).

[71] W. H. C. de Krom, *J. Lightwave Technol.* **9**, 641 (1991).

[72] Y.-H. Lee, C.-C. Kuo, and H.-W. Tsao, *Microwave Opt. Tech. Lett.* **5**, 168 (1992).

[73] H. Van de Stadt, *Astron. Astrophys.* **36**, 341 (1974).

[74] G. L. Abbas, V. W. Chan, and T. K. Yee, *J. Lightwave Technol.* **3**, 1110 (1985).

[75] B. L. Kasper, C. A. Burrus, J. R. Talman, and K. L. Hall, *Electron. Lett.* **22**, 413 (1986).

[76] S. B. Alexander, *J. Lightwave Technol.* **5**, 523 (1987).

[77] T. Okoshi, *J. Lightwave Technol.* **3**, 1232 (1985).

[78] T. G. Hodgkinson, R. A. Harmon, and D. W. Smith, *Electron. Lett.* **23**, 513 (1987).

[79] M. W. Maeda and D. A. Smith, *Electron. Lett.* **27**, 10 (1991).

[80] P. Poggiolini and S. Benedetto, *IEEE Trans. Commun.* **42**, 2105 (1994).

[81] S. Benedetto and P. Poggiolini, *IEEE Trans. Commun.* **42**, 2915 (1994).

[82] G. P. Agrawal, *Quantum Semiclass. Opt.* **8**, 383 (1996).

[83] B. Glance, *J. Lightwave Technol.* **5**, 274 (1987).

[84] D. Kreit and R. C. Youngquist, *Electron. Lett.* **23**, 168 (1987).

[85] T. Okoshi and Y. C. Cheng, *Electron. Lett.* **23**, 377 (1987).

[86] A. D. Kersey, A. M. Yurek, A. Dandridge, and J. F. Weller, *Electron. Lett.* **23**, 924 (1987).

[87] S. Ryu, S. Yamamoto, and K. Mochizuki, *Electron. Lett.* **23**, 1382 (1987).

[88] M. Kavehrad and B. Glance, *J. Lightwave Technol.* **6**, 1386 (1988).

[89] I. M. I. Habbab and L. J. Cimini, *J. Lightwave Technol.* **6**, 1537 (1988).

[90] B. Enning, R. S. Vodhanel, E. Dietrich, E. Patzak, P. Meissner, and G. Wenke, *J. Lightwave Technol.* **7**, 459 (1989).

[91] N. G. Walker and G. R. Walker, *Electron. Lett.* **3**, 290 (1987); *J. Lightwave Technol.* **8**, 438 (1990).

[92] R. Noé, H. Heidrich, and D. Hoffman, *J. Lightwave Technol.* **6**, 1199 (1988).

[93] T. Pikaar, K. Van Bochove, A. Van Rooyen, H. Frankena, and F. Groen, *J. Lightwave Technol.* **7**, 1982 (1989).

[94] H. W. Tsao, J. Wu, S. C. Yang, and Y. H. Lee, *J. Lightwave Technol.* **8**, 385 (1990).

[95] T. Imai, *J. Lightwave Technol.* **9**, 650 (1991).

[96] Y. H. Cheng, T. Okoshi, and O. Ishida, *J. Lightwave Technol.* **7**, 368 (1989).

[97] R. Langenhorst, W. Pieper, M. Eiselt, D. Rhode, and H. G. Weber, *IEEE Photon. Technol. Lett.* **3**, 80 (1991).

[98] A. R. Chraplyvy, R. W. Tkaeh, L. L. Buhl, and R. C. Alferness, *Electron. Lett.* **22**, 409 (1986).

[99] K. Tajima, *J. Lightwave Technol.* **6**, 322 (1988).

[100] A. A. Elrefaie, R. E. Wagner, D. A. Atlas, and D. G. Daut, *J. Lightwave Technol.* **6**, 704 (1988).

[101] K. Nosu and K. Iwashita, *J. Lightwave Technol.* **6**, 686 (1988).

[102] R. G. Priest and T. G. Giallorenzi, *Opt. Lett.* **12**, 622 (1987).

[103] N. Takachio and K. Iwashita, *Electron. Lett.* **24**, 108 (1988).

[104] G. S. Mosehytz and P. Horn, *Active Filter Design Handbook*, Wiley, New York, 1981.

[105] L. J. Cimini, Jr., L. J. Greenstein, and A. A. M. Saleh, *IEEE Photon. Technol. Lett.* **2**, 200 (1990).

[106] A. H. Gnauck, L. J. Cimini, Jr., J. Stone, and L. W. Stulz, *IEEE Photon. Technol. Lett.* **2**, 585 (1990).

[107] T. G. Pratt and M. A. Ingram, *Proc. SPIE* **1787**, 390 (1992).

[108] C. De Angelis, A. Galtarossa, C. Campanile, and F. Matera, *J. Opt. Commun.* **16**, 173 (1995).

[109] L. Kazovsky, *Electron. Lett.* **24**, 522 (1988).

[110] F. N. Farokhrooz and J. P. Raina, *Int. J. Optoelectron.* **10**, 115 (1995).

[111] G. P. Agrawal, *Nonlinear Fiber Optics*, 2nd ed., Academic Press, San Diego, CA, 1995.

[112] Y. Aoki, K. Tajima, and I. Mito, *J. Lightwave Technol.* **6**, 710 (1988).

[113] T. Sugie, *J. Lightwave Technol.* **9**, 1145 (1991); *Opt. Quantum Electron.* **27**, 643 (1995).

[114] R. A. Linke, B. L. Kasper, N. A. Olsson, and R. C. Alferness, *Electron. Lett.* **22**, 30 (1986).

[115] N. A. Olsson, M. G. Oberg, L. A. Koszi, and G. Przyblek, *Electron. Lett.* **24**, 36 (1988).

[116] A. H. Gnauck, K. C. Reichmann, J. M. Kahn, S. K. Korotky, J. J. Veselka, and T. L. Koch, *IEEE Photon. Technol. Lett.* **2**, 908 (1990).

[117] R. Noé, E. Meissner, B. Borchert, and H. Rodler, *IEEE Photon. Technol. Lett.* **4**, 1151 (1992).

[118] J. J. O. Pires and J. R. F. de Rocha, *J. Lightwave Technol.* **10**, 1722 (1992).

[119] E. A. Swanson, J. C. Livas, and R. S. Bondurant, *IEEE Photon. Technol. Lett.* **6**, 263 (1994).

[120] I. Bar-David, *IEE Proc.* **141**, Pt. J, 38 (1994).

[121] G. Jacobsen, *Electron. Lett.* **28**, 254 (1992); *IEEE Photon. Technol. Lett.* **5**, 105 (1993).

[122] A. Murat, P. A. Humblet, and J. S. Young, *J. Lightwave Technol.* **11**, 290 (1993).

[123] C. P. Kaiser, P. J. Smith, and M. Shafi, *J. Lightwave Technol.* **13**, 525 (1995).

[124] P. J. Smith, M. Shafi, and C. P. Kaiser, *IEEE J. Sel. Areas Commun.* **13**, 557 (1995).

[125] S. R. Chinn, D. M. Bronson, and J. C. Livas, *J. Lightwave Technol.* **14**, 370 (1996).

[126] S. Saito, T. Imai, and T. Ito, *J. Lightwave Technol.* **9**, 161 (1991).

[127] E. Iannone, F. S. Locati, F. Matera, M. Romagnoli, and M. Settembre, *Electron. Lett.* **28**, 645 (1992); *J. Lightwave Technol.* **11**, 1478 (1993).

[128] J. Farre, G. Jacobsen, E. Bodtker, and K. Stubkjaer, *J. Opt. Commun.* **14**, 65 (1993).

[129] F. Matera, M. Settembre, B. Daino, and G. De Marchis, *Opt. Commun.* **119**, 289 (1995).

[130] A. Schoepflin, *Electron. Lett.* **26** 255 (1990).

[131] L. G. Kazovsky and D. A. Atlas, *IEEE Photon. Technol. Lett.* **2**, 431 (1990).

[132] L. G. Kazovsky, D. A. Atlas, and R. W. Smith, *IEEE Photon. Technol. Lett.* **2**, 589 (1990); D. A. Atlas and L. G. Kazovsky, *Electron. Lett.* **26**, 1032 (1990).

[133] T. Chikama, S. Watanabe, T. Naito, H. Onaka, T. Kiyonaga, Y. Onada, H. Miyata, M. Suyuma, M. Seimo, and H. Kuwahara, *J. Lightwave Technol.* **8**, 309 (1990).

[134] D. J. Maylon, D. W. Smith, and R. Wyatt, *Electron. Lett.* **22**, 421 (1986)

[135] L. G. Kazovsky, *J. Lightwave Technol.* **4**, 182 (1986); L. G. Kazovsky and D. A. Atlas, *IEEE Photon. Technol. Lett.* **1**, 395 (1989).

[136] J. M. Kahn, B. L. Kasper, and K. J. Pollock, *Electron. Lett.* **25**, 626 (1989).

[137] S. Norimatsu, K. Iwashita, and K. Sato, *IEEE Photon. Technol. Lett.* **2**, 374 (1990).

[138] C.-H. Shin, and M. Ohtsu, *IEEE Photon. Technol. Lett.* **2**, 297 (1990); *IEEE J. Quantum Electron.* **29**, 374 (1991).

[139] S. Norimatsu, *J. Lightwave Technol.* **13**, 2183 (1995).

[140] J. M. Kahn, *IEEE Photon. Technol. Lett.* **1**, 340 (1989).

[141] A. Schoepflin, S. Kugelmeier, G. Gottwald, D. Feicio, and G. Fischer, *Electron. Lett.* **26**, 395 (1990).

[142] D. A. Atlas and L. G. Kazovsky, *IEEE Photon. Technol. Lett.* **2**, 367 (1990).

[143] J. M. Kahn, A. H. Gnauck, J. J. Veselka, S. K. Korotky, and B. L. Kasper, *IEEE Photon. Technol. Lett.* **2**, 285 (1990).

[144] S. Norimatsu, K. Iwashita, and K. Noguchi, *Electron. Lett.* **26**, 648 (1990).

[145] B. Wandernorth, *Electron. Lett.* **27**, 1693 (1991); **28**, 387 (1992).

[146] S. L. Miller, *IEE Proc.* **139**, Pt. J, 215 (1992).

[147] K. Yonenaga and N. Takachio, *IEEE Photon. Technol. Lett.* **5**, 949 (1993).

[148] K. Yonenaga and S. Norimatsu, *IEEE Photon. Technol. Lett.* **7**, 929 (1995).

[149] C.-C. Huang and L. Wang, *J. Lightwave Technol.* **13**, 1963 (1995).

[150] S. Huang and L. Wang, *J. Lightwave Technol.* **14**, 661 (1996).

[151] S. Ryu, S. Yamamoto, Y. Namahira, K. Mochizuki, and M. Wakabayashi, *Electron. Lett.* **24**, 399 (1988).

[152] M. C. Brain, M. J. Creaner, R. Steele, N. G. Walker, G. R. Walker, J. Mellis, S. Al-Chalabi, J. Davidson, M. Rutherford, and I. C. Sturgess, *J. Lightwave Technol.* **8**, 423 (1990).

[153] T. W. Cline, J. M. P. Delavaux, N. K. Dutta, P. V. Eijk, C. Y. Kuo, B. Owen, Y. K. Park, T. C. Pleiss, R. S. Riggs, R. E. Tench, Y. Twu, L. D. Tzeng, and E. J. Wagner, *IEEE Photon. Technol. Lett.* **2**, 425 (1990).

[154] T. Imai, Y. Hayashi, N. Ohkawa, T. Sugie, Y. Ichihashi, and T. Ito, *Electron. Lett.* **26**, 1407 (1990).

[155] S. Ryu, S. Yamamoto, Y. Namihira, K. Mochizuki, and H. Wakabayashi, *J. Lightwave Technol.* **9**, 675 (1991).

[156] T. Imai, N. Ohkawa, Y. Hayashi, and Y. Ichihashi, *J. Lightwave Technol.* **9**, 761 (1991).

[157] T. L. Koch and U. Koren, *IEEE J. Quantum Electron.* **27**, 641 (1991).

[158] D. Trommer, A. Umbach, W. Passenberg, and G. Unterbörsch, *IEEE Photon. Technol. Lett.* **5**, 1038 (1993).

[159] F. Ghirardi, J. Brandon, F. Huet, M. Carré, A. Bruno, and A. Carenco, *IEEE Photon. Technol. Lett.* **6**, 814 (1994).

[160] U. Hilbk, T. Hermes, P. Meissner, C. Jacumeit, R. Stentel, and G. Unterbörsch, *IEEE Photon. Technol. Lett.* **7**, 129 (1995).

[161] F. Ghirardi, A. Bruno, B. Mersali, J. Brandon, L. Giraudet, A. Scavennec, and A. Carenco, *J. Lightwave Technol.* **13**, 1536 (1995).

[162] E. C. M. Pennings, D. Schouten, G.-D. Khoe, R. A. van Gils, and G. F. G. Depovere, *J. Lightwave Technol.* **13**, 1985 (1995).

[163] M. Hamacher, D. Trommer, K. Li, H. Schroeter-Janssen, W. Rehbein, and H. Heidrich, *IEEE Photon. Technol. Lett.* **8**, 75 (1996).

[164] E.-J. Bachus, T. Almeida, P. Demeester, G. Depovere, A. Ebberg, M. R. Ferreira, G.-D. Khoe, O. Koning, R. Marsden, J. Rawsthorne, and N. Wauters, *J. Lightwave Technol.* **14**, 1309 (1996).

Chapter 7

MULTICHANNEL LIGHTWAVE SYSTEMS

In principle, the signal bandwidth in optical communication systems can exceed 1 THz because of a large carrier frequency associated with the optical carrier. In practice, however, the bit rate is often limited to 10 Gb/s or less because of the limitations imposed by fiber dispersion, fiber nonlinearity, and the speed of electronic components. The transmission of multiple optical channels over the same fiber provides a simple way for making use of the unprecedented capacity offered by optics. Channel multiplexing can be done in either the time or frequency domain, leading to *time-division multiplexing* (TDM) in the former case and *frequency-division multiplexing* (FDM) in the latter case. The concept of TDM and FDM was introduced in Section 1.2.2. Indeed, as explained there, TDM in the electrical domain is invariably used even for the single-channel lightwave systems. In this chapter we focus on the TDM and FDM techniques used in the optical domain. To make the distinction explicit, it is common to refer to the two optical-domain techniques as *optical* TDM (OTDM) and *wavelength-division multiplexing* (WDM), respectively. Lightwave systems making use of such techniques are referred to as multichannel communication systems. The development of such systems has attracted considerable attention during the 1990s. Indeed, WDM lightwave systems became available commercially by 1996.

This chapter is organized as follows. Sections 7.1 to 7.3 are devoted to the development of WDM lightwave systems by considering in different sections the architectural aspects of such systems, the optical components needed for their implementation, and the performance issues such as interchannel crosstalk. Subcarrier multiplexing, a scheme in which FDM is implemented in the microwave domain, is discussed in Section 7.4. In Section 7.5 we consider the basic concepts behind OTDM systems and issues related to the practical implementation of such systems. Finally, in Section 7.6 we discuss a new kind of multiplexing scheme known as code-division multiplexing.

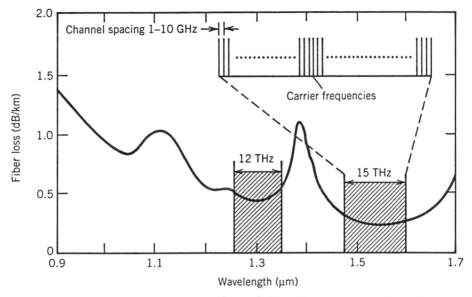

Figure 7.1 Low-loss transmission windows of silica fibers in the wavelength regions near 1.3 and 1.55 μm. The inset shows schematically multichannel operation in the 1.55-μm transmission window.

7.1   WDM LIGHTWAVE SYSTEMS

WDM corresponds to the scheme in which multiple optical carriers at different wavelengths are modulated by using independent electrical bit streams (which can themselves use TDM and FDM techniques in the electrical domain) and are then transmitted over the same fiber. The optical signal at the receiver is demultiplexed into separate channels by using an optical technique. WDM has the potential for exploiting the large bandwidth offered by the optical fiber. For example, hundreds of 10-Gb/s channels can be transmitted over the same fiber if the channel spacing is reduced to 40–50 GHz. Figure 7.1 shows the low-loss transmission windows of optical fibers centered near 1.3 and 1.55 μm. Each window covers a bandwidth of more than 10 THz, indicating that the total capacity of WDM systems may exceed 10 Tb/s.

The concept of WDM has been pursued since the first commercial lightwave system became available in 1980. In its simplest form, WDM is used to transmit two channels in different transmission windows of the optical fiber. For example, an existing 1.3-μm lightwave system can be upgraded in capacity by adding another channel near 1.55 μm, resulting in a channel spacing of 250 nm. Considerable attention was directed during the 1980s toward reducing the channel spacing, and multichannel systems with a channel spacing of less than 0.1 nm had been demonstrated by 1990 [1]–[3]. However, it was during the decade of 1990s that WDM systems were developed aggressively. By 1996, WDM systems operating with a total capacity of 40 Gb/s became available commercially. A

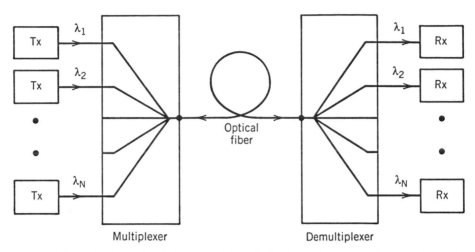

Figure 7.2 Multichannel point-to-point fiber link. Separate transmitter-receiver pairs are used to send and receive the signal at different wavelengths

transpacific system (TPC-6) operating at 100 Gb/s by the year 2000 has also been planned. This section focuses on the design and architecture of WDM systems by classifying them into three categories introduced in Section 5.1.

7.1.1 High-Capacity Point-to-Point Links

For long-haul point-to-point fiber links, the single-channel bit rate is often limited to 10 Gb/s or less because of fiber dispersion. The role of WDM in such systems is simply to increase the total bit rate. This is quite different from the role of WDM in optical networks, where information is distributed by using switching techniques. Figure 7.2 shows schematically a WDM fiber link. The output of several transmitters, each operating at its own carrier frequency (or wavelength), is multiplexed together. The multiplexed signal is launched into the optical fiber for transmission to the other end, where a demultiplexer sends each channel to its own receiver. When N channels at bit rates B_1, B_2,..., and B_N are transmitted simultaneously over a fiber of length L, the total bit rate–distance product, BL, becomes

$$BL = (B_1 + B_2 + \cdots + B_N)L. \qquad (7.1.1)$$

For equal bit rates, the system capacity is enhanced by a factor of N. An early experiment in 1985 demonstrated the BL product of 1.37 (Tb/s)-km at 1.5 μm by transmitting 10 channels at 2 Gb/s over 68.3 km of standard fiber [4] with a channel spacing of 1.35 nm. Such a performance would be impossible for single-channel systems, as fiber dispersion limits BL to values considerably below 1 (Tb/s)-km (see Fig. 5.4).

The ultimate capacity of WDM fiber links depends on how closely channels can be packed in the wavelength domain. The minimum *channel spacing* is limited by the extent of interchannel crosstalk, an issue covered in Section 7.3. Typically, channel spacing should exceed four times the bit rate. The low-loss region of the state-of-the-art optical fibers extends over 120 nm in the wavelength region near 1.55 μm (see Fig. 7.1). The minimum channel spacing can be as small as 80 GHz or 0.6 nm for 20-Gb/s channels. Since 200 channels can be accommodated over the 120-nm bandwidth, the resulting effective bit rate can be as large as 4 Tb/s. If we assume that the WDM signal can be transmitted over 150 km without the need for electronic regeneration or optical amplification, the effective BL product exceeds 600 (Tb/s)-km with the use of WDM technology. This should be contrasted with third-generation commercial lightwave systems, which transmit a single channel over 80 km or so at a bit rate of up to 2.5 Gb/s, resulting in BL values of at most 0.2 (Tb/s)-km. Clearly, the use of WDM has the potential of improving the performance of fourth-generation lightwave systems by a factor of more than 1000.

In practice, many factors limit the use of the entire low-loss window extending over 120 nm. For example, optical amplifiers are often used to avoid electronic regeneration of the WDM signal. The number of channels is then limited by the bandwidth over which amplifiers can provide nearly uniform gain. The bandwidth of erbium-doped fiber amplifiers is limited to 30–35 nm even with the use of gain-flattening techniques (see Chapter 8). Among other factors that limit the number of channels are (1) stability and tunability of distributed feedback (DFB) semiconductor lasers, (2) signal degradation during transmission because of various nonlinear effects, and (3) interchannel crosstalk during demultiplexing. The commercialization of high-capacity WDM fiber links requires the development of many high-performance components, such as transmitters integrating multiple DFB lasers, channel multiplexers and demultiplexers with add-drop capability, and large-bandwidth constant-gain amplifiers. The design and operation of WDM components are discussed in Section 7.2 and Section 7.3 covers linear and nonlinear crosstalk issues.

Experimental results on WDM links can be divided into two groups based on whether the transmission distance is ~ 100 km or exceeds 1000 km. Since the 1985 experiment in which ten 2-Gb/s channels were transmitted over 68.3 km [4], both the number of channels and the bit rate of individual channels have increased considerably. By 1995, a capacity of 340 Gb/s was demonstrated by transmitting 17 channels, each operating at 20 Gb/s, over 150 km [5]. This record was broken within a year by three experiments that used WDM to realize the total bit rate of 1 Tb/s or more. In one experiment, 55 channels, spaced 0.8 nm apart and each operating at 20 Gb/s, were transmitted over 150 km using two in-line amplifiers, resulting in a total bit rate of 1.1 Tb/s and a BL product of 165 (Tb/s)-km [6]. In another experiment, 50 channels, each operating at 20 Gb/s, were transmitted over 55 km [7] using WDM in combination with *polarization-division multiplexing* (PDM), a technique discussed in Section 10.4.2 in the context of solitons. In the third experiment, ten 100-Gb/s channels were transmitted over 40 km [8] using a combination of WDM and OTDM (each 100 Gb/s was obtained through TDM of ten 10-Gb/s channels). By the end of 1996, a bit rate of 2.64 Tb/s was demonstrated in a 132-channel

Table 7.1 Record-setting WDM transmission experiments

Channels N	Bit rate B (Gb/s)	Capacity NB (Gb/s)	Distance L (km)	$\overset{\bullet}{N}BL$ Product [(Tb/s)-km]
10	100	1000	40	40
16	10	160	531	85
32	10	320	640	205
32	5	160	9300	1488
50	20	1000	55	55
55	20	1100	150	165
132	20	2640	120	317

WDM experiment using 0.27 nm channel spacing [9]. Table 7.1 lists several record-setting WDM transmission experiments performed after 1995.

The second group of WDM experiments, with a transmission distance of more than 1000 km [10], [11] can be divided into two categories, depending on whether a linear fiber link or a recirculating fiber loop was used. A 1994 experiment realized transmission at 40 Gb/s over 1420 km of a linear fiber link by multiplexing sixteen 2.5-Gb/s channels while maintaining an amplifier spacing of about 100 km [12]. It was followed by many experiments that increased either the transmission distance or the bit rate. In one test-bed experiment, a transmission distance of 6000 km at 20 Gb/s (8 channels at 2.5 Gb/s) has been realized with an amplifier spacing of 75 km [13]. In another test-bed experiment, the bit rate could be extended to 40 Gb/s (8 channels at 5 Gb/s), but the transmission distance was only 4500 km even when the RZ format was used to improve the SNR [14]. In another approach, a segment of the existing transpacific cable (TPC-5) was used to demonstrate transmission over 6600 km by multiplexing several 5 or 10 Gb/s, resulting in a bit rate as large as 25 Gb/s [15]. On the high-bit-rate end, a 1996 experiment multiplexed sixteen 10-Gb/s channels to realize transmission at 160 Gb/s, but the link length was only 531 km [16]. By 1997, another experiment demonstrated 320-Gb/s transmission over 640 km by multiplexing 32 channels (spaced apart by 0.8 nm and operating at 10 Gb/s) and using broadband, gain-flattened, fiber amplifiers [17]. In contrast, a fiber-loop experiment demonstrated that 100-Gb/s transmission (20 channels at 5 Gb/s) over a transoceanic distance of 9100 km is possible using the polarization-scrambling and forward-error-correction techniques [11]. The number of channels was later increased to 32, resulting in a 160-Gb/s transmission over 9300 km [18].

The development of WDM fiber links has led to the advent of the fourth generation of lightwave systems, which make use of the WDM technology to increase the bit rate and in-line optical amplifiers to increase the transmission distance. Four-channel WDM links, each channel operating at 2.5 Gb/s, became available commercially in 1995. By 1996, WDM systems with a capacity of 40 Gb/s (16 channels at 2.5 Gb/s or 4 channels at 10 Gb/s) were commercial-

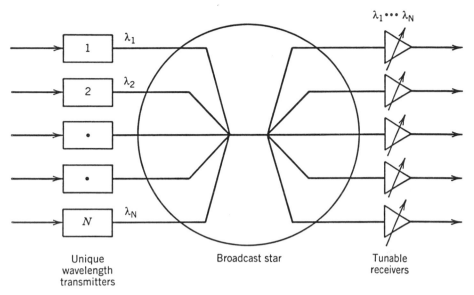

Figure 7.3 Distribution of multiple channels through a broadcast star coupler.

ized [19]. The 16-channel system covers a wavelength range of about 12 nm in the 1.55-μm region with a channel spacing of 0.8 nm. Each channel can operate at various levels of SONET or SDH specification (see Section 1.2.2), ranging from 155 Mb/s (STM-1 or OC-3) to 2.5 Gb/s (STM-16 or OC-48). WDM fiber links operating at 80 Gb/s are likely to appear in 1997, followed by 160-Gb/s systems designed by multiplexing sixteen 10 Gb/s channels. Clearly, the use of WDM has revolutionized the design of fiber-optic communication systems.

7.1.2   Broadcast and Distribution Networks

As discussed in Section 5.1.2, some networks broadcast or distribute a multi-channel WDM signal to a group of subscribers and let the receiver select each channel through demultiplexing. An example is provided by the cable-television (CATV) networks, which distribute multiple video channels through an *optical bus*. Although analog CATV networks often use a subcarrier-multiplexing scheme (see Section 7.4), digital video networks (e.g., high-definition television or HDTV) can benefit by the WDM approach. Figure 7.3 shows an example of the WDM distribution network based on the use of a *broadcast star*. Each channel is transmitted using a unique optical carrier frequency. The output of all transmitters is combined in a passive star and distributed to all receivers equally. Each subscriber receives all channels and can select one of them by using a tuning scheme. Such a network is also known as a *broadcast-and-select network*.

Many experiments have demonstrated the potential of broadcast-and-select networks [20]–[28]. An early experiment [20] used an 8 × 8 broadcast star to distribute seven channels spaced by 15 nm. Each receiver used a mechanically tunable filter of 10-nm bandwidth and 400-nm tuning range to detect the 280-Mb/s signal. In a 10-channel experiment employing coherent detection [21], each channel operated at 70 Mb/s with a channel spacing of only 6 GHz. The experiment used a partial 128×128 broadcast star to demonstrate the potential for serving a large number of subscribers. These initial experiments were followed by many other experiments with improved performance [22]–[26]. In a 16-channel experiment [23], the channel bit rate was as high as 2 Gb/s. In another [26], the channel bit rate was relatively low ($B = 622$ Mb/s), but 100 channels were distributed by using a 128×128 broadcast star.

The ultimate throughput of a broadcast network is limited by the distribution and insertion losses. The distribution loss of an optical bus increases so rapidly with increasing N [see Eq. (5.1.1)] that such an architecture is limited to rather small values of N (below 100) unless optical amplifiers are used to compensate for distribution losses. In the case of a broadcast star, distribution losses are much smaller. For an $N \times N$ star, the average power $\bar{P}_R$ received by each subscriber can be written as [see Eq. (5.1.2)]

$$\bar{P}_R = (\bar{P}_T/N)(1 - \delta)^{\log_2 N}(1 - C_L), \qquad (7.1.2)$$

where $\bar{P}_T$ is the average transmitted power and δ the insertion loss of each directional coupler within the broadcast star. The parameter C_L includes other losses, such as fiber loss and connector loss. The power received should exceed the receiver sensitivity $\bar{P}_{\text{rec}}$ for satisfactory operation of the network. If we express $\bar{P}_{\text{rec}}$ in terms of the average number of photons/bit $\bar{N}_p$, $\bar{P}_{\text{rec}} = \bar{N}_p h\nu B$. By using $\bar{P}_R = \bar{P}_{\text{rec}}$ in Eq. (7.1.2), the effective bandwidth, governed by the product BN, is given by

$$BN = \left(\frac{\bar{P}_T}{\bar{N}_p h\nu}\right)(1 - \delta)^{\log_2 N}(1 - C_L). \qquad (7.1.3)$$

We can use Eq. (7.1.3) to estimate the capacity of broadcast networks. A representative value of $\bar{N}_p$ for coherent receivers is 100 (see Section 6.4). The transmitted power is typically limited to 1 mW. The photon energy $h\nu = 0.8$ eV for 1.55-μm lightwave systems. For a lossless star coupler ($\delta = 0$) and lossless transmission ($C_L = 0$), Eq. (7.1.3) provides $BN = 78$ Tb/s. This value reduces to 7.8 Tb/s for the case of direct detection if we use $\bar{N}_p = 1000$ as a representative value. In practice, BN is considerably lower because of insertion and other losses ($C_L \neq 0$), which reduce the power transmitted. Figure 7.4 shows the dependence of BN on the number of channels for direct and coherent-detection schemes by taking into account the dependence of the receiver sensitivity on the bit rate. The BN product is assumed to be 10 Tb/s for an ideal lossless star coupler (the solid horizontal line). The dotted line shows the decrease in BN occurring for $\delta = 0.05$ (0.2-dB coupler loss). The solid curves show the BN product for realistic receivers. The dashed lines correspond to constant bit rates. For $B = 622$ Mb/s, BN approaches 1 Tb/s for $N \approx 1600$ when direct

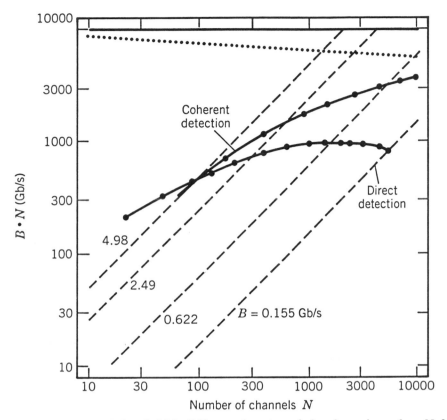

Figure 7.4 Network bandwidth BN as a function of the channel number N for direct- and coherent-detection schemes. The solid horizontal line on top shows the ideal lossless case by assuming a receiver sensitivity of 100 photons/bit. The dotted line illustrates the effect of 0.2 dB loss per coupler in the broadcast star. The dashed lines correspond to constant bit rates. (After Ref. [3]. ©1990 IEEE. Reprinted with permission.)

detection is used. This value increases by a factor of 3 ($BN \approx 3$ Tb/s) with $N \approx 4800$ for the case of coherent detection.

The distribution-loss problem can be overcome by the use of optical amplifiers. In a 1990 experiment [27], two erbium-doped fiber amplifiers (see Chapter 8) were used to demonstrate a broadcast network capable of reaching 39.5 million subscribers. The experiment demonstrated the concept by using 12 DFB lasers with a wavelength spacing of 2 nm. Each laser could be modulated at 2.2 Gb/s. The whole system was capable of transmitting 384 digital video channels to 39.5 million subscribers. The role of amplifiers was to compensate for the distribution losses to ensure that each subscriber was able to get enough power to demodulate the signal locally. This concept has been extended to demonstrate a WDM broadcast network capable of reaching 43.8 million subscribers while operating at 39.81 Gb/s over a 507-km range [28].

7.1.3   Multiple-Access WDM Networks

Multiple-access networks differ from broadcast networks in one important aspect: They offer a random bidirectional access to each subscriber [29]–[33]. Each user should be able not only to receive but also to transmit information to any other user of the network. Telephone service provides one example of such networks, generally referred to as *subscriber-loop* or *local-loop* networks. Another example is provided by networks used for connecting multiple computers (e.g., the *Internet*). In 1996 both the local-loop and computer networks were using electrical techniques to provide bidirectional multiple access through *circuit* or *packet switching*. As discussed in Section 5.1.3, it is necessary to establish protocol rules for satisfactory operation of a network. Examples of common protocols include Ethernet TCP/IP (transmission control protocol/Internet protocol) and *asynchronous transfer mode*(ATM). The main limitation of such techniques is that each node on the network must be capable of processing the entire network traffic. Since it is difficult to achieve electronic processing speeds in excess of 10 Gb/s, such networks are inherently limited by the electronics.

The use of WDM permits a novel approach in which the channel wavelength itself can be used for switching, routing, or distributing each channel to its destination, resulting in an *all-optical network* (AON). Since wavelength is used for multiple access, such a WDM approach is referred to as *wavelength-division multiple access* (WDMA). A considerable amount of research and development work was done during the decade of the 1990s to develop WDMA networks [34]–[37]. Broadly speaking, WDMA networks can be classified into two categories, called *single-hop* and *multihop* AONs [38]. As the name itself implies, every node is directly connected to all other nodes in a single-hop AON, resulting in a fully connected network. In contrast, multihop AONs are only partially connected such that an optical signal sent by one node may require several hops through intermediate nodes before reaching its destination. In each category, transmitters and receivers can have their operating frequencies either fixed or tunable, resulting in four possible combinations of devices.

Several architectures have been proposed for multihop AONs [32]. Hypercube architecture provides one example; it has been used for interconnecting multiple processors in a supercomputer [39]. The hypercube configuration can be easily visualized in three dimensions such that 8 nodes are located at 8 corners of a simple cube. In general, the number of nodes N must be of the form 2^m, where m is the dimensionality of the hypercube. Each node is connected to m different nodes. The maximum number of hops is limited to m, while the average number of hops is about $m/2$ for large N. Each node requires m receivers. The number of receivers can be reduced by using a variant, known as the *deBruijn network* [29], but it requires more than $m/2$ hops on average. Another example of multihop WDM network is provided by the *shuffle network* or its bidirectional equivalent, the *Banyan network* [40].

An example of a single-hop AON is provided by the *Lambdanet architecture*, shown schematically in Fig. 7.5. A broadcast star is used to distribute the signal to all nodes. The new feature of the Lambdanet is that each node is equipped

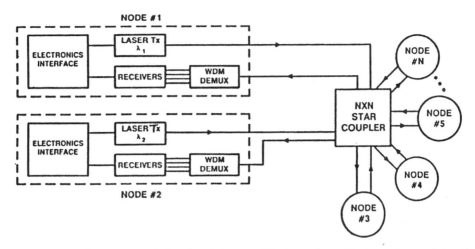

Figure 7.5 Schematic of the Lambdanet with N nodes. Each node consists of one transmitter and N receivers. (After Ref. [41]. ©1990 IEEE. Reprinted with permission.)

with one transmitter emitting at a unique wavelength and N receivers operating at the N wavelengths, where N is the number of nodes (or users). Each node receives the entire network traffic. This feature creates a completely connected, nonblocking network whose capacity and connectivity can be reconfigured electronically depending on the application. The network is also transparent to the bit rate or the modulation format. Different users can transmit data at different bit rates with different modulation formats. The network at a given wavelength can use analog or digital transmission as long as the receivers corresponding to that wavelength are suitably chosen. The flexibility of the Lambdanet makes it suitable for many applications. It can be used for voice-traffic transport in an interoffice network. Since transmitters and receivers operate at a fixed wavelength and do not require tuning (a relatively slow operation, in practice), the Lambdanet can also be used for networks using packet switching. A 1987 experimental demonstration [41] used 18 channels operating at 1.5 Gb/s, resulting in a network capacity of 27 Gb/s. The channels could be transmitted over 57.8 km without much power penalty. The main drawback of the Lambdanet is that the number of users is limited by the number of available wavelengths. Moreover, each node requires many receivers (equal to the number of nodes), resulting in a considerable investment in hardware costs.

A tunable receiver can reduce the cost and complexity of the Lambdanet. This is the approach adopted for the *Rainbow network* [42], which can support up to 32 nodes, each of which can transmit 1-Gb/s signals over 10–20 km. It makes use of a central passive star (see Fig. 7.5) together with the *high-performance parallel interface* (HIPPI) to connect multiple computers. A tunable optical filter is used to select the unique wavelength associated with each node. The main shortcoming of the Rainbow network is that tuning of a receiver is a

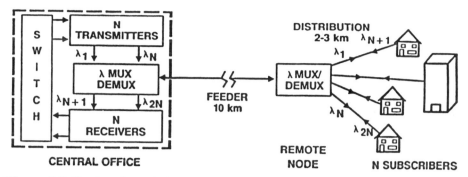

Figure 7.6 Passive photonic loop for local-loop applications. (After Ref. [44]. ©1988 IEE. Reprinted with permission.)

relatively slow process, making it difficult to use packet switching. An example of the WDM computer network that uses packet switching is provided by the *Starnet* [43]. It makes use of coherent detection to transmit data at bit rates of up to 1.25 Gb/s per node.

WDM networks making use of a passive star coupler are often called *passive optical networks* (PONs) since each node receives all traffic, eliminating the need for active switching. PONs have the potential for bringing the optical fiber to the home (or at least to the curb). In one scheme, called a *passive photonic loop* [44], multiple wavelengths are used for routing signals in the local loop. Figure 7.6 shows a block diagram of such a network. The central office contains N transmitters emitting at wavelengths $\lambda_1, \lambda_2, \ldots, \lambda_N$ and N receivers operating at wavelengths $\lambda_{N+1}, \ldots, \lambda_{2N}$ for a network of N subscribers. The signals to each subscriber are carried on separate wavelengths in each direction. A remote node multiplexes signals from the subscribers to send the combined signal to the central office. It also demultiplexes signals for individual subscribers. The remote node is passive and requires little maintenance if passive WDM components (see Section 7.2) are used. A switch at the central office routes signals depending on their wavelengths.

The design of PONs for local-loop applications was still evolving in 1996, and a large number of schemes were under consideration [45]–[47]. The goal is to provide broadband access to each user and to deliver audio, video, and data channels on demand, while keeping the cost down. A technique known as *spectral slicing* uses the broad emission spectrum of an LED to provide multiple WDM channels inexpensively. A novel WDM component, known as the *waveguide-grating router* (WGR) is used for wavelength routing [45]. Spectral slicing and WGR are discussed in Section 7.2. Figure 7.7 shows how WGRs and LEDs can be used to realize potentially low-cost PONs capable of bidirectional communication between a user and a central office.

Another approach is based on the *hybrid fiber-coaxial* (HFC) technology [46], [47]. It makes use of the existing CATV coaxial networks while delivering simultaneously telephony and video services in the local loop. Optical fibers are used to deliver the broadband signal to the local serving area, where the

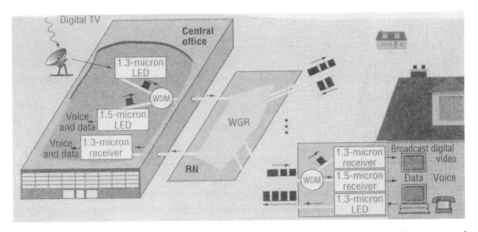

Figure 7.7 WDM passive optical network for local-loop applications. Remote node (RN) contains a waveguide-grating router (WGR) for routing voice, video, and data channels. (After Ref. [45]. ©1996 PennWell. Reprinted with permission.)

signal is converted to the electrical domain and distributed to each subscriber through conventional means. The HFC network is not really a WDM network since channel multiplexing is done in the electrical domain (see Section 7.4). Nonetheless, it is likely to be commercialized first because of its low-cost nature. The largest bandwidth for each subscriber is provided by an architecture known as the *passive cable network*. Its use permits delivery of 10-MHz bandwidth in the forward path (to the subscriber) and a 1.5-Mb/s data channel in the reverse path [47].

PONs can be used only for local- and *metropolitan-area networks* (MANs) since the number of nodes is typically limited to 100 or so because of the need to assign a unique wavelength to each node. Several MANs can be connected to form a *wide-area network* (WAN) that can form the backbone of a national wideband WDMA network. This is the approach adopted by the AON Consortium [48]. The network consists of three levels. At the LAN level, a broadcast star is used in a manner similar to Fig. 7.5 to combine multiple channels. At the next level, several LANs are connected to a MAN by using passive wavelength routing. At the highest level, several MANs connect to a WAN whose nodes are interconnected in a mesh topology. At the WAN level, the network makes extensive use of switches and wavelength-shifting devices so that it is dynamically configurable. The available wavelengths are divided into three sets. One set is reserved for LANs, another for MANs, and the third for the WAN. In a test-bed implementation of such a network, the set of 20 wavelengths was divided into three sets of 10, 5, and 5, with the largest set used for LANs.

A transport network, called a *multiwavelength optical network* (MONET), was in the test-bed stage in 1996 in the northeastern part of the United States. It is designed to incorporate the existing diverse switching technologies (SONET or SDH, ATM, FDDI, etc.) into a long-haul AON by using the WDM cross-

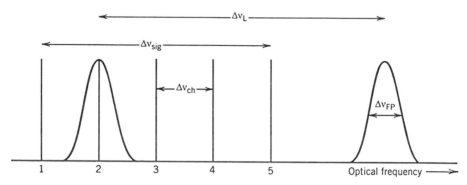

Figure 7.8 Channel selection through a tunable optical filter.

connect technology [49]. A set of eight standard wavelengths has been established (MONET-compliant wavelengths) in the 1.55-μm region with a channel spacing of 200 GHz (about 1.6 nm). In another test-bed demonstration, a configurable AON was formed by using DFB laser arrays, multichannel OEIC receivers, tunable filters, and optical cross-connect switches [50]. Similar transport networks are also under development in Europe [51] and Japan [52]. Clearly, the WDM technique has revolutionized the state of the art of lightwave technology during the 1990s.

7.2 WDM COMPONENTS

The implementation of WDM technology for fiber-optic communication systems requires several new optical components. Among them are multiplexers, which combine the output of several transmitters and launch it into an optical fiber (see Fig. 7.2); demultiplexers, which split the received multichannel signal into individual channels destined to different receivers (see Fig. 7.2); star couplers, which mix the output of several transmitters and broadcast the mixed signal to multiple receivers (see Fig. 7.3); tunable optical filters, which filter out one channel at a specific wavelength that can be changed by tuning the passband of the optical filter; multiwavelength optical transmitters with tunable semiconductor lasers, whose wavelength can be tuned over a few nanometers; wavelength routers which can distribute the WDM signal to different ports; and wavelength shifters which switch the channel wavelength. In this section we consider various WDM components separately.

7.2.1 Tunable Optical Filters

It is helpful to consider optical filters first since they are often the building blocks of more complex WDM components, and thus help to understand the underlying operating mechanism. The role of a tunable optical filter in a WDM system is to select a desired channel at the receiver. Channel selection can be made in the optical or electrical domain. The electric-domain selection technique is suitable for coherent detection. The optical-domain channel selection requires a tunable optical filter that is placed just before the receiver. Figure 7.8 shows the selection technique schematically. The filter bandwidth must be large enough to transmit the desired channel but, at the same time, small enough to block the neighboring channels to avoid the crosstalk.

All optical filters require a wavelength-selective mechanism and can be classified into two broad categories depending on whether optical interference or diffraction is the underlying physical mechanism. Each category can be further subdivided according to the implementation scheme adopted. In this section we consider four kinds of optical filters; Fig. 7.9 shows an example of each kind. The desirable properties of a tunable optical filter include (1) wide tuning range to maximize the number of channels that can be selected, (2) negligible crosstalk to avoid interference from adjacent channels, (3) fast tuning speed to minimize the access time, (4) small insertion loss, (5) polarization insensitivity, (6) stability against environmental changes (humidity, temperature, vibrations, etc.), and (7) last but not the least, low cost.

Fabry–Perot Filters

A Fabry–Perot (FP) interferometer, a cavity formed by two highly reflecting mirrors [see Fig. 7.9(a)], can act as a tunable optical filter if its length is controlled electronically by using a piezoelectric transducer. The operating mechanism can be understood from the discussion in Section 3.3.2. The transmittivity of a FP filter peaks at wavelengths that correspond to the longitudinal-mode frequencies given by Eq. (3.3.5). Hence, the frequency spacing between two successive transmission peaks, known as the *free spectral range*, is given by

$$\Delta\nu_L = c/2n_gL, \tag{7.2.1}$$

where n_g is group index of the intracavity material for a FP filter of length L.

If the filter is designed to pass a single channel (see Fig. 7.8), the combined bandwidth of the multichannel signal, $\Delta\nu_{\text{sig}} = NS_{\text{ch}}B$, must be less than $\Delta\nu_L$, where N is the number of channels, S_{ch} is the normalized channel spacing ($S_{\text{ch}} = \Delta\nu_{\text{ch}}/B$), and B is the bit rate. At the same time, the filter bandwidth $\Delta\nu_{\text{FP}}$, the width of the transmission peak in Fig. 7.8, should be large enough to pass the entire frequency contents of the selected channel. Typically, $\Delta\nu_{\text{FP}} \sim B$. The channel number N is thus limited by

$$N < \frac{\Delta\nu_L}{S_{\text{ch}}\Delta\nu_{\text{FP}}} = \frac{F}{S_{\text{ch}}}, \tag{7.2.2}$$

where $F = \Delta\nu_L/\Delta\nu_{\text{FP}}$ is the *finesse* of the FP filter. The concept of finesse is well known in the theory of FP interferometers [53]. If the internal loss is neglected, the finesse is governed by the mirror reflectivity R, assumed to be the same for both mirrors, and is given by [53]

$$F = \frac{\pi\sqrt{R}}{1 - R}. \tag{7.2.3}$$

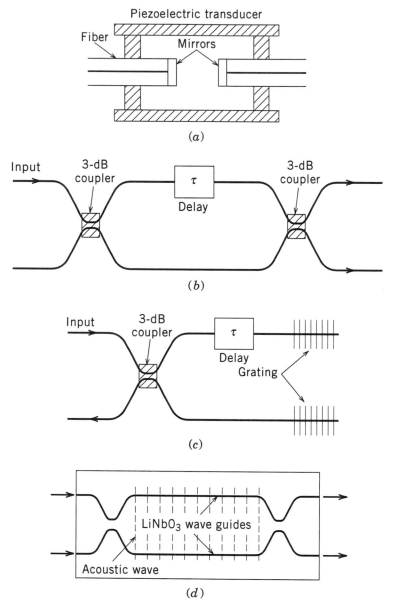

Figure 7.9 Four kinds of filters based on various interferometric and diffractive devices: (a) Fabry–Perot filter; (b) Mach–Zehnder filter; (c) grating-based Michelson filter; (d) acousto-optic filter. The shaded area represents a surface acoustic wave.

Equation (7.2.2) provides a remarkably simple condition for the number of channels that a FP filter can resolve. The channel spacing can be as small as $3\Delta\nu_{FP}$ while keeping the crosstalk below 10% (-10 dB). By using the limiting value $S_{ch} = 3$ and F from Eq. (7.2.3), the maximum number of channels for a

FP filter is restricted by

$$N < \frac{\pi\sqrt{R}}{3(1-R)}. \tag{7.2.4}$$

As an example, a FP filter with 99% reflecting mirrors can be used to select up to 104 channels. Channel selection is made by changing the filter length L electronically. The length needs to be changed by only a fraction of the wavelength to tune the filter and select different channels. The filter length L itself is determined from Eq. (7.2.1) together with the condition $\Delta\nu_L > \Delta\nu_{\text{sig}}$. As an example, for a 10-channel WDM signal with 0.8-nm channel spacing, $\Delta\nu_{\text{sig}} \approx 1$ THz if the operating wavelength is near 1.55 μm. If $n_g = 1.5$ is used as a representative value for the group index, L should be smaller than 100 μm. Such a short length together with the requirement of high mirror reflectivities underscores the complexity of the design of FP filters for WDM applications.

The most practical design of FP filters uses an air gap between two pieces of an optical fiber (see Fig. 7.9). The two fiber ends forming the gap are coated to act as high-reflectivity mirrors, making an air-filled FP cavity [54]. The entire structure is enclosed in a piezoelectric chamber so that the gap length can be changed electronically for tuning and selecting a specific channel. The advantage of fiber FP filters is that they can be integrated within the system without incurring coupling losses. Such filters are available commercially and were used in commercial WDM fiber links starting in 1995. The number of channels is limited in the range 50–100 because of a limited finesse of a practical FP filter ($F = 100$ for 97% mirror reflectivity). But N can be increased by an order of magnitude by using two FP filters in tandem, a scheme that increases the effective finesse to $F \sim 1000$. Tuning speed is relatively slow because of the mechanical nature of the tuning mechanism, but switching times ~ 100 μs can be obtained with piezoelectric transducers.

Tunable FP filters can also be made by using a *liquid crystal*, the same material used for flat-panel displays. The operating mechanism makes use of the anisotropic nature of liquid crystals that makes it possible to change its refractive index electronically. A FP cavity is still formed by enclosing the liquid crystal material within two high-reflectivity mirrors, but the tuning is done by changing the refractive index rather than the cavity length. Such FP filters can provide high finesse ($F \sim 200$–400) with a bandwidth of 0.2–0.3 nm [55]. They can be tuned electrically over 50 nm, but switching time is typically ~ 1 ms or more when nematic liquid crystals are used. It can be reduced to below 10 μs by using smectic liquid crystals [56].

Mach–Zehnder Filters

A chain of MZ interferometers can also be used for making a tunable optical filter. A MZ interferometer can be constructed simply by connecting the two output ports of a 3-dB coupler to the two input ports of another 3-dB coupler [see Fig. 7.9(b)]. The first coupler splits the input signal equally into two parts, which acquire different phase shifts (if the arm lengths are made different) before they interfere at the second coupler. Since the relative phase shift is wavelength dependent, the transmittivity $T(\nu)$ is also wavelength dependent. In fact, it is simply given by $T(\nu) = \cos^2(\pi \nu \tau)$, where ν is the frequency and τ is the relative delay in the two arms of the MZ interferometer. A cascaded chain of such MZ interferometers with relative delays adjusted suitably acts as an optical filter that can be tuned by changing the arm lengths slightly. Mathematically, the transmittivity of a chain of M MZ interferometers is given by

$$T(\nu) = \prod_{m=1}^{M} \cos^2(\pi \nu \tau_m), \tag{7.2.5}$$

where τ_m is the relative delay for the mth member of the chain.

A commonly used method implements the relative delays τ_m such that each MZ stage blocks the alternate channels successively. This scheme requires $\tau_m = (2^m \Delta \nu_{ch})^{-1}$ for a channel spacing of $\Delta \nu_{ch}$. The resulting transmittivity of a 10-stage MZ chain has a channel selectivity as good as that offered by a FP filter having a finesse of 1600 [29]. Moreover, such a chain is capable of selecting closely spaced channels. The MZ chain can be built by using fiber couplers or by using silica waveguides on a silicon substrate. The *silica-on-silicon technology* was exploited extensively during the 1990s to make many WDM components. Such devices are often referred to as *planar lightwave circuits* because of the use of planar optical waveguides formed on a silicon substrate [57]–[59]. The underlying technology is sometimes called the *silicon optical-bench technology* [60]. In the case of a MZ chain formed in a planar lightwave circuit, tuning is realized through a chromium heater deposited on one arm of each MZ interferometer. Since the tuning mechanism is thermal, it results in a slow response with a switching time of about 1 ms.

Grating-Based Filters

A separate class of tunable optical filters makes use of the wavelength selectivity provided by a Bragg grating [53]. Ruled optical gratings are rarely used in practice because of their bulky nature and size incompatibility with fiber-optic systems. Rather, an index grating is formed within an optical waveguide. Fiber Bragg gratings provide a simple example of grating-based optical filters [61]. The operating mechanism of such gratings is discussed in Section 9.6. In its simplest form, a fiber grating acts as a reflection filter whose central wavelength can be controlled by changing the grating period, and whose bandwidth can be tailored by changing the grating strength and by chirping the grating period slightly. The reflective nature of the fiber grating is often a limitation in practice and requires the use of an *optical circulator*. A phase shift in the middle of the grating can convert a fiber grating into a narrowband transmission filter [62]. Many other schemes can be used to make transmission filters by using fiber gratings. In one approach, fiber gratings are used as mirrors of a FP filter, resulting in transmission filters whose free spectral range can vary over a wide range 0.1–10 nm [63]. In another design, a grating is inserted in each arm of a MZ interferometer to provide a transmission filter [64]. Other kinds of interferometers, such as the *Sagnac* and *Michelson interferometers*, can also be used to realize transmission filters. Figure 7.9(c) shows an example of the Michelson interferometer made by using a 3-dB fiber coupler and two fiber gratings acting as mirrors for the two arms of the Michelson interferometer [65]. Most of these schemes can also be implemented in the form of a planar lightwave circuit by using silica waveguides on a silicon substrate.

In another grating-based scheme, borrowed from the DFB semiconductor-laser technology, the InGaAsP/InP material system is used to form planar waveguides functioning near 1.55 μm. The wavelength selectivity is provided by a built-in grating whose Bragg wavelength is tuned electrically by using the phenomenon of electrorefraction [66]. A phase-control section, similar to that used for multisegment DFB lasers, have also been used to tune DBR filters. Multiple gratings, each tunable independently, can also be used to make tunable filters [67]. Such filters can be tuned quickly (in a few nanoseconds) and can be designed to provide net gain since one or more amplifiers can be integrated with the filter. They can also be integrated within the receiver, as they use the same semiconductor material. These two properties of InGaAsP/InP filters make them quite attractive for WDM applications.

Gratings etched permanently into silica or InGaAsP waveguides suffer from a limited tuning range. In another class of filters, the grating is formed dynamically by using acoustic waves. Such filters, called *acousto-optic filters*, exhibit a wide tuning range (> 100 nm) and have been extensively developed. The physical mechanism behind the operation of acousto-optic filters is the *photo-elastic effect* [29] through which an acoustic wave propagating through an acousto-optic material creates periodic changes in the refractive index corresponding to the regions of local compression and rarefaction. In effect, the acoustic wave creates a periodic index grating that can diffract an optical beam. The wavelength selectivity stems from this acoustically induced grating. An incident wave at the frequency ν and having a propagation vector $\mathbf{k}$ is diffracted from this grating and appears at a slightly different frequency $\nu' = \nu \pm \nu_a$ in a direction determined by the *phase-matching condition* $\mathbf{k}' = \mathbf{k} \pm \mathbf{K_a}$, where ν_a and $\mathbf{K_a}$ are the frequency and the wave vector of the acoustic wave, respectively. These relations can be easily understood in terms of energy and momentum conservation if Bragg diffraction is viewed as a photon-phonon collision.

Acousto-optic tunable filters can be made by using bulk components as well as waveguides, and both kinds are available commercially. For WDM applications, the LiNbO$_3$ waveguide technology is often used since it can produce compact, polarization-independent, acousto-optic filters having a bandwidth of about 1 nm and a tuning range over 100 nm [68]. The basic design, shown schematically in Fig. 7.9(d), uses two polarization beam splitters, two LiNbO$_3$ waveguides, a surface-acoustic-wave transducer, all integrated on the same substrate. The incident WDM signal is split into its orthogonally polarized components by the first beam splitter. The channel whose wavelength satisfies the Bragg condition is directed to a different output port by the second beam splitter because of an acoustically induced change in its polarization direction; all other channels go to the other output port. Tuning is relatively fast because of its electronic nature, resulting in a switching time of less than 10 μs. Acousto-optic tunable filters appear to be a practical candidate for wavelength routing and optical cross-connect applications in dense WDM systems [68], [69].

Amplification-Based Filters

Another category of tunable optical filters operates on the principle of amplification of a selected channel. Any amplifier with a gain bandwidth smaller than the channel spacing can be used as an optical filter. Tuning is achieved by changing the wavelength at which the gain peak occurs. Stimulated Brillouin scattering (SBS) occurring naturally in silica fibers [70] can be used for selective amplification of one channel [71], since the gain bandwidth is quite small ($\sim$ 100 MHz). As discussed in Section 8.4, SBS also involves interaction between the optical and acoustic waves and is governed by the same energy and momentum conservation laws as those for acousto-optic filters. In particular, SBS occurs only in the backward direction and involves a frequency shift of about 10 GHz, corresponding to the acoustic-wave frequency in silica fibers in the 1.55-μm region.

To use the SBS amplification as a tunable optical filter, a CW pump beam is launched at the receiver end of the optical fiber in a direction opposite to that of the multichannel signal, and the pump wavelength is tuned to select the channel. The pump beam transfers a part of its energy to a channel down-shifted from the pump frequency by exactly the Brillouin shift (about 10 GHz at 1.55 μm). A tunable pump laser is a prerequisite for this scheme. The bit rate of each channel is limited to about 100 MHz when a narrow-linewidth pump laser is used. It can be increased by broadening the pump spectrum as the SBS-gain bandwidth then increases as well. The potential of this scheme was demonstrated in an experiment [72] in which a 128-channel WDM network was simulated by using two 8 × 8 star couplers in series. A specific channel operating at 150 Mb/s could be selected with a channel separation as small as 1.5 GHz while maintaining bit-error rates below 10^{-10}.

Semiconductor laser amplifiers (see Chapter 8) can also be used for channel selection provided that a DFB structure is used to narrow the gain bandwidth [73]. A built-in grating can easily provide a filter bandwidth below 1 nm. Tuning is achieved using a *phase-control section* in combination with a shift of Bragg wavelength through electrorefraction. In fact, such amplifiers are nothing but multisection semiconductor lasers (discussed in Section 3.3.5) whose facet reflectivities have been made negligible by using antireflection coatings. In one experimental demonstration [74], two channels operating at 1 Gb/s and separated by 0.23 nm could be separated by selective amplification (> 10 dB) of one channel. The tuning range of such amplifiers is limited to 4–5 nm, restricting the number of channels to about 20. A tunable FP laser or amplifier can also be used for channel selection [75]. A tuning range of 188 GHz with 23-dB gain and 5-GHz bandwidth has been achieved by using a two-section geometry.

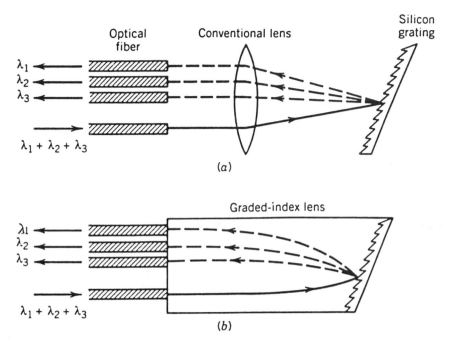

Figure 7.10 Grating-based demultiplexer making use of (a) a conventional lens and (b) a graded-index lens.

Four-wave mixing in a semiconductor laser amplifier can also be used to form a tunable filter whose center wavelength is determined by the pump laser [76].

7.2.2 Multiplexers and Demultiplexers

Multiplexers and demultiplexers are the essential components of a WDM system. Similar to the case of optical filters, demultiplexers require a wavelength-selective mechanism and can be classified into two broad categories. *Diffraction-based demultiplexers* use an angularly dispersive element, such as a diffraction grating, which disperses incident light spatially into various wavelength components. *Interference-based demultiplexers* make use of devices such as optical filters and directional couplers. In both cases, the same device can be used as a multiplexer or a demultiplexer, depending on the direction of propagation because of the inherent reciprocity of optical waves in dielectric media.

Grating-based demultiplexers use the phenomenon of Bragg diffraction from an optical grating. Figure 7.10 shows the design of "bulk" grating-based demultiplexers schematically. The input WDM signal is focused onto a reflection grating which splits various wavelength components spatially, and a lens focuses them onto individual fibers. Use of a graded-index lens simplifies alignment and provides a relatively compact device. The focusing lens can be eliminated altogether by using a concave grating. For a compact design, the concave grating can be integrated within a silicon slab waveguide [1]. In a different approach [77]

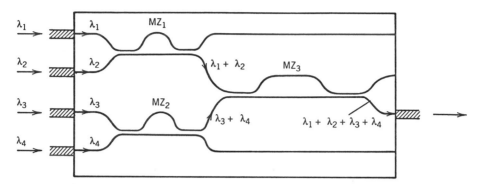

Figure 7.11 Layout of an integrated four-channel waveguide multiplexer based on Mach–Zehnder interferometers. (After Ref. [80]. ©1988 IEEE. Reprinted with permission.)

directly etched elliptical Bragg gratings have been used to realize waveguide multiplexers with the silicon technology. The idea behind this approach is simple. If the input and output fibers are placed at the two foci of the elliptical grating, and the grating period L is adjusted to a specific wavelength λ_0 by using the *Bragg condition* $2\Lambda n_{\text{eff}} = \lambda_0$, where n_{eff} is the effective index of the waveguide mode, the grating would selectively reflect that wavelength and focus it onto the output fiber. Multiple gratings need to be etched, as each grating reflects only one wavelength. In practice, the gratings are ordered such that shorter wavelengths are reflected first. The scheme has been used to demonstrate a four-channel multiplexer [77].

A problem with grating demultiplexers is that their bandpass characteristics depend on the dimensions of the input and output fibers. In particular, the core size of output fibers must be large to ensure a flat passband and low insertion losses. For this reason, most early design of multiplexers used multimode fibers. In a 1991 design [78], a microlens array was used to solve this problem and to demonstrate a 32-channel multiplexer for single-mode fiber applications. The fiber array was produced by fixing single-mode fibers in V-shaped grooves etched into a silicon wafer. The microlens transforms the relatively small mode diameter of fibers ($\sim 10\ \mu$m) into a much wider diameter (about $80\ \mu$m) just beyond the lens. This scheme provides a multiplexer that can work with channels spaced by only 1 nm in the wavelength region near $1.55\ \mu$m while accommodating a channel bandwidth of 0.7 nm [78].

Filter-based demultiplexers use the phenomenon of optical interference to select the wavelength [1]. Demultiplexers based on the MZ filter have attracted the most attention. Similar to the case of an optical filter, several MZ interferometers can be combined to form a WDM demultiplexer [79]–[82]. A 128-channel multiplexer fabricated with the silicon waveguide technology was demonstrated by 1989 [81]. Figure 7.11 illustrates the basic concept by showing the layout of a four-channel multiplexer. It consists of three MZ interferometers. One arm of each MZ interferometer is made longer than the other to provide a wavelength-dependent phase shift between the two arms. The path-length difference is chosen such that the total input power from two input ports at different wavelengths appears at only one output port. The whole structure can be fabricated on a silicone substrate by using SiO_2 waveguides in the form of a planar lightwave circuit.

Fiber Bragg gratings can also be used for making all-fiber demultiplexers. In one approach [62], a $1 \times N$ fiber coupler is converted into a demultiplexer by forming a *phase-shifted grating* at the end of each output port, opening a narrowband transmission window (~ 0.1 nm) within the stop band. The position of the transmission window is varied by changing the amount of phase shift so that each arm of the $1 \times N$ fiber coupler transmits only one channel. The fiber-grating technology has been applied to form photo-induced Bragg gratings directly on a planar silica waveguide [83]. This approach has attracted attention since it permits integration of Bragg gratings within planar lightwave circuits. Such gratings were incorporated as early as 1993 in a MZ interferometer with equal arm lengths to form a bandpass optical filter [64]. Their use in an asymmetric MZ interferometer (unequal arm lengths) results in a compact multiplexer [84].

It is possible to construct multiplexers by using directional couplers which transfer power from one waveguide to another through evanescent mode coupling [85]. The basic scheme is similar to Fig. 7.11 but simpler since MZ interferometers are not used. Furthermore, one can make an all-fiber multiplexer by using fiber couplers [85], thus avoiding the coupling losses that occur whenever light is coupled into or out of the fiber. A *fused biconical taper* can also be used to make the fiber coupler [86]. Multiplexers based on fiber couplers are attractive for WDM. However, their use requires a relatively large channel spacing (> 10 nm), making them unsuitable for dense WDM applications.

From the standpoint of system design, integrated multiplexers with low insertion losses are preferred. Even grating-based multiplexers based on silicon technology suffer from the coupling-loss problem since multiple fibers should be connected to the input and output ports of the device. This problem can be solved if the grating is made by using InGaAsP/InP technology since it can then be integrated within a receiver. Although the fabrication of an *etched grating* is not a simple step, such integrated receivers have been made and are discussed later in this section.

Another integrated approach to demultiplexing uses a *phased array* of optical waveguides that acts as a grating. Such gratings are often called *waveguide gratings* and have attracted considerable attention [87]–[90] since they can be fabricated by using InGaAsP/InP technology, permitting their integration within a WDM transmitter or receiver. The basic idea is quite simple [87]. The incoming WDM signal is coupled into an array of planar waveguides after passing through a coupling section. During its propagation, the signal in each waveguide experiences a different phase shift because of different lengths of waveguides. Moreover, the phase shifts are wavelength dependent because of the frequency dependence of the mode-propagation constant. As a result, different channels focus to different spatial spots when the output of waveguides diffracts through another coupling section. In essence, such a phased array of waveguides acts as a conventional diffraction grating. The efficiency of a waveguide grating can be close to 100% with a proper design [91]. A *waveguide-grating demulti-plexer* capable of resolving 16 channels with a 1.8-nm spacing in the wavelength range 1535–1565 nm was demonstrated in 1994 [88]. Polarization sensitivity becomes a critical issue for such devices, but polarization-insensitive performance can be achieved with a proper design [89], [90]. A waveguide grating can also be made by using the silica-on-silicon (SiO_2/Si) technology. In fact, such gratings are useful for many WDM components (such as a waveguide-grating router) as discussed later in this section. In another variant, a reflective waveguide-grating demultiplexer was fabricated by using $LiNbO_3$ technology [92].

The performance of multiplexers is judged mainly by the amount of insertion (or coupling) loss for each channel. The performance criterion for demultiplexers is more stringent. First, the performance of a demultiplexer should be insensitive to the polarization of the incident WDM signal. Second, a demultiplexer should separate each channel without any leakage from the neighboring channels. In practice, some power leakage is likely to occur, particularly in the case of dense WDM systems with small interchannel spacing. Such power leakage is referred to as crosstalk and should be quite small (< -20 dB) for a satisfactory system performance. The issue of interchannel crosstalk is discussed in Section 7.3.

7.2.3   Add/Drop Multiplexers and Filters

Add/drop multiplexers are needed for WDM networks in which one or more channels need to be dropped or added while preserving the integrity of other channels. One can think of such a WDM device as a demultiplexer–multiplexer pair since its operation requires demultiplexing of the input WDM signal, changing the data content of one or more specific wavelength channels and then multiplexing the entire signal back again. Figure 7.12(a) shows such a generic add/drop multiplexer schematically Since a demultiplexer can act as a multiplexer in the reverse direction, an add/drop multiplexer uses a pair of demultiplexers arranged suitably. Any demultiplexer design discussed in the preceding subsection can be used to make add/drop multiplexers. It is even possible to amplify the WDM signal and equalize the channel powers at the add/drop multiplexer since each channel can be individually controlled [93].

If a single channel needs to be demultiplexed and no active control of individual channels is required, one can use a multiport device that sends a single channel to one port while all other channels are transferred to another port, thereby avoiding the need for demultiplexing all channels. Such devices are often called add/drop filters since they filter out a specific channel without affecting the WDM signal. If only a small portion of the channel power is filtered out, such a device acts as an "optical tap," as it leaves the contents of the WDM signal intact.

Several kinds of multiport add/drop filters have been developed. The simplest scheme uses a series of interconnected directional couplers, forming a MZ chain similar to that of a MZ filter discussed earlier. However, in contrast with

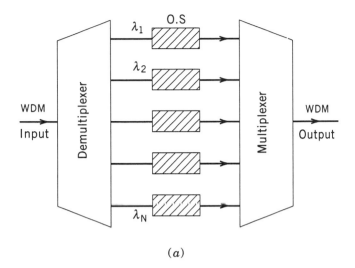

(a)

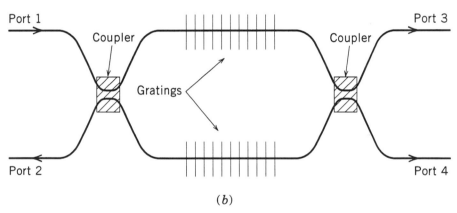

(b)

Figure 7.12 (a) Generic add/drop multiplexer using optical switches (O.S.); (b) add/drop filter made with a Mach–Zehnder interferometer and two identical fiber gratings.

the MZ filter of Section 7.2.1, the relative delay τ_m in Eq. (7.2.5) is made the same for each MZ interferometer. Such a device is sometimes referred to as a *resonant coupler* ince it resonantly couples out a specific-wavelength channel to one output port while the remainder of the channels appear at the other output port. Its performance can be optimized by controlling the coupling ratios of various directional couplers [94]. Although resonant couplers can be implemented in an all-fibers configuration by using fiber couplers, the silica-on-silicon waveguide technology provides a compact alternative to designing such add/drop filters [95].

The wavelength selectivity of Bragg gratings can also be used to make add/drop filters. In one device, referred to as a *grating-assisted* or *grating-folded* directional coupler, a Bragg grating is fabricated in the middle of a directional coupler [96]. In another, two identical Bragg gratings are formed on the two arms of a MZ interferometer [97]. The operation of both devices can be understood from Fig. 7.12(b), where the MZ configuration is shown schematically. A single channel, whose wavelength λ_g falls within the stop band of the Bragg grating, is totally reflected and appears at port 2. The remaining channels are not affected by the gratings and appear at port 4. The same device can add a channel at the wavelength λ_g if the signal at that wavelength is injected from port 3. If the add and drop operations are performed simultaneously, it is important to make the gratings highly reflecting (close to 100%) to minimize the crosstalk.

Grating-folded directional couplers have been made by using InGaAsP/InP waveguides [96]. This approach permits the integration of add/drop filter with a detector or a laser [98]. An all-fiber design of add/drop filters is also attractive since it avoids the coupling loss that invariably occurs when semiconductor-based add/drop filters are used. The MZ configuration of the add/drop filter has been used to make an all-fiber add/drop filter [99] that exhibited extraction efficiency of more than 99%, while keeping the crosstalk below 1%. Several other schemes use gratings to make add/drop filters. In one scheme, a waveguide with a built-in phase-shifted grating is used to add or drop one channel from a WDM signal propagating in a neighboring waveguide [100]. In another, two identical waveguide gratings are combined with amplifiers to realize a channel-dropping filter [101].

7.2.4 Broadcast Star Couplers

The role of a star coupler, as seen in Fig. 7.3, is to combine the optical signals entering from its multiple input ports and divide it equally among its output ports. In contrast with demultiplexers, star couplers do not contain wavelength-selective elements, as they do not attempt to separate individual channels. The number of input and output ports need not be the same. For example, in the case of video distribution, a relatively small number of video channels (say 100) are broadcast to thousands or even millions of subscribers [27]. The number of input and output ports is generally the same for LAN applications. Such a passive star coupler is referred to as an $N \times N$ broadcast star, where N is the number of input (or output) ports. A reflection star is sometimes used for LAN applications by reflecting the combined signal back to its input ports. For instance, a mirror placed at the center of the star in Fig. 7.3 would convert the transmission star into a reflection star. Such a geometry saves considerable fiber when users are distributed over a large geographical area.

Several kinds of star couplers have been developed. An early approach made use of multiple 3-dB fiber couplers [102]. Each fiber coupler is capable of mixing two input signals and dividing it equally between its two output ports, the same operation required for a 2×2 star coupler. Higher-order $N \times N$ stars can be formed by combining several 2×2 couplers [102] as long as N is a multiple of two. Figure 7.13 shows such a combination scheme for an 8×8 star, requiring

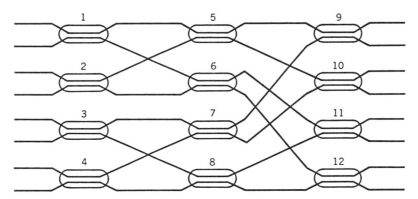

Figure 7.13 A 8 × 8 star coupler formed by using twelve 2 × 2 single-mode fiber couplers.

interconnection among 12 fiber couplers. Clearly, the complexity of such star couplers grows enormously with the number of ports.

A solution is offered by fused biconical-taper couplers [86] which can be used to form compact, monolithic star couplers. Figure 7.14 shows schematically the transmission and reflection stars formed by the use of this technique. Star couplers of 100 × 100 size were made as early as 1979 by using the biconical-taper scheme [103]. The idea is to fuse together a large number of fibers and elongate the fused portion to form a biconically tapered structure. In the tapered portion, signals from each fiber mix together and are shared almost equally among its output ports. Such a scheme works relatively well for multimode fibers. In the case of single-mode fibers it is limited to fusing of a few fibers only. Fused 2 × 2 couplers by using single-mode fibers were made quite early [86]. They can also be designed to operate over a wide wavelength range. One can then form higher-order stars by using the combinatorial scheme shown in Fig. 7.12 for an 8 × 8 star coupler [104]. Fused star couplers are preferred in practice because of their relatively compact nature [105]–[107].

A common approach for fabricating a compact broadcast star makes use of the silica-on-silicon technology in which two arrays of planar SiO_2 waveguides, separated by a central slab region, are formed on a silicon substrate. Such a star coupler was first demonstrated in 1989 in a 19 × 19 configuration [108]. The SiO_2 channel waveguides were 200 μm apart at the input end, but the final spacing near the central region was only 8 μm. The 3-cm-long star coupler had an efficiency of about 55%. Recently, the silicon-on-insulator technology has been used for making star couplers. A 5 × 9 star made by using silicon rib waveguides exhibited low loss (1.3 dB) with a relatively uniform coupling [109].

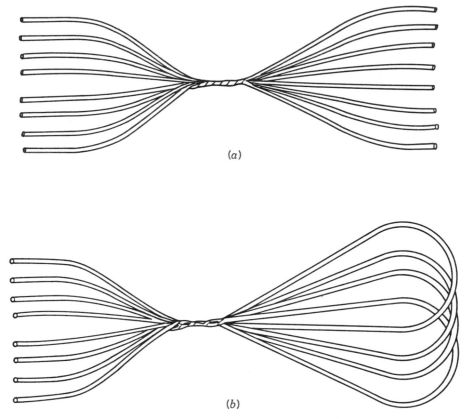

Figure 7.14 (a) Transmission and (b) reflection star couplers formed by using the fused biconical tapering method.

7.2.5 Wavelength Routers

An important WDM component is an $N \times N$ router, a device that combines the functionality of a star coupler with multiplexing and demultiplexing operations. Figure 7.15(a) shows the operation of such a wavelength router schematically for $N = 4$. The WDM signal entering from each input port is split into N parts directed toward the N output ports of the router. Such a device is an example of a passive router since its use does not involve any active element requiring electrical power. It is also called a *static router* since the routing topology is not dynamically reconfigurable. Despite its static nature, such a WDM device has many potential applications in WDM networks.

An $N \times N$ multiplexer can be used as a wavelength router. One design uses two $N \times M$ star couplers such that M output ports of one star coupler are connected with M input ports of another star coupler through an array of M waveguides that acts as a waveguide grating [110]. Such a device, called a *waveguide-grating router* (WGR), is shown schematically in Fig. 7.15(b). The first $N \times M$ star coupler distributes the power of N input channels equally to

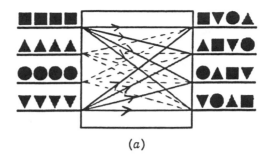

(a)

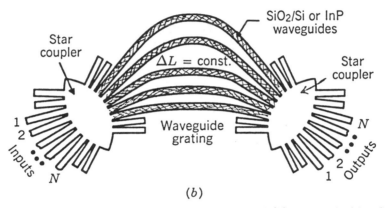

(b)

Figure 7.15 Schematic of (a) a wavelength router and (b) its practical implementation through a waveguide-grating router making use of two star couplers and an array of planar waveguides.

its M output ports. The grating formed with M waveguides separates different channels according to their wavelengths. The second $M \times N$ star coupler distributes the demultiplexed signals to its output ports. The net result is that the input WDM signals coming from N different nodes are routed to another set of N different nodes, and the distribution is made based solely on the wavelengths of incoming channels.

The waveguide grating is the most important part of a WGR since it performs the demultiplexing operation. The waveguide array is designed such that the length difference remains constant from one waveguide to next, resulting in a constant phase difference. Because of this property, the amplitude transmission T_{pq} from the pth input port to the qth output port is given by [110]

$$T_{pq} = \eta_p \eta_q \frac{\sum_{m=1}^{M} P_m \exp[imb(2\pi/\lambda)(p-q)]}{\sum_{m=1}^{M} P_m}, \qquad (7.2.6)$$

where η_p and η_q are the coupling efficiencies of the two star couplers, b is a constant that depends on the angular separation among the waveguide-array ports, and P_m is the power in the mth waveguide. It is clear from Eq. (7.2.6) that the transmission spectrum of a WGR is periodic. For a given value of p, $|T_{pq}|^2$ peaks at a different wavelength as q is varied, and the device acts as a $1 \times N$ demultiplexer. Thus, a WGR can be thought of as N demultiplexers working in parallel with the following property. If the WDM signal from the first input port is distributed to N output ports in the order λ_1, $\lambda_2 \ldots, \lambda_N$, the WDM signal from the second input port will be distributed as λ_N, $\lambda_1, \ldots, \lambda_{N-1}$, and the same cyclic pattern is followed for other input ports. The optimization of a WGR to reduce the crosstalk and to maximize the coupling efficiency requires precise control of many design parameters, and several design rules have been established [110]. As an example, the number of waveguides forming the waveguide grating should be close to $2N$, to allow a relatively small channel spacing with a minimum of crosstalk.

Despite the complexity of the design, WGRs have been fabricated by using both silica-on-silicon technology [111]–[113] and InGaAsP/InP technology [114], resulting in an integrated, compact device (~ 1 cm^2). In early attempts, the number of input and output ports was limited to below 16, but 128×128 WGRs having 128 input and output ports were available by 1996 in the form of a planar lightwave circuit [115]. They can operate on WDM signals with a channel spacing as small as 0.2 nm while maintaining crosstalk below 16 dB. WGRs are needed for wavelength routing in WDM networks, but they can also be used for applications other than wavelength routing [116], [117]. They have been used for making multichannel transmitters and receivers (discussed later in this section), tunable add-drop optical filters [118], [119], and add/drop multiplexers [120]. Their use has led to a novel technique, called *spectral slicing*, that permits the use of an LED as a low-cost multiwavelength source for local-loop applications [121]; an example is shown in Fig 7.7. The basic idea is quite simple. If the output of an LED is modulated to impose a signal and then connected to a WGR, the broad spectrum of the LED is sliced into as many parts as the number of WGR output ports. As a result, the modulated signal is distributed to many users at different wavelengths determined by WGR through spectral slicing.

7.2.6   Optical Cross-Connects

The development of wide-area WDM networks [48]–[51] requires dynamic wavelength routing that can reconfigure the network while maintaining its non-blocking (transparent) nature. This functionality is provided by an *optical cross-connect* (OXC) which performs the same function as that provided by electronic digital switches in telephone networks. The use of dynamic routing also solves the problem of a limited number of available wavelengths through the wavelength-reuse technique. The design and fabrication of OXCs were a major topic of research during the decade of 1990s [122]–[127].

Figure 7.16 shows the design of an OXC schematically. It contains N input ports, each port receiving a WDM signal consisting of M wavelengths. Demultiplexers split the signal into individual wavelengths and distribute each wavelength to M switches, each switch receiving N input signals at the same

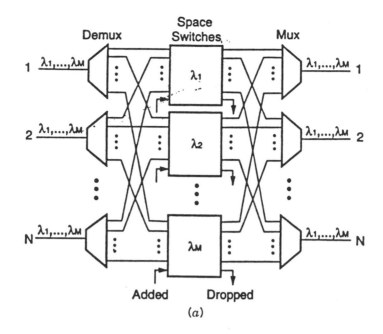

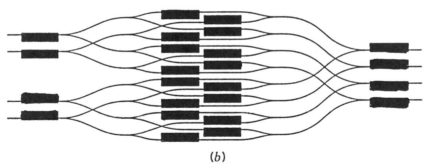

Figure 7.16 (a) Schematic of an optical cross-connect making use of space-division switches; (b) Example of such a 4 × 4 switch based on semiconductor laser amplifiers.

wavelength. An extra input and output port is added to the switch to allow dropping or adding of a specific channel. Switches send their output to N multiplexers which combine their M inputs to form the WDM signal. Such an OXC needs N multiplexers, N demultiplexers, and $M(N+1) \times (N+1)$ optical switches. Switches used by an OXC are the space-division switches since they operate on the basis of the spatial location of their inputs.

Many schemes have been developed for performing the switching operation. In one scheme [122], a delivery-and-coupling type optical switch was used to form a 8 × 16 OXC by using the planar lightwave circuit technology. Since 128 optical paths, each carrying a 2.5-Gb/s signal, were cross-connected in a non-blocking manner, the OXC had a throughput of 320 Gb/s. By 1997, such an OXC was packaged in the form of switch boards of a standard dimension (33×33 cm^2) [127]. In another scheme [128], shown schematically in Fig. 7.16 for a 4×4 switch, a semiconductor laser amplifier (SLA) is used as a gate switch (see Section 8.2). Each input is divided into several branches using 3-dB splitters, and each branch passes through an SLA which either blocks it through absorption or transmits it while amplifying the signal simultaneously. Such *space-division switches* have the advantage that all components can be integrated using the InGaAsP/InP technology while providing low insertion loss or even net gain because of the SLAs. They can operate at high bit rates, and operation at bit rates as high as 2.5 Gb/s has been demonstrated within an installed fiber network [128].

The main drawback of the OXC shown in Fig. 7.16 is the large number of components and interconnections required that grows exponentially as the number of nodes N and the number of wavelengths M increase. In an alternative scheme, the signal wavelength itself is used for switching by making use of wavelength-division switches. Such a scheme makes use of static wavelength routers such as a WGR (see Fig. 7.15) together with a new WDM component—the *wavelength converter*. The next subsection discusses the operation of wavelength converters.

7.2.7   Wavelength Converters

A wavelength converter changes the input wavelength to a new wavelength without modifying the data content of the signal. Many schemes were developed during the 1990s to realize wavelength converters [129]; four among them are shown schematically in Fig. 7.17.

A conceptually simple scheme uses an optoelectronic regenerator [see Fig. 7.17(a)] in which a receiver first converts the optical signal at the input wavelength λ_1 to electric current $I(t)$, followed by a transmitter that uses $I(t)$ to generate the optical signal at the desired wavelength λ_2. Such a scheme is relatively easy to implement since it uses standard components. Its other advantages include insensitivity to input polarization and possibility of net amplification. Among its disadvantages are limited transparency to bit rate and data format, speed limited by electronics, and a relatively high cost.

Several all-optical techniques for wavelength conversion make use of semiconductor laser amplifiers [130], [131] (see Section 8.2). The simplest one is based on gain saturation in a SLA and is shown in Fig. 7.17(b). The incident signal at the wavelength λ_1 is launched into the SLA together with a CW beam at the wavelength λ_2 at which the converted signal is desired. Both beams saturate the amplifier. However, the CW beam is amplified by a large amount during 0 bits, but by a much smaller amount during 1 bits. As a result, the bit pattern of the incident signal is transferred to the new wavelength, although in reverse. This technique has been used in many experiments and can work at bit rates as high as 20 Gb/s. It can provide net gain due to amplification and can be made nearly polarization insensitive. Its main disadvantages are (1) requirement of a

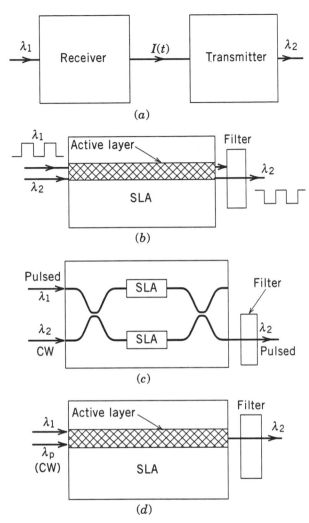

Figure 7.17 Four schemes for wavelength conversion: (a) optoelectronic regenerator; (b) gain saturation in a semiconductor laser amplifier (SLA); (c) phase modulation in a SLA placed in one arm of a Mach-Zehnder interferometer; (d) four-wave mixing inside a SLA.

rapidly tunable CW laser, (2) signal degradation due to spontaneous emission, and (3) phase distortion due to frequency chirping that invariable occurs (see Section 3.3.7).

Frequency chirping can be used to advantage by using a SLA in one or both arms of a MZ interferometer. Figure 7.17(c) shows the case in which two SLAs are used. The pulsed signal at the wavelength λ_1 and the CW signal at the wavelength λ_2 propagate simultaneously in the two arms. However, an additional phase shift is introduced for each 1 bit in the signal. The MZ interferometer is designed such that the two waves interfere constructively or destructively, depending on this additional phase shift. As a result, the CW wave is directed to different output ports of the MZ interferometer depending on the bit pattern. The output from one port is an exact replica of the incident signal at the new wavelength. Other types of interferometers (such as Michelson and Sagnac interferometers) can also be used, with similar results. A MZ interferometer is often used in practice since it can be easily integrated by using SiO_2/Si or InGaAsP/InP waveguides, resulting in a compact device. Such a device can operate at high bit rates (10 Gb/s or more), offers a large contrast, and degrades the signal relatively little, although spontaneous emission does affect the SNR. Its main disadvantage is a narrow dynamic range of the input power since the phase induced by the amplifier depends on it.

Another scheme uses the SLA as a nonlinear medium for *four-wave mixing* (FWM), the same nonlinear phenomenon (see Section 2.6.3) that is a major source of interchannel crosstalk in WDM systems (see Section 7.3.2). This technique is also known as *optical phase conjugation* and is discussed in Section 9.7 in the context of dispersion compensation. As seen in Fig. 7.17(d), its use requires a CW pump beam that is launched with the signal whose wavelength needs to be converted. If ν_1 and ν_2 are the frequencies of the input signal and the converted signal, the pump frequency ν_p is chosen such that $\nu_p = (\nu_1 + \nu_2)/2$. At the amplifier output, a replica of the input signal appears at the carrier frequency ν_2 since FWM requires the presence of both the pump and signal. One can understand the process physically as scattering of two pump photons of energy $2h\nu_p$ into two photons of energy $h\nu_1$ and $h\nu_2$. The nonlinearity responsible for the FWM process has its origin in a fast intraband relaxation process occurring at a time scale of 0.1 ps [132]. As a result, frequency shifts as large as 10 THz, corresponding to wavelength conversion over a range of 80 nm, are possible. For the same reason, this technique can work at bit rates as high as 100 Gb/s and is transparent to both the bit rate and the data format. Because of the gain provided by the amplifier, conversion efficiency can be quite high, resulting even in a net gain [133]. An added advantage of this technique is the reversal of the frequency chirp since its use inverts the signal spectrum (see Section 9.7). In practice, the system performance can also be improved by using two SLAs in a tandem configuration [134].

The main disadvantage of a wavelength-conversion technique using a semiconductor laser amplifier is that it requires a tunable laser source whose light should be coupled into the amplifier, typically resulting in large coupling losses. An alternative is to integrate the functionality of a wavelength converter within a tunable semiconductor laser. Several such devices have been developed [130]. In the simplest scheme, the signal whose wavelength needs to be changed is injected into a tunable laser directly. The change in the laser threshold resulting from injection translates into modulation of the laser output, mimicking the bit pattern of the injected signal. Such a scheme requires relatively large input powers. A low-power scheme uses the input signal to produce a frequency shift (typically, 10 GHz/mW) in the laser output for each "1" bit. The resulting frequency-modulated CW signal can be converted into amplitude modulation by using a MZ interferometer. This scheme can work at bit rates as high as the relaxation-oscillation frequency of the tunable laser (up to 25 GHz). Another scheme uses FWM inside the cavity of a tunable semiconductor laser which also plays the role of the pump laser. A phase-shifted DFB laser has provided wavelength conversion over a range of 30 nm with this technique [135].

7.2.8   WDM Transmitters and Receivers

Although WDM transmitters and receivers attracted attention throughout the decade of 1980s, it was only during the 1990s that monolithically integrated WDM transmitters and receivers, operating near 1.55 μm with a channel spacing of 1 nm or less, were developed by using the InP-based *optoelectronic integrated-circuit* (OEIC) technology [136]–[147]. In a different approach, planar light-wave circuits fabricated with the silica-on-silicon technology are used to develop hybrid-integrated transmitters and receivers [148]. Such OEIC components are quite important for implementation of WDM technology.

Many schemes have been used to make monolithic WDM transmitters [137]. In one approach, the output of several DFB or DBR semiconductor lasers, independently tunable through Bragg gratings, is combined by using passive waveguides [136]. A built-in amplifier boosts the power of the multiplexed signal to increase the transmitted power. In a 1993 experiment, a WDM transmitter not only integrated 16 DBR lasers with 0.8-nm wavelength spacing, but an electro-absorption modulator was also integrated with each laser [138]. In another experiment, 16 gain-coupled DFB lasers were integrated [139], and their wavelengths were controlled by changing the width of the ridge waveguides together with a fine tuning over a 1-nm range through a thin-film resistor. Vertical-cavity surface-emitting lasers (VCSELs) provide another unique approach to WDM transmitters that have the potential of low cost while integrating a two-dimensional array of lasers covering a wide wavelength span [140].

In a different approach, a grating is integrated within the laser cavity to realize lasing at several wavelengths simultaneously. The *multistripe-array grating-integrated cavity* (MAGIC) laser made use of an etched grating [141]. A waveguide grating can also be integrated within the laser cavity. In a 1996 demonstration, simultaneous operation at 18 wavelengths, spaced apart by about 0.8 nm, was realized using an intracavity WGR [142]. Figure 7.18 shows the laser cavity schematically. Spontaneous emission of the amplifier on the left side is demultiplexed into 18 spectral bands by the WGR through spectral slicing. The amplifier array on the right side selectively amplifies the set of 18 wavelengths, resulting in a laser emitting at all wavelengths simultaneously. This approach has provided simultaneous operation at up to 32 wavelengths [143].

Monolithic WDM receivers integrate a photodiode array with a demultiplexer. In one approach, a planar concave-grating demultiplexer is integrated with a photodetector array [144]–[146]. In another approach, a WGR is integrated with a photodiode array. Even electronic amplifiers can be integrated within the same chip. The design of such a monolithic receiver is similar to the transmitter shown in Fig. 7.18 except that no cavity is formed and the amplifier array is replaced with a photodiode array. Such a WDM receiver

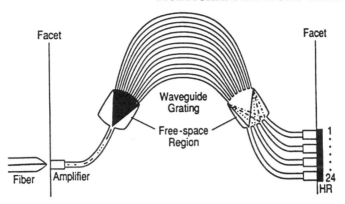

Figure 7.18 Schematic of a multiwavelength laser source using a waveguide grating as a demultiplexer inside the laser cavity. (After Ref. [142]. ©1996 IEEE. Reprinted with permission.)

was demonstrated in 1995 [147]. It integrated an 8-channel WGR (with 0.8-nm channel spacing), eight p-i-n photodiodes, and eight preamplifiers by using heterojunction-bipolar transistors. The WDM receiver was capable of detecting 8 channels, each operating at 2.5 Gb/s.

A unique approach to WDM sources converts the large spectral bandwidth of ultrashort optical pulses into multiple WDM channels. In one implementation of this idea [149], picosecond pulses from a mode-locked fiber laser are first broadened spectrally (to 200-nm bandwidth) through supercontinuum generation in an optical fiber [70]. Spectral slicing of the output by a demultiplexer (a WGR) then produces many WDM channels (up to 200) with a channel spacing of 1 nm or less. In another experiment, femtosecond pulses with a spectral width of 50 nm were chirped and stretched to a duration of about 15 ns by propagating them through 15 km of standard telecommunication fiber [150]. Spectral slicing of the output by a demultiplexer can again provide many channels, each of which can be modulated independently. This technique also permits simultaneous modulation of all channels by using a single modulator before the demultiplexer if the modulator is driven by a suitable electrical bit stream composed through TDM. A 32-channel WDM source was demonstrated by using this method [150]. Such WDM sources are appropriate when the return-to-zero (RZ) format is used for signal transmission or a device capable of RZ-to-NRZ conversion is used.

7.3   SYSTEM PERFORMANCE ISSUES

The most important issue in the design of WDM lightwave systems is the *interchannel crosstalk*. The system performance degrades whenever crosstalk leads to a transfer of power from one channel to another. Such a transfer can occur because of the nonlinear effects in optical fibers, a phenomenon referred to as nonlinear crosstalk as it depends on the nonlinear nature of the communication channel. However, some crosstalk occurs even in a perfectly linear channel because of the imperfect nature of various WDM components such as optical filters, demultiplexers, and switches. In this section we discuss linear and non-linear crosstalk in separate subsections and then consider other performance and design issues relevant for WDM systems.

7.3.1 Linear Crosstalk

Linear crosstalk has been studied extensively [151]–[162]. It can be classified into two categories, depending on its origin. Optical filters and demultiplexers often let leak a fraction of the signal power from neighboring channels that interferes with the detection process. Such crosstalk is called *heterowavelength* or *out-of-band crosstalk* and is less of a problem because of its incoherent nature than the *homowavelength* or *in-band crosstalk* that occurs during routing of the WDM signal from multiple nodes. The difference between the two kinds of crosstalk is best understood by considering an example of each.

Filter-Induced Crosstalk

Consider the case in which a tunable optical filter is used to select a single channel among the N channels incident on it. If the optical filter is set to pass the mth channel, the optical power reaching the photodetector can be written as

$$P = P_m + \sum_{n \neq m}^{N} T_{mn} P_n, \qquad (7.3.1)$$

where P_m is the power in the mth channel and T_{mn} is the filter transmittivity for channel n when channel m is selected. Crosstalk occurs if $T_{mn} \neq 0$ for $n \neq m$. Such crosstalk is out-of-band crosstalk since it belongs to the signals lying outside the spectral band occupied by the channel detected. Its incoherent nature is also apparent in Eq. (7.3.1) since it depends only on the power of the signal in neighboring channels.

To evaluate the impact of the crosstalk on system performance, one should consider the power penalty, defined as the additional power required at the receiver to counteract the effect of crosstalk. The photocurrent generated in response to the incident optical power is given by

$$I = R_m P_m + \sum_{n \neq m}^{N} R_n T_{mn} P_n = I_{ch} + I_X, \qquad (7.3.2)$$

where $R_m = \eta_m q / h \nu_m$ is the photodetector responsivity for channel m at the optical frequency ν_m and η_m is the quantum efficiency, which may be different for different channels. The second term I_X in Eq. (7.3.2) denotes the crosstalk contribution to the receiver current I. Its value depends on the bit pattern and becomes maximum when all interfering channels carry 1 bits simultaneously. This is referred to as the worst case.

A simple approach to calculating the *crosstalk power penalty* is based on the eye closure (see Section 4.3.3) occurring as a result of the crosstalk [151]. The maximum *eye closure* happens in the worst case for which I_X is maximum. In practice, I_{ch} is increased to maintain system performance. If I_{ch} needs to be increased by a factor δ_X, the peak current corresponding to the top of the eye is $I_1 = \delta_X I_{ch} + I_X$. The decision threshold is set at $I_D = I_1/2$. The *eye opening* from I_D to the top level would be maintained at its original value $I_{ch}/2$ if [151]

$$(\delta_X I_{ch} + I_X) - I_X - \tfrac{1}{2}(\delta_X I_{ch} + I_X) = \tfrac{1}{2}I_{ch}, \qquad (7.3.3)$$

or when

$$\delta_X = 1 + I_X/I_{ch}. \qquad (7.3.4)$$

The quantity δ_X is just the power penalty for the mth channel. By using I_X and I_{ch} from Eq. (7.3.2), δ_X can be written (in dB) as

$$\delta_X = 10 \log_{10}\left(1 + \frac{\sum_{n \neq m}^{N} R_n T_{mn} P_n}{R_m P_m}\right), \qquad (7.3.5)$$

where the powers correspond to their on-state values. If the peak power is assumed to be the same for all channels, the crosstalk penalty becomes power independent. Further, if the photodetector responsivity is nearly the same for all channels ($R_m \approx R_n$), δ_X is well approximated by

$$\delta_X \approx 10 \log_{10}(1 + X), \qquad (7.3.6)$$

where $X = \sum_{n \neq m}^{N} T_{mn}$ is a measure of the out-of-band crosstalk since it represents the fraction of total power leaked into a specific channel from all other channels. The numerical value of X depends on the transmission characteristics of the specific optical filter. For a FP filter, X can be obtained in a closed form [152].

The preceding analysis of crosstalk penalty is based on the eye closure rather than the bit-error rate (BER). One can obtain an expression for the BER [152] if I_X is treated as a random variable in Eq. (7.3.2). For a given value of I_X, the BER is obtained by using the analysis of Section 4.5.1. In particular, the BER is given by Eq. (4.5.6) with the on- and off-state currents given by $I_1 = I_{ch} + I_X$ and $I_0 = I_X$ if we assume that $I_{ch} = 0$ in the off-state. The decision threshold is set at $I_D = I_{ch}(1 + X)/2$, which corresponds to the worst-case situation in which all neighboring channels are in the on-state. The final BER is obtained by averaging over the distribution of the random variable I_X. The distribution of I_X has been calculated for a FP filter and is generally far from being Gaussian. The crosstalk power penalty δ_X can be calculated by finding the increase in I_{ch} needed to maintain a certain value of BER. Figure 7.19 shows the calculated penalty [152] for several values of BER plotted as a function of N/F, where the FP-filter finesse F was taken to be 100. The solid curve corresponds to the error-free case (BER $= 0$). The power penalty can be kept below 0.2 dB to maintain a BER of 10^{-9} for values of N/F as large as 0.33. From Eq. (7.2.2) the channel spacing can be as little as three times the bit rate for such FP filters. Similar results are obtained for acousto-optic filters [153].

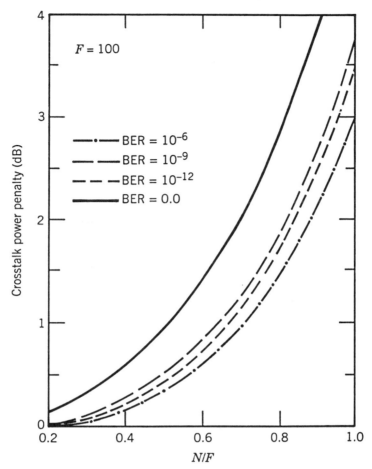

Figure 7.19 Crosstalk power penalty for four different values of the BER calculated for a FP filter of finesse $F = 100$. (After Ref. [152]. ©1990 IEEE. Reprinted with permission.)

WGR-Induced Crosstalk

The origin of in-band crosstalk [158], [159] can be understood by considering a static wavelength router such as a WGR (see Fig. 7.15). For an $N \times N$ router, there exist N^2 combinations through which N-wavelength WDM signals can be split. Consider the output at one wavelength, say λ_m. Among the $N^2 - 1$ interfering signals that can accompany the desired signal, $N - 1$ signals have the same carrier wavelength λ_m, while the remaining $N(N - 1)$ belong to different carrier wavelengths and are likely to be eliminated as they pass through other WDM components. The $N - 1$ crosstalk signals at the same wavelengths (in-band crosstalk) originate from the incomplete filtering by the WGR because of a partial overlap among its N transmission peaks. The total optical field, including only the in-band crosstalk, can be written as

$$E_m(t) = \left(E_m + \sum_{n \neq m}^{N} E_n \right) \exp(-i\omega_m t), \qquad (7.3.7)$$

where E_m is the desired signal and $\omega_m = 2\pi c/\lambda_m$. The coherent nature of the in-band crosstalk is obvious from Eq. (7.3.7).

To see the impact of the in-band crosstalk on the system performance, one should evaluate the power penalty. However, the receiver current $I = R|E_m(t)|^2$ in this case contains interference or beat terms because of the coherent nature of the crosstalk. One can identify two types of beat terms, signal-crosstalk beating with terms like $E_m E_n$ and crosstalk-crosstalk beating with terms like $E_k E_n$, where both k and $n \neq m$. The latter terms are negligible in practice and can be ignored. The receiver current is then given approximately by

$$I(t) \approx RP_m(t) + 2R \sum_{n \neq m}^{N} \sqrt{P_m(t)P_n(t)} \cos[\phi_m(t) - \phi_n(t)], \qquad (7.3.8)$$

where $P_n = |E_n|^2$ is the power and $\phi_n(t)$ is the phase. In practice, $P_n << P_m$ for $n \neq m$ because a WGR is built to reduce the crosstalk. Since phases are likely to fluctuate randomly, one can write Eq. (7.3.8) as $I(t) = R(P_m + \Delta P)$ and treat the crosstalk as intensity noise and use the approach of Section 4.6.2 to calculate the power penalty. In fact, the result is the same as in Eq. (4.6.11), i.e.,

$$\delta_X = -10 \log_{10}(1 - r_X^2 Q^2), \qquad (7.3.9)$$

where

$$r_X^2 = \langle (\Delta P)^2 \rangle / P_m^2 = X(N - 1), \qquad (7.3.10)$$

and $X = P_n/P_m$ is the crosstalk level defined as the fraction of power leaking through the WGR and is taken to be the same for all $N - 1$ sources of coherent in-band crosstalk by assuming equal powers. An average over the phases was performed by replacing $\cos^2 \theta = 1/2$. In addition, r_X^2 was multiplied by another factor of $1/2$ to account for the fact that P_n is zero on average for half of the bits for each crosstalk channel. Experimental measurements of power penalty for a WGR agree with this simple model [158].

The impact of in-band crosstalk can be estimated from Fig. 4.19, where power penalty δ_X is plotted as a function of r_X. To keep the power penalty below 2 dB, $r_X < 0.07$ is required, a condition that limits $X(N - 1)$ to below -23 dB from Eq. (7.3.10). Thus, the crosstalk level X must be below -38 dB for $N = 16$ and below -43 dB for $N = 100$, rather stringent requirements.

The calculation of crosstalk penalty for the case of dynamic wavelength routing through optical cross-connects becomes quite complicated because of a large number of crosstalk elements that a signal can pass through in such WDM networks [160]. The worst-case analysis predicts a large power penalty (> 3 dB) when the number of crosstalk elements becomes more than 25 even if the crosstalk level of each component is only -40 dB. Clearly, the linear crosstalk is of primary concern in the design of WDM networks and should be controlled. Each WDM component must be designed to reduce the crosstalk as much as possible. Linear crosstalk can also be reduced through its compensation at the receiver [161].

7.3.2 Nonlinear Crosstalk

Several nonlinear effects in optical fibers [70] can lead to interchannel crosstalk such that the intensity and the phase of one channel are influenced by other neighboring channels. Section 2.6 discussed such nonlinear effects and their origin from a physical point of view. This section shows how the performance of WDM lightwave systems is affected by them [163]–[166].

Stimulated Raman Scattering

As discussed in Section 2.6, stimulated Raman scattering (SRS) is generally not of concern for single-channel systems because of its relatively high threshold (about 500 mW near 1.55 μm). The situation is quite different for WDM systems for which the fiber acts as a Raman amplifier such that the long-wavelength channels are amplified by the short-wavelength channels as long as the wavelength difference is within the bandwidth of the Raman gain [167]. Fiber Raman amplifiers are discussed in Section 8.3. The Raman gain spectrum of silica fibers is so broad that the amplification can occur for channels spaced as far apart as 200 nm. The shortest-wavelength channel is most depleted, as it can pump many channels simultaneously. Such an energy transfer among channels can be detrimental for system performance as it depends on the bit pattern: Amplification occurs only when 1 bits are present in both channels simultaneously. The signal-dependent amplification leads to enhanced power fluctuations which add to receiver noise and degrade receiver performance.

The *Raman crosstalk* can be avoided if channel powers are made so small that Raman amplification is negligible over the fiber length. It is thus important to estimate the limiting value of the channel power. A simple model [167] considers the depletion of the highest-frequency channel under the worst case in which all channels transmit the bit 1 simultaneously. The amplification factor for each channel is $G_i = \exp(g_i L_{\text{eff}})$, where $L_{\text{eff}} = [1 - \exp(-\alpha L)]/\alpha$ is the effective interaction length [see Eq. (2.6.2)] and $g_i = g_R(\Omega_i)P_{\text{ch}}/A_{\text{eff}}$ is the Raman gain at $\Omega_i = \omega_0 - \omega_i$. For $g_i L_{\text{eff}} \ll 1$, the highest-frequency channel at ω_0 is depleted by a fraction $g_i L_{\text{eff}}$ due to Raman amplification of the ith channel [70]. The total depletion (fractional loss) is given by

$$D = \sum_{i=2}^{N} g_R(\Omega_i)P_{\text{ch}}L_{\text{eff}}/A_{\text{eff}}. \qquad (7.3.11)$$

The summation in Eq. (7.3.11) can be carried out analytically if the Raman gain spectrum (see Fig. 8.11) is approximated by a triangular profile such that

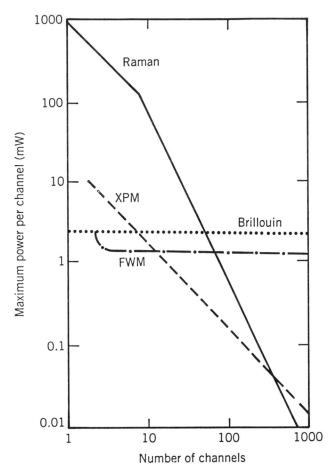

Figure 7.20 Limitations on the channel power imposed by the four nonlinear effects. The system is assumed to operate near 1.55 μm with a fiber loss of 0.2 dB/km. (After Ref. [163]. ©1990 IEEE. Reprinted with permission.)

$g_R(\Omega)$ increases linearly for Ω up to 15 THz and then drops to zero. The result is [167]

$$D = \frac{\Delta\nu_{\mathrm{ch}} g_R^{\max} P_{\mathrm{ch}} L_{\mathrm{eff}} N(N-1)}{6 \times 10^{13} A_{\mathrm{eff}}}, \qquad (7.3.12)$$

where all channels were assumed to have a constant channel spacing $\Delta\nu_{\mathrm{ch}}$, and $g_R^{\max}$ is the peak value of the Raman gain. The gain was reduced by a factor of 2 to account for polarization scrambling occurring inside optical fibers. The power penalty (in dB) is related to D as $\delta_R = -10\log_{10}(1-D)$ since the input channel power must be increased by a factor of $(1-D)^{-1}$ to maintain the same system performance. D should be less than 5% to keep the power penalty below 0.5 dB. Equation (7.3.12) provides the limiting channel power by using $D = 0.05$. This limit is plotted in Fig. 7.20 for a fiber of 8-μm core diameter and 0.2-dB loss near 1.55 μm. SRS limits the channel power below 1 mW when the number of WDM channels exceeds 80.

The simple model has neglected the fact that signals in each channel consist of a random sequence of 0 and 1 bits. A statistical analysis shows that the Raman crosstalk is lower by about a factor of 2 when signal modulation is taken into account [168]. A more complete model should consider the depletion of each channel through Raman pumping of longer-wavelength channels and its own amplification because of pumping from short-wavelength channels [169]. Periodic amplification of the WDM signal in a long-haul fiber link can also magnify the impact of SRS-induced degradation. The reason is that in-line amplifiers add noise which experiences less Raman loss than the signal itself, resulting in degradation of the SNR. In fact, the total capacity of WDM systems is then limited to below 100 Gb/s for a transmission distance of 5000 km or more [170].

Stimulated Brillouin Scattering

Stimulated Brillouin scattering (SBS) can also transfer energy from a high-frequency channel to a low-frequency one when the channel spacing equals the Brillouin shift. However, in contrast with the case of SRS, such an energy transfer is easily avoided with the proper design of multichannel communication systems. The reason is that the Brillouin-gain bandwidth is extremely narrow ($\sim$ 50 MHz) compared with the Raman-gain bandwidth ($\sim$ 5 THz). Thus, the channel spacing must match almost exactly the *Brillouin shift* (about 10 GHz in the 1.55-μm region) for SBS to occur; such an exact match is easily avoided. Furthermore, as discussed in Section 2.6, the two channels must be counter-propagating for Brillouin amplification to occur.

Although SBS does not induce interchannel crosstalk when all channels propagate in the forward direction, it nonetheless limits the channel powers. The reason is that a part of the channel power can be transferred to a backward-propagating Stokes wave generated from noise when the threshold condition [see Eq. (2.6.4)] is satisfied. This condition is independent of the number and the presence of other channels. However, the *Brillouin threshold* for each channel can be reached for a channel power as small as 2–3 mW (see Section 2.6.1). For a fiber with $A_{\text{eff}} = 50$ μm^2 and $\alpha = 0.2$ dB/km, Eq. (2.6.4) predicts a threshold power of only 2.4 mW when the fiber length is long enough ($>$ 20 km) that L_{eff} can be replaced by $1/\alpha$. This value is shown in Fig. 7.20 by a horizontal dotted line. The estimate above neglects the effects of signal modulation. In general, the Brillouin threshold depends on the modulation format as well as on the ratio of the bit rate to the Brillouin-gain bandwidth [171]–[173]. Moreover, it can be increased to close to 100 mW by modulating the phase of the optical carrier at frequencies 0.2–0.5 GHz. Such a phase modulation broadens the laser linewidth, effectively suppressing the SBS.

Cross-Phase Modulation

An important mechanism of crosstalk in multichannel lightwave systems is the nonlinear phenomenon of *cross-phase modulation* (XPM) [70]. As discussed in Section 2.6.2, XPM originates from the intensity dependence of the refractive index which is responsible for an intensity-dependent phase shift as the signal propagates through the optical fiber. The phase shift for a specific channel depends not only on the power of that channel but also on the power of other channels. The total phase shift for the jth channel is given by Eq. (2.6.9) or by

$$\phi_j^{NL} = \frac{\bar{\gamma}}{\alpha} \left(P_j + 2 \sum_{m \neq j}^{N} P_m \right), \qquad (7.3.13)$$

where L_{eff} was replaced by $1/\alpha$ by assuming that $\alpha L \gg 1$. The parameter $\bar{\gamma}$ is defined in Eq. (2.6.6) and is typically $\sim 1 \ W^{-1}km^{-1}$. For IM/DD systems, the nonlinear phase shift depends on the bit pattern of various channels and can vary from zero to its maximum value $(\bar{\gamma}/\alpha)(2N - 1)P_j$, if we assume equal channel powers. However, it does not affect the system performance if the GVD effects are negligible and the receiver responds only to the incident power.

The situation is dramatically different for coherent multichannel systems because of the phase-sensitive nature of the receiver. The effect of XPM depends on the modulation format. The worst case occurs for the case of ASK since the phase shift depends on the bit pattern of various channels. The effect of XPM can be made negligible by choosing the channel power P_{ch} such that the maximum phase shift

$$(\bar{\gamma}/\alpha)(2N - 1)P_{\text{ch}} \ll 1. \qquad (7.3.14)$$

The channel power is then limited to below 1 mW even for 10 channels.

The impact of XPM on coherent systems is less severe for the FSK and PSK formats since the channel power does not depend on the bit pattern. In fact, XPM would be harmless if the channel powers were constant, as a constant phase shift of Eq. (7.3.13) does not affect system performance. In practice, channel powers fluctuate because of the intensity noise associated with the transmitter lasers. XPM converts intensity fluctuations into phase fluctuations via Eq. (7.3.13). If σ_p^2 is the variance of power fluctuations, assumed to be the same for all channels, the phase variance σ_ϕ^2 can be obtained by adding individual variances since the power in each channel fluctuates independently. Therefore, σ_ϕ for large N is given approximately by

$$\sigma_\phi \approx (2\bar{\gamma}/\alpha)\sigma_p \sqrt{N}. \qquad (7.3.15)$$

Typically, $\sigma_p = 5 \times 10^{-3} P_{\text{ch}}$, where P_{ch} is the average channel power. Even for $P_{\text{ch}} = 100$ mW and $N = 100$, σ_ϕ is below 0.1 rad, a value that leads to negligible crosstalk penalty.

Much larger crosstalk penalty can occur when phase or frequency modulation is accompanied by residual amplitude modulation. This can happen when semiconductor lasers are directly modulated or when fiber dispersion is large enough to convert phase modulation into amplitude variations. The value of σ_p in that case can be as large as $0.2P_{\text{ch}}$. Moreover, σ_ϕ grows linearly with N rather than $\sqrt{N}$. To limit the power penalty below 1 dB, the average channel power (in mW) should be $< 21/N$ for typical values of α and $\bar{\gamma}$ [163]. This condition is plotted in Fig. 7.20 and is compared with the power limitations imposed by SRS and SBS. XPM becomes the dominant crosstalk mechanism for WDM systems with 10 or more channels. It limits the channel power to below 0.1 mW for $N > 100$.

Four-Wave Mixing

FWM becomes a major source of nonlinear crosstalk [174]–[178] whenever the channel spacing and fiber dispersion are small enough to satisfy the phase-matching condition [70]. Its impact is most severe for coherent multichannel systems, for which the channel spacing is typically ~ 10 GHz. For noncoherent WDM systems, FWM becomes of major concern when the channel wavelengths are close to the zero-dispersion wavelength of the fiber. This is the case for 1.55-μm systems when dispersion-shifted fibers are used. In fact, optical fibers are often designed such that their dispersion is small enough to minimize the dispersive effects but, at the same time, large enough that the FWM-induced crosstalk is also minimized.

The physical origin of FWM-induced crosstalk and the resulting system degradation can be understood by noting that FWM generates a new wave at the frequency $\omega_{ijk} = \omega_i + \omega_j - \omega_k$, whenever three waves at frequencies ω_i, ω_j, and ω_k copropagate inside the fiber. For an N-channel system, i, j, and k can vary from 1 to N, resulting in a large combination of new frequencies generated by FWM. In the case of equally spaced channels, the new frequencies coincide with the existing frequencies, leading to coherent in-band crosstalk. When channels are not equally spaced, most FWM components fall in between the channels and lead to incoherent out-of-band crosstalk. In both cases the system performance is degraded because of a loss in the channel power, but the coherent crosstalk degrades system performance much more severely. For this reason, WDM systems are sometimes designed with unequal channel spacings, and channel wavelengths are chosen such that all FWM-generated frequencies fall outside the signal spectra [178].

The FWM efficiency varies as P_{ch}^3 if we assume equal power in each channel [174]–[178]. It also depends on the channel spacing through the phase-matching requirement. For a given channel spacing, P_{ch} should be reduced below a certain value to reduce the impact of FWM on system performance. Figure 7.20 shows this limit for the case of 10-GHz channel spacing. A similar limitation occurs for WDM systems with a channel spacing of 100 GHz when dispersion-shifted fibers are used. Typically, FWM limits the channel powers to below 1 mW unless unequal channel spacings are used. Experimental measurements confirm the advantage of unequal channel spacings [178]. However, this is often not a practical solution since many WDM components (such as a FP filter or a WGR) make use of equal channel spacings. An alternative is offered by the dispersion-management technique discussed in Section 9.9.

To conclude, several nonlinear effects in optical fibers limit the WDM system performance drastically and require the minimization of nonlinear crosstalk through a proper system design. Figure 7.20 can be used as a guideline for designing the WDM lightwave systems.

7.3.3 Other Design and Performance Issues

The design of WDM communication systems requires careful consideration of many transmitter and receiver characteristics. An important issue concerns the stability of the carrier frequency (or wavelength) associated with each channel. The frequency of light emitted from DFB or DBR semiconductor lasers can change considerably because of changes in the operating temperature ($\sim$ 10 GHz/°C). Similar changes can also occur with aging of lasers [179]. Such frequency changes are generally not of concern for single-channel communication systems. In the case of WDM lightwave systems it is important that the carrier frequency of each channel remain stable, at least relatively, so that the channel spacing does not fluctuate with time.

A number of techniques have been used for frequency stabilization [180]–[182]. A common technique uses *electrical feedback* provided by a frequency discriminator using an atomic or molecular resonance to lock the laser frequency to the resonance frequency. For example, one can use ammonia, krypton, or acetylene for semiconductor lasers operating in the 1.55-μm region, as all three have resonances near that wavelength. Frequency stability to within 1 MHz can be achieved by this technique. Another technique makes use of the *optogalvanic effect* to lock the laser frequency to an atomic or molecular resonance. A phase-locked loop can also be used for frequency stabilization. In another scheme, a Michelson interferometer, calibrated by using a frequency-stabilized master DFB laser, provides a set of equally spaced reference frequencies [182].

An important issue in the design of WDM systems is related to the loss of signal power due to insertion, distribution, and transmission losses that invariably occur. Fiber amplifiers offer a simple solution by boosting the signal power periodically, although only at the expense of degrading the SNR through spontaneous emission. Many channels can be amplified simultaneously because of a relatively large bandwidth of fiber amplifiers (see Section 8.5). However, not all channels are amplified by the same factor because of the nonuniformity of the gain spectrum. As a result, channel powers can deviate by a large amount (by 10 dB or more) when the signal passes through many in-line amplifiers before being detected. Several gain-equalization techniques can be used to flatten the gain spectrum (discussed in Section 8.6.4) of fiber amplifiers and to reduce the undesirable nonuniformity of channel powers. Alternatively, one can control the loss of individual channels (through selective attenuation) at each node within a WDM network to make the channel powers nearly uniform. The issue of power management in WDM networks is quite complex and requires attention to many details [183]. The buildup of the amplifier noise can also become a limiting factor when the WDM signal passes through a large number of amplifiers.

Another major issue in the design of WDM systems concerns the role of fiber dispersion. As discussed in Sections 2.4 and 5.2, the performance of single-channel systems is often limited by the group-velocity dispersion (GVD). These limitations also apply to WDM lightwave systems and set the maximum bit rate of each channel. One can use a *dispersion-compensation scheme* to reduce the impact of GVD as discussed in Chapter 9. However, it becomes difficult to compensate GVD for all channels simultaneously. Moreover, the nonlinear effects in optical fibers are also affected by GVD, and the two cannot be controlled without considering their mutual interplay [184]. An example is provided by FWM, whose impact increases rapidly if the system is designed to operate close to the zero-dispersion wavelength of the fiber. It is common to employ a *dispersion-management* scheme if the system designer has the liberty of choosing fibers with different dispersion characteristics (see Section 9.9). The underlying idea consists of mixing fibers with positive and negative GVDs such that not only is the total dispersion reduced for all channels but the nonlinear effects are minimized simultaneously. In a 1995 experiment, a 340-Gb/s WDM signal (17 channels at 20 Gb/s) was transmitted over 150 km using dispersion management [185]. In another experiment [186], dispersion management was used to transmit a 100-Gb/s WDM signal over 560 km with 80-km amplifier spacing. A fiber-loop experiment in 1996 demonstrated that 100-Gb/s transmission (20 channels at 5 Gb/s) over 9100 km is possible with a suitable design [11].

7.4 TIME-DIVISION MULTIPLEXING

As discussed in Section 1.2.2, TDM is commonly performed in the electrical domain to obtain digital hierarchies for telecommunication systems. In this sense, even single-channel lightwave systems carry multiple TDM channels. This scheme becomes difficult to implement for bit rates above 10 Gb/s because of limitations imposed by high-speed electronics and the direct-modulation capability of semiconductor lasers. A solution is offered by the *optical* TDM (OTDM), a scheme that has the potential of increasing the bit rate for a single optical carrier to a value as high as 1 Tb/s. Although not yet commercialized, OTDM systems were studied extensively during the decade of the 1990s [187]–[191]. Their development requires new types of optical transmitters and receivers employing all-optical multiplexing and demultiplexing techniques. In this section we first discuss these new techniques and then focus on the design and performance issues related to OTDM lightwave systems.

7.4.1 Channel Multiplexing

In OTDM lightwave systems, several optical signals modulated at the bit rate B using the same carrier frequency are multiplexed optically to form a composite optical signal at the bit rate NB, where N is the number of multiplexed optical

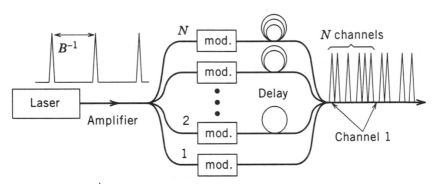

Figure 7.21 Schematic of an OTDM transmitter capable of multiplexing N channels. Pulses shorter than $(NB)^{-1}$ are separated into N branches, modulated, selectively delayed, and then recombined to form the composite bit stream at the bit rate NB.

channels. Figure 7.21 shows the basic design of an OTDM transmitter. It requires a laser capable of generating a periodic pulse train at the repetition rate equal to the single-channel bit rate B. Moreover, it should produce pulses of width T_p such that $T_p < (NB)^{-1}$ to ensure that the pulse will fit within its allocated time slot. The laser output is split equally into N branches, after amplification if necessary, and a modulator in each branch blocks the pulse for every 0 bit, creating N independent bit streams at the bit rate B.

Multiplexing of N bit streams is achieved by a delay technique that can be implemented optically in a simple manner. The modulated bit stream in the nth branch is delayed by an amount $(n-1)/(NB)$, where $n = 1, \ldots, N$, and the output of all branches is combined to form a composite signal. It should be clear that the multiplexed bit stream has a bit slot $T_B = (NB)^{-1}$. Furthermore, N consecutive bits in each interval of duration B^{-1} belong to N different channels, as required by the TDM scheme (see Section 1.2.2). The entire multiplexer (except for modulators which require LiNbO$_3$ or semiconductor waveguides) can be built by using single-mode fibers. Splitting and recombining of signals in N branches can be accomplished with $1 \times N$ fused fiber couplers that are available commercially. The optical delay can be implemented by using fiber segments of controlled lengths. As an example, a 1-mm fiber length introduces a delay of about 5 ps. The main point to note is that the relative delay in each branch must be precisely controlled to ensure the proper alignment of bits belonging to different channels. For a precision of 0.1 ps, typically required for a 40-Gb/s OTDM signal, the delay length should be controlled to within 20 μm.

An alternative approach makes use of planar lightwave circuits fabricated with the silica-on-silicon technology [57]–[60]. Such devices can be made polarization insensitive while providing a precise control of the delay lengths. However, it is difficult to build the entire multiplexer on a planar lightwave circuit since modulators cannot be integrated with this technology. Nonetheless, their use reduces the physical size and improves the stability and robustness of the multiplexer.

An important difference between the OTDM and WDM techniques should be apparent from Fig. 7.21. OTDM requires the use of the RZ format (see Section 1.2.3), whereas the NRZ format is used commonly for lightwave systems. In this respect, OTDM is similar to soliton communication systems (covered in Chapter 10), which also must use the RZ format. In fact, mode-locked semiconductor or fiber lasers are needed in both cases; such optical sources are discussed in Section 10.2.4.

7.4.2   Channel Demultiplexing

Demultiplexing of individual channels from an OTDM signal requires electro-optical or all-optical techniques. Several schemes have been developed, each having its own merits and drawbacks [192]. Figure 7.22 shows three schemes discussed in this section. All demultiplexing techniques require a *clock signal*, a periodic pulse train at the bit rate of a single channel. The clock signal is in the electrical form for electro-optic demultiplexing but consists of an optical pulse train for all-optical demultiplexing.

An electro-optical technique uses several MZ-type LiNbO$_3$ modulators whose design is shown schematically in Fig. 6.3(a). Each modulator halves the bit rate by rejecting alternate bits in the incoming signal. Thus, an 8-channel OTDM system requires three modulators connected in series [Fig. 7.22(a)], driven by the same clock signal, but with different voltages equal to $4V_0$, $2V_0$, and V_0, where V_0 is the voltage required for π phase shift in one arm of the MZ interferometer. Different channels can be selected by changing the phase of the clock signal. The main advantage of this technique is that it uses well-developed components that are commercially available. However, it also has several disadvantages. It requires a large number of expensive components, some of which need high drive voltage. It is also limited by the speed of modulators because of its electro-optic nature. Nonetheless, LiNbO$_3$ modulators operating at speeds as high as 20 Gb/s are available, making this technique a viable option.

Several all-optical techniques make use of a *nonlinear optical-loop mirror* (NOLM) constructed by using a fiber loop whose ends are connected to the two input ports of a 3-dB fiber coupler [see Fig. 7.22(b)]. Such a device is an all-optical implementation of the Sagnac interferometer. NOLM is called a mirror since it reflects its input entirely when the counterpropagating waves experience the same phase shift over one round trip. However, if the symmetry is broken to introduce a relative phase shift of π between them, the signal is fully transmitted by the NOLM. The operation of NOLM is based on XPM [70], the same nonlinear phenomenon that normally introduces crosstalk in WDM systems (see Section 7.3.2). The clock signal, consisting of a train of optical pulses at the single-channel bit rate is injected into the loop such that it propagates only in the clockwise direction. The OTDM signal enters the NOLM after being equally split into counterpropagating directions by the 3-dB coupler. The clock signal introduces a phase shift through XPM for pulses belonging to a specific channel within the OTDM signal. In the simplest case, optical fiber itself introduces XPM. The power of the optical signal and the loop length are made large

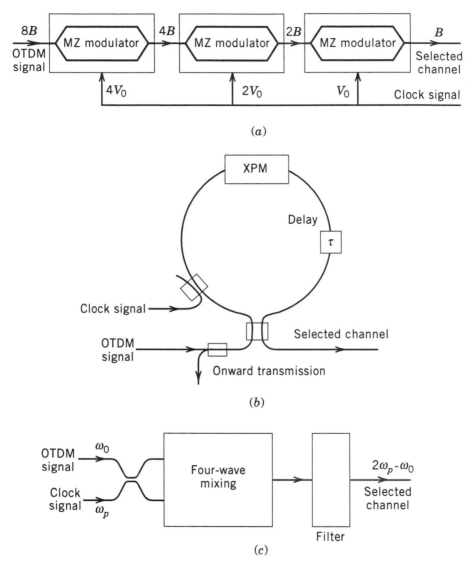

Figure 7.22 Three demultiplexing schemes for an OTDM signal: (a) electro-optic technique using LiNbO$_3$ modulators; (b) XPM in a nonlinear optical-loop mirror; (c) FWM in a nonlinear medium.

enough to introduce a relative phase shift of π. As a result, a single channel is demultiplexed by the NOLM. In this sense, a NOLM is the TDM counterpart of the WDM add/drop multiplexers discussed in Section 7.2.3. All channels can be demultiplexed simultaneously by using several NOLMs in parallel [192]. Fiber nonlinearity is fast enough that such a device can respond at femtosecond time scales. Demultiplexing of a 6.3-Gb/s channel from a 100-Gb/s OTDM signal was demonstrated in 1993 [193].

The main limitation of a NOLM, making use of XPM in a fiber loop, stems from the weak fiber nonlinearity. The loop length should be 10 km or more in order to introduce a phase shift of π with practical power levels of the clock signal. In a variant of the basic idea, a SLA is inserted within the fiber loop and its nonlinearity is used for demultiplexing. The SLA induces a phase shift because of gain saturation in a way similar to the wavelength-conversion scheme shown in Fig. 7.17(c). This phase shift can be changed for the data bits belonging to a specific channel by injecting the clock signal, resulting in demultiplexing of that channel. The nonlinearity of the SLA is large enough that a relative phase shift of π can be introduced by a SLA of length 1 mm. However, this nonlinearity is relatively slow because it requires recombination of electron–hole pairs within the active region of the SLA and is governed by the carrier lifetime ($\sim$ 1 ns). By injecting a CW signal with the clock signal (at a different wavelength), the carrier lifetime can be reduced to below 100 ps. Even a faster response can be realized by placing the SLA asymmetrically within the loop. Such a device [194] is referred to as the *terahertz optical asymmetrical demultiplexer* (TOAD) because it can respond at time scales of 1 ps. Its operation at bit rates as high as 250 Gb/s has been demonstrated [195].

The third scheme for demultiplexing, shown schematically in Fig. 7.22(c), makes use of FWM in a nonlinear medium and works in a way similar to the wavelength-conversion scheme discussed in Section 7.2.5 [see Fig. 7.17(d)]. The OTDM signal is launched together with the clock signal (at a different wavelength) into a nonlinear medium. The clock signal acts as a pump for the FWM process. A new pulse train is generated at a third wavelength (see Section 2.6.3) that is an exact replica of the channel that needs to be demultiplexed. An optical filter is used to separate the demultiplexed channel from the OTDM and clock signals. A polarization-preserving fiber is often used as the nonlinear medium in which FWM takes place because of the ultrafast nature of the nonlinearity and its ability to preserve the state of polarization despite various environmental fluctuations. Indeed, error-free demultiplexing of a 10-Gb/s channel from a 500-Gb/s OTDM signal has been demonstrated by using clock pulses of about 1 ps duration [196]. Alternatively, the nonlinearity of an SLA can be used for FWM; a 200 Gb/s signal has been demultiplexed by such a device [197]. Polarization sensitivity is a major issue when SLAs are used for demultiplexing. This problem can be solved by using two identical pieces of polarization-preserving fibers before and after the SLA such that their major axes are orthogonal.

Demultiplexing of an OTDM signal requires the recovery of a clock signal at the bit rate of a single channel. An all-optical scheme is needed because of the high bit rates associated with an OTDM signal. An optical phase-locked loop based on the FWM process is commonly used for this purpose [198]. Other schemes based on a NOLM or a fiber laser can also be used [190]. The FWM scheme can make use of any nonlinear medium. Optical fibers are often used because of their fast response. The use of SLAs is attractive because of their compactness and permits clock recovery at bit rates $\sim$ 100 Gb/s. The FWM component is used to generate a phase-difference signal that is fed back into a voltage-controlled oscillator to realize phase locking [198].

7.4.3   System Performance

The transmission distance L of OTDM signals is limited in practice by fiber dispersion because of the use of short optical pulses ($\sim$ 1 ps) dictated by relatively high bit rates. In fact, since use of an OTDM signal carrying N channels at the bit rate B is equivalent to transmitting a single channel at the composite bit rate of NB, the bit rate–distance product NBL is restricted by the dispersion limits indicated in Sections 2.4.3 and 5.2.2. As an example, it is evident from Fig. 2.13 that a 200-Gb/s system is limited to $L < 50$ km even when the system is designed to operate exactly at the zero-dispersion wavelength of the fiber. Thus, the implementation of such systems requires not only dispersion-shifted fibers but also the use of a dispersion-compensation technique (see Section 9.8) to reduce the impact of higher-order dispersive effects.

Considerable progress was made during the 1990s in realizing high-speed point-to-point fiber links using the OTDM technique. In one experiment the 100-Gb/s OTDM signal, consisting of 16 channels at 6.3 Gb/s, was transmitted over 560 km by using several 80-km-spaced optical amplifiers together with dispersion management [199]. The laser source in this experiment was a mode-locked fiber laser producing pulses of about 3 ps width at a repetition rate of 6.3 GHz, the bit rate of a single channel. A multiplexing scheme similar to that shown in Fig. 7.21 was used to generate the 100-Gb/s OTDM signal. The total bit rate was later extended to 400 Gb/s by using a *supercontinuum* pulse source producing 1-ps-wide pulses [200]. Such short pulses are needed since the bit slot is only 2.5-ps wide at 400 Gb/s. It was necessary to compensate for the dispersion slope (third-order dispersion β_3) since 1-ps pulses were severely distorted with oscillatory tails extending to beyond 5 ps (typical characteristic of the third-order dispersion) in the absence of such a compensation. Even then, the transmission distance was limited to 40 km.

A simple technique to realize lightwave systems operating at bit rates above 1 Tb/s is to combine the TDM and WDM techniques. For example, a WDM signal consisting of M separate optical carriers such that each carrier carries N OTDM channels at the bit rate B, has the total capacity $B_{\text{tot}} = MNB$. The dispersion limitations of such a system are set by the OTDM-signal bit rate of NB. In a 1996 experiment, this approach was used to realize the total capacity of 1 Tb/s by using $M = 10$, $N = 10$, and $B = 10$ Gb/s [8]. The channels were spaced apart by 400 GHz (about 3.2 nm) to avoid overlap between neighboring WDM channels at the 100-Gb/s bit rate. The total capacity of such systems can exceed 5 Tb/s in principle, although in practice many factors, such as the amplifier bandwidth, nonlinear effects in fibers, and the practicality of dispersion compensation over a wide bandwidth, may limit their practical realization [201].

OTDM has also been used for designing transparent optical networks capable of connecting multiple nodes for random bidirectional access [188]–[191]. Similar to the case of WDM networks, both single-hop and multihop architectures have been considered. Single-hop OTDM networks use passive star couplers to distribute the signal from one node to all other nodes. In contrast, multihop OTDM networks require signal processing at each node to route the traffic. The

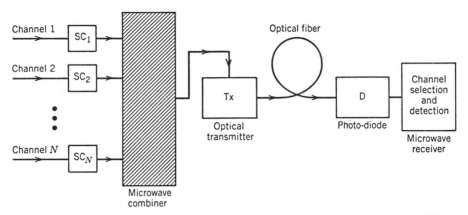

Figure 7.23 Multichannel SCM lightwave system. Microwave subcarriers (SC) are modulated by using analog or digital modulation. The composite electrical signal obtained from the microwave combiner is used to modulate the optical carrier.

packet-switching technique is commonly used for such networks. Considerable effort was under way in 1996 for developing OTDM networks.

7.5 SUBCARRIER MULTIPLEXING

The basic concept behind *subcarrier multiplexing* (SCM) is borrowed from microwave communication technology, which employs multiple microwave carriers for transmission of multiple channels (electrical FDM) over coaxial cables or free space. The total bandwidth is limited to well below 1 GHz when coaxial cables are used to transmit the multichannel microwave signal. However, if the multichannel microwave signal is transmitted optically by using optical fibers, the signal bandwidth can easily exceed 10 GHz for a single optical carrier. Such a scheme is referred to as SCM, since multiplexing is done by using microwave subcarriers rather than the optical carrier. It has attracted considerable attention because of its commercial use by the cable-television (CATV) industry. It can also be combined with TDM or WDM. A combination of SCM and WDM has the potential for achieving a bandwidth in excess of 1 THz.

Figure 7.23 shows a block diagram of a SCM lightwave system designed with a single optical carrier. The main advantage of SCM is the flexibility and the upgradability offered by it in the design of broadband networks. One can use analog or digital modulation, or a combination of the two, to transmit multiple voice, data, and video signals to a large number of users. Each user can be served by a single subcarrier, or the multichannel signal can be distributed to all users as done commonly by the CATV industry. The SCM technique has been studied extensively because of its wide-ranging practical applications. Its use for video distribution had reached the commercial stage by 1992. In this section we describe both analog and digital SCM lightwave systems, with emphasis on their design and performance.

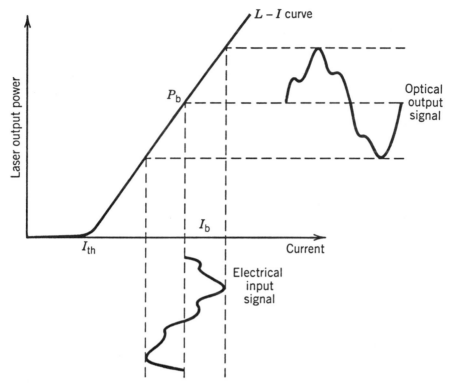

Figure 7.24 Analog-modulation scheme for semiconductor lasers. The laser is biased at the bias current I_b and modulated by adding the modulation current to I_b. The linear nature of the L–I curve produces an optical replica of the electrical analog signal.

7.5.1   Analog SCM Lightwave Systems

The emphasis in Chapters 4–6 was on digital lightwave systems because digital modulation techniques are employed almost universally in the design of fiber-optic communication systems. An exception occurs in the case of SCM systems designed for video distribution. Most CATV networks distribute television channels by using an analog techniques [202]–[208] based on *frequency modulation* (FM) or *amplitude modulation with vestigial sideband* (AM-VSB) since its use permits the use of existing commercial microwave components. However, since the wave form of an analog signal must be preserved during transmission, analog SCM systems require a high SNR at the receiver and impose strict linearity requirements on the optical source and the communication channel.

Figure 7.23 shows the basic design of an SCM lightwave system. Each microwave subcarrier is modulated, and the output of all modulated subcarriers is summed by using a microwave power combiner. The composite signal is used to modulate the intensity of a semiconductor laser directly by adding it to the bias current. Figure 7.24 shows how the laser output replicates the electrical signal when the L–I characteristics of the semiconductor laser are linear. Any deviation from linearity distorts the analog optical signal and affects system performance. The transmitted power can be written as

$$P(t) = P_b \left[1 + \sum_{j=1}^{N} m_j a_j \cos(2\pi f_j t + \phi_j) \right], \qquad (7.5.1)$$

where P_b is the output power at the bias level and m_j, a_j, f_j, and ϕ_j are, respectively, the *modulation index*, amplitude, frequency, and phase associated with the jth microwave subcarrier; a_j, f_j, or ϕ_j is modulated to impose the signal, depending on whether the AM, FM, or PM technique is used.

The power received is also given by Eq. (7.5.1), except for an overall power reduction occurring because of transmission losses if the communication channel is perfectly linear. In practice, the signal is distorted because of a departure from linearity. Such a distortion is referred to as *intermodulation distortion* (IMD) and is similar in nature to the FWM distortion discussed in Section 7.3. Any nonlinearity in the response of the semiconductor laser or in the propagation characteristics of fibers generates new frequencies of the form $f_i + f_j$ and $f_i + f_j \pm f_k$, some of which lie within the transmission bandwidth and distort the analog signal. The new frequencies are referred to as the *intermodulation products* (IMPs). If all subcarriers lie within an octave of bandwidth, only third-order IMPs can appear within the transmission bandwidth. These are further subdivided as *two-tone* IMPs and *triple-beat* IMPs, depending on whether two frequencies coincide or the three frequencies are distinct. The triple-beat IMPs tend to be a major source of third-order IMD because of their large number. An N-channel SCM system generates $N(N-1)(N-2)/2$ triple-beat terms compared with $N(N-1)$ two-tone, third-order terms. The second-order IMD must also be considered whenever subcarriers occupy more than an octave of bandwidth. It often dominates system performance, depending on the number of IMPs that contribute to second-order IMD.

IMD has its origin in several distinct nonlinear mechanisms. The dynamic response of semiconductor lasers is governed by the rate equations (see Section 3.3.6), which are intrinsically nonlinear. The solutions of these equations provide expressions for the second- and third-order IMPs originating from this intrinsic nonlinearity [203]. Their contribution is largest whenever the IMP frequency falls near the relaxation-oscillation frequency. A second source of IMD is the nonlinearity of the light-current (L–I) curve (see Fig. 3.21). The magnitude of IMPs can be calculated by expanding the output power in a Taylor series around the bias power [202]. In both cases, the second- and third-order IMPs increase with the modulation index m as m^2 and m^3, respectively. Several other mechanisms, such as fiber dispersion, frequency chirp, and mode-partition noise can cause IMD, and their impact on the SCM systems has been studied extensively [209]–[214].

The IMD-induced degradation of the system performance depends on the interchannel interference created by IMPs. Depending on the channel spacing among microwave subcarriers, some of the IMPs fall within the bandwidth of a specific channel and affect the signal recovery. It is common to introduce *composite second-order* (CSO) and *composite triple-beat* (CTB) distortion values by adding the power for all IMPs that fall within the passband of a specific channel [215]. The CSO and CTB distortion values are normalized to the carrier power of that channel and expressed in dBc units, where the "c" in "dBc" denotes normalization with respect to the carrier power. Typically, both CSO and CTB distortion values should be below -60 dBc for negligible impact on the system performance; they increase rapidly with an increase in the modulation index [215].

System performance depends on the SNR associated with the demodulated signal. In the case of SCM systems, the *carrier-to-noise ratio* (CNR) is often used in place of SNR, as the two are related to each other. The CNR is defined as the ratio of RMS carrier power to RMS noise power at the output of the photodetector. It can be written as

$$\text{CNR} = \frac{(mR\bar{P})^2/2}{\sigma_s^2 + \sigma_T^2 + \sigma_I^2 + \sigma_{\text{IMD}}^2}, \tag{7.5.2}$$

where m is the channel modulation index, R is the detector responsivity, $\bar{P}$ is the average received optical power, and σ_s, σ_T, σ_I, and σ_{IMD} are the RMS values of the noise currents associated with the shot noise, thermal noise, intensity noise, and IMD, respectively. The expressions for σ_s^2 and σ_T^2 can be obtained from Section 4.4.1 and are given by

$$\sigma_s^2 = 2q(R\bar{P} + I_d)\Delta f, \tag{7.5.3}$$

$$\sigma_T^2 = 4k_B T F_n \Delta f / R_L, \tag{7.5.4}$$

where the notation of Chapter 4 is used. The RMS value σ_I of the intensity noise can be obtained from Eq. (4.6.6) in Section 4.6.2. If we assume that the relative intensity noise (RIN) of the laser is nearly uniform within the receiver bandwidth,

$$\sigma_I^2 = (\text{RIN})(R\bar{P})^2\Delta f. \tag{7.5.5}$$

The RMS value of σ_{IMD} depends on the CSO and CTB distortion values.

The CNR requirements of SCM systems depend on the modulation format. In the case of AM-VSB format, the CNR should typically exceed 50 dB for satisfactory performance. Such large values can be realized only by increasing the received optical power $\bar{P}$ to a relatively large value (> 0.1 mW). This requirement has two effects. First, the power budget of AM-analog SCM systems is extremely limited unless the transmitter power is increased above 10 mW. Second, the intensity-noise contribution to the receiver noise dominates the system performance since σ_I^2 increases quadratically with $\bar{P}$ compared with the linear dependence for the shot noise in Eq. (7.5.3). In fact, the CNR becomes independent of the received optical power when σ_I dominates. From Eqs. (7.5.2) and (7.5.5) the CNR is then given by

$$\text{CNR} \approx \frac{m^2}{2(\text{RIN})\Delta f}. \tag{7.5.6}$$

As an estimate, the RIN of the transmitter laser should be below -150 dB/Hz to realize a CNR of 50 dB if $m = 0.1$ and $\Delta f = 50$ MHz are used as the representative values. Larger values of RIN can be tolerated only by increasing the modulation index m or by decreasing the receiver bandwidth. Indeed, DFB lasers with low values of the RIN have been developed for CATV applications. In general, the DFB laser is biased high above threshold to provide a bias power P_b in excess of 5 mW, since the RIN decreases as P_b^{-3}. The high values of the bias power also permit an increase in the modulation index m.

The intensity noise can become a problem even when the transmitter laser is selected with a low RIN value to provide a large CNR in accordance with Eq. (7.5.6). The reason is that the RIN can be enhanced during signal transmission in optical fibers. One such mechanism is provided by multiple reflections between two reflecting surfaces along the transmission path [216]. As discussed in Section 5.4.5, the two reflective surfaces act as a FP interferometer which can convert the laser-frequency noise into intensity noise. The *reflection-induced* RIN depends on both the laser linewidth and the spacing between reflecting surfaces. It can be avoided by using fiber components (splices and connectors) with negligible parasitic reflections (< -40 dB) and by using lasers with a narrow linewidth (< 1 MHz). Another mechanism for the RIN enhancement is provided by the dispersive fiber itself [217]. Because of fiber dispersion, different frequency components travel at slightly different speeds. As a result, frequency fluctuations are converted into intensity fluctuations during signal transmission. The *dispersion-induced* RIN depends on the laser linewidth and increases quadratically with the fiber length. Fiber dispersion also enhances CSO and CTB distortion when the link length approaches 100 km [218]. It may be necessary to use dispersion-compensation techniques (see Chapter 9) for such SCM systems. In a 1996 experiment [219], the use of a chirped fiber grating for dispersion compensation reduced the RIN by more than 30 dB for fiber spans of 30 and 60 km.

The high-CNR requirement of AM-analog SCM systems results in a tight power budget and limits the transmission distance and the number of distribution sites to relatively small values. A solution is offered by optical amplifiers, which can boost the signal power as in-line amplifiers (see Chapter 8). Another solution is to change the modulation format to FM. The bandwidth of a FM subcarrier is considerably larger than in the AM case (30 MHz in place of 4 MHz). However, the required CNR at the receiver is much lower (about 16 dB in place of 50 dB) because of the so-called FM advantage, which can provide a studio-quality video signal (> 50-dB SNR) with only 16-dB CNR [202]. As a result, the optical power received can be as small as 10 μW. The RIN is not much of a problem for such systems as long as the RIN value is below -135 dB/Hz. In fact, the receiver noise of FM systems is generally dominated by the thermal noise.

Several experiments performed during 1988–1991 demonstrated the potential of analog SCM lightwave systems for video transmission over optical fibers [202]–[207]. In one AM-VSB experiment, 42 video channels with 6-MHz channel spacing were transmitted using microwave subcarriers in the frequency range

50–400 MHz [203]. In another experiment [204], 19 AM-VSB video channels were transmitted using an erbium-doped fiber amplifier, which increased the transmitter power to 8 dBm (6.3 mW). The FM video transmission has also been demonstrated. In a 120-channel experiment [202], microwave subcarriers occupied the frequency range 2.7–7.5 GHz with 40-MHz channel spacing. The modulation depth of each channel was about 2.3%, with a total $m_{\rm RMS}$ of 25%. In another experiment [220], 60 FM-video channels were transmitted together with a 100-Mb/s digital baseband (without microwave subcarrier) signal. The baseband digital channel and SCM analog video channels were transmitted by modulating the same laser source. This experiment demonstrates the flexibility of SCM, since analog and digital channels can be mixed together for transmission over the same fiber by using the same semiconductor laser.

The main limitation of AM analog SCM systems results from their tight power budget because of the high CNR requirement. However, it is not a major practical concern with the advent of optical amplifiers. In a 1991 experiment [221], a fiber amplifier boosted the output power to 11 dBm (12.6 mW), resulting in a 16-dB power budget for the 35-channel SCM system making use of the AM-VSB format. Amplified analog SCM systems have attracted considerable attention [222].

A fundamental limit on analog SCM systems is imposed by the $L-I$ characteristics of the semiconductor laser itself [223]. As seen in Fig. 7.24, the laser power drops to near zero whenever the drive current is below threshold: Thus, the optical signal would be a replica of the electrical signal only if the modulation current I_m is kept below $I_b - I_{th}$. For a multichannel SCM system the instantaneous value of I_m depends on the relative phases of microwave subcarriers and can drop below $I_b - I_{th}$ if the total modulation index mN for N channels exceeds 1. The SCM systems are designed to keep the RMS value $(m_{\rm RMS} = m\sqrt{N})$ of the total modulation index below 1. Since mN can exceed 1, especially for large values of N, some distortion is likely to occur as a result of clipping of the optical power at the laser threshold. This distortion occurs even when the $L-I$ curve is perfectly linear and sets the fundamental limit on the performance of analog SCM systems. It can be made negligible by keeping $mN < 40\%$. This restriction limits the modulation index for each channel below 4% for a 100-channel SCM system. Although this value does not limit FM systems, it becomes a limiting factor for AM systems, which often require large values of m to achieve the CNR in excess of 50 dB in accordance with Eq. (7.5.6). *Clipping distortion* has been studied extensively because of its practical impact [224], [225].

7.5.2   Digital SCM Lightwave Systems

During the 1990s, the emphasis of SCM systems shifted from analog to digital modulation. One application of digital SCM systems is related to the distribution of high-definition television (HDTV) channels. It was realized early that digital SCM systems can benefit from the coherent-detection technology discussed in Chapter 6 in two ways. First, the 10–15 dB improvement in receiver sensitivity would increase their power budget by the same amount. Second, digital modulation of subcarriers together with coherent detection [226], [227] would improve the signal quality with a low requirement for the CNR. In this section we consider the CNR of digital SCM lightwave systems.

The FSK format is commonly used for modulating microwave subcarriers digitally, as it is simplest to implement using voltage-controlled oscillators. The channel spacing is typically 200 MHz for a 100-Mb/s bit rate. The FSK-modulated subcarriers are combined with a power combiner (see Fig. 7.23), amplified, and then fed to a phase modulator, which modulates the phase of a CW optical signal obtained from a narrow-linewidth laser. At the receiver, the optical signal is mixed coherently with the output of a local oscillator, as discussed in Section 6.1. To avoid the phase-locking requirement, heterodyne detection is used with an intermediate frequency ν_{IF}. The photocurrent generated at the receiver can be written as

$$I(t) = R\{P_s + P_{LO} + 2\sqrt{P_s P_{LO}}\cos[2\pi\nu_{IF}t + \phi_m(t)]\}, \qquad (7.5.7)$$

where P_{LO} is the local-oscillator power and P_s is the signal power. The phase $\phi_m(t)$ contains the composite multichannel microwave signal and is given by

$$\phi_m(t) = \sum_{j=1}^{N} \beta_j \cos[2\pi f_j t + \alpha_j(t)], \qquad (7.5.8)$$

where β_j is the PM index of jth channel with the microwave subcarrier frequency f_j, and $\alpha_j(t)$ is the subcarrier phase whose time dependence is due to FSK modulation and contains the digital information being transmitted. A BPF together with a tunable local-oscillator permits channel selection.

To obtain the signal current of a specific channel, one can use the Bessel-function expansion for the cosine function in Eq. (7.5.7) and obtain [226]

$$I(t) = 2\sqrt{P_s P_{LO}} \sum_{k_1} \cdots \sum_{k_N} J_{k_1}(\beta_1) \cdots J_{k_N}(\beta_N) \times$$
$$\cos[2\pi\nu_{IF}t + k_1(2\pi f_1 t + \alpha_1) + \cdots + k_N(2\pi f_N t + \alpha_N)], \quad (7.5.9)$$

where the direct-current (dc) term in Eq. (7.5.7) is ignored. The PM indices β_j $(j = 1, N)$ are generally small enough that only terms with low values of $k_1, \ldots, k_N$ contribute to the sum in Eq. (7.5.9). The dominant term in the current generated for the jth channel is obtained by setting $k_j = -1$ and remaining indices to zero so that

$$I_j(t) = 2\sqrt{P_s P_{LO}} J_1(\beta)[J_0(\beta)]^{N-1}\cos[2\pi(\nu_{IF} - f_j)t - \alpha_j], \qquad (7.5.10)$$

where the PM indices are taken to be the same for all channels. The jth channel is selected by tuning the local oscillator such that the BPF lets the signal centered at $\nu_{IF} - f_j$ pass. The CNR is obtained by averaging $I_j^2(t)$ and dividing it by the current-noise variance. The result is

$$CNR = \frac{2R^2 P_{LO} P_s J_1^2(\beta) J_0^{2N-2}(\beta)}{\sigma_s^2 + \sigma_T^2 + \sigma_I^2 + \sigma_{IMD}^2 + \sigma_X^2}, \qquad (7.5.11)$$

where σ_s^2, σ_T^2, and σ_I^2 are obtained from Eqs. (7.5.3)–(7.5.5) with $\bar{P}$ replaced by P_{LO}. The term σ_{IMD}^2 takes into account the contribution of IMD to receiver noise. An additional term σ_X^2 has been added to include interchannel crosstalk occurring owing to leakage of energy from neighboring channels through the bandpass filter. Both σ_{IMD} and σ_X should be negligible. The intensity noise can be avoided using a balanced heterodyne receiver (see Section 6.5.2). The thermal noise is also negligible in the limit of large local-oscillator power. The CNR in the shot-noise limit is thus obtained by using σ_s^2 from Eq. (7.5.3) in Eq. (7.5.11) and is given by

$$\mathrm{CNR} = \frac{R P_s J_1^2(\beta) J_0^{2N-2}(\beta)}{q \Delta f}, \qquad (7.5.12)$$

where the contribution of dark current is neglected. The PM index β for each channel is generally small enough ($\beta < 0.2$) that $J_0(\beta)$ and $J_1(\beta)$ can be approximated by 1 and $\beta/2$, respectively. By using $R = \eta q/h\nu$, CNR is given by

$$\mathrm{CNR} = \frac{\eta P_s \beta^2}{4 h\nu \Delta f}. \qquad (7.5.13)$$

This expression should be compared with Eq. (7.5.6) obtained for analog SCM systems in the intensity-noise limit. The role of modulation index m is played by the PM index β in the case of digital SCM systems. By using $\beta = 0.1$, $h\nu = 0.8$ eV, and $\Delta f = 120$ MHz, Eq. (7.5.13) shows that a CNR of 20 dB can be obtained for an average signal power of only 10 nW. In practice, the required value of CNR depends on the demodulation scheme (see Chapter 6) but is generally in the range 16–20 dB.

Several experiments have demonstrated the potential of digital coherent SCM lightwave systems. An early experiment [228] used digital modulation for 20 FSK video channels, operating at 100 Mb/s with 200-MHz channel spacing, but used direct detection of the optical signal at the receiver. It was followed by a 5-channel SCM system [229] that used FSK modulation of microwave subcarriers together with heterodyne detection at the receiver of the PM optical signal. The experiment was later extended to demonstrate the transmission of 20 channels [226] in the frequency range 2–6 GHz by using microwave subcarriers at 2.1, 2.3,..., 5.9 GHz. This choice of subcarrier frequencies reduces the second-order IMPs, as they fall in between the subcarriers. The use of coherent detection improved receiver sensitivity by 14 dB over that of the direct-detection case [228]. In a 1990 experiment [230] two 560-Mb/s channels were transmitted over 122 km of fiber sing ASK modulation of two microwave subcarriers at 3.2 and 5.0 GHz. This scheme can be used to transmit multiple channels over long distances using a single laser. The real potential of SCM systems lies in broadcasting multiple channels.

A practical problem with the use of coherent detection for broadcasting networks is that each receiver should be equipped with a laser acting as a local oscillator, a rather expensive scheme for LAN applications. This problem can be solved by superimposing the local-oscillator output on the multiplexed signal

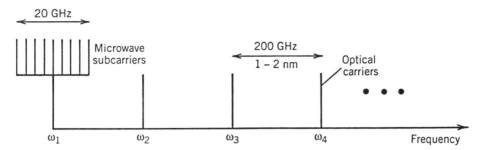

Figure 7.25 Frequency allocation in a multiwavelength SCM network.

at the transmitter itself [230]. The transmission loss can be tolerated for LANs, as the transmission distance is typically less than 10 km. The advantage is that a single local oscillator is shared by all subscribers, making this approach cost-effective. Another advantage is that the relative polarization of the signal and local oscillator is maintained during transmission, and a polarization-diversity receiver is not needed. The sharing approach makes coherent SCM systems quite attractive for broadcasting digital video channels to a large number of subscribers [231].

The use of coherent detection is not always necessary for digital SCM systems. However, their high-bandwidth requirement was of some concern initially. A single digital video channel requires a bit rate of more than 100 Mb/s, especially if the HDTV format is used, in contrast with the analog channel occupying a bandwidth of only 5–8 MHz. This disadvantage has practically disappeared with the emergence of compression standards (such as MPEG-2). In fact, several compressed digital video channels can be packed within the bandwidth of a single analog channel using novel modulation techniques such as quadrature AM, carrier-less AM/PM, quadrature PSK, and orthogonal FSK. To increase the capacity of an SCM system, it is common to employ hybrid techniques by mixing analog (AM) and digital (quadrature AM) formats. Clipping noise becomes a major issue in such hybrid systems, and many techniques have been developed to reduce its impact [232], [233].

7.5.3 Multiwavelength SCM Lightwave Systems

The combination of WDM and SCM provides the potential of designing broadband passive optical networks capable of providing integrated services (audio, video, data, etc.) to a large number of subscribers. In this scheme, shown in Fig. 7.25, multiple optical carriers are launched into the same optical fiber by using the WDM techniques discussed in Section 7.1. Each optical carrier carries multiple SCM channels by using several microwave subcarriers. One can mix analog and digital signals by using different subcarriers or different optical carriers. Such networks are extremely flexible and easy to upgrade as the demand grows. In a 1990 demonstration [234], 16 DFB lasers with wavelength spacing of 2 nm in the 1.55-μm band were modulated with 100 analog FM video channels and six 622-Mb/s digital channels. Video channels were multiplexed using the SCM technique such that one DFB laser carried 10 FM video channels over the bandwidth 300–700 MHz. Therefore, 10 DFB lasers were used for 100 FM video channels and the remaining six lasers for six digital channels, each operating at 622 Mb/s. The combined signal was split into 4096 parts using star couplers and a single fiber amplifier capable of providing 17–24 dB gain over the 34-nm spectral range.

The networking capabilities of multiwavelength SCM systems have attracted considerable attention [235]–[238] because of several practical advantages over the WDM scheme of Section 7.1. Such systems can provide multiple services (telephone, analog and digital TV channels, computer data, etc.) with only one optical transmitter and one optical receiver per user if different services use different microwave subcarriers, thus lowering the cost of terminal equipment. Microwave subcarriers can be spaced closely (< 1 GHz) and are much easier to stabilize in frequency. Different services can be offered without the need for synchronization. Finally, microwave subcarriers can be processed using commercial electronic components. In one scheme [235], each user is assigned a unique wavelength for transmitting multiple SCM messages, but the user can receive multiple wavelengths. Optical filters can be used to select optical wavelengths electronically at the site for network control. In another scheme [239], each user is assigned a unique wavelength and a unique subcarrier frequency to receive and transmit the signal. The main advantage of multiwavelength SCM is that the network can serve NM users, where N is the number of optical wavelengths and M is the number of microwave carriers, by using only N distinct transmitter wavelengths, thereby increasing the network capacity. The optical wavelengths can be relatively far apart (a few nanometers), making it possible to use direct detection, thereby reducing the cost of the terminal equipment.

A limiting factor for multiwavelength SCM networks is the phenomenon of *optical beat interference* [239], [240]. It is a result of two or more users transmitting simultaneously on the same optical channel by using different subcarrier frequencies. Since the optical carrier frequencies are then slightly different, their beating would produce a beat note in the photocurrent, as the user receives all channels transmitted at one wavelength. If the beat-note frequency overlaps an active subcarrier channel, an interference signal would limit the detection process in a way similar to IMD. Statistical models have been used [239] to estimate the probability of channel outage because of optical beat interference. The results show that as M is increased by adding more subcarriers, network capacity increases sublinearly with M and eventually saturates. The nonlinear effects in optical fibers such as SRS and XPM (see Section 7.3.2) also affect the performance of multiwavelength SCM networks [241], [242].

As discussed in Section 7.1.3 in the context of local-loop applications, the multiwavelength SCM systems have advanced in 1996 close to the commercial stage [45]–[47]. In one approach [243], digital SCM video channels are multiplexed with the audio and data traffic through WDM (see Fig. 7.7). In another approach adopted by the CATV industry [47], the hybrid fiber/coaxial (HFC) technology is used to provide broadband integrated services to the sub-

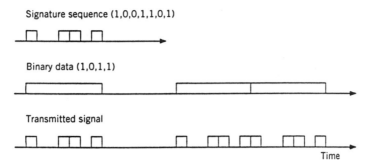

Figure 7.26 Coding of binary data by using a multibit signature sequence for CDM lightwave systems.

scriber. Digital video transport systems (e.g., DV6000), operating at 10 Gb/s by combining the WDM and SCM techniques, were available commercially by 1996 [244]. The use of WDM and SCM for *personal communication networks* is also attracting considerable attention [245].

7.6   CODE-DIVISION MULTIPLEXING

The WDM and SCM techniques discussed so far in this chapter can be classified as scheduled multiple-access techniques, as different users use the network according to a fixed assignment. Their major advantage is the simplicity of data routing among users. This simplicity, however, is achieved through an inefficient utilization of channel bandwidth. This drawback can be overcome by using a random multiple-access technique which allows users to access any channel randomly at an arbitrary time. A multiplexing scheme that is attracting attention makes use of the *spread-spectrum technique* [246]. It is referred to as *code-division multiplexing* (CDM), as each channel is coded in such a way that its spectrum spreads over a much wider region than occupied by the original signal. CDM has been studied extensively in the context of microwave communications and is known to provide the most flexibility in a multiuser environment. Its use for fiber-optic communications has attracted considerable attention during the 1990s [247]–[258]. In this section we describe the CDM technique with emphasis on potential applications in lightwave systems.

In the spread-spectrum method [246], the signal spectrum is spread over a considerably wider region than the minimum bandwidth necessary for transmission. Spectrum spreading is accomplished by means of a code that is independent of the signal itself. The receiver uses the same code for compressing the signal spectrum and recovering the data. An advantage of the spread-spectrum method is that it is difficult to jam or intercept the signal because of its coded nature. The CDM technique is thus useful when security of the data is of concern. The spectrum-spreading code is called a *signature sequence* or a key. Several methods can be used for coding including *direct sequence, time hopping,* and *frequency hopping*. Figure 7.26 shows an example of direct-sequence coding for optical CDM systems. Each bit of data is coded using a signature sequence which consists of 7 bits $(1, 0, 0, 1, 1, 0, 1)$. The effective bit rate increases by a factor of 7 because of coding. As a result, the spectrum is spread over a much wider region. Different signature sequences are assigned to different users. All users share the same optical bandwidth. Transmitters are allowed to transmit messages at arbitrary times. The receivers recover messages by decoding the received signal using the same signature sequence that was used to code the message. Such a technique can be used to multiplex several digital video channels by encoding the signal through pulse-position modulation and decoding it by using optical-correlation techniques [251].

Spectrum spreading can also be accomplished by the technique of frequency hopping, where the carrier frequency is shifted periodically according to a certain preassigned code. The situation differs from WDM in the sense that a fixed frequency is not assigned to a given channel. Rather, all channels share the entire bandwidth by using different carrier frequencies at different times according to a code. A *frequency-hop signal* can be represented in a matrix form, shown in Fig. 7.27, such that its rows correspond to assigned frequencies and its columns correspond to time slots. A matrix element m_{ij} equals 1 if and only if the frequency ω_i is transmitted in the interval t_j. Different users are assigned different frequency-hop patterns (or codes) to ensure that two users do not transmit at the same frequency during the same time slot. The code sequences that satisfy this property are said to be *orthogonal sequences*. In the case of asynchronous transmission, complete orthogonality cannot be ensured. Such systems make use of *pseudo-orthogonal codes* with maximum autocorrelation and minimum cross correlation to make the BER as low as possible. In general, the BER of CDM systems is relatively high (typically, $> 10^{-6}$). They can, nonetheless, be useful for some applications by using forward-error correction schemes.

Frequency-hopping CDM lightwave systems [249] can be implemented by changing the carrier frequency in the optical or electrical domain. A tunable semiconductor laser is required for optical-domain implementation. The alternative approach consists of hopping the frequency of a microwave subcarrier and then using the SCM technique to transmit the CDM signal. This approach has the advantage that coding and decoding is done in the electrical domain, where the existing commercial microwave components can be used. A hybrid approach in which each 1 bit is transmitted at a different wavelength has been proposed for enhanced security [253].

In another approach, referred to as *coherence multiplexing* [254], a broadband optical source, in combination with an unbalanced MZ interferometer which introduces in one of its branches a delay longer than the *oherence time*, is used to multiplex several channels. Such CDM systems rely on coherence to discriminate among channels and are affected severely by the optical beat noise. In a 1995 demonstration of this technique [254], four 1-Gb/s channels were multiplexed. The optical source was a SLA operating below threshold with a bandwidth of 17 nm. A differential-detection technique was used to reduce the impact of optical beat noise. Indeed, bit-error rates below 10^{-9} could be

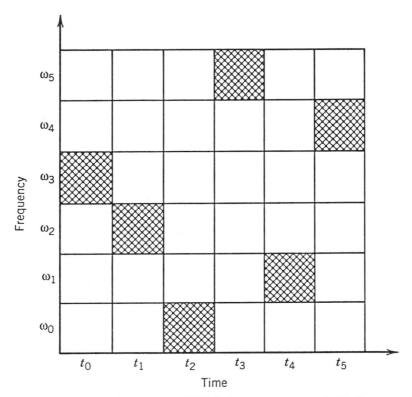

Figure 7.27 Frequency hopping in CDM lightwave systems. A filled square shows the frequency corresponding to a specific time slot. A specific frequency-hop sequence (3, 2, 0, 5, 1, 4) is shown.

achieved by using differential detection even when all four channels were operating simultaneously. Coherence multiplexing can be combined with CDM to provide enhanced security [257].

PROBLEMS

7.1 Explain the difference between optical TDM and WDM, using diagrams whenever necessary.

7.2 The low-loss region of a silica fiber extends from 1.5 to 1.6 μm. How many channels can be transmitted by using WDM when the channel spacing is 10 GHz? If each channel is operated at 2 Gb/s with a power budget of 30 dB allocated to fiber loss, calculate the effective bit rate–distance product of the multichannel system by assuming a loss of 0.2 dB/km.

7.3 A 128 × 128 broadcast star is made by using 2 × 2 directional couplers, each having an insertion loss of 0.2 dB. Each channel transmits 1 mW of average power and requires 1 μW of average received power for operation at 1 Gb/s. What is the maximum transmission distance for each channel? Assume a cable loss of 0.25 dB/km and a loss of 3 dB from connectors and splices.

7.4 Consider a FP filter of length L, refractive index n, and equal mirror reflectivities R for the two facets. Derive an expression for the transmittivity by considering multiple reflections. Use it to show that the finesse F is given by Eq. (7.2.3).

7.5 A FP filter is used to select 100 channels spaced apart by 0.1 nm. What should be the length and the mirror reflectivities of the filter? Assume a refractive index of 1.5 and an operating wavelength of 1.55 μm.

7.6 Explain how a Mach–Zehnder interferometer works by using diagrams as necessary. Prove that the transmission of a chain of such interferometers acts as a filter whose transmittivity is given by Eq. (7.2.5).

7.7 How can you use multiple Mach–Zehnder interferometers to make a demultiplexer? Show the design of a 4×4 demultiplexer schematically.

7.8 You have been given ten 3-dB fiber couplers. Design a 4×4 demultiplexer with as few couplers as possible.

7.9 Use a Mach–Zehnder interferometer and two fiber gratings to design an add/drop filter. Explain the operation of such a device.

7.10 Explain how an array of planar waveguides can be used for wavelength routing in WDM lightwave systems.

7.11 Use a waveguide-grating router to design an integrated WDM transmitter. How would the design change for a WDM receiver?

7.12 What is meant by the in-band linear crosstalk? Derive an expression for the power penalty induced by the in-band crosstalk when a waveguide-grating router is used.

7.13 Explain how stimulated Raman scattering can cause crosstalk in a multichannel lightwave system. Derive Eq. (7.3.12) by approximating the Raman gain by a triangular profile.

7.14 Derive an expression for the power level at which the threshold for stimulated Brillouin scattering is reached.

7.15 Derive Eq. (7.3.13) by considering nonlinear phase change induced by both self- and cross-phase modulation. Use it to obtain Eq. (7.3.15).

7.16 Derive an expression for the CNR of analog SCM lightwave systems by including thermal noise, shot noise, and intensity noise. Plot it as a function of the average optical power and show that CNR saturates to a constant value at high power levels.

7.17 Consider an analog SCM lightwave system operating at 1.55 μm. It uses a receiver of 90% quantum efficiency, 10 nA dark current, and thermal-noise RMS current of 0.1 mA over a 50-MHz bandwidth. The RIN of the transmitter laser is -150 dB/Hz. Calculate the average received power necessary to obtain 50-dB CNR for an AM–VSB system with a modulation index of 0.2.

7.18 The analog SCM lightwave system of Problem 7.1o is operated by using an FM format requiring 16-dB CNR. What is the required average optical power at the receiver if the modulation index is only 0.02?

7.19 Complete the derivation of Eq. (7.5.11) by starting from Eq. (7.5.7) for the photocurrent generated at the coherent receiver. Plot the CNR as a function of the local-oscillator power P_{LO} and show that it becomes maximum for an optimum value of P_{LO}. Neglect the contribution of crosstalk and IMD to the receiver noise.

7.20 Derive an expression for the optimum value of the local-oscillator power for which the CNR of coherent SCM receivers is maximum. Neglect the contribution of crosstalk and IMD to the receiver noise.

7.21 Calculate the local-oscillator power required to achieve 18-dB CNR by using the receiver parameters given in Problem 7.11. Use Eq. (7.5.11) by neglecting both σ_X and σ_{IMD}. Assume a PM index of 0.1 and $P_s = 1$ mW.

REFERENCES

[1] H. Ishio, J. Minowa, and K. Nosu, *J. Lightwave Technol.* **2**, 448 (1984).

[2] G. Winzer, *J. Lightwave Technol.* **2**, 369 (1984).

[3] C. A. Brackett, *IEEE J. Sel. Areas Commun.* **8**, 948 (1990).

[4] N. A. Olsson, J. Hegarty, R. A. Logan, L. F. Johnson, K. L. Walker, L. G. Cohen, B. L. Kasper, and J. C. Campbell, *Electron. Lett.* **21**, 105 (1985).

[5] A. R. Chraplyvy, A. H. Gnauck, R. W. Tkach, R. M. Derosier, C. R. Giles, B. M. Nyman, G. A. Ferguson, J. W. Sulhoff, and J. L. Zyskind, *IEEE Photon. Technol. Lett.* **7**, 98 (1995).

[6] H. Onaka, H. Miyata, G. Ishikawa, K. Otsuka, H. Ooi, Y. Kai, S. Kinoshita, M. Seino, H. Nishimoto, and T. Chikama, Paper PD19, *Proc. Optical Fiber Commun. Conf.*, Optical Society of America, Washington, DC, 1996.

[7] A. R. Chraplyvy, A. H. Gnauck, R. W. Tkach, J. L. Zyskind, J. W. Sulhoff, A. J. Lucero, Y. Sun, R. M. Jopson, F. Forghieri, R. M. Derosier, C. Wolf, and A. R. McCormick, *IEEE Photon. Technol. Lett.* **8**, 1264 (1996).

[8] T. Morioka, H. Takara, S. Kawanishi, O. Kamatani, K. Takiguchi, K. Uchiyama, M. Saruwatari, H. Takahashi, M. Yamada, T. Kanamori, and H. Ono, *Electron. Lett.* **32**, 906 (1996).

[9] Y. Yano et al., Proc. Europ. Conf. Opt. Commun., Oslo (1996).

[10] H. Taga, *J. Lightwave Technol.* **14**, 1287 (1996).

[11] N. S. Bergano and C. R. Davidson, *J. Lightwave Technol.* **14**, 1287 (1996).

[12] A. R. Chraplyvy, J.-M. Delavaux, R. M. Derosier, G. A. Ferguson, D. A. Fishman, C. R. Giles, J. A. Nagel, B. M. Nyman, J. W. Sulhoff, R. E. Tench, R. W. Tkach, and J. L. Zyskind, *IEEE Photon. Technol. Lett.* **6**, 1371 (1994).

[13] O. Gautheron, G. Bassier, V. Letellier, G. Grandpierre, and P. Bollaert, *Electron. Lett.* **32**, 1019 (1996).

[14] M. Suyama, H. Iwata, S. Harasawa, and T. Naito, Paper PD26, *Proc. Optical Fiber Commun. Conf.*, Optical Society of America, Washington, DC, 1996.

[15] J. C. Feggeler, C.-C. Chen, Y. C. Chen, J.-M. P. Delavaux, F. L. Heisman, G. M. Homsey, H. D. Kidorf, T. M. Kissel, J. L. Miller, Jr., J. A. Nagel, R. Nuyts, Y.-K. Park, and W. E. Patterson, *Electron. Lett.* **32**, 1314 (1996).

[16] S. Artigaud, M. Chbat, P. Nouchi, F. Chiquet, D. Bayart, L. Hamon, A. Pitel, F. Goudeseune, P. Bousselet, and J.-L. Beylat, *Electron. Lett.* **32**, 1389 (1996).

[17] A. K. Srivastava, J. B. Judkins, Y. Sun, J. L. Zyskind, J. W. Sulhoff, C. Wolf, J. Zhou, R. P. Espindola, A. M. Vengsarkar, A. R. Chraplyvy, L. Garett, R. M. Derosier, A. H. Gnauck, and R. W. Tkach, Paper PD18, *Proc. Optical Fiber Commun. Conf.*, Optical Society of America, Washington, DC, 1997.

[18] N. S. Bergano, C. R. Davidson, M. A. Mills, P. C. Corbett, S. G. Evangelides, B. Pederson, R. Menges, J. L. Zyskind, J. W. Sulhoff, A. K. Srivastava, C. Wolf, and J. B. Judkins, Paper PD16, *Proc. Optical Fiber Commun. Conf.*, Optical Society of America, Washington, DC, 1997.

[19] G. Kotelelly, *Lightwave* **13** (8), 7 (1996).

[20] D. B. Payne and J. R. Stern, *J. Lightwave Technol.* **4**, 864 (1986).

[21] E.-J. Bachus, R.-P. Braun, C. Caspar, E. Grossman, H. Foisel, K. Hermes, H. Lamping, B. Strebel, and F. J. Westphal, *Electron. Lett.* **22**, 1002 (1986).

[22] B. S. Glance, J. Stone, K. J. Pollock, P. J. Fitzergald, C. A. Burrus, Jr., B. L. Kasper, and L. W. Stulz, *J. Lightwave Technol.* **6**, 1770 (1988).

[23] C. Lin, H. Kobrinski, A. Frenkel, and C. A. Brackett, *Electron. Lett.* **24**, 1215 (1988).

[24] I. P. Kaminow, P. P. Iannone, J. Stone, and L. W. Stulz, *J. Lightwave Technol.* **6**, 1406 (1988).

[25] R. Welter, W. B. Sessa, M. W. Maeda, R. E. Wagner, L. Curtis, J. Young, T. P. Lee, K. Nanduri, H. Kodera, Y. Koga, and J. R. Barry, *J. Lightwave Technol.* **7**, 1438 (1989).

[26] H. Toba, K. Oda, K. Nakanishi, N. Shibata, K. Nosu, N. Takato, and M. Fukuda, *J. Lightwave Technol.* **8**, 1396 (1990).

[27] A. M. Hill, R. Wyatt, J. F. Massicott, K. J. Blyth, D. S. Forrester, R. A. Lobbett, P. J. Smith, and D. B. Payne, *Electron. Lett.* **26**, 1882 (1990).

[28] D. S. Forrester, A. M. Hill, R. A. Lobbett, R. Wyatt, and S. F. Carter, *Electron. Lett.* **27**, 2051 (1991).

[29] P. E. Green, Jr., *Fiber-Optic Networks*, Prentice-Hall, Upper Saddle River, NJ, 1993.

[30] D. W. Smith, *Optical Network Technology*, Chapman & Hall, New York, 1995.

[31] D. J. G. Mestdagh, *Fundamentals of Multiaccess Optical Fiber Networks*, Artec House, Boston, 1995.

[32] M. M.-K. Liu, *Principles and Applications of Optical Communications*, Irwin, Chicago, 1996.

[33] K. Nosu *Optical FDM Network Technologies*, Artec House, Boston, 1997.

[34] Special issue on broadband optical networks, *J. Lightwave Technol.* **11** (6) (1993).

[35] Special issue on multiwavelength optical technology and networks, *J. Lightwave Technol.* **14** (6) (1996).

[36] Special issue on optical networks, *IEEE J. Sel. Areas Commun.* **14** (1996).

[37] P. E. Green, Jr., *IEEE J. Sel. Areas Commun.* **14**, 764 (1996).

[38] B. Mukherjee, *IEEE Network Mag.* **6** (3), 12 (1992); **6** (4), 20 (1992).

[39] P. W. Dowd, *IEEE Trans. Comput.* **41**, 1223 (1992).

[40] K. Tang, *J. Lightwave Technol.* **12**, 2023 (1994).

[41] M. S. Goodman, H. Kobrinski, M. P. Vecchi, R. M. Bulley, and J. L. Gimlett, *IEEE J. Sel. Areas Commun.* **8**, 995 (1990).

[42] E. Hall, J. Kravitz, R. Ramaswami, M. Halvorson, S. Tenbrink, and R. Thomsen, *IEEE J. Sel. Areas Commun.* **14**, 814 (1996).

[43] T.-K. Chinag, S. K. Agrawal, D. T. Mayweather, D. Sadot, C. F. Berry, M. Hickey, and L. G. Kazovsky, *IEEE J. Sel. Areas Commun.* **14**, 824 (1996).

[44] S. S. Wagner, H. Kobrinski, T. J. Robe, H. L. Lemberg, and L. S. Smoot, *Electron. Lett.* **24**, 344 (1988).

[45] P. P. Iannone and N. J. Frigo, *Lightwave* **13** (8), 44 (1996).

[46] P. W. Shumate, *Lightwave* **13** (8), 52 (1996).

[47] C. Podlesny, *Lightwave* **13** (8), 58 (1996).

[48] I. P. Kaminow, C. R. Doerr, C. Dragone, T. Koch, U. Koren, A. A. M. Saleh, A. J. Kirby, C. M. Özveren, B. Schofield, R. E. Thomas, R. A. Barry, D. M. Castagnozzi, V. W. S. Chan, B. R. Hemenway, D. Marquis, S. A. Parikh, M. L. Stevens, E. A. Swanson, S. G. Finn, and R. G. Gallager, *IEEE J. Sel. Areas Commun.* **14**, 780 (1996).

[49] R. E. Wagner, R. C. Alferness, A. A. M. Saleh, and M. S. Goodman, *J. Lightwave Technol.* **14**, 1349 (1996).

[50] G.-K. Chang, G. Elinas, J. K. Gamelin, M. Z. Iqbal, and C. A. Brackett, *J. Lightwave Technol.* **14**, 1320 (1996).

[51] S. Johansson, *J. Lightwave Technol.* **14**, 1341 (1996).

[52] S. Okamoto, A. Watanabe, and K.-I. Sato, *J. Lightwave Technol.* **14**, 1410 (1996).

[53] M. Born and E. Wolf, *Principles of Optics*, 6th ed., Pergamon Press, Oxford, 1980.

[54] J. Stone and L. W. Stulz, *Electron. Lett.* **23**, 781 (1987).

[55] K. Hirabayashi, H. Tsuda, and T. Kurokawa, *IEEE Photon. Technol. Lett.* **3**, 741 (1991); *J. Lightwave Technol.* **11**, 2033 (1993).

[56] A. Sneh and K. M. Johnson, *J. Lightwave Technol.* **14**, 1067 (1996).

[57] Y. Hibino, F. Hanawa, H. Nakagome, M. Ishii, and N. Takato, *J. Lightwave Technol.* **13**, 1728 (1995).

[58] S. Mino, K. Yoshino, Y. Yamada, T. Terui, M. Yasu, and K. Moriwaki, *J. Lightwave Technol.* **13**, 2320 (1995).

[59] M. Kawachi, *IEE Proc.* **143**, 257 (1996).

[60] Y. P. Li and C. H. Henry, *IEE Proc.* **143**, 263 (1996).

[61] I. Bennion, J. A. R. Wiliams, L. Zhang, and K. Sugden, *Opt. Quantum Electron.* **28**, 93 (1996).

[62] G. P. Agrawal and S. Radic, *IEEE Photon. Technol. Lett.* **6**, 995 (1994).

[63] G. E. Town, K. Sugde, J. A. R. Williams, I. Bennion, and S. B. Poole, *IEEE Photon. Technol. Lett.* **7**, 78 (1995).

[64] R. Kashyap, G. D. Maxwell, and B. J. Ainslie, *IEEE Photon. Technol. Lett.* **5**, 191 (1993).

[65] F. Bilodeau, K. O. Hill, B. Malo, D. C. Johnson, and J. Albert, *IEEE Photon. Technol. Lett.* **6**, 80 (1994).

[66] T. Numai, S. Murata, and I. Mito, *Appl. Phys. Lett.* **53**, 83 (1988); **54**, 1859 (1989).

[67] J.-P. Weber, B. Stoltz, and M. Dasler, *Electron. Lett.* **31**, 220 (1995).

[68] J. L. Jackel, M. S. Goodman, J. E. Baran, W. J. Tomlinson, G.-K. Chang, M. Z. Iqbal, G. H. Song, K. Bala, C. A. Brackett, D. A. Smith, R. S. Chakravarthy, R. H. Hobbs, D. J. Fritz, R. W. Ade, and K. M. Kissa, *J. Lightwave Technol.* **14**, 1056 (1996).

[69] D. A. Smith, R. S. Chakravarthy, Z. Bao, J. E. Baran, J. L. Jackel, A. d'Alessandro, D. J. Fritz, S. H. Huang, X. Y. Zou, S.-M. Hwang, A. E. Willner, and K. D. Li, *J. Lightwave Technol.* **14**, 1005 (1996).

[70] G. P. Agrawal, *Nonlinear Fiber Optics*, 2nd ed., Academic Press, San Diego, CA, 1995.

[71] A. R. Chraplyvy and R. W. Tkach, *Electron. Lett.* **22**, 1084 (1986).

[72] R. W. Tkach, A. R. Chraplyvy, and R. M. Derosier, *IEEE Photon. Technol. Lett.* **1**, 111 (1989).

[73] K. Margari, H. Kawaguchi, K. Oe, Y. Nakano, and M. Fukuda, *Appl. Phys. Lett.* **51**, 1974 (1987); *IEEE J. Quantum Electron.* **24**, 2178 (1988).

[74] T. Numai, S. Murata, and I. Mito, *Appl. Phys. Lett.* **53**, 1168 (1988).

[75] T. Numai, *IEEE Photon. Technol. Lett.* **2**, 401 (1990); L. Kazovsky and J. Werner, *Appl. Opt.* **28**, 553 (1989).

[76] S. Dubovitsky and W. H. Steier, *J. Lightwave Technol.* **14**, 1020 (1996).

[77] C. H. Henry, R. F. Kazarinov, Y. Shani, R. C. Kistler, V. Pol, and K. J. Orlowsky, *J. Lightwave Technol.* **8**, 748 (1990).

[78] D. R. Wisely, *Electron. Lett.* **27**, 520 (1991).

[79] H. Toba, K. Oda, N. Nakato, and K. Nosu, *Electron. Lett.* **19**, 583 (1987).

[80] B. H. Verbeek, C. H. Henry, N. A. Olsson, K. J. Orlowsky, R. F. Kazarinov, and B. H. Johnson, *J. Lightwave Technol.* **6**, 1011 (1988).

[81] K. Oda, N. Tokato, T. Kominato, and H. Toba, *IEEE Photon. Technol. Lett.* **1**, 137 (1989).

[82] N. Takato, T. Kominato, A. Sugita, K. Jinguji, H. Toba, and M. Kawachi, *IEEE J. Sel. Areas Commun.* **8**, 1120 (1990).

[83] G. D. Maxwell, R. Kashyap, and B. J. Ainslie, *Electron. Lett.* **28**, 2100 (1992).

[84] Y. Hibino, T. Kitagawa, K. O. Hill, F. Bilodeau, B. Malo, J. Albert, and D. C. Johnson, *IEEE Photon. Technol. Lett.* **8**, 84 (1996).

[85] M. J. F. Digonnet and H. J. Shaw, *Appl. Opt.* **22**, 484 (1983).

[86] B. S. Kawasaki, K. O. Hill, and R. G. Gaumont, *Opt. Lett.* **6**, 327 (1981).

[87] A. R. Vellekoop and M. K. Smit, *J. Lightwave Technol.* **9**, 310 (1991).

[88] H. Bissessur, F. Gaborit, B. Martin, P. Pagnod-Rossiaux, J.-L. Peyre, and M. Renaud, *Electron. Lett.* **30**, 336 (1994).

[89] H. Bissessur, F. Gaborit, B. Martin, and G. Ripoche, *Electron. Lett.* **31**, 1372 (1995).

[90] L. H. Spiekman, M. R. Amersfoort, A. H. de Vreede, F. P. G. M. van Ham, A. Kuntze, J. W. Pedersen, P. Dremeester, and M. K. Smit, *J. Lightwave Technol.* **14**, 991 (1996).

[91] C. Dragone, *J. Lightwave Technol.* **7**, 479 (1989); *IEEE Photon. Technol. Lett.* **1**, 238 (1989).

[92] H. Okayama, M. Kawahara, and T. Kamijoh, *J. Lightwave Technol.* **14**, 985 (1996).

[93] F. Shehadeh, R. S. Vodhanel, M. Krain, C. Gibbons, R. E. Wagner, and M. Ali, *IEEE Photon. Technol. Lett.* **7**, 1075 (1995).

[94] M. Kuznetsov, *J. Lightwave Technol.* **12**, 226 (1994).

[95] H. H. Yaffe, C. H. Henry, M. R. Serbin, and L. G. Cohen, *J. Lightwave Technol.* **12**, 1010 (1994).

[96] R. C. Alferness, L. L. Buhl, M. J. R. Martyak, M. D. Divino, C. H. Joyner, and A. G. Dentai, *Electron. Lett.* **24**, 150 (1988).

[97] D. C. Johnson, K. O. Hill, F. Bilodeau, and S. Faucher, *Electron. Lett.* **23**, 668 (1987).

[98] C. M. Ragdale, T. J. Reid, D. C. J. Reid, A. C. Carter, and P. J. Williams, *Electron. Lett.* **28**, 712 (1992).

[99] F. Bilodeau, D. C. Johnson, S. Thériault, B. Malo, J. Albert, and K. O. Hill, *IEEE Photon. Technol. Lett.* **7**, 388 (1995).

[100] H. A. Haus and Y. Lai, *J. Lightwave Technol.* **10**, 57 (1992).

[101] M. Zirngibl, C. H. Joyner, and B. Glance, *IEEE Photon. Technol. Lett.* **6**, 513 (1994).

[102] M. E. Marhic, *Opt. Lett.* **9**, 368 (1984).

[103] E. G. Rawson and M. D. Bailey, *Electron. Lett.* **15**, 432 (1979).

[104] D. B. Mortimore, *Electron. Lett.* **21**, 502 (1985); **21**, 742 (1985); **22**, 1205 (1986).

[105] J. W. Arkwright and D. B. Mortimore, *Electron. Lett.* **26**, 1534 (1989).

[106] D. B. Mortimore and J. W. Arkwright, *Appl. Opt.* **30**, 650 (1991).

[107] J. W. Arkwright, D. B. Mortimore, and R. M. Adams, *Electron. Lett.* **27**, 737 (1991).

[108] C. Dragone, C. H. Henry, I. P. Kaminow, and R. C. Kistler, *IEEE Photon. Technol. Lett.* **1**, 241 (1989).

[109] P. D. Trinh, S. Yegnanaraynan, and B. Jalali, *IEEE Photon. Technol. Lett.* **8**, 794 (1996).

[110] C. Dragone, *IEEE Photon. Technol. Lett.* **3**, 812 (1991).

[111] C. Dragone, C. A. Edwards, and R. C. Kistler, *IEEE Photon. Technol. Lett.* **3**, 896 (1991).

[112] H. Takahashi, Y. Hibino, and I. Nishi, *Opt. Lett.* **17**, 499 (1992).

[113] R. Adar, C. H. Henry, C. Dragone, R. C. Kistler, and M. A. Milbrodt, *J. Lightwave Technol.* **11**, 212 (1993).

[114] M. Zirngibl, C. Dragone, and C. H. Joyner, *IEEE Photon. Technol. Lett.* **4**, 1250 (1992).

[115] K. Okamoto, K. Syuto, H. Takahashi, and Y. Ohmori, *Electron. Lett.* **32**, 1474 (1996).

[116] B. Glance, I. P. Kaminow, and R. W. Wilson, *J. Lightwave Technol.* **12** 957 (1994).

[117] Y. Tachikawa, Y. Inoue, M. Ishii, and T. Nozawa, *J. Lightwave Technol.* **14**, 977 (1996).

[118] Y. Tachikawa and Y. Inoue, *Electron. Lett.* **31**, 2029 (1995).

[119] B. Glance, *IEEE Photon. Technol. Lett.* **8**, 245 (1996).

[120] K. Okamoto, M. Okuno, A. Himeno, and Y. Ohmori, *Electron. Lett.* **32**, 1471 (1996).

[121] M. Zirngibl, C. R. Doerr, and L. W. Stulz, *IEEE Photon. Technol. Lett.* **8**, 721 (1996).

[122] M. Koga, Y. Hamazumi, A. Watanabe, S. Okamoto, H. Obara, K.-I. Sato, M. Okuno, and S. Suzuki, *J. Lightwave Technol.* **14**, 1106 (1996).

[123] Y. Jin and M. Kavehrad, *J. Lightwave Technol.* **14**, 1183 (1996).

[124] A. Jourdan, F. Masetti, M. Garnot, G. Soulage, and M. Sotom, *J. Lightwave Technol.* **14**, 1198 (1996).

[125] S. Okamoto, A. Watanabe, and K.-I. Sato, *J. Lightwave Technol.* **14**, 1410 (1996).

[126] W. D. Zhong, J. P. R. Lacey, and R. S. Tucker, *J. Lightwave Technol.* **14**, 1613 (1996).

[127] A. Watanabe, S. Okamoto, M. Koga, K. Sato, and M. Okuno, *Electron. Lett.* **33**, 67 (1997).

[128] E. Almström, C. P. Larsen, L. Gillner, W. H. van Berlo, M. Gustavsson, and E. Berglind, *J. Lightwave Technol.* **14**, 996 (1996).

[129] S. J. B. Yoo, *J. Lightwave Technol.* **14**, 955 (1996).

[130] G.-H. Duan, in *Semiconductor Lasers: Past, Present, and Future*, G. P. Agrawal, Ed., AIP Press, Woodbury, NY, 1995, Chap. 10.

[131] T. Derhuus, B. Mikkelsen, C. Joergensen, S. L. Danielsen, and K. E. Stubkjaer, *J. Lightwave Technol.* **14**, 942 (1996).

[132] G. P. Agrawal, *Appl. Phys. Lett.* **51**, 302 (1987).

[133] A. D'Ottavi, F. Martelli, P. Spano, A. Mecozzi, S. Scotti, R. Dall'Ara, J. Eckner, and G. Guekos, *Appl. Phys. Lett.* **68**, 2186 (1996).

[134] D. Nesset, D. D. Marcenac, and A. E. Kelly, *Electron. Lett.* **33**, 148 (1997).

[135] H. Kuwatsuka, H. Shoji, M. Matsuda, and H. Ishikawa, *Electron. Lett.* **31**, 2108 (1995).

[136] T. L. Koch and U. Koren, *IEE Proc.* **138**, Pt. J, 139 (1991); *IEEE J. Quantum Electron.* **27**, 641 (1991).

[137] T. P. Lee, C. E. Zah, R. Bhat, W. C. Young, B. Pathak, F. Favire, P. S. D. Lin, N. C. Andreadakis, C. Caneau, A. W. Rahjel, M. Koza, J. K. Gamelin, L. Curtis, D. D. Mahoney, and A. Lepore, *J. Lightwave Technol.* **14**, 967 (1996).

[138] M. G. Young, U. Koren, B. I. Miller, M. A. Newkirk, M. Chien, M Zirngibl, C. Dragone, B. Tell, H. M. Presby, and G. Raybon, *IEEE Photon. Technol. Lett.* **5**, 908 (1993).

[139] G. P. Li, T. Makino, A. Sarangan, and W. Huang, *IEEE Photon. Technol. Lett.* **8**, 22 (1996).

[140] C. J. Chang-Hasnain, in *Semiconductor Lasers: Past, Present, and Future*, G. P. Agrawal, Ed., AIP Press, Woodbury, NY, 1995, Chap. 5.

[141] K. R. Poguntke, J. B. D. Soole, A. Scherer, H. P. LeBlanc, R. Bhat, and M. A. Koza, *Appl. Phys. Lett.* **62**, 2034 (1993).

[142] M. Zirngibl, C. H. Joyner, C. R. Doerr, L. W. Stulz, and H. M. Presby, *IEEE Photon. Technol. Lett.* **8**, 870 (1996).

[143] Y. Tachikawa and K. Okamoto, *Electron. Lett.* **31**, 1665 (1995).

[144] J. B. D. Soole, A. Scherer, H. P. LeBlanc, N. C. Andreadakis, R. Bhat, and M. A. Koza, *Appl. Phys. Lett.* **58**, 1949 (1991).

[145] C. Cremer, N. Emeis, M. Schier, G. Heise, G. Ebbinghaus, and L. Stoll, *IEEE Photon. Technol. Lett.* **4**, 108 (1992).

[146] M. Fallahi, K. A. McGreer, A. Delage, I. M. Templeton, F. Chatenoud, and R. Barber, *IEEE Photon. Technol. Lett.* **5**, 794 (1993).

[147] S. Chandrasekhar, M. Zirngibl, A. G. Dentai, C. H. Joyner, F. Storz, C. A. Burrus, and L. M. Lunardi, *IEEE Photon. Technol. Lett.* **7**, 1342 (1995).

[148] I. Ikushima, S. Himi, T. Hamaguchi, M. Suzuki, N. Maeda, H. Kodera, and K. Yamashita, *J. Lightwave Technol.* **13**, 517 (1995).

[149] T. Morioka, K. Uchiyama, S. Kawanishi, S. Suzuki, and M. Saruwatari, *Electron. Lett.* **31**, 1064 (1995).

[150] M. C. Nuss, W. H. Knox, and U. Koren, *Electron. Lett.* **32**, 1311 (1996).

[151] P. A. Rosher and A. R. Hunwicks, *IEEE J. Sel. Areas Commun.* **8**, 1108 (1990).

[152] P. A. Humblet and W. M. Hamdy, *IEEE J. Sel. Areas Commun.* **8**, 1095 (1990).

[153] F. Tian and H. Hermann, *J. Lightwave Technol.* **13**, 1146 (1995).

[154] M. Fukutoku, K. Oda, and H. Toba, *J. Lightwave Technol.* **13**, 2224 (1995).

[155] Y. D. Jin, Q. Jiang, and M. Kavehrad, *IEEE Photon. Technol. Lett.* **7**, 1210 (1995).

[156] D. J. Blumenthal, P. Granestrand, and L. Thylen, *IEEE Photon. Technol. Lett.* **8**, 284 (1996).

[157] G. Murtaza and J. M. Senior, *IEEE Photon. Technol. Lett.* **8**, 440 (1996).

[158] H. Takahashi, K. Oda, and H. Toba, *J. Lightwave Technol.* **14**, 1097 (1996).

[159] C.-S. Li and F. Tong, *J. Lightwave Technol.* **14**, 1120 (1996).

[160] J. Zhou, R. Cadeddu, E. Casaccia, C. Cavazzoni, and M. J. O'Mahony, *J. Lightwave Technol.* **14**, 1423 (1996).

[161] K.-P. Ho and J. M. Kahn, *J. Lightwave Technol.* **14**, 1127 (1996); K.-P. Ho and C. Lin, *Electron. Lett.* **32**, 1119 (1996).

[162] L. A. Buckman, L. P. Chen, and K. Y. Lau, *IEEE Photon. Technol. Lett.* **9**, 250 (1997).

[163] A. R. Chraplyvy, *J. Lightwave Technol.* **8**, 1548 (1990).

[164] R. G. Waarts, A. A. Frisem, E. Lichtman, H. H. Yaffe, and R. P. Braun, *Proc. IEEE* **78**, 1344 (1990).

[165] N. Shibata, K. Nosu, K. Iwashita, and Y. Azuma, *IEEE J. Sel. Areas Commun.* **8**, 1068 (1990).

[166] F. Forghieri, *IEEE Photon. Technol. Lett.* **8**, 1400 (1996).

[167] A. R. Chraplyvy, *Electron. Lett.* **20**, 58 (1984).

[168] F. Forghieri, R. W. Tkach, and A. R. Chraplyvy, *IEEE Photon. Technol. Lett.* **7**, 101 (1995).

[169] Y. Zhao, J. S. Wang, W. Zhou, H. Y. Tam, and M. S. Demokan, *Microwave Opt. Tech. Lett.* **12**, 111 (1996).

[170] A. R. Chraplyvy and R. W. Tkach, *IEEE Photon. Technol. Lett.* **5**, 666 (1993).

[171] Y. Aoki, K. Tajima, and I. Mito, *J. Lightwave Technol.* **6**, 710 (1988).

[172] E. Lichtman, R. G. Waarts, and A. A. Friesem, *J. Lightwave Technol.* **7**, 171 (1989).

[173] D. A. Fishman and J. A. Nagel, *J. Lightwave Technol.* **11**, 1721 (1993).

[174] R. G. Waarts and R. P. Braun, *Electron. Lett.* **22**, 873 (1986).

[175] M. W. Maeda, W. B. Sessa, W. I. Way, A. Yi-Yan, L. Curtis, R. Spicer, and R. I. Laming, *J. Lightwave Technol.* **8**, 1402 (1990).

[176] K. Inoue, K. Nakanishi, K. Oda, and H. Toba, *J. Lightwave Technol.* **12**, 1423 (1994).

[177] K. Inoue and H. Toba, *J. Lightwave Technol.* **13**, 88 (1995).

[178] F. Forghieri, R. W. Tkach, and A. R. Chraplyvy, *J. Lightwave Technol.* **13**, 889 (1995).

[179] Y. C. Chung, J. Jeong, and L. S. Cheng, *IEEE Photon. Technol. Lett.* **6**, 792 (1994).

[180] T. Ikegami, S. Sudo, and Y. Sakai, *Frequency Stabilization of Semiconductor Laser Diodes*, Artec House, Boston, 1995.

[181] M. Kourogi and M. Ohtsu, in *Semiconductor Lasers: Past, Present, and Future*, G. P. Agrawal, Ed., AIP Press, Woodbury, NY, 1995, Chap. 3.

[182] M. Guy, B. Villeneuve, C. Latrasse, and M. Têtu, *J. Lightwave Technol.* **14**, 1136 (1996).

[183] J. Zhou and M. J. O'Mahony, *IEE Proc.* **143**, 178 (1996).

[184] D. Marcuse, A. R. Chraplyvy, and R. W. Tkach, *J. Lightwave Technol.* **12**, 885 (1994).

[185] A. R. Chraplyvy, A. H. Gnauck, R. W. Tkach, R. M. Derosier, C. R. Giles, B. M. Nyman, G. A. Ferguson, J. W. Sulhoff, and J. L. Zyskind, *IEEE Photon. Technol. Lett.* **7**, 98 (1995).

[186] S. Kawanishi, H. Takara, O. Kamatani, T. Morioka, and M. Saruwatari, *Electron. Lett.* **32**, 470 (1996).

[187] D. M. Spirit, A. D. Ellis, and P. E. Barnsley, *IEEE Commun. Mag.* **32** (12), 56 (1994).

[188] M. Saruwatari, *IEEE Commun. Mag.* **32** (9), 98 (1994).

[189] A. D. Ellis, D. M. Patrick, D. Flannery, R. J. Manning, D. A. O. Davies, and D. M. Spirit, *J. Lightwave Technol.* **13**, 761 (1995).

[190] R. A. Barry, V. W. S. Chan, K. L. Hall, E. S. Kintzer, J. D. Moores, K. A. Rauschenbach, E. A. Swanson, L. E. Adams, C. R. Doerr, S. G. Finn, H. A. Haus, E. P. Ippen, and W. S. Wong, *IEEE J. Sel. Areas Commun.* **14**, 999 (1996).

[191] S.-W. Seo, K. Bergman, and P. R. Prucnal, *IEEE J. Sel. Areas Commun.* **14**, 1039 (1996).

[192] E. Bødtker and J. E. Bowers, *J. Lightwave Technol.* **13**, 1809 (1995).

[193] K. Uchiyama, H. Takara, S. Kawanishi, T. Morioka, M. Saruwatari, and T. Kitoh, *Electron. Lett.* **29**, 1870 (1993).

[194] J. P. Sokoloff, P. R. Prucnal, and M. Kane, *IEEE Photon. Technol. Lett.* **5**, 787 (1993).

[195] J. Glesk, J. P. Sokoloff, and P. R. Prucnal, *Electron. Lett.* **30**, 339 (1994).

[196] T. Morioka, H. Takara, S. Kawanishi, T. Kitoh, and M. Saruwatari, *Electron. Lett.* **32**, 833 (1996).

[197] T. Morioka, H. Takara, S. Kawanishi, K. Uchiyama, and M. Saruwatari, *Electron. Lett.* **32**, 840 (1996).

[198] O. Kamatani and S. Kawanishi, *J. Lightwave Technol.* **14**, 1757 (1996).

[199] S. Kawanishi, H. Takara, O. Kamatani, T. Morioka, and M. Saruwatari, *Electron. Lett.* **32**, 470 (1996).

[200] S. Kawanishi, H. Takara, T. Morioka, O. Kamatani, K. Takaguchi, T. Kitoh, and M. Saruwatari, *Electron. Lett.* **32**, 916 (1996).

[201] K. H. Kim, H. K. Lee, S. Y. Park, and E.-H. Lee, *J. Lightwave Technol.* **13**, 1597 (1995).

[202] R. Olshansky, V. A. Lanzisera, and P. M. Hill, *J. Lightwave Technol.* **7**, 1329 (1989).

[203] T. E. Darcie, *J. Lightwave Technol.* **5**, 1103 (1987); *IEEE J. Sel. Areas Commun.* **8**, 1240 (1990).

[204] W. I. Way, *J. Lightwave Technol.* **7**, 1806 (1989).

[205] M. Maeda and M. Yamamoto, *IEEE J. Sel. Areas Commun.* **8**, 1257 (1990).

[206] R. Olshansky, R. Gross, and M. Schmidt, *IEEE J. Sel. Areas Commun.* **8**, 1268 (1990).

[207] T. Darcie and G. Bodeep, *IEEE Trans. Microwave Theory Tech.* **38**, 524 (1990).

[208] M. Nazarathy, J. Berger, A. J. Ley, I. M. Levi, and Y. Kagan, *J. Lightwave Technol.* **11**, 82 (1993).

[209] C. S. Ih and W. Gu, *IEEE J. Sel. Areas Commun.* **8**, 1296 (1990).

[210] C. R. Phillips, T. E. Darcie, D. Marcuse, G. E. Bodeep, and N. J. Frigo, *IEEE Photon. Technol. Lett.* **3**, 481 (1991).

[211] C. Y. Kuo, *J. Lightwave Technol.* **11**, 7 (1993).

[212] G. J. Meslener, *J. Lightwave Technol.* **12**, 118 (1994).

[213] M. T. Abuelmaatti, *Microwave Opt. Tech. Lett.* **11**, 202 (1996).

[214] A. J. Rainal, *J. Lightwave Technol.* **14**, 474 (1996).

[215] J. H. Angenent, *Electron. Lett.* **26**, 2049 (1990).

[216] W. I. Way, C. Lin, C. E. Zah, L. Curtis, R. Spicer, and W. C. Young, *IEEE Photon. Technol. Lett.* **2**, 360 (1990).

[217] K. Petermann, *Electron. Lett.* **26**, 2097 (1990).

[218] G. K. Gopalkrishnan, T. J. Brophy, and C. Breverman, *Electron. Lett.* **32**, 1309 (1996).

[219] J. Marti, A. Montero, J. Capmany, J. M. Fuster, and D. Pastor, *Electron. Lett.* **32**, 1605 (1996).

[220] R. Olshansky, V. Lanzisera, and P. M. Hill, *Electron. Lett.* **24**, 1234 (1988).

[221] P. M. Gabla, V. Lamaire, H. Krimmel, J. Otterbach, J. Augé, and A. Dursin, *IEEE Photon. Technol. Lett.* **3**, 56 (1991).

[222] E. Yoneda, K. Suto, K. Kikushima, and H. Yoshinaga, *J. Lightwave Technol.* **11**, 128 (1993).

[223] A. A. M. Saleh, *Electron. Lett.* **25**, 776 (1989).

[224] N. J. Frigo, M. R. Phillips, and G. E. Bodeep, *J. Lightwave Technol.* **11**, 138 (1993).

[225] K.-P. Ho and J. M. Kahn, *IEEE Photon. Technol. Lett.* **8**, 125 (1996).

[226] R. Gross and R. Olshansky, *J. Lightwave Technol.* **8**, 406 (1990).

[227] S. Watanabe, T. Terahara, I. Yokata, T. Naito, T. Chikama, and H. Kuwabara, *J. Lightwave Technol.* **11**, 116 (1993).

[228] P. M. Hill and R. Olshansky, *Electron. Lett.* **24**, 892 (1988).

[229] R. Gross, R. Olshansky, and P. Hill, *IEEE Photon. Technol. Lett.* **1**, 179 (1989).

[230] M. S. Kao and J. Wu, *Electron. Lett.* **26**, 1680 (1990).

[231] R. W. Gross, W. Rideout, R. Olshansky, and G. R. Joyce, *J. Lightwave Technol.* **9**, 524 (1991).

[232] S. S. Wagner, T. E. Chapuran, and R. C. Menendez, *IEEE Photon. Technol. Lett.* **8**, 275 (1996).

[233] Q. Pan and R. J. Green, *IEEE Photon. Technol. Lett.* **8**, 278 (1996); *IEEE Photon. Technol. Lett.* **8**, 1079 (1996).

[234] W. I. Way, S. S. Wagner, M. M. Choy, C. Lin, R. C. Menendez, H. Tohme, A. Yi-Yan, A. C. von Lehman, R. E. Spicer, M. Andrejco, M. Saifi, and H. L. Lemberg, *IEEE Photon. Technol. Lett.* **2**, 665 (1990).

[235] S. C. Liew and K.-W. Cheung, *J. Lightwave Technol.* **7**, 1825 (1989).

[236] D. W. Faulkner, D. B. Payne, J. R. Stern, and J. W. Ballance, *J. Lightwave Technol.* **7**, 1741 (1989).

[237] R. Olshansky, V. A. Lanzisera, S.-F. Su, R. Gross, A. M. Forcucci, and A. H. Oakes, *J. Lightwave Technol.* **11**, 60 (1993).

[238] W.-P. Lin, *Microwave Opt. Tech. Lett.* **12**, 277 (1996).

[239] C. Desem, *IEEE J. Sel. Areas Commun.* **8**, 1290 (1990); *IEEE Photon. Technol. Lett.* **3**, 387 (1991).

[240] S. L. Woodward, X. Lu, T. E. Darcie, and G. E. Bodeep, *IEEE Photon. Technol. Lett.* **8**, 694 (1996).

[241] A. Li, C. J. Mahon, Z. Wang, G. Jacobsen, and E. Bodtker, *Electron. Lett.* **31**, 1538 (1995).

[242] Z. Wang, E. Bodtker, and G. Jacobsen, *Electron. Lett.* **31**, 1591 (1995).

[243] P. P. Iannone, K. C. Reichmann, and N. J. Frigo, *IEEE Photon. Technol. Lett.* **8**, 930 (1996).

[244] G. Lawton, *Lightwave* **13** (10), 14 (1996).

[245] O. K. Tonguz and H. Jung, *J. Lightwave Technol.* **14**, 1400 (1996).

[246] R. E. Ziemer, R. L. Peterson, and D. E. Borth, *An Introduction to Spread Spectrum Communications*, Prentice Hall, Upper Saddle River, NJ, 1995.

[247] J. A. Salehi, *IEEE Trans. Commun.* **37**, 824 (1989); J. A. Salehi and C. A. Brackett, *IEEE Trans. Commun.* **37**, 834 (1989).

[248] J. A. Salehi, A. M. Weiner, and J. P. Heritage, *J. Lightwave Technol.* **8**, 478 (1990).

[249] K. Kiaseleh, *IEEE Photon. Technol. Lett.* **3**, 173 (1991).

[250] W. C. Kwong, P. A. Perrier, and P. R. Prucnal, *IEEE Trans. Commun.* **39**, 1625 (1991).

[251] R. M. Gagliardi, A. J. Mendez, M. R. Dale, and E. Park, *J. Lightwave Technol.* **11**, 20 (1993).

[252] F. Khaleghi and M. Kavehrad, *IEEE Trans. Commun.* **43**, 75 (1995).

[253] L. Tančevski, I. Andonovic, and J. Budin, *IEEE Photon. Technol. Lett.* **7**, 573 (1995).

[254] G. J. Pendock and D. D. Sampson, *IEEE Photon. Technol. Lett.* **7**, 1504 (1995).

[255] A. Gameiro, *Proc. SPIE* **2602**, 63 (1996).

[256] K. Iversen and D. Hampicke, *Proc. SPIE* **2614**, 110 (1996).

[257] N. Karafolas, G. C. Gupta, and D. Uttamchandani, *Opt. Commun.* **123**, 11 (1996).

[258] D. J. G. Mestdagh, *Opt. Fiber Technol.* **2**, 7 (1996).

Chapter 8

OPTICAL AMPLIFIERS

As discussed in Chapter 5, the transmission distance of a fiber-optic communication system is limited by fiber loss and dispersion. For long-haul lightwave systems, the loss limitation has traditionally been overcome using optoelectronic repeaters in which the optical signal is first converted into an electric current and then regenerated using a transmitter. Such regenerators become quite complex and expensive for multichannel lightwave systems. An alternative approach makes use of optical amplifiers, which amplify the optical signal directly without requiring its conversion to electric domain. Several kinds of optical amplifiers were studied and developed during the 1980s, and the use of optical amplifiers for long-haul lightwave systems became widespread during the 1990s. By 1996, optical amplifiers were a part of the fiber-optic cables laid across the Atlantic and Pacific oceans.

This chapter is devoted to optical amplifiers. In Section 8.1 we discuss general concepts common to all optical amplifiers. Semiconductor laser amplifiers are considered in Section 8.2. Fiber Raman and Brillouin amplifiers are covered in Sections 8.3 and 8.4, respectively. Sections 8.5 and 8.6 are devoted to fiber amplifiers made by doping the fiber core with a rare-earth element. The emphasis is on the erbium-doped fiber amplifiers, which are used almost exclusively for 1.55-μm lightwave systems. Indeed, such fiber amplifiers have been used in commercial lightwave systems since 1996.

8.1 BASIC CONCEPTS

Optical amplifiers amplify incident light through stimulated emission, the same mechanism as that used by lasers (see Section 3.1). Indeed, an optical amplifier is nothing but a laser without feedback. Its main ingredient is the *optical gain* realized when the amplifier is pumped (optically or electrically) to achieve *population inversion*. The optical gain, in general, depends not only on the frequency (or wavelength) of the incident signal, but also on the local beam intensity at any point inside the amplifier. Details of the frequency and intensity dependence of the optical gain depend on the amplifier medium. To illustrate the general concepts, let us consider the case in which the gain medium is modeled as a homogeneously broadened two-level system. The *gain coefficient* of such a medium can be written as [1]

$$g(\omega) = \frac{g_0}{1 + (\omega - \omega_0)^2 T_2^2 + P/P_s},$$

(8.1.1)

where g_0 is the peak value of the gain, ω is the optical frequency of the incident signal, ω_0 is the atomic transition frequency, and P is the optical power of the signal being amplified. The saturation power P_s depends on gain-medium parameters such as the fluorescence time T_1 and the transition cross section; its expression for different kinds of amplifiers is given in the following sections. The parameter T_2 in Eq. (8.1.1), known as the *dipole relaxation time*, is typically quite small (< 1 ps). The fluorescence time T_1, also called the *population relaxation time*, varies in the range 100 ps to 10 ms, depending on the gain medium. Equation (8.1.1) can be used to discuss important characteristics of optical amplifiers, such as the gain bandwidth, amplification factor, and output saturation power.

8.1.1   Gain Spectrum and Bandwidth

Consider the unsaturated regime in which $P/P_s \ll 1$ throughout the amplifier. By neglecting the term P/P_s in Eq. (8.1.1), the gain coefficient becomes

$$g(\omega) = \frac{g_0}{1 + (\omega - \omega_0)^2 T_2^2}.$$

(8.1.2)

This equation shows that the gain is maximum when the incident frequency ω coincides with the atomic transition frequency ω_0. The gain reduction for $\omega \neq \omega_0$ is governed by a Lorentzian profile that is a characteristic of homogeneously broadened two-level systems [1]. As discussed later, the gain spectrum of actual amplifiers can deviate considerably from the Lorentzian profile. The gain bandwidth is defined as the full width at half maximum (FWHM) of the gain spectrum $g(\omega)$. For the Lorentzian spectrum, the gain bandwidth is given by $\Delta\omega_g = 2/T_2$, or by

$$\Delta\nu_g = \frac{\Delta\omega_g}{2\pi} = \frac{1}{\pi T_2}.$$

(8.1.3)

As an example, $\Delta\nu_g \sim 3$ THz for semiconductor laser amplifiers for which $T_2 \sim 0.1$ ps. Amplifiers with a relatively large bandwidth are preferred for optical communication systems, since the gain is then nearly constant over the entire bandwidth of even a multichannel signal.

The concept of *amplifier bandwidth* is commonly used in place of the gain bandwidth. The difference becomes clear when one considers the amplifier gain G, known as the *amplification factor* and defined as

$$G = P_{\text{out}}/P_{\text{in}},$$

(8.1.4)

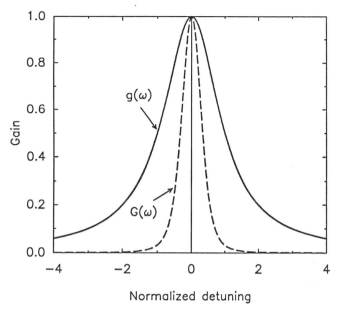

Figure 8.1 Lorentzian gain profile $g(\omega)$ and the corresponding amplifier-gain spectrum $G(\omega)$ for a two-level gain medium.

where P_{in} and P_{out} are the input and output powers of the continuous-wave (CW) signal being amplified. We can obtain an expression for G by using

$$\frac{dP}{dz} = gP, \qquad (8.1.5)$$

where $P(z)$ is the optical power at a distance z from the input end. A straightforward integration with the initial condition $P(0) = P_{\text{in}}$ shows that the signal power grows exponentially as

$$P(z) = P_{\text{in}} \exp(gz). \qquad (8.1.6)$$

By noting that $P(L) = P_{\text{out}}$ and using Eq. (8.1.4), the amplification factor for an amplifier of length L is given by

$$G(\omega) = \exp[g(\omega)L], \qquad (8.1.7)$$

where the frequency dependence of both G and g is shown explicitly. Both the amplifier gain $G(\omega)$ and the gain coefficient $g(\omega)$ are maximum when $\omega = \omega_0$ and decrease with the signal detuning $\omega - \omega_0$. However, $G(\omega)$ decreases much faster than $g(\omega)$. The amplifier bandwidth $\Delta\nu_A$ is defined as the FWHM of $G(\omega)$ and is related to the gain bandwidth $\Delta\nu_g$ as

$$\Delta\nu_A = \Delta\nu_g \left(\frac{\ln(2)}{\ln(G_0/2)} \right)^{1/2}, \qquad (8.1.8)$$

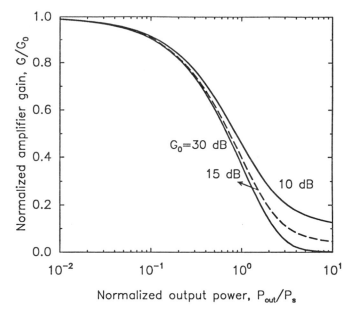

Figure 8.2 Saturated amplifier gain G as a function of the output power (normalized to the saturation power) for several values of the unsaturated amplifier gain G_0.

where $G_0 = \exp(g_0 L)$. Figure 8.1 shows the gain profile $g(\omega)$ and the amplification factor $G(\omega)$ by plotting g/g_0 and G/G_0 as a function of $(\omega - \omega_0)T_2$. The amplifier bandwidth is smaller than the gain bandwidth, and the difference depends on the amplifier gain itself.

8.1.2   Gain Saturation

The origin of gain saturation lies in the power dependence of the $g(\omega)$ in Eq. (8.1.1). Since g is reduced when P becomes comparable to P_s, the amplification factor G decreases with an increase in the signal power. This phenomenon is called gain saturation. Consider the case in which incident signal frequency is exactly tuned to the gain peak ($\omega = \omega_0$). The detuning effects can be incorporated in a straightforward manner. By substituting g from Eq. (8.1.1) in Eq. (8.1.5), we obtain

$$\frac{dP}{dz} = \frac{g_0 P}{1 + P/P_s}. \tag{8.1.9}$$

This equation can easily be integrated over the amplifier length. By using the initial condition $P(0) = P_{\text{in}}$ together with $P(L) = P_{\text{out}} = G P_{\text{in}}$, we obtain the following implicit relation for the large-signal amplifier gain:

$$G = G_0 \exp\left(-\frac{G-1}{G}\frac{P_{\text{out}}}{P_s}\right). \tag{8.1.10}$$

Equation (8.1.10) shows that the amplification factor G decreases from its unsaturated value G_0 when P_{out} becomes comparable to P_s. Figure 8.2 shows the saturation characteristics by plotting G as a function of P_{out}/P_s for several values of G_0. A quantity of practical interest is the output saturation power P_{out}^s, defined as the output power for which the amplifier gain G is reduced by a factor of 2 (or by 3 dB) from its unsaturated value G_0. By using $G = G_0/2$ in Eq. (8.1.10),

$$P_{out}^s = \frac{G_0 \ln(2)}{G_0 - 2} P_s. \tag{8.1.11}$$

P_{out}^s is smaller than P_s by about 30%. Indeed, by noting that $G_0 \gg 2$ in practice ($G_0 = 1000$ for 30-dB amplifier gain), $P_{out}^s \approx (\ln 2)P_s \approx 0.69\, P_s$. As seen in Fig. 8.2, P_{out}^s becomes nearly independent of G_0 for $G_0 > 20$ dB.

8.1.3   Amplifier Noise

All amplifiers degrade the signal-to-noise ratio (SNR) of the amplified signal because of spontaneous emission that adds noise to the signal during its amplification. The SNR degradation is quantified through a parameter F_n, called the *amplifier noise figure* in analogy with the electronic amplifiers (see Section 4.4.1) and defined as [2]

$$F_n = \frac{(SNR)_{in}}{(SNR)_{out}}, \tag{8.1.12}$$

where SNR refers to the electric power generated when the optical signal is converted into an electric current. In general, F_n depends on several detector parameters that govern thermal noise associated with the detector (see Section 4.4.1). A simple expression for F_n can be obtained by considering an ideal detector whose performance is limited by shot noise only [2].

Consider an amplifier with the gain G such that the output and input powers are related by $P_{out} = GP_{in}$. The SNR of the input signal is given by

$$(SNR)_{in} = \frac{\langle I \rangle^2}{\sigma_s^2} = \frac{(RP_{in})^2}{2q(RP_{in})\Delta f} = \frac{P_{in}}{2h\nu\Delta f}, \tag{8.1.13}$$

where $\langle I \rangle = RP_{in}$ is the average photocurrent, $R = q/h\nu$ is the responsivity of an ideal photodetector with unit quantum efficiency (see Section 4.1), and

$$\sigma_s^2 = 2q(RP_{in})\Delta f \tag{8.1.14}$$

is obtained from Eq. (4.4.5) for the shot noise by setting the dark current $I_d = 0$. Here Δf is the detector bandwidth. To evaluate the SNR of the amplified signal, one should add the contribution of spontaneous emission to the receiver noise.

The spectral density of *spontaneous-emission-induced noise* is nearly constant (white noise) and can be written as [2]

$$S_{sp}(\nu) = (G - 1)n_{sp}h\nu, \tag{8.1.15}$$

where ν is the optical frequency. The parameter n_{sp} is called the *spontaneous-emission factor* or *population-inversion factor* and is given by

$$n_{sp} = N_2/(N_2 - N_1), \tag{8.1.16}$$

where N_1 and N_2 are the atomic populations for the ground and excited states, respectively. The effect of spontaneous emission is to add fluctuations to the amplified power which are converted to current fluctuations during the photodetection process.

It turns out that the dominant contribution to the receiver noise comes from beating of spontaneous emission with the signal [2]. This beating phenomenon is similar to heterodyne detection (see Section 6.1) in the sense that the spontaneously emitted radiation mixes with the amplified signal at the photodetector and produces a heterodyne component of the photocurrent. The beating of spontaneous emission with the signal produces a noise current $\Delta I = 2R(GP_{\text{in}}P_{\text{sp}})^{1/2}\cos\theta$, where $P_{\text{sp}} = 2S_{\text{sp}}\Delta f$ is the spontaneous-emission power within the receiver bandwidth and θ is a random phase difference. If all other noise sources are neglected, the variance of the photocurrent can be written as [2]

$$\sigma^2 \approx 4(RGP_{\text{in}})(RS_{\text{sp}})\Delta f, \qquad (8.1.17)$$

where $\cos^2\theta$ was replaced by its average value $1/2$. The SNR of the amplified signal is thus given by

$$(\text{SNR})_{\text{out}} = \frac{\langle I \rangle^2}{\sigma^2} = \frac{(RGP_{\text{in}})^2}{\sigma^2} \approx \frac{GP_{\text{in}}}{4S_{\text{sp}}\Delta f}. \qquad (8.1.18)$$

The amplifier noise figure can now be obtained by substituting Eqs. (8.1.13) and (8.1.18) in Eq. (8.1.12). If we also use Eq. (8.1.15) for S_{sp},

$$F_n = 2n_{\text{sp}}(G - 1)/G \approx 2n_{\text{sp}}. \qquad (8.1.19)$$

This equation shows that the SNR of the amplified signal is degraded by 3 dB even for an ideal amplifier for which $n_{\text{sp}} = 1$. For most practical amplifiers, F_n exceeds 3 dB and can be as large as 6–8 dB. For its application in optical communication systems, an optical amplifier should have F_n as low as possible.

8.1.4 Amplifier Applications

Optical amplifiers can serve several purposes in the design of fiber-optic communication systems: four applications are shown schematically in Fig. 8.3. An important application for long-haul systems consists of using amplifiers as in-line amplifiers which replace electronic regenerators. Such a replacement can be carried out as long as the system performance is not limited by the cumulative effects of fiber dispersion, fiber nonlinearity, and amplifier noise. The use of optical amplifiers is particularly attractive for multichannel lightwave systems since they can amplify all channels simultaneously.

Another way to use optical amplifiers is to increase the transmitter power by placing an amplifier just after the transmitter. Such amplifiers are called power amplifiers or *power boosters*, as their main purpose is to boost the power transmitted. A power amplifier can increase the transmission distance by 100 km or more depending on the amplifier gain and the fiber loss. Transmission distance

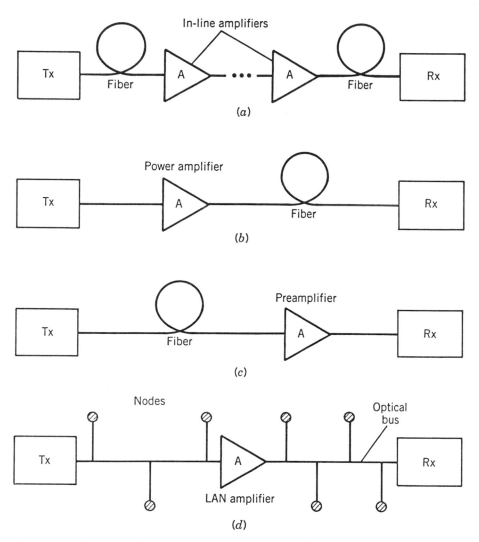

Figure 8.3 Four possible applications of optical amplifiers in lightwave systems: (a) as in-line amplifiers; (b) as a booster of transmitter power; (c) as a preamplifier to the receiver; (d) for compensation of distribution losses in local-area networks.

can also be increased by putting an amplifier just before the receiver to boost the received power. Such amplifiers are called *optical preamplifiers* and are commonly used to improve the receiver sensitivity. Another application of optical amplifiers is to use them for compensating distribution losses in local-area networks. As discussed in Sections 5.1 and 7.1, distribution losses often limit the number of nodes in a network. Many other applications of optical amplifiers were discussed in Section 7.2.

8.2   SEMICONDUCTOR LASER AMPLIFIERS

All lasers act as amplifiers close to but before reaching threshold, and semiconductor lasers are no exception. Indeed, research on semiconductor laser amplifiers (SLAs) started soon after the invention of semiconductor lasers in 1962. However, it was only during the 1980s that SLAs were developed for practical applications, largely motivated by their potential applications in lightwave systems [3]–[9]. In this section we discuss the amplification characteristics of SLAs from the standpoint of their applications in fiber-optic communication systems.

8.2.1   Amplifier Design

The amplifier characteristics discussed in Section 8.1 were for an optical amplifier without feedback. Such amplifiers are called *traveling-wave* (TW) amplifiers to emphasize that the amplified signal travels in the forward direction only. Semiconductor lasers experience a relatively large feedback because of reflections occurring at the cleaved facets (32% reflectivity). They can be used as amplifiers when biased below threshold, but multiple reflections at the facets must be included by considering a Fabry–Perot (FP) cavity. Such amplifiers are called *FP amplifiers*. The amplification factor is obtained by using the standard theory of FP interferometers and is given by [5]

$$G_{\mathrm{FP}}(\nu) = \frac{(1 - R_1)(1 - R_2)G(\nu)}{(1 - G\sqrt{R_1 R_2})^2 + 4G\sqrt{R_1 R_2} \sin^2[\pi(\nu - \nu_m)/\Delta\nu_L]}, \qquad (8.2.1)$$

where R_1 and R_2 are the facet reflectivities, ν_m represents the cavity-resonance frequencies [see Eq. (3.3.5)], and $\Delta\nu_L$ is the longitudinal-mode spacing, also known as the free spectral range of the FP cavity. The single-pass amplification factor G corresponds to that of a TW amplifier and is given by Eq. (8.1.7) when gain saturation is negligible. Indeed, G_{FP} reduces to G when $R_1 = R_2 = 0$.

As evident from Eq. (8.2.1), $G_{\mathrm{FP}}(\nu)$ peaks whenever ν coincides with one of the cavity-resonance frequencies and drops sharply in between them. The amplifier bandwidth is thus determined by the sharpness of the cavity resonance. One can calculate the amplifier bandwidth from the detuning $\nu - \nu_m$ for which G_{FP} drops by 3 dB from its peak value. The result is given by

$$\Delta\nu_A = \frac{2\Delta\nu_L}{\pi} \sin^{-1}\left(\frac{1 - G\sqrt{R_1 R_2}}{(4G\sqrt{R_1 R_2})^{1/2}}\right). \qquad (8.2.2)$$

To achieve a large amplification factor, $G\sqrt{R_1 R_2}$ should be quite close to 1. As seen from Eq. (8.2.2), the amplifier bandwidth is then a small fraction of the free spectral range of the FP cavity (typically, $\Delta\nu_L \sim 100$ GHz and $\Delta\nu_A < 10$ GHz). Such a small bandwidth makes FP amplifiers unsuitable for most lightwave system applications.

TW-type SLAs can be made if the reflection feedback from the end facets is suppressed. A simple way to reduce the reflectivity is to coat the facets with an *anti-reflection coating*. However, it turns out that the reflectivity must be

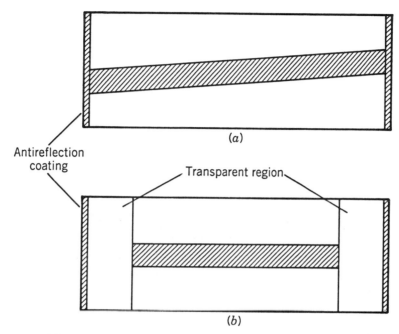

Figure 8.4 (a) Tilted-stripe and (b) and buried-facet structures for nearly traveling-wave (TW) semiconductor laser amplifiers.

extremely small ($< 0.1\%$) for the SLA to act as a TW amplifier. Furthermore, the minimum reflectivity depends on the amplifier gain itself. One can estimate the tolerable value of the facet reflectivity by considering the maximum and minimum values of G_{FP} from Eq. (8.2.1) near a cavity resonance. It is easy to verify that their ratio is given by

$$\Delta G = \frac{G_{\mathrm{FP}}^{\mathrm{max}}}{G_{\mathrm{FP}}^{\mathrm{min}}} = \left(\frac{1 + G\sqrt{R_1 R_2}}{1 - G\sqrt{R_1 R_2}}\right)^2. \tag{8.2.3}$$

If ΔG exceeds 3 dB, the amplifier bandwidth is set by the cavity resonances rather than by the gain spectrum. To keep $\Delta G < 2$, the facet reflectivities should satisfy the condition

$$G\sqrt{R_1 R_2} < 0.17. \tag{8.2.4}$$

It is customary to characterize the SLA as a TW amplifier when Eq. (8.2.4) is satisfied. A SLA designed to provide 30-dB amplification factor ($G = 1000$) should have facet reflectivities such that $\sqrt{R_1 R_2} < 1.7 \times 10^{-4}$.

Considerable effort is required to produce antireflection coatings with reflectivities less than 0.1%. Even then, it is difficult to obtain low facet reflectivities in a predictable and regular manner. For this reason, alternative techniques have been developed to reduce the reflection feedback in SLAs. In one method, the active-region stripe is tilted from the facet normal, as shown in Fig. 8.4(a).

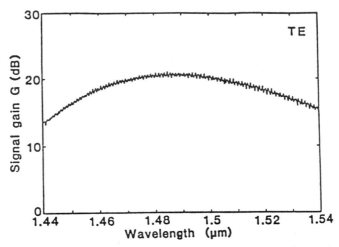

Figure 8.5 Amplifier gain versus signal wavelength for a semiconductor laser amplifier whose facets are coated to reduce reflectivity to about 0.04%. (After Ref. [4]. ©1987 IEEE. Reprinted with permission.)

Such a structure is referred to as the *angled-facet* or *tilted-stripe* structure [10]. The reflected beam at the facet is physically separated from the forward beam because of the angled facet. Some feedback can still occur, as the optical mode spreads beyond the active region in all semiconductor laser devices. In practice, the combination of an antireflection coating and the tilted stripe can produce reflectivities below 10^{-3} (as small as 10^{-4} with design optimization). In an alternative scheme [11] a transparent window region is inserted between the active-layer ends and the facets [see Fig. 8.4(b)]. The optical beam spreads in this window region before arriving at the semiconductor–air interface. The reflected beam spreads even further on the return trip and does not couple much light into the thin active layer. Such a structure is called *buried-facet* or *window-facet* structure and has provided reflectivities as small as 10^{-4} when used in combination with antireflection coatings.

8.2.2 Amplifier Characteristics

The amplification factor of SLAs is given by Eq. (8.2.1). Its frequency dependence results mainly from the frequency dependence of $G(\nu)$ when condition (8.2.4) is satisfied. The measured amplifier gain exhibits ripples reflecting the effects of residual facet reflectivities. Figure 8.5 shows the wavelength dependence of the amplifier gain measured for a SLA with the facet reflectivities of about 4×10^{-4}. Condition (8.2.4) is well satisfied as $G\sqrt{R_1 R_2} \approx 0.04$ for this amplifier. Gain ripples are negligibly small because the SLA operated in a nearly TW mode. The 3-dB amplifier bandwidth is about 70 nm because of a relatively broad gain spectrum of SLAs (see Section 3.3.1).

To discuss gain saturation, consider the peak gain and assume that it increases linearly with the carrier population N as (see Section 3.3.1)

$$g(N) = (\Gamma\sigma_g/V)(N - N_0), \tag{8.2.5}$$

where Γ is the confinement factor, σ_g is the differential gain, V is the active volume, and N_0 is the value of N required at transparency. The gain has been reduced by Γ to account for spreading of the waveguide mode outside the gain region of SLAs. The carrier population N changes with the injection current I and the signal power P as indicated in Eq. (3.3.28). By expressing the photon number in terms of the optical power, this equation can be written as

$$\frac{dN}{dt} = \frac{I}{q} - \frac{N}{\tau_c} - \frac{\sigma_g(N - N_0)}{\sigma_m h\nu}P, \tag{8.2.6}$$

where τ_c is the carrier lifetime and σ_m is the cross-section area of the waveguide mode. In the case of a CW beam, or pulses much longer than τ_c, the steady-state value of N can be obtained by setting $dN/dt = 0$ in Eq. (8.2.6). When the solution is substituted in Eq. (8.2.5), the optical gain is found to saturate as

$$g = \frac{g_0}{1 + P/P_s}, \tag{8.2.7}$$

where the small-signal gain g_0 is given by

$$g_0 = (\Gamma\sigma_g/V)(I\tau_c/q - N_0), \tag{8.2.8}$$

and the saturation power P_s is defined as

$$P_s = h\nu\sigma_m/(\sigma_g\tau_c). \tag{8.2.9}$$

A comparison of Eqs. (8.1.1) and (8.2.7) shows that the SLA gain saturates in the same way as that for a two-level system. Thus, the output saturation power P_{out}^s is obtained from Eq. (8.1.11) with P_s given by Eq. (8.2.9). Typical values of P_{out}^s are in the range 5–10 mW.

The noise figure F_n of SLAs is larger than the minimum value of 3 dB for several reasons. The dominant contribution comes from the spontaneous-emission factor n_{sp}. For SLAs, n_{sp} is obtained from Eq. (8.1.16) by replacing N_2 and N_1 by N and N_0, respectively. An additional contribution results from internal losses (such as free-carrier absorption or scattering loss) which reduce the available gain from g to $g - \alpha_{\text{int}}$. By using Eq. (8.1.19) and including this additional contribution, the noise figure can be written as [7]

$$F_n = 2\left(\frac{N}{N - N_0}\right)\left(\frac{g}{g - \alpha_{\text{int}}}\right). \tag{8.2.10}$$

Residual facet reflectivities increase F_n by an additional factor that can be approximated by $1 + R_1 G$, where R_1 is the reflectivity of the input facet [3]. In most TW amplifiers, $R_1 G \ll 1$, and this contribution can be neglected. Typical values of F_n for SLAs are in the range 5–7 dB.

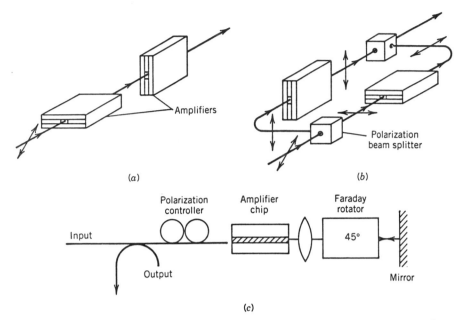

Figure 8.6 Three configurations used to reduce the polarization sensitivity of semi-conductor laser amplifiers: (a) twin amplifiers in series; (b) twin amplifiers in parallel; and (c) double pass through a single amplifier.

An undesirable characteristic of SLAs is their *polarization sensitivity*. The amplifier gain G differs for the TE and TM modes (see Section 3.3.4) by as much as 5–8 dB simply because both G and σ_g are different for the two or-thogonally polarized modes. This feature makes the amplifier gain sensitive to the polarization state of the input beam, a property undesirable for lightwave systems in which the state of polarization changes with propagation along the fiber (unless polarization-preserving fibers are used). Several schemes have been devised to reduce the polarization sensitivity [11]–[16]. In one scheme, the am-plifier is designed such that the width and the thickness of the active region are comparable. A gain difference of less than 1.3 dB between TE and TM polar-izations has been realized by making the active layer 0.26 μm thick and 0.4 μm wide [11]. Another scheme makes use of a large-optical-cavity structure; a gain difference of less than 1 dB has been obtained with such a structure [12].

Several other schemes reduce the polarization sensitivity by using two am-plifiers or two passes through the same amplifier. Figure 8.6 shows three such configurations. In Fig. 8.6(a), the TE-polarized signal in one amplifier becomes TM polarizedl in the second amplifier, and vice versa. If both amplifiers have identical gain characteristics, the twin-amplifier configuration provides signal gain that is independent of the signal polarization. A drawback of the *series configuration* is that residual facet reflectivities lead to mutual coupling between the two amplifiers. In the *parallel configuration* shown in Fig. 8.6(b) the inci-dent signal is split into a TE- and a TM-polarized signals, each of which is amplified by separate amplifiers. The amplified TE and TM signals are then combined to produce the amplified signal with the same polarization as that of the input beam [13]. The *double-pass configuration* of Fig. 8.6(c) passes the signal through the same amplifier twice, but the polarization is rotated by 90° between the two passes [14]. Since the amplified signal propagates in the backward direction, a 3-dB fiber coupler is needed to separate it from the incident signal. Despite a 6-dB loss occurring at the fiber coupler (3 dB for the input signal and 3 dB for the amplified signal) this configuration provides high gain from a single amplifier, as the same amplifier supplies gain on the two passes.

8.2.3 Multichannel Amplification

As mentioned in Section 8.1.4, one of the advantages of using optical amplifiers is that they can be used to amplify several channels simultaneously as long as the carrier frequencies of multiple channels lie within the amplifier bandwidth. Ideally, the signal in each channel should be amplified by the same amount. In practice, several nonlinear phenomena in SLAs induce *interchannel crosstalk*, an undesirable feature that should be minimized for practical lightwave systems. Two such nonlinear phenomena are *cross saturation* and *four-wave mixing* (FWM). They both originate from the stimulated recombination term in the carrier rate equation. In the case of multichannel amplification, the power P in Eq. (8.2.6) corresponds to

$$P = \frac{1}{2} \left| \sum_{j=1}^{M} A_j \exp(-i\omega_j t) + \text{c.c.} \right|^2, \tag{8.2.11}$$

where c.c. stands for complex conjugate, M is the number of channels, A_j is the amplitude, and ω_j is the carrier frequency for the jth channel. Because of the coherent addition of individual channel fields, Eq. (8.2.11) contains time-dependent terms resulting from beating of the signal in different channels, i.e.,

$$P = \sum_{j=1}^{M} P_j + \sum_{j=1}^{M} \sum_{k \neq j}^{M} 2\sqrt{P_j P_k} \cos(\Omega_{jk} t + \phi_j - \phi_k), \tag{8.2.12}$$

where $A_j = \sqrt{P_j} \exp(i\phi_j)$ was assumed together with $\Omega_{jk} = \omega_j - \omega_k$. When Eq. (8.2.12) is substituted in Eq. (8.2.6), the carrier population is also found to oscillate at the beat frequency Ω_{jk}. Since the gain and the refractive index both depend on N, they are also modulated at the frequency Ω_{jk}; such a modulation creates gain and index gratings, which induce interchannel crosstalk by scattering a part of signal from one channel to another. This phenomenon can also be viewed as four-wave mixing [17].

The origin of cross saturation is also evident from Eq. (8.2.12). The first term on the right side shows that the power P in Eq. (8.2.7) should be replaced by the total power in all channels. Thus, the gain of a specific channel is saturated not only by its own power but also by the power of neighboring channels, a phenomenon known as cross saturation. As discussed in Section 7.2, cross saturation can be used for wavelength conversion in WDM systems. However, it is undesirable for signal amplification since the amplifier gain changes with time depending on the bit pattern of neighboring channels. As a result, the amplified signal appears to fluctuate more or less randomly. Such fluctuations degrade the effective SNR at the receiver. Such an interchannel crosstalk occurs regardless of the extent of channel spacing. It can be avoided only by operating SLAs in the unsaturated regime. It is also absent in coherent systems making use of the FSK or the PSK format, since the power in each channel, and hence the total power, remains constant with time.

Interchannel crosstalk induced by FWM, on the other hand, can occur for all multichannel lightwave systems irrespective of the modulation format used [18]– [21]. Its impact is most severe in coherent systems [20] because of their relatively small channel spacing. FWM can occur even for widely spaced channels through intraband nonlinearities [22] occurring at fast time scales (< 1 ps). In fact, such highly nondegenrate FWM is used for wavelength conversion as discussed in Section 7.2. In general, cross saturation and FWM make it difficult to use SLAs as in-line amplifiers.

8.2.4 Pulse Amplification

The large bandwidth of TW-type SLAs suggests that they are capable of amplifying ultrashort optical pulses (as short as a few picoseconds) without significant pulse distortion. However, when the pulse width τ_p becomes shorter than the carrier lifetime τ_c, gain dynamics play an important role, since both N and g in Eq. (8.2.5) become time dependent.

One can adapt the formulation developed in Section 2.4 for pulse propagation in optical fibers to the case of SLAs by making a few changes. The dispersive effects are not important for SLAs because of negligible material dispersion and a short amplifier length (< 1 mm in most cases). The amplifier gain can be included by adding the term $gA/2$ on the right side of Eq. (2.4.7). By setting $\beta_2 = \beta_3 = 0$, the amplitude $A(z, t)$ of the pulse envelope then evolves as [23]

$$\frac{\partial A}{\partial z} + \frac{1}{v_g}\frac{\partial A}{\partial t} = \frac{1}{2}(1 - i\beta_c)gA, \qquad (8.2.13)$$

where carrier-induced index changes are included through the linewidth enhancement factor β_c (see Section 3.3.7). The time dependence of g is governed by Eqs. (8.2.5) and (8.2.6). The two equations can be combined to yield

$$\frac{\partial g}{\partial t} = \frac{g_0 - g}{\tau_c} - \frac{gP}{E_{\text{sat}}}, \qquad (8.2.14)$$

where the saturation energy E_{sat} is defined as

$$E_{\text{sat}} = h\nu(\sigma_m/\sigma_g), \qquad (8.2.15)$$

and g_0 is given by Eq. (8.2.8).

Equations (8.2.13) and (8.2.14) govern amplification of optical pulses in SLAs. They can be solved analytically for pulses whose duration is short compared with the carrier lifetime ($\tau_p \ll \tau_c$). The first term on the right side of Eq. (8.2.14) can then be neglected during pulse amplification. By introducing the reduced time $\tau = t - z/v_g$ together with $A = \sqrt{P}\exp(i\phi)$, Eqs. (8.2.13) and (8.2.14) can be written as [23]

$$\frac{\partial P}{\partial z} = g(z,\tau)P(z,\tau), \tag{8.2.16}$$

$$\frac{\partial \phi}{\partial z} = -\tfrac{1}{2}\beta_c g(z,\tau), \tag{8.2.17}$$

$$\frac{\partial g}{\partial \tau} = -g(z,\tau)P(z,\tau)/E_{\text{sat}}. \tag{8.2.18}$$

Equation (8.2.16) can easily be integrated over the amplifier length L to yield

$$P_{\text{out}}(\tau) = P_{\text{in}}(\tau)\exp[h(\tau)], \tag{8.2.19}$$

where $P_{\text{in}}(\tau)$ is the input power and $h(\tau)$ is the total integrated gain defined as

$$h(\tau) = \int_0^L g(z,\tau)\,dz. \tag{8.2.20}$$

If Eq. (8.2.18) is integrated over the amplifier length after replacing gP by $\partial P/\partial z$, $h(\tau)$ satisfies [23]

$$\frac{dh}{d\tau} = -\frac{1}{E_{\text{sat}}}[P_{\text{out}}(\tau) - P_{\text{in}}(\tau)] = -\frac{P_{\text{in}}(\tau)}{E_{\text{sat}}}(e^h - 1). \tag{8.2.21}$$

Equation (8.2.21) can easily be solved to obtain $h(\tau)$. The amplification factor $G(\tau)$ is related to $h(\tau)$ as $G = \exp(h)$ and is given by [1]

$$G(\tau) = \frac{G_0}{G_0 - (G_0 - 1)\exp[-E_0(\tau)/E_{\text{sat}}]}, \tag{8.2.22}$$

where G_0 is the unsaturated amplifier gain and

$$E_0(\tau) = \int_{-\infty}^{\tau} P_{\text{in}}(\tau)\,d\tau \tag{8.2.23}$$

is the partial energy of the input pulse defined such that $E_0(\infty)$ equals the input pulse energy E_{in}.

The solution (8.2.22) shows that the amplifier gain is different for different parts of the pulse. The leading edge experiences the full gain G_0 as the amplifier is not yet saturated. The trailing edge experiences the least gain since the whole pulse has saturated the amplifier gain. The final value G_f of $G(\tau)$ after passage of the pulse is obtained from Eq. (8.2.22) by replacing $E_0(\tau)$ by E_{in}. The intermediate values of the gain depend on the pulse shape. Figure 8.7 shows the shape dependence of $G(\tau)$ for super-Gaussian input pulses by using

$$P_{\text{in}}(t) = P_0 \exp[-(\tau/\tau_p)^{2m}], \tag{8.2.24}$$

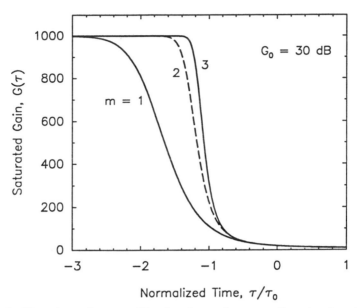

Figure 8.7 Time-dependent amplification factor for super-Gaussian input pulses of input energy such that $E_{in}/E_{sat} = 0.1$. The unsaturated value G_0 is 30 dB in all cases. The input pulse is Gaussian for $m = 1$ but becomes nearly rectangular as m increases.

where m is the shape parameter. The input pulse is Gaussian for $m = 1$ but becomes nearly rectangular as m increases. For comparison purposes, the input energy is held constant for different pulse shapes by choosing $E_{in}/E_{sat} = 0.1$. The shape dependence of the amplification factor $G(\tau)$ implies that the output pulse is distorted, and distortion is itself shape dependent.

As seen from Eq. (8.2.17), gain saturation leads to a time-dependent phase shift across the pulse. This phase shift is found by integrating Eq. (8.2.17) over the amplifier length and is given by

$$\phi(\tau) = -\tfrac{1}{2}\beta_c \int_0^L g(z,\tau)\, dz = -\tfrac{1}{2}\beta_c h(\tau) = -\tfrac{1}{2}\beta_c \ln[G(\tau)]. \qquad (8.2.25)$$

Since the pulse modulates its own phase through gain saturation, this phenomenon is referred to as *saturation-induced* self-phase modulation [23]. The frequency chirp is related to the phase derivative as (see Section 2.4.2)

$$\Delta\nu_c = -\frac{1}{2\pi}\frac{d\phi}{d\tau} = \frac{\beta_c}{4\pi}\frac{dh}{d\tau} = -\frac{\beta_c P_{in}(\tau)}{4\pi E_{sat}}[G(\tau) - 1], \qquad (8.2.26)$$

where Eq. (8.2.21) was used. Figure 8.8 shows the chirp profiles for several input pulse energies when a Gaussian pulse is amplified in a SLA with 30-dB unsaturated gain. The frequency chirp is larger for more energetic pulses simply because gain saturation sets in earlier for such pulses.

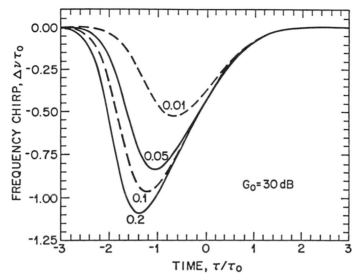

Figure 8.8 Frequency chirp imposed across the amplified pulse for several values of E_{in}/E_{sat}. A Gaussian input pulse is assumed together with $G_0 = 30$ dB and $\beta_c = 5$. (After Ref. [24]. ©1989 IEEE. Reprinted with permission.)

Self-phase modulation and the associated frequency chirp can affect lightwave systems considerably. The spectrum of amplified pulse becomes considerably broad and contains several peaks of different amplitudes [23]. The dominant peak is shifted toward the red side and is broader than the input spectrum. It is also accompanied by one or more satellite peaks. Figure 8.9 shows the expected shape and spectrum of amplified pulses when a Gaussian pulse of energy such that $E_{in}/E_{sat} = 0.1$ is amplified by a SLA. The temporal and spectral changes depend on amplifier gain and are quite significant for $G_0 = 30$ dB. The experiments performed by using picosecond pulses from mode-locked semiconductor lasers confirm this behavior [23]. In particular, the spectrum of amplified pulses is found to be shifted toward the red side by 50–100 GHz, depending on the amplifier gain. Spectral distortion in combination with the frequency chirp would affect the transmission characteristics when amplified pulses are propagated through optical amplifiers.

It turns out that the frequency chirp imposed by the SLA is opposite in nature compared with that imposed by directly modulated semiconductor lasers. If we also note that the chirp is nearly linear over a considerable portion of the amplified pulse (see Fig. 8.8), it is easy to understand that the amplified pulse would pass through an initial compression stage when it propagates in the anomalous-dispersion region of optical fibers (see Section 2.4.2). Such a compression was observed in an experiment [24] in which 40-ps optical pulses were first amplified in a 1.52-μm SLA and then propagated through 18 km of single-mode fiber with $\beta_2 = -18$ ps^2/km. This compression mechanism can be used to design fiber-optic communication systems in which in-line SLAs

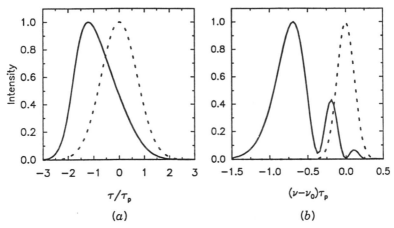

Figure 8.9 (a) Shape and (b) spectrum at the output of a semiconductor laser amplifier with $G_0 = 30$ dB and $\beta_c = 5$ for a Gaussian input pulse of energy $E_{in}/E_{sat} = 0.1$. The dashed curves show for comparison input shape and spectrum.

are used to compensate simultaneously for both fiber loss and dispersion by operating SLAs in the saturation region so that they impose frequency chirp on the amplified pulse. The basic concept was demonstrated in 1989 in an experiment [25] in which a 16-Gb/s signal was transmitted over 70 km by using an SLA. In the absence of the SLA or when the SLA was operated in the unsaturated regime, the system was dispersion limited to the extent that signal could not be transmitted over more than 20 km. The use of SLAs for dispersion compensation is discussed further in Chapter 9.

8.2.5 System Applications

The use of SLAs as a preamplifier to the receiver is attractive since it permits monolithic integration of the SLA with the receiver. As seen in Fig. 8.3(c), in this application the signal is optically amplified before it falls on the receiver. The preamplifier boosts the signal to such a high level that the receiver performance is limited by shot noise rather than by thermal noise. The basic idea is similar to the case of avalanche photodiodes (APDs), which amplify the signal in the electrical domain. However, just as APDs add additional noise (see Section 4.4.3), preamplifiers also degrade the SNR through spontaneous-emission noise. A relatively large noise figure of SLAs ($F_n = 5$–6 dB) makes them less than ideal as a preamplifier. Nonetheless, they can improve the receiver sensitivity considerably. SLAs can also be used as power amplifiers to boost the transmitter power. It is, however, difficult to achieve powers in excess of 10 mW because of a relatively small value (~ 5 mW) of the output saturation power.

SLAs were used as in-line amplifiers in several system experiments before 1990. In one experiment [26], a signal at 1 Gb/s was transmitted over 313 km by using four cascaded SLAs. In another experiment making use of coherent detection [27], four cascaded SLAs were used to transmit 420-Mb/s FSK signal over 370 km. SLAs have also been employed to overcome distribution losses in the local-area network (LAN) applications. In one experiment [28], an SLA was used as a dual-function device. It amplified five FSK channels multiplexed by using the subcarrier multiplexing (SCM) technique (see Section 7.5), improving the power budget by 11 dB. At the same time, it was used to monitor the network performance through a baseband control channel. The 100-Mb/s baseband control signal modulated the carrier density of the amplifier, which in turn produced a corresponding electric signal that was used for monitoring.

SLAs suffer from several drawbacks which make their use as in-line amplifiers impractical. A few among them are polarization sensitivity, interchannel crosstalk, and a large coupling loss. Since fiber amplifiers do not suffer from these problems, they are used almost exclusively for signal amplification in 1.55-μm lightwave systems. The situation is somewhat different near the 1.3-μm wavelength for which the properties of fiber amplifiers are less than satisfactory. SLAs provide an alternative for 1.3-μm systems and are often used in practice. Moreover, as was seen in Section 7.2, SLAs have found many other applications. They can be used as wavelength converter and as a fast switch for wavelength routing in WDM networks, They can also be used as a nonlinear element for clock recovery and demultiplexing in TDM systems (see Section 7.4).

8.3   FIBER RAMAN AMPLIFIERS

A fiber Raman amplifier uses *stimulated Raman scattering* (SRS) occurring in silica fibers when an intense pump beam propagates through it [29]–[31]. The main features of SRS were discussed in Sections 2.6.1 and 7.3.2. SRS differs from stimulated emission in one fundamental aspect. Whereas in the case of stimulated emission an incident photon stimulates emission of another identical photon without losing its energy, in the case of SRS the incident pump photon gives up its energy to create another photon of reduced energy at a lower frequency (inelastic scattering); the remaining energy is absorbed by the medium in the form of molecular vibrations (optical phonons). Thus fiber Raman amplifiers must be pumped optically to provide gain, in contrast with SLAs (discussed in Section 8.2), which can be pumped electrically. An important difference from the case of SLA is that population inversion is not required for fiber Raman amplifiers. In fact, SRS is a nonresonant nonlinear phenomenon that does not require population transfer among energy levels. Figure 8.10 shows schematically a fiber Raman amplifier and the energy-level scheme. The pump and the signal beams at frequencies ω_p and ω_s are injected into the fiber through a WDM fiber coupler. The energy is transferred from the pump beam to the signal beam through SRS as the two beams copropagate along the fiber. The pump and signal beams can also be injected in such a way that they counterpropagate inside the fiber.

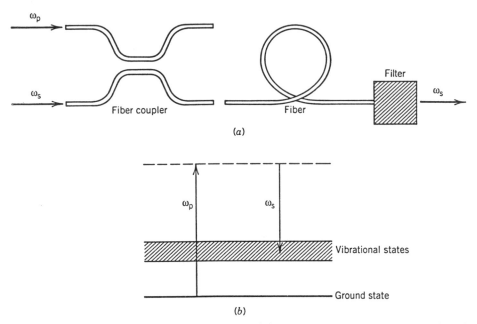

Figure 8.10 (a) Fiber Raman amplifier and (b) energy-level scheme associated with SRS.

8.3.1 Raman Gain and Bandwidth

The frequency difference $\Omega_R = \omega_p - \omega_s$, known as the *Stokes shift*, plays an important role in the SRS process. The vibrational energy levels of molecules dictate the value of Ω_R and the spread in Ω_R that can be tolerated for SRS to occur. Fortunately, vibrational energy levels of silica molecules merge together to form a band because of the amorphous nature of glass. As a result, the signal frequency ω_s can differ from ω_p over a wide range (up to 20 THz) and still experience considerable amplification through SRS. Figure 8.11 shows the Raman gain spectrum of silica fibers. The Raman gain coefficient g_R scales linearly with ω_p (or inversely with the pump wavelength λ_p). Since the optical gain g is proportional to the pump intensity I_p, it is obtained by using $g = g_R I_p$. In terms of the pump power P_p, the gain can be written as

$$g(\omega) = g_R(\omega)(P_p/a_p), \qquad (8.3.1)$$

where a_p is the cross-section area of the pump beam inside the fiber. As evident from Fig. 8.11, the gain spectrum of fiber Raman amplifiers is far from being Lorentzian. The gain peaks at a Stokes shift of about 13.2 THz. The gain bandwidth $\Delta\nu_g$ is about 6 THz if we define it as the FWHM of the dominant peak in Fig. 8.11. The large bandwidth of fiber Raman amplifiers makes them attractive for fiber-optic communication applications. However, a relatively large pump power is required to realize a large amplification factor. For example, if we use Eq. (8.1.7) by assuming operation in the unsaturated region, $gL \approx 6.7$ is required for $G = 30$ dB. By using $g_R = 6 \times 10^{-14}$ m/W at the gain peak

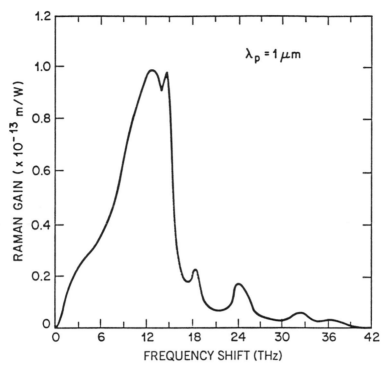

Figure 8.11 Measured Raman gain spectrum for fused silica. Raman gain scales inversely with the pump wavelength λ_p. (After Ref. [29]. ©1972 AIP. Reprinted with permission.)

at 1.55 μm and $a_p = 50$ μm^2, the required pump power is more than 5 W for 1-km-long fiber. The required power can be reduced for longer fibers, but then fiber loss must be included. In the following section we include both fiber loss and pump depletion.

8.3.2   Gain Saturation

The origin of gain saturation in fiber Raman amplifiers is quite different from SLAs. Since the pump supplies energy for signal amplification, it begins to deplete as the signal power P_s increases. A decrease in the pump power P_p reduces the optical gain as seen from Eq. (8.3.1). This reduction in gain is referred to as gain saturation. Variations of pump and signal powers along the amplifier length can be studied by solving two coupled equations [31]:

$$dP_s/dz = -\alpha_s P_s + (g_R/a_p)P_p P_s, \qquad (8.3.2)$$
$$dP_p/dz = -\alpha_p P_p - (\omega_p/\omega_s)(g_R/a_p)P_s P_p, \qquad (8.3.3)$$

where α_s and α_p represent the fiber loss at the signal and pump frequencies ω_s and ω_p, respectively.

Consider first the case of small-signal amplification for which pump depletion can be neglected [the last term in Eq. (8.3.3)]. By substituting $P_p(z) = P_p(0)\exp(-\alpha_p z)$ in Eq. (8.3.2), the signal intensity at the output of an amplifier of length L is given by

$$P_s(L) = P_s(0)\exp(g_R P_0 L_{\text{eff}}/a_p - \alpha_s L), \qquad (8.3.4)$$

where $P_0 = P_p(0)$ is the input pump power and L_{eff} is defined as

$$L_{\text{eff}} = [1 - \exp(-\alpha_p L)]/\alpha_p. \qquad (8.3.5)$$

Because of fiber loss at the pump wavelength, the effective length of the amplifier is less than the actual length L; $L_{\text{eff}} \approx 1/\alpha_p$ for $\alpha_p L \gg 1$. Since $P_s(L) = P_s(0)\exp(-\alpha_s L)$ in the absence of Raman amplification, the amplifier gain is given by

$$G_A = \frac{P_s(L)}{P_s(0)\exp(-\alpha_s L)} = \exp(g_0 L), \qquad (8.3.6)$$

where the small-signal gain g_0 is defined as

$$g_0 = g_R \left(\frac{P_0}{a_p}\right)\left(\frac{L_{\text{eff}}}{L}\right) \approx \frac{g_R P_0}{a_p \alpha_p L}. \qquad (8.3.7)$$

The last relation holds for $\alpha_p L \gg 1$. The amplification factor G_A becomes length independent for large values of $\alpha_p L$. Figure 8.12 shows the experimentally observed variation of G_A with P_0 for several values of the input signal powers for a 1.3-km fiber Raman amplifier operating at 1.064 μm by using a 1.017-μm pump. The amplification factor G_A increases exponentially with P_0 initially but then starts to deviate for $P_0 > 1$ W because of gain saturation. Deviations become larger with an increase in $P_s(0)$ as gain saturation sets in earlier along the amplifier length. The solid lines in Fig. 8.12 are obtained by solving Eqs. (8.3.2) and (8.3.3) numerically to include pump depletion.

An approximate expression for the saturated amplifier gain G_s can be obtained by assuming that $\alpha_s = \alpha_p$ in Eqs. (8.3.2) and (8.3.3). The result is [31]

$$G_s = \frac{1 + r_0}{r_0 + G_A^{-(1+r_0)}}, \qquad (8.3.8)$$

where r_0 is related to the signal-to-pump power ratio at the amplifier input as

$$r_0 = \frac{\omega_p}{\omega_s}\frac{P_s(0)}{P_p(0)}. \qquad (8.3.9)$$

Figure 8.13 shows the saturation characteristics by plotting G_s/G_A as a function of $G_A r_0$ for several values of G_A. The amplifier gain is reduced by 3 dB when $G_A r_0 \approx 1$. This condition is satisfied when the power of the amplified signal becomes comparable to the input pump power P_0. In fact, P_0 is a good measure of the saturation power. Since typically $P_0 \sim 1$ W, the saturation power of fiber Raman amplifiers is much larger than that of SLAs.

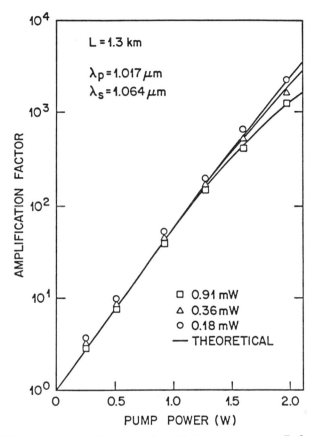

Figure 8.12 Variation of amplifier gain G_0 with the pump power P_0 for a 1.3-km-long fiber Raman amplifier. Different symbols show the experimental data for three values of the input signal power. Solid lines show the theoretical prediction. (After Ref. [32]. ©1981 Elsevier. Reprinted with permission.)

8.3.3 Amplifier Performance

As seen in Fig. 8.12, fiber Raman amplifiers can provide 30-dB gain at a pump power of about 1.5 W for a 1.3-km-long fiber. For the optimum performance, the frequency difference between the pump and signal beams should correspond to the peak of the Raman gain spectrum in Fig. 8.11 (occurring at about 13 THz). In the near-infrared region, the most practical pump source is a diode-pumped Nd:YAG laser operating at 1.06 μm. For such a pump laser, the maximum gain occurs for signal wavelengths near 1.12 μm, respectively. However, the signal wavelengths of most interest for fiber-optic communication systems are near 1.3 and 1.5 μm. A Nd:YAG laser can still be used if a higher-order Stokes line, generated through *cascade SRS*, is used as a pump. For instance, the third-order Stokes line at 1.24 μm generated by using SRS can act as a pump for amplifying the 1.3-μm signal. Amplifier gains of up to 20 dB were measured in

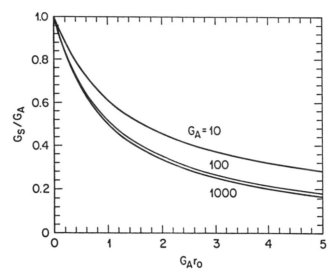

Figure 8.13 Gain–saturation characteristics of fiber Raman amplifiers for several values of the unsaturated amplifier gain G_A.

1984 by using this technique [33].

An early application of fiber Raman amplifiers was as a preamplifier for improving the receiver sensitivity [34]. The improvement was limited by amplifier noise associated with *spontaneous Raman scattering* that invariably accompanies the amplification process. Even though the noise figure of such amplifiers is close to its ideal value (3 dB), receiver noise is enhanced due to an increase in the shot noise simply because spontaneous Raman scattering occurs over a wide frequency range (> 5 THz). The performance can be improved by passing the amplified signal through an optical filter. The broad bandwidth of fiber Raman amplifiers is useful for amplifying several channels simultaneously. In one experiment [35], signals from three DFB semiconductor lasers operating in the range 1.57–1.58 μm were amplified simultaneously using a 1.47-μm pump. This experiment used a multimode semiconductor laser as a pump source. An amplifier gain of 5 dB was obtained at a pump power of only 60 mW.

The broad bandwidth of fiber Raman amplifiers makes them suitable for amplification of short optical pulses. In particular, they can be used to overcome the fiber loss in soliton-based communication systems. This aspect is discussed in Chapter 10. A novel feature of this approach is that the same fiber used for signal transmission is also used for signal amplification. This property is sometimes referred to as *distributed amplification*, as opposed to lumped amplification provided by other amplifiers.

The main drawback of fiber Raman amplifiers from the standpoint of lightwave system applications is that a high-power laser is required for pumping. The experiments near 1.55 μm often use tunable color-center lasers as a pump; such lasers are too bulky for communication applications. Indeed, with the advent of the erbium-doped fiber amplifier (covered in Section 8.5), Raman amplifiers are rarely used in the 1.55-μm wavelength region. The situation may change with the availability of compact high-power semiconductor lasers. In an interesting 1992 experiment [36], a fiber-Raman amplifier was pumped by a 1.55-μm semiconductor laser whose output was amplified using an erbium-doped fiber amplifier. The 140-ns pump pulses had 1.4-W peak power at the 1-kHz repetition rate and were capable of amplifying 1.66-μm signal pulses by more than 23 dB through SRS in a 20-km-long dispersion-shifted fiber. The 200-mW peak power of 1.66-μm pulses was large enough for their use for optical time-domain reflection measurements, a technique that is commonly used for supervising and maintaining fiber-optic networks [37].

The use of fiber Raman amplifiers in the 1.3-μm region has attracted considerable attention during the 1990s [38]–[42] simply because the properties of 1.3-μm doped-fiber amplifiers are often less than satisfactory. In one approach [38], three pairs of fiber gratings (see Section 9.6) are inserted within the fiber used for Raman amplification. The Bragg wavelengths of these gratings are chosen such that they form three cavities for three Raman lasers operating at wavelengths 1.117, 1.175, and 1.24 μm that correspond to first, second, and third-order Stokes line of a 1.06-μm pump. All three lasers are pumped by using a diode-pumped Nd-fiber laser through cascade SRS. The 1.24-μm laser then pumps the fiber Raman amplifier to provide signal amplification in the 1.3-μm region. The same idea of cascaded SRS was used to obtain 39-dB gain at 1.3 μm by using WDM couplers in place of fiber gratings [39]. In a different approach, the core of silica fiber is doped heavily with germania. Such a fiber can be pumped to provide the 30-dB gain at a pump power of only 350 mW [40]. Such pump powers can be obtained by using two or more semiconductor lasers.

Cascaded-SRS-based 1.3-μm Raman amplifiers have been developed with many applications in mind, including their use by the CATV industry. They can exhibit gains as high as 40 dB with a noise figure of about 4 dB, and are quite suitable as an optical preamplifier for high-speed optical receivers. In a 1996 experiment, such a receiver yielded the sensitivity of 151 photons/bit at a bit rate of 10-Gb/s [41]. The 1.3-μm Raman amplifiers can also be used to upgrade the transmission capacity of existing fiber links from 2.5 to 10 Gb/s [42].

8.4   FIBER BRILLOUIN AMPLIFIERS

The operating principle behind fiber Brillouin amplifiers [31] is essentially the same as for fiber Raman amplifiers except that the optical gain is provided by *stimulated Brillouin scattering* (SBS) instead of SRS [43]. Fiber Brillouin amplifiers are also pumped optically, and a part of the pump power is transferred to the signal through SBS. Physically, each pump photon of energy $\hbar\omega_p$ uses most of its energy to create a signal photon of energy $\hbar\omega_s$, while the remaining energy is used to excite an *acoustic phonon*. Classically, the pump beam scatters from an acoustic wave moving through the medium at the speed of sound. Despite a formal similarity between SBS and SRS, SBS differs from SRS in three important aspects which affect the operation of fiber Brillouin amplifiers considerably: (1) amplification occurs only when the signal beam propagates in a direction opposite to that of the pump beam (backward-pumping configuration); (2) the Stokes shift for SBS is smaller ($\sim$ 10 GHz) by three orders of magnitude compared with that of SRS and depends on the pump frequency; and (3) the Brillouin gain spectrum is extremely narrow, with a bandwidth < 100 MHz. The origin of these differences lies in a relatively small value of the ratio v_A/c ($\sim 10^{-5}$), where v_A is the acoustic velocity in silica and c is the velocity of light. The narrow bandwidth of fiber Brillouin amplifiers renders them unsuitable for amplifications of optical channels in lightwave systems. The same feature can, however, be exploited for some novel applications in coherent and multichannel communication systems. In this section we describe fiber Brillouin amplifiers with emphasis on lightwave system applications [44]–[46].

8.4.1   Brillouin Gain and Amplifier Bandwidth

Since SBS can be viewed as scattering of a pump wave from an acoustic wave, the Stokes shift $\Omega = \omega_p - \omega_s$ corresponds to the acoustic frequency and must satisfy the dispersion relation [31]

$$\Omega = |k_A|v_A = 2v_A|k_p|\sin(\theta/2), \qquad (8.4.1)$$

where the acoustic wave vector $\mathbf{k}_A = \mathbf{k}_p - \mathbf{k}_s$ in order to satisfy the phase-matching condition, $\mathbf{k}_p$ and $\mathbf{k}_s$ are the wave vectors of the pump and signal beams, respectively, and θ is the angle between them. In obtaining Eq. (8.4.1), $|k_p| \approx |k_s|$ was used, as is also the case in practice. Equation (8.4.1) shows that Ω vanishes in the forward direction ($\theta = 0$) and is maximum in the backward direction ($\theta = \pi$). In single-mode fibers, light travels only in the forward and backward directions. Thus, SBS occurs only in the backward direction with a frequency shift $\Omega_B = 2v_A|k_p|$. By using $k_p = 2\pi\bar{n}/\lambda_p$, where λ_p is the pump wavelength, the *Brillouin shift* is given by

$$\nu_B = \Omega_B/2\pi = 2\bar{n}v_A/\lambda_p, \qquad (8.4.2)$$

where $\bar{n}$ is the mode index. Using $v_A = 5.96$ km/s and $\bar{n} = 1.45$ as typical values for silica fibers, $\nu_B = 11.1$ GHz at $\lambda_p = 1.55$ μm. Equation (8.4.2) shows that ν_B scales inversely with the pump wavelength.

The gain spectrum [44] of SBS is determined by the damping time of acoustic waves or, equivalently, by the lifetime T_B of acoustic phonons. If the acoustic wave is assumed to decay as $\exp(-t/T_B)$, the Brillouin gain has a Lorentzian spectral profile given by

$$g_B(\Omega) = \frac{g_B(\Omega_B)}{1 + (\Omega - \Omega_B)^2 T_B^2}. \qquad (8.4.3)$$

A comparison of Eq. (8.4.3) with Eq. (8.1.2) reveals that the Brillouin gain spectrum has the same form as that of a two-level system. In fact, the FWHM

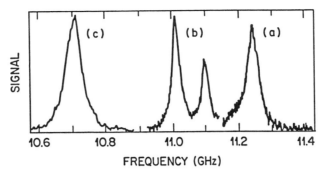

Figure 8.14 Brillouin-gain spectra measured at the pump wavelength $\lambda_p = 1.525$ μm for three fibers with different structures and germania doping: (a) silica-core fiber; (b) depressed-cladding fiber; (c) dispersion-shifted fiber. Vertical scale is arbitrary. (After Ref. [45]. ©1986 IEE. Reprinted with permission.)

is given by a relation similar to Eq. (8.1.3) as $\Delta\nu_B = (\pi T_B)^{-1}$. Typically, $T_B > 10$ ns and $\Delta\nu_B < 100$ MHz. Figure 8.14 shows the measured Brillouin gain spectra at $\lambda_p = 1.525$ μm for three different kinds of single-mode silica fibers. Both the Brillouin shift ν_B and the gain bandwidth $\Delta\nu_B$ can vary from fiber to fiber because of the guided nature of light and the presence of dopants in the fiber core. The fiber labeled (a) in Fig. 8.14 has a core of nearly pure silica (germania concentration of about 0.3% per mole). The measured Brillouin shift $\nu_B = 11.25$ GHz is in agreement with Eq. (8.4.2). The Brillouin shift is reduced for fibers (b) and (c) with nearly inverse dependence on the germania concentration in the fiber core. Fiber (b) has a double-peak structure resulting from an inhomogeneous distribution of germania within the core.

The gain bandwidth in Fig. 8.14 is larger than that expected for bulk silica ($\Delta\nu_B \approx 17$ MHz at $\lambda_p = 1.525$ μm). A part of the increase is due to the guided nature of acoustic modes in optical fibers. However, most of the increase in bandwidth is attributable to inhomogeneities in the core cross section along the fiber length. Because such inhomogeneities are specific to each fiber, the Brillouin gain bandwidth is generally different for different fibers and can exceed 100 MHz, although typical values are 50–60 MHz for λ_p near 1.55 μm.

The peak value of the gain in Eq. (8.4.3) occurs for $\Omega = \Omega_B$ and is given by [47]

$$g_B(\Omega_B) = \frac{2\pi n^7 p_{12}^2}{c\lambda_p^2 \rho_0 v_A \Delta\nu_B}, \tag{8.4.4}$$

where p_{12} is the longitudinal *elasto-optic coefficient* and ρ_0 is material density. Equation (8.4.4) shows that g_B depends on both λ_p and $\Delta\nu_B$. However, the bandwidth $\Delta\nu_B$ is found to scale with λ_p as λ_p^{-2}. Since $\lambda_p^2\Delta\nu_B$ is then nearly a constant, the peak gain is independent of the pump wavelength (in contrast with SRS where g_R varies as λ_p^{-1}). By using the parameter values typical of fused silica, g_B is estimated to be about 5×10^{-11} m/W. This value is larger by nearly three orders of magnitude than the Raman gain at $\lambda_p = 1.55$ μm. The optical gain, given by Eq. (8.3.1) by replacing g_R by g_B, is also larger by the same amount for a given value of the pump power P_p. Because of such a large value of the Brillouin gain, fiber Brillouin amplifiers can amplify the signal by 30–35 dB with pump powers $\sim$ 5 mW [44].

The expression for the Brillouin gain, Eq. (8.4.4), assumes CW pumping by a laser whose linewidth $\Delta\nu_p \ll \Delta\nu_B$. The gain is reduced when the pump linewidth exceeds the Brillouin linewidth [46] and is given by

$$\bar{g}_B(\Omega_B) = \frac{\Delta\nu_B}{\Delta\nu_B + \Delta\nu_p} g_B(\Omega_B). \tag{8.4.5}$$

For this reason fiber Brillouin amplifiers need to be pumped by narrow-linewidth semiconductor lasers ($\Delta\nu_p < 10$ MHz). In particular, FP semiconductor lasers are not suitable for pumping because of their multimode nature.

8.4.2 Gain Saturation

Similar to the case of SRS, the gain of fiber Brillouin amplifiers saturates when the amplified signal begins to deplete the pump. In fact, gain saturation is governed by a pair of nonlinear equations similar to Eqs. (8.3.2) and (8.3.3). If we assume for simplicity that $\alpha_p \approx \alpha_s = \alpha$, $\omega_p \approx \omega_s$, and $a_p \approx a_s$ (because of a small frequency shift between pump and signal waves), these equations become

$$dP_s/dz = \alpha P_s - (g_B/a_p)P_pP_s. \tag{8.4.6}$$
$$dP_p/dz = -\alpha P_p - (g_B/a_p)P_pP_s, \tag{8.4.7}$$

where the signal is assumed to propagate in the backward direction.

In the case of small-signal amplification ($P_s \ll P_p$), the last term in Eq. (8.4.7) can be neglected. The signal is then found to grow exponentially in the backward direction as

$$P_s(0) = P_s(L)\exp(g_B P_0 L_{\text{eff}}/a_p - \alpha L), \tag{8.4.8}$$

where $P_0 = P_p(0)$ is the pump power at $z = 0$ and L_{eff} is given by Eq. (8.3.5). The amplification factor becomes $G_A = \exp(g_0 L)$ with the small-signal gain g_0 defined as

$$g_0 = g_B \left(\frac{P_0}{a_p}\right)\left(\frac{L_{\text{eff}}}{L}\right) \approx \frac{g_B P_0}{a_p \alpha_p L}, \tag{8.4.9}$$

where the last relation holds for $\alpha_p L \gg 1$. The amplified signal grows exponentially with an increase in the pump power, similar to the case of fiber Raman amplifiers, as long as $P_s \ll P_0$ so that pump depletion can be neglected.

Gain saturation is included by solving Eqs. (8.4.6) and (8.4.7). An analytic solution is possible only for $\alpha = 0$ [47]. The amplified signal $P_s(0)$ is related to the input signal $P_s(L)$ as [31]

$$P_s(L) = \frac{(1-b_0)P_s(0)}{G_A^{(1-b_0)} - b_0}, \tag{8.4.10}$$

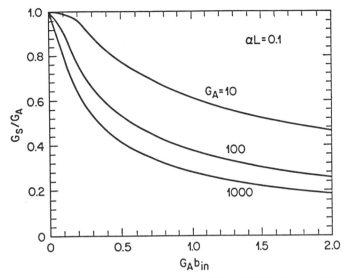

Figure 8.15 Gain–saturation characteristics of fiber Brillouin amplifiers for several values of unsaturated amplifier gain G_A.

where b_0 itself depends on the amplified signal as $b_0 = P_s(0)/P_0$. The saturated gain G_s can be written as $G_s = b_0/b_{in}$, where $b_{in} = P_s(L)/P_0$ is the ratio of the input signal power to the pump power. Figure 8.15 shows the saturation characteristics of fiber Brillouin amplifiers by plotting G_s/G_A as a function of $G_A b_{in}$ for several values of the small-signal amplification factor $[G_A = \exp(g_0 L)]$. The saturated gain is reduced by 3 dB when $G_A b_{in} \approx 0.5$ for G_A in the range 20–30 dB. This condition is satisfied when the amplified signal power becomes about 50% of the input pump power. Since typical pump powers are a few milliwatts, the saturation pump power of fiber Brillouin amplifiers is quite small (~ 1 mW) in contrast with fiber Raman amplifiers, whose saturation power is ~ 1 W.

8.4.3 Amplifier Performance

Although SBS in optical fibers was observed [43] in 1972 and was later studied [44] as a nonlinear phenomenon limiting the signal power in fiber-optic communication systems (see Section 7.4.2), its use for making fiber Brillouin amplifiers did not attract much attention because of the narrow gain bandwidth associated with such amplifiers. Since the pump and signal frequencies should differ almost exactly by the Brillouin shift (about 11 GHz at 1.55 μm), which varies from fiber to fiber (see Fig. 8.14), a tunable pump laser with a narrow linewidth (~ 1 MHz) is required. In one experiment [48], two external-cavity semiconductor lasers (linewidths < 10 kHz) were used as pump and signal lasers. Both lasers operated continuously and were tunable over a wide frequency range near 1.5 μm. Figure 8.16 shows the experimental setup schematically. The pump beam was coupled into a 37.5-km-long fiber through a 3-dB coupler. The probe

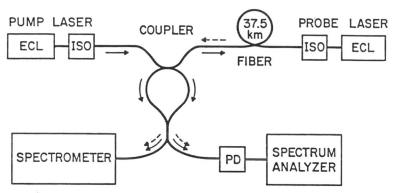

Figure 8.16 Fiber Brillouin amplifier. ECL, ISO, and PD stand, respectively, for external-cavity semiconductor laser, optical isolator, and photodetector. Solid and dashed arrows show the optical path of the pump and signal beams, respectively. (After Ref. [48]. ©1986 IEE. Reprinted with permission.).

laser provided a weak input signal (about 10 mW power) at the other end of the fiber. The measured amplifier gain varied exponentially with the pump power, as expected from Eq. (8.4.8). An amplification factor of 16 dB ($G_A = 40$) was obtained at a pump power of 3.7 mW.

Fiber Brillouin amplifiers can be used as a preamplifier to improve the receiver sensitivity. In a 1987 experiment [49], the receiver sensitivity improved by 16 dB when a 90-Mb/s signal was transmitted over 30 km of fiber and amplified simultaneously by injecting 2.9 mW of pump power at the receiver end. The sensitivity improvement nearly matched the amplifier gain of 16.5 dB, indicating little penalty from amplifier noise. This performance was possible because the signal power inside the fiber was considerably larger than the noise generated by spontaneous Brillouin scattering. The noise figure of fiber Brillouin amplifiers is quite large (> 15 dB) because of a large population of the acoustic phonons at room temperature and limits their usefulness as a preamplifier. Their use as power amplifiers or in-line amplifiers is also limited because of a narrow bandwidth and a small saturation power.

The narrow bandwidth of fiber Brillouin amplifiers can be used to advantage in coherent and multichannel lightwave systems. In the case of coherent communication systems, it is possible to perform homodyne detection without requiring a phase-locked local oscillator at the receiver. The basic idea [50] is to amplify the optical carrier selectively while leaving the modulation sidebands unamplified; the amplified carrier acts as a local oscillator whose phase is automatically locked to the phase of the transmitted carrier. Brillouin amplification is ideal for this purpose because of its narrow bandwidth. The scheme should work well at bit rates > 100 Mb/s because the modulation sidebands then fall outside the amplifier bandwidth, and the optical carrier can be amplified selectively. In a demonstration of this scheme [51], the carrier was amplified by 30 dB more than the modulation sidebands at bit rates as low as 80 Mb/s.

Under ideal conditions, the expected improvement in the receiver sensitivity is about $\sqrt{G_A}$, where G_A is the amplification factor. Thus, a sensitivity improvement of up to 15 dB seems feasible by a proper design. The limiting factor is the nonlinear phase shift induced by the pump on the signal (a kind of cross-phase modulation) when the pump frequency and the carrier frequency do not match the Brillouin shift exactly. The calculations show [52] that the deviation from the Brillouin shift should be < 100 kHz for a phase stability of 0.1 rad. The nonlinear phase shift can also lead to undesirable amplitude modulation of a frequency-modulated signal [53].

Another application of narrow-bandwidth amplifiers is for channel selection in a multichannel communication system [54]–[57]. As discussed in Section 7.2.3, a channel can be selectively amplified through Brillouin amplification by launching a pump beam at the receiver end so that it propagates inside the fiber in a direction opposite to that of the multichannel optical signal. The pump frequency is adjusted in such a way that it is higher than that of the selected channel by exactly the Brillouin shift. Different channels can be selectively amplified by tuning the pump laser. The main drawback of this channel-selection scheme is that the bit rate is limited to about 100 Mb/s, as the channel spectrum should fall within the bandwidth of the Brillouin-gain spectrum. It can be increased by broadening the pump spectrum, since the gain bandwidth then increases, although the peak gain is reduced [see Eq. (8.4.5)]. The potential of this channel-selection scheme was demonstrated in an experiment [55] in which a 128-channel WDM network was stimulated by using two 8×8 star couplers in series. A specific channel operating at 150 Mb/s could be selected with a channel separation as small as 1.5 GHz. It is also possible to amplify and demodulate the FSK signal simultaneously at bit rates of up to 250 Mb/s by using fiber Brillouin amplifiers [56]. In a similar approach, a fiber Brillouin amplifier can be used to selectively amplify multiple subcarrier-multiplexed (SCM) channels carried by a single optical carrier within a WDM signal [57]. In another application, Brillouin amplification within the transmission fiber is used to amplify a backward-propagating supervisory signal [58]. This scheme is unique in the sense that a 2.5-Gb/s data channel, and not a CW beam, was used to pump a 5-kb/s supervisory channel.

8.5 DOPED-FIBER AMPLIFIERS

An important class of fiber amplifiers makes use of *rare-earth elements* as a gain medium by doping the fiber core during the manufacturing process (see Section 2.7). Although doped-fiber amplifiers were studied as early as 1964 [59], their use became practical only 25 years later, after the techniques for fabrication and characterization of low-loss doped fibers were perfected [60]. Amplifier characteristics such as the operating wavelength and the gain bandwidth are determined by the dopants rather than by the silica fiber, which plays the role of a host medium. Many different rare-earth ions, such as erbium, holmium, neodymium, samarium, thulium, and ytterbium, can be used to realize fiber am-

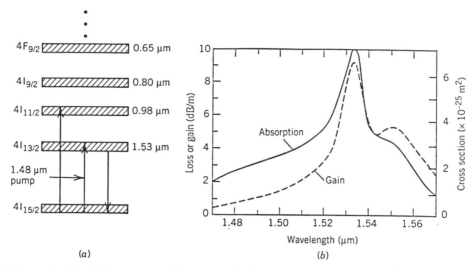

Figure 8.17 (a) Energy-level diagram of erbium ions in silica fibers; (b) absorption and gain spectra of an EDFA whose core was codoped with germania to increase the refractive index. (After Ref. [71]. ©1991 IEEE. Reprinted with permission.)

plifiers operating at different wavelengths covering a wide region extending over 0.5–3.5 μm. Erbium-doped fiber amplifiers (EDFAs) have attracted the most attention [61]–[63] simply because they operate near 1.55 μm, the wavelength region in which the fiber loss is minimum. Their development has revolutionized the design of lightwave systems. EDFAs were used for video distribution by 1992. The transatlantic and transpacific fiber-optic cables, using EDFAs as in-line amplifiers while operating at 5 Gb/s, became operational in 1996. EDFAs are also used routinely for terrestrial WDM systems. This section is devoted to their operating characteristics, with emphasis on potential applications.

8.5.1 Pumping Requirements

The gain characteristics of EDFAs depend on the pumping scheme as well as on other dopants, such as germania and alumina, that are present within the fiber core. The amorphous nature of silica broadens the energy levels of erbium ions (Er^{3+}) into bands. Figure 8.17(a) shows a few energy levels of Er^{3+} in silica glasses. Many transitions can be used to pump the EDFA. Early experiments used for pumping the visible radiation emitted from high-power lasers, such as argon-ion, Nd:YAG, and dye lasers even though such pumping schemes are relatively inefficient. From a practical standpoint, the use of a semiconductor laser is preferred.

Efficient EDFA pumping is possible using semiconductor lasers operating near 0.98- and 1.48-μm wavelengths. Indeed, the development of such lasers was fueled by the need for suitable pump lasers for EDFAs. It is possible to obtain high amplifier gains in the range 30–40 dB with only 10–20 mW of absorbed pump power when EDFAs are pumped by using such semiconductor lasers. Efficiencies as high as 11 dB/mW have been achieved with 0.98-μm pumping [64]. The pumping transition $4I_{15/2}$–$4I_{9/2}$ can use high-power GaAs lasers to pump EDFAs. The *pumping efficiency* of about 1 dB/mW can be obtained at 820 nm [65]. Even though the efficiency is smaller than that of 0.98–μm lasers by one order of magnitude, the 30-dB gain can be realized with 40–50 mW of pump power, power levels readily obtained from GaAs semiconductor lasers. The pump power can be reduced by using silica fibers doped with aluminum and phosphorus or by using *fluorophosphate fibers* [66]. With the availability of visible semiconductor lasers, EDFAs can also be pumped in the wavelength range 0.6–0.7 μm. In one experiment [67], 33-dB gain was realized at 27 mW of pump power obtained from an AlGaInP laser operating at 670 nm. The pumping efficiency was as high as 3 dB/mW at low pump powers.

EDFAs can be designed to operate in such a way that the pump and signal beams propagate in opposite directions, a configuration referred to as backward pumping to distinguish it from the forward-pumping configuration in which both beams propagate along the same direction. The performance is nearly the same in the two pumping configurations when the signal power is small enough for the amplifier to remain unsaturated. In the saturation regime, the power-conversion efficiency is generally better in the backward-pumping configuration, mainly because of the important role played by the amplified spontaneous emission [68]. In the bidirectional pumping configuration, the amplifier is pumped in both directions simultaneously by using two semiconductor lasers located at the two fiber ends. This configuration requires two pump lasers but has the advantage that the population inversion, and hence the small-signal gain, is relatively uniform along the entire amplifier length.

8.5.2 Gain Spectrum

The gain spectrum of EDFAs is affected considerably by the amorphous nature of silica and by the presence of other codopants within the fiber core such as germania and alumina [69]–[71]. The gain profile of the erbium ions alone (in isolation) is homogeneously broadened and its bandwidth is determined by the dipole relaxation time T_2 in accordance with Eq. (8.1.2). However, it is considerably broadened by the presence of silica glass. Structural disorders lead to inhomogeneous broadening of the gain profile, whereas *Stark splitting* of various energy levels is responsible for homogeneous broadening. Mathematically, the gain $g(\omega)$ of Eq. (8.1.2) should be averaged over the distribution of atomic transition frequencies ω_0 so that

$$g_{\text{eff}}(\omega) = \int_{-\infty}^{\infty} g(\omega, \omega_0) f(\omega_0) \, d\omega_0, \qquad (8.5.1)$$

where $f(\omega_0)$ is the distribution function, whose form depends on the presence of codopants within the fiber core. Figure 8.17(b) shows the gain and absorption spectra of an EDFA whose core was doped with germania [71]. The gain

spectrum is quite broad (FWHM > 10 nm), with a double-peak structure. The
addition of alumina to the core broadens the gain spectrum even more. Attempts
have been made to isolate the contributions of homogeneous and inhomogeneous
broadening through measurements of *spectral hole burning* [1]. For germania-
doped EDFAs the relative contributions of homogeneous and inhomogeneous
broadening are found to be 4 and 8 nm, respectively [70]. By contrast, the gain
spectrum of aluminosilicate glasses has roughly equal contributions from homogeneous and inhomogeneous broadening mechanisms. The gain bandwidth of
such EDFAs typically exceeds 30 nm.

   Gain spectrum of EDFAs can vary from amplifier to amplifier even when
core composition is the same, because it also depends on the amplifier length.
The reason is that the gain depends on both the absorption and emission cross
sections, which have different spectral characteristics. The local inversion or
local gain varies along the fiber length because of pump power variations. The
total gain is obtained by integrating over the amplifier length. This feature can
be used to obtain a relatively flat gain spectrum by optimizing the fiber length.

8.5.3   Gain Characteristics

The gain of EDFAs depends on a large number of device parameters such as
erbium-ion concentration, amplifier length, core radius, and pump power. Considerable effort has been made to develop an understanding of the gain characteristics through theoretical modeling [71]–[73]. A three-level rate-equation
model commonly used for some lasers [1] (e.g., a Ruby laser) can be adapted
for EDFAs for all pumping wavelengths, since stimulated emission terminates
in the ground state in each case [see Fig. 8.17(a)]. It is sometimes necessary
to add a fourth level to include the *excited-state absorption*. In general, the
resulting equations must be solved numerically. Much insight can be gained by
using a simple two-level model valid when amplified spontaneous emission and
excited-state absorption are neglected [73]. The model assumes that pump level
3 of the *three-level system* remains nearly empty because of a rapid transfer of
the pumped population to the excited state 2. If N_2 is the population of the
excited state, it satisfies the rate equation [1]

$$\frac{\partial N_2}{\partial t} = W_p N_1 - W_s (N_2 - N_1) - \frac{N_2}{T_1}, \qquad (8.5.2)$$

where $N_1 = N_t - N_2$ is the ground-state population (N_t is the total atomic
density) and T_1 is the spontaneous lifetime of the excited state. W_p and W_s are
the transition rates for the pump and signal waves such that

$$W_p = \sigma_p P_p / a_p h \nu_p \qquad (8.5.3)$$

together with a similar expression for W_s. Here σ_p is the transition cross section
at the pump frequency ν_p, a_p is the cross-section area of the pump mode inside
the fiber, and P_p is the pump power. The steady-state solution of Eq. (8.5.2) is

given by

$$N_2 = \frac{(P_p' + P_s')N_t}{1 + 2P_s' + P_p'},$$ (8.5.4)

where $P_p' = P_p/P_p^{\text{sat}}$ and $P_s' = P_s/P_s^{\text{sat}}$, and the saturation powers are defined as

$$P_p^{\text{sat}} = \frac{a_p h\nu_p}{\sigma_p T_1}, \qquad P_s^{\text{sat}} = \frac{a_s h\nu_s}{\sigma_s T_1}.$$ (8.5.5)

The pump and signal powers vary along the amplifier length because of absorption, stimulated emission, and spontaneous emission. If the contribution of spontaneous emission is neglected, P_p and P_s satisfy

$$\frac{dP_s}{dz} = \sigma_s(N_2 - N_1) - \alpha P_s, \qquad \frac{dP_p}{dz} = \sigma_p N_1 - \alpha' P_p,$$ (8.5.6)

where $\alpha_s = \sigma_s N_t$ and $\alpha_p = \sigma_p N_t$ account for the dopant-induced absorption, while α and α' take into account the fiber loss, at the signal and pump wavelengths, respectively. The loss of silica fibers can be neglected for typical amplifier lengths of 10–20 m. However, it must be included in the case of distributed amplification discussed later. By substituting $N_1 = N_t - N_2$ and N_2 from Eq. (8.5.4), the signal and pump powers are found to satisfy

$$\frac{dP_s}{dz} = \frac{(P_p' - 1)\alpha_s P_s}{1 + 2P_s' + P_p'} - \alpha P_s,$$ (8.5.7)

$$\frac{dP_p}{dz} = -\frac{(P_s' + 1)\alpha_p P_p}{1 + 2P_s' + P_p'} - \alpha' P_p,$$ (8.5.8)

These equations govern the growth of signal power inside the EDFA and can be used to study both small-signal and large-signal amplification characteristics. Their predictions are in good agreement with experiments [73] as long as amplified spontaneous emission remains negligible.

A drawback of the foregoing model is that the absorption and emission cross sections are taken to be the same for both the pump and signal beams. As seen in Fig. 8.17(b), these cross sections are generally different. It is easy to extend the model to include this feature. Figure 8.18 shows the calculated small-signal gain at 1.55 μm as a function of the pump power and the amplifier length by using typical parameter values. For a given amplifier length L, the amplifier gain initially increases exponentially with the pump power, but the increase becomes much smaller when the pump power exceeds a certain value [corresponding to the "knee" in Fig. 8.18(a)]. For a given pump power, the amplifier gain becomes maximum at an optimum value of L and drops sharply when L exceeds this optimum value. The reason is that the latter portion of the amplifier remains unpumped and absorbs the amplified signal.

Since the optimum value of L depends on the pump power P_p, it is necessary to choose both L and P_p appropriately. Figure 8.18(b) shows that for 1.48-μm pumping a 35-dB gain can be realized at a pump power of 5 mW for $L = 30$ m. It is possible to design amplifiers such that high gain is obtained for

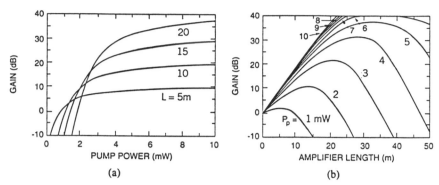

Figure 8.18 Small-signal gain as a function of (a) pump power and (b) amplifier length for an EDFA assumed to be pumped at 1.48 μm. (After Ref. [71]. ©1991 IEEE. Reprinted with permission.)

amplifier lengths as short as a few meters. The qualitative features shown in Fig. 8.18 are observed in all EDFAs; the agreement between theory and experiment is generally quite good [73]. The saturation characteristics of EDFAs are similar to those shown in Figs. 8.13 and 8.15 for Raman and Brillouin amplifiers, respectively. In general, the output saturation power is smaller than the output pump power expected in the absence of signal. It can vary over a wide range depending on the EDFA design, with typical values $\sim$ 10 mW. For this reason the output power levels of EDFAs are generally limited to below 100 mW, although powers as high as 250 mW have been obtained with a proper design [74].

The analysis above assumes that both pump and signal waves are in the form of CW beams. In practice, EDFAs are pumped by using CW semiconductor lasers, but the signal is in the form of a pulse train (containing a random sequence of 1 and 0 bits), and the duration of individual pulses is inversely related to the bit rate. All pulses should experience the same gain. Fortunately, this occurs naturally in EDFAs for pulses shorter than a few microseconds. The reason is related to a relatively large value of the fluorescence time associated with the excited erbium ions ($T_1 \sim$ 10 ms). When the time scale of signal-power variations is much shorter than T_1, erbium ions are unable to follow such fast variations. Since single-pulse energies are typically much below the saturation energy ($\sim$ 10 μJ), EDFAs respond to average power. As a result, gain saturation is governed by the average signal power, and amplifier gain does not vary from pulse to pulse.

In some applications such as packet-switched networks, signal power may vary on a time scale comparable to T_1. Amplifier gain in that case is likely to become time dependent, an undesirable feature from the standpoint of system performance. It is possible to implement a built-in gain-control mechanism that keeps the amplifier gain pinned at a constant value. The basic idea is to make the EDFA oscillate at a controlled wavelength outside the range of interest (typically below 1.5 μm). Since the gain remains clamped at the threshold value for a laser,

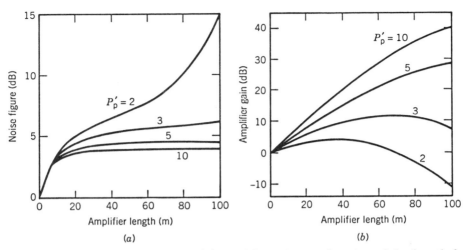

Figure 8.19 (a) noise figure and (b) amplifier gain as a function of the length for several pumping levels. (After Ref. [79]. ©1990 IEE. Reprinted with permission.)

the signal is amplified by the same factor despite variations in the signal power. In one implementation of this scheme, an EDFA was forced to oscillate at 1.48 μm by fabricating two fiber Bragg gratings acting as high-reflectivity mirrors at the two ends of the amplifier [75].

8.5.4 Amplifier Noise

Since amplifier noise is the ultimate limiting factor for system applications, it has been studied extensively [76]–[79]. As discussed in Section 8.1.3, amplifier noise is quantified through the noise figure F_n given by $F_n = 2n_{sp}$. The spontaneous emission factor n_{sp} depends on the relative populations N_1 and N_2 of the ground and excited states as $n_{sp} = N_2/(N_2 - N_1)$. Since EDFAs operate on the basis of a three-level pumping scheme, $N_1 \neq 0$ and $n_{sp} > 1$. Thus, the noise figure of EDFAs is expected to be larger than the ideal value of 3 dB.

The spontaneous-emission factor for EDFAs can be calculated by using the three-level rate-equation model discussed in Section 8.5.3. However, one should take into account the fact that both N_1 and N_2 vary along the fiber length because of their dependence on the pump and signal powers [see Eq. (8.5.4)]; hence n_{sp} should be averaged along the amplifier length. As a result, the noise figure depends both on the amplifier length L and the pump power P_p, just as the amplifier gain does. Figure 8.19(a) shows the variation of F_n with the amplifier length for several values of P_p/P_p^{sat} when a 1.53-μm signal is amplified with an input power of 1 mW. The amplifier gain under the same conditions is also shown in Fig. 8.19(b). The results show that a noise figure close to 3 dB can be obtained for a high-gain amplifier pumped such that $P_p \gg P_p^{sat}$ [76].

The experimental results confirm that F_n close to 3 dB can be realized in ED-FAs. A noise figure of 3.2 dB was measured [77] in a 30-m-long EDFA pumped at 0.98 μm with 11 mW of power. A similar value was measured in another experiment [78] with only 5.8 mW of pump power at 0.98 μm. In general, it is difficult to achieve high gain, low noise, and high pumping efficiency simultaneously. The main limitation is imposed by amplified spontaneous emission traveling backward toward the pump and depleting the pump power. Incorporation of an internal isolator alleviates this problem to a large extent. In one implementation, 51-dB gain was realized with a 3.1-dB noise figure at a pump power of only 48 mW [80].

The measured values of F_n are generally larger for EDFAs pumped at 1.48 μm. A noise figure of 4.1 dB was obtained for a 60-m-long EDFA when pumped at 1.48 μm with 24 mW of pump power [77]. The reason for a larger noise figure for 1.48-μm-pumped EDFAs can be understood from Fig. 8.17(a), which shows that the pump level and the excited level lie within the same band for 1.48-μm pumping. It is difficult to achieve complete population inversion ($N_1 \approx 0$) under such conditions. It is nonetheless possible to realize $F_n < 3.5$ dB for pumping wavelengths near 1.46 μm.

Relatively low noise levels of EDFAs make them an ideal choice for lightwave system applications. In spite of low noise, the performance of long-haul fiber-optic communication systems employing multiple EDFAs is often limited by the amplifier noise. The noise problem is particularly severe when the system operates in the anomalous-dispersion region of the fiber since a nonlinear phenomenon known as the modulation instability [31] enhances the amplifier noise [81] and degrades the signal spectrum [82]. Amplifier noise also affects the performance of soliton communication systems, where it leads to the problem of timing jitter. This issue is discussed in Section 10.3.

8.5.5 Multichannel Amplification

An advantage of optical amplifiers is that they can be used to amplify several communication channels simultaneously as long as the bandwidth of the multichannel signal is smaller than the amplifier bandwidth. The bandwidth of EDFAs is comparable to that of SLAs (1–5 THz), and both of them can be used for multichannel amplification. As discussed in Section 8.2.3, SLAs suffer from the interchannel crosstalk induced by carrier-density modulation occurring at the beat frequency Ω of the neighboring optical carriers. Such crosstalk does not occur when $\Omega\tau_c \gg 1$, where τ_c is the carrier lifetime. In the case of EDFAs, the role of τ_c is played by T_1 in Eq. (8.5.1). Since T_1 is much larger (about 10 ms) compared with τ_c (about 0.5 ns), the condition $\Omega T_1 \gg 1$ is satisfied for a channel spacing as small as 10 kHz. As a result, the gain-modulation effects are negligible when EDFAs are used for amplification of a multichannel signal. This property makes them suitable for WDM lightwave systems.

A second source of interchannel crosstalk is cross-saturation occurring because the gain of a specific channel is saturated not only by its own power (self-saturation) but also by the power of neighboring channels. This mechanism of crosstalk is common to all optical amplifiers, and EDFAs are no exception [83]–[85]. It can be avoided by operating the amplifier in the unsaturated region.

Experimental results support this conclusion. In one experiment [83] negligible power penalty was observed when an EDFA was used to amplify two channels operating at 2 Gb/s and separated by 2 nm as long as the channel powers were low enough to avoid gain saturation.

A third source of interchannel crosstalk is related to the *spectral nonuniformity* of the amplifier gain in EDFAs. Even though the gain spectrum of an EDFA is relatively broad, as seen in Fig. 8.17, the gain is far from bring uniform (or flat) over a wide wavelength range. As a result, different channels of a WDM signal are amplified by different amounts. This problem becomes quite severe in long-haul systems employing a cascaded chain of EDFAs and is discussed in Section 8.6.4. It also leads to distortion in CATV networks [86] transmitting multiple video channels through subcarrier multiplexing (see Section 7.5). The EDFA-induced distortion results from FM-to-AM conversion of a chirped signal because of the *gain tilt* $(dg/d\omega \neq 0)$ at the operating wavelength [87]. The gain tilt of EDFAs used for CATV applications can be minimized through a suitable design [88].

8.5.6 Ultrashort Pulse Amplification

A relatively large bandwidth ($\sim$ 30 nm) of EDFAs suggests that they can be used to amplify optical pulses of duration 1 ps or shorter without distortion. In a 1990 experiment [89], 9-ps optical pulses were amplified by 30 dB without significant changes in the pulse shape or width. Gain saturation was negligible at low repetition rates but became important when the repetition rate exceeded 100 MHz. However, the amount of saturation was determined by the average power, resulting in the same gain for all pulses. These properties of EDFAs are quite useful for their use in soliton communication systems (see Chapter 10).

As discussed in Section 7.4, the use of time-division multiplexing forces the use of femtosecond pulses as the bit rate approaches 1 Tb/s. In several experiments [90]–[93], EDFAs were used to amplify femtosecond optical pulses. The spectral bandwidth of such ultrashort pulses is large enough that gain dispersion (spectral nonuniformity of gain) must be considered. It is also important to include the *intrapulse* stimulated Raman scattering [94], a phenomenon through which high-frequency components of an ultrashort pulse can pump the low-frequency components of the same pulse. The amplification of ultrashort optical pulses is also influenced by the group-velocity dispersion (GVD) and the intensity dependence of the refractive index of the silica fiber acting as a host to the erbium ions, which provide gain [31]. Numerical simulations show that an input pulse can be amplified and compressed simultaneously by an EDFA when its peak power exceeds a certain value [94]. It can also split into several subpulses. Such features have been observed experimentally [91]–[93]. From the standpoint of lightwave system applications, the potential of EDFAs lies in their ability to amplify picosecond pulses without distortion.

8.5.7   Pr-Doped Fiber Amplifiers

EDFAs can only be used for lightwave systems operating near 1.55 μm. However, the worldwide telecommunication network contains more than 50 million kilometers of the standard fiber whose performance is optimized for 1.3-μm operation. Moreover, video distribution by the CATV industry is mostly carried out at 1.3 μm. Clearly, such lightwave systems would benefit if fiber amplifiers operating near 1.3 μm were available. Considerable effort has been directed toward developing broadband optical amplifiers providing gain in this wavelength region. Doping of silica fibers with neodymium ions provides a fiber amplifier capable of amplifying in the spectral region 1.30–1.36 μm. However, undesirable effects such as excited-state absorption and competing radiative transitions have limited the performance of such amplifiers. It turns out that the performance is improved considerably if Nd^{3+} ions are doped in a fluoride fiber [95]. The fiber commonly used for this purpose is known as ZBLAN, an acronym formed by using the first letters of the materials used to fabricate it (ZrF_4–BaF_2–LaF_3–AlF_3–NaF).

Because of the relatively poor performance of Nd-doped amplifiers, fiber amplifiers operating at 1.3 μm have been developed by doping ZBLAN fibers with praseodymium ions (Pr^{3+}), which have a resonance at 1.32 μm. These amplifiers require pumping in the wavelength region near 1.01 μm. Amplifier gains of up to 38 dB and saturation output powers of up to 20 mW were demonstrated in 1991 using a Ti:sapphire laser for pumping [96]–[99]. The pumping efficiency of 1.3-μm amplifiers is quite small (typically, < 0.2 dB/mW) compared with EDFAs (up to 11 dB/mW).

Pump powers needed to obtain 30-dB gain from Pr-doped amplifiers are quite large ($\sim$ 1 W). For this reason, most early experiments used bulky solid-state or ion lasers for optical pumping. A diode-pumped solid-state laser provides an alternative to direct semiconductor-laser pumping. A Nd:YLF laser operating at 1.047 μm was used as a pump laser in a 1993 experiment [100]. The absorption coefficient at 1.047 μm is only 30% of its peak value occurring near 1.017 μm. Nonetheless, the absorption coefficient of 0.36 dB/m was large enough that 29.5-dB gain could be achieved in a 30-m-long fiber amplifier at a pump power of 1.8 W. The amplifier was capable of providing 10-dB gain at a pump power of 325 mW. The output saturation power was quite large ($\sim$ 0.2 W). Indeed, the 250-mW output at 1.3 μm was obtained by launching only 1.2 mW of the signal power.

Strained-layer quantum-well semiconductor lasers operating at 1.017 μm were developed to meet the pumping requirement of Pr-doped fiber amplifiers. In a 1993 demonstration [101], such an amplifier provided 28.3-dB gain at a pump power of 280 mW. Such high pump powers were obtained by using a bidirectional pumping scheme, with two orthogonally polarized semiconductor lasers on each side. Each laser was capable of delivering 100 mW of output power at a drive current of 250 mA. The maximum gain was limited not by the available pump power but by the lasing threshold of the amplifier. Figure 8.20 shows the amplifier gain as a function of the pump power and the drive current

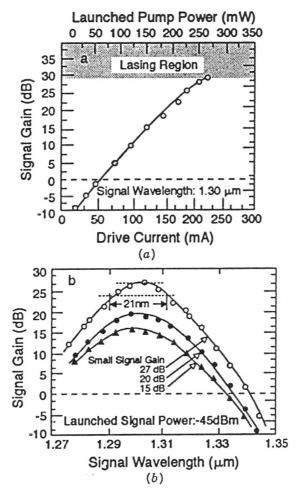

Figure 8.20 (a) Measured signal gain as a function of the drive current, or the launched pump power, for a Pr-doped fiber amplifier; (b) measured signal gain as a function of the signal wavelength for the same amplifier. (After Ref. [101]. ©1993 IEEE. Reprinted with permission.)

when a 1.3-μm signal with 32-nW input power is injected into the amplifier. The wavelength dependence of the signal is also shown for several values of the small-signal gain [101]. The amplifier was capable of delivering more than 25-dB gain over a wide bandwidth (> 20 nm). In a later experiment making use of Nd:YLF lasers, a 40-dB gain and 5-dB noise figure were realized by cascading two Pr-doped fiber amplifiers [102]. The development of Pr-doped fiber amplifiers reached the maturity stage by 1995 when such amplifiers became available commercially. However, a relatively high cost and the practical difficulty of working with fluoride fibers hinders their wide acceptance.

8.5.8   Distributed-Gain Amplifiers

Most doped-fiber amplifiers provide 20-to-25 dB amplification over a length
~ 10 m through a relatively high density of dopants (~ 500 parts per million).
Since such EDFAs compensate for a loss accumulated over 100 km in a relatively
short distance of 10–20 m, they are often referred to as the lumped amplifiers.
An alternative technique compensates for the fiber loss through *distributed amplification.* In this approach, the transmission fiber itself is lightly doped (dopant
density ~ 50 parts per billion) to provide the gain distributed over the entire
fiber length that locally compensates for the fiber loss, resulting in a virtually
transparent ("lossless") fiber at a specific wavelength when the fiber is pumped
at another suitable wavelength by using a bidirectional pumping configuration.
The scheme is similar to that discussed in Section 8.3 for distributed Raman
amplifiers, except that the dopants provide the gain instead of the nonlinear
phenomenon of SRS and has the potential advantage that it may require less
pump power than that required for SRS. Considerable attention has been paid
during the 1990s to the design of distributed EDFAs [103]–[110].

Ideally, one would like to compensate for the fiber loss in such a way that
the signal energy does not change during propagation. Such a performance
is, however, never realized in practice because of the pump-power variations
occurring due to both the pump absorption by the dopants and the fiber loss at
the pump wavelength. The optimal pumping wavelength for distributed EDFAs
is 1.48 μm, since fiber loss at this wavelength is minimal (about 0.25 dB/km)
compared to that at other pump wavelengths (such as 0.98 μm). Even then,
the total pump loss typically exceeds 0.4 dB/km, indicating a loss of 10 dB for
a fiber length of only 25 km. If the fiber is pumped unidirectionally by injecting
the pump beam from one end, nonuniform pumping leads to large variations in
the signal power. A *bidirectional pumping configuration* is therefore commonly
used in which the fiber is pumped from both ends by using two 1.48 μm lasers.
The amplification characteristics of such a distributed EDFA are obtained by
integrating Eqs. (8.5.7) and (8.5.8) subject to the boundary conditions that P_p
equals P_1 and P_2 at $z = 0$ and $z = L$, respectively, where P_1 and P_2 are the
pump powers injected at the two ends of the fiber.

In general, variations in the signal power due to nonuniform pumping can
be kept to a minimum for a relatively short fiber length of 10–15 km. Indeed,
in a 1990 experiment [103], transmission over a 10-km length exhibited *ultralow*
signal-power excursion. For practical reasons, it is important to increase the
fiber length close to 50 km or more so that the pumping stations could be spaced
that far apart. In a 1995 experiment [108], 11.5-ps pulses were transmitted over
93.4 km of a distributed EDFA by injecting up to 90 mW of pump power from
each end. The signal power was estimated to vary by a factor of more than 10
because of the nonuniform pumping. A potential application of a distributed
EDFA is for soliton transmission, a topic covered in Chapter 10.

The performance of distributed EDFAs depends on the signal wavelength
since both the noise figure and the pump power required to achieve transparency
change with the signal wavelength [107]. The noise figure is typically large

(> 6 dB) compared with a lumped EDFA because of a low value of the small-signal gain throughout the fiber length [105]. In general, one should consider the effect of SRS in distributed EDFAs pumped at 1.48 μm since the pump-signal wavelength difference lies within the Raman-gain bandwidth (see Fig. 8.11). As a result, the signal experiences not only the gain provided by the dopants but also the gain provided by SRS (see Section 8.3). SRS increases the net gain and reduces the noise figure for a given amount of pump power.

Distributed EDFAs have been used for system experiments. In a 1996 experiment, a 40-Gb/s RZ signal was transmitted over 68 km by using 7.8-ps optical pulses [109]. Computer simulations show that the use of distributed amplification for NRZ systems has the potential of doubling the pump-station spacing in comparison with the spacing for the lumped amplifiers [110].

8.6   SYSTEM APPLICATIONS

Fiber amplifiers have found many applications and have become an integral part of many fiber-optic communication systems because of their excellent amplification characteristics, such as low insertion loss, high gain, large bandwidth, low noise, and low crosstalk. They have been used for all four applications shown in Fig. 8.3. In this section we focus on the design and performance issues for lightwave systems making use of fiber amplifiers.

8.6.1   Optical Preamplifiers

Fiber amplifiers are used routinely for improving the receiver sensitivity by preamplifying the incident optical signal before it falls on the detector. The basic idea is similar to coherent detection: Preamplification of the optical signal makes it strong enough that thermal noise becomes negligible compared with shot noise. As a result, the receiver sensitivity can be improved 10–20 dB by using an EDFA as a preamplifier [111]–[116]. In a 1990 experiment [111], only 152 photons/bit were needed for an IM/DD system operating at bit rates in the range 0.6–2.5 Gb/s. In another experiment [114], a receiver sensitivity of −37.2 dBm (147 photons/bit) was achieved at the high bit rate of 10 Gb/s. It is even possible to use two preamplifiers in series; the receiver sensitivity improved by 18.8 dB by using this technique [113]. An experiment in 1992 demonstrated a record sensitivity of −38.8 dBm (102 photons/bit) at 10 Gb/s by using two EDFAs [115]. Sensitivity degradation was limited to below 1.2 dB when the signal was transmitted over 45 km of dispersion-shifted fiber.

An important performance issue in the design of lightwave systems making use of optical preamplifiers is the contamination of the amplified signal by spontaneous emission. Because of the incoherent nature of spontaneous emission, the amplified signal is noisier than the input signal. In this section we consider the SNR, the BER, and the receiver sensitivity for direct-detection and coherent-detection lightwave systems.

Direct-Detection Lightwave Systems

As discussed in Section 8.1.3, the amplified signal incident at the photodetector can be written as

$$P_{\text{amp}} = GP_s + P_{\text{sp}}, \tag{8.6.1}$$

where G is the amplifier gain, P_s is the input optical signal, and P_{sp} is the spontaneous-emission noise power added to the signal such that

$$P_{\text{sp}} = S_{\text{sp}}\Delta\nu_{\text{sp}}. \tag{8.6.2}$$

The spectral density S_{sp} is given by Eq. (8.1.15) and $\Delta\nu_{\text{sp}}$ is the effective bandwidth of spontaneous emission; $\Delta\nu_{\text{sp}}$ can be approximated by the amplifier bandwidth in practice.

The photocurrent generated at the detector can be written as

$$I = I_p + \Delta I, \tag{8.6.3}$$

where $I_p = RP_{\text{amp}}$ is the signal and ΔI represents current fluctuations originating from the effects of shot noise, thermal noise, and spontaneous-emission noise. The variance $\sigma^2 = \langle(\Delta I)^2\rangle$ of current fluctuations can be written as [3], [6]

$$\sigma^2 = \sigma_T^2 + \sigma_s^2 + \sigma_{sp-sp}^2 + \sigma_{sig-sp}^2 + \sigma_{s-sp}^2, \tag{8.6.4}$$

where σ_T^2 is the thermal noise (see Section 4.4.1) and the remaining four terms are given by [117]

$$\sigma_s^2 = 2q[R(GP_s + P_{\text{sp}}) + I_d]\Delta f, \tag{8.6.5}$$

$$\sigma_{sp-sp}^2 = 4R^2 S_{\text{sp}}^2 \Delta\nu_{\text{opt}}\Delta f, \tag{8.6.6}$$

$$\sigma_{sig-sp}^2 = 4R^2 GP_s S_{\text{sp}}\Delta f, \tag{8.6.7}$$

$$\sigma_{s-sp}^2 = 4qRS_{\text{sp}}\Delta\nu_{\text{opt}}\Delta f, \tag{8.6.8}$$

where $R = \eta q/h\nu$ is the photodetector responsivity. The shot-noise term σ_s^2 is the same as in Section 4.4.1 except that P_{sp} has been added to GP_s to account for the shot noise generated by spontaneous emission.

The three contributions σ_{sp-sp}^2, σ_{sig-sp}^2, and σ_{s-sp}^2 originate from *beating* of spontaneous emission against itself, signal, and shot noise, respectively. Their origin has been discussed extensively by using both classical and quantum-mechanical approaches [3]. Physically, the signal and spontaneous emission are not at the same optical frequency, and the two can beat with each other. Spontaneous emission can also beat against itself because it spans a wide frequency range governed by $\Delta\nu_{\text{sp}}$. All such beating terms generate current fluctuations and contribute to the receiver noise. The expression for σ_{sig-sp}^2 was obtained in Section 8.1.3. Other beating terms can be obtained in a similar manner. In Eqs. (8.6.5)–(8.6.8), Δf is the receiver (electrical) bandwidth, whereas $\Delta\nu_{\text{opt}}$ is the optical bandwidth of the spontaneous-emission noise that can be different than $\Delta\nu_{\text{sp}}$ if an optical filter is placed before the photodetector to reduce the amount of spontaneous emission. In that case, $\Delta\nu_{\rm opt}$ is related to the filter bandwidth. For a FP filter with a Lorentzian passband, $\Delta\nu_{\rm opt} = (\pi/4)\Delta\nu_f$, where $\Delta\nu_f$ is the filter bandwidth [117].

The BER can be obtained by following the analysis of Section 4.5.1. In fact, it is given by Eq. (4.5.9) or

$$\text{BER} = \tfrac{1}{2}\text{erfc}(Q/\sqrt{2}), \qquad (8.6.9)$$

where

$$Q = \frac{I_1 - I_0}{\sigma_1 + \sigma_0} = \frac{RG(2\bar{P}_{\rm rec})}{\sigma_1 + \sigma_0}. \qquad (8.6.10)$$

Equation (8.6.10) is obtained by assuming zero extinction ratio ($I_0 = 0$) so that $I_1 = RGP_1 = RG(2\bar{P}_{\rm rec})$, where $\bar{P}_{\rm rec}$ is the receiver sensitivity for a given value of BER ($Q = 6$ for BER $= 10^{-9}$). The RMS noise currents σ_1 and σ_0 are obtained from Eqs. (8.6.4)–(8.6.8) by setting $P_s = P_1 = 2\bar{P}_{\rm rec}$ and $P_s = 0$, respectively. Equation (8.6.10) is solved to obtain the receiver sensitivity $\bar{P}_{\rm rec}$ and its dependence on various amplifier parameters.

The analysis can be simplified considerably by comparing the magnitude of various terms in Eqs. (8.6.4). For this purpose it is useful to substitute $S_{\rm sp}$ from Eq. (8.1.15), use $R = \eta q/h\nu$ and Eq. (8.1.19), and write Eqs. (8.6.5)–(8.6.8) in terms of the noise figure F_n as

$$\sigma_s^2 = 2q^2\eta GP_s\Delta f/h\nu, \qquad (8.6.11)$$
$$\sigma_{sp-sp}^2 = (q\eta GF_n)^2\Delta\nu_{\rm opt}\Delta f, \qquad (8.6.12)$$
$$\sigma_{sig-sp}^2 = 2(q\eta G)^2 F_n P_s\Delta f/h\nu, \qquad (8.6.13)$$
$$\sigma_{s-sp}^2 = 2q^2\eta GF_n\Delta\nu_{\rm opt}\Delta f, \qquad (8.6.14)$$

where $RP_{\rm sp}$ and I_d were neglected in Eq. (8.6.5) as they contribute negligibly to the shot noise. A comparison of Eqs. (8.6.11) and (8.6.13) shows that σ_s^2 can be neglected in comparison with σ_{sig-sp}^2, as it is smaller by a large factor ηGF_n. Similarly, a comparison of Eqs. (8.6.12) and (8.6.14) shows that σ_{s-sp}^2 can be neglected in comparison with σ_{sp-sp}^2. The thermal noise σ_T^2 can also be neglected in comparison with the dominant terms. The noise currents σ_1 and σ_0 are thus well approximated by

$$\sigma_1 = (\sigma_{sig-sp}^2 + \sigma_{sp-sp}^2)^{1/2}, \qquad \sigma_0 = \sigma_{sp-sp}. \qquad (8.6.15)$$

The receiver sensitivity is obtained by substituting Eq. (8.6.15) in Eq. (8.6.10), using Eqs. (8.6.12) and (8.6.13) with $P_s = 2\bar{P}_{\rm rec}$, and solving for $\bar{P}_{\rm rec}$. The result is

$$\bar{P}_{\rm rec} = h\nu F_n\Delta f[Q^2 + Q(\Delta\nu_{\rm opt}/\Delta f)^{1/2}]. \qquad (8.6.16)$$

The receiver sensitivity can also be written in terms of the average number of photons/bit, $\bar{N}_p$, by using $\bar{P}_{\rm rec} = \bar{N}_p h\nu B$. By taking $\Delta f = B/2$ as a typical value of the receiver bandwidth, $\bar{N}_p$ is given by

$$\bar{N}_p = \tfrac{1}{2}F_n[Q^2 + Q(2\Delta\nu_{\rm opt}/B)^{1/2}]. \qquad (8.6.17)$$

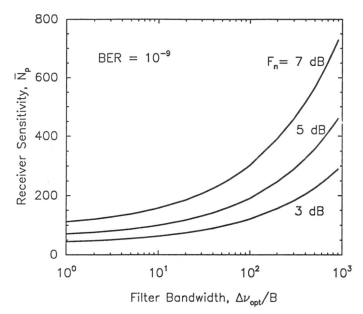

Figure 8.21 Receiver sensitivity versus optical-filter bandwidth for several values of the noise figure F_n when an optical amplifier is used for preamplification of the received signal.

Equation (8.6.17) is a remarkably simple expression for the receiver sensitivity. It clearly shows why amplifiers with a small noise figure must be used; the receiver sensitivity degrades as F_n increases. It also shows how optical filters can improve the receiver sensitivity by reducing $\Delta\nu_{\rm opt}$. Figure 8.21 shows $\bar{N}_p$ as a function of $\Delta\nu_{\rm opt}/B$ for several values of the noise figure F_n by using $Q = 6$, a value required to achieve a BER of 10^{-9}. The minimum optical bandwidth is equal to the bit rate to avoid blocking the signal. The minimum value of F_n is 2 for an ideal amplifier (see Section 8.1.3). Thus, by using $Q = 6$, the best receiver sensitivity from Eq. (8.6.17) is $\bar{N}_p = 44.5$ photons/bit. This value should be compared with $\bar{N}_p = 10$ for an ideal receiver (see Section 4.5.3) operating in the quantum-noise limit. Of course, $\bar{N}_p = 10$ is never realized in practice because of thermal noise; typically, $\bar{N}_p$ exceeds 1000 for p–i–n receivers without optical amplifiers. The analysis of this section shows that $\bar{N}_p < 100$ can be realized when optical amplifiers are used to preamplify the signal received despite the degradation caused by spontaneous-emission noise. The effect of a finite laser linewidth on the receiver sensitivity has also been included with similar conclusions [118].

Coherent Lightwave Systems

The analysis of Section 8.6.1 can easily be extended to the case of coherent receivers. The SNR of heterodyne receivers is obtained by following the analysis of Section 6.1.4. In fact, Eq. (6.1.12) can be used after adding the contributions resulting from spontaneous-emission noise. The result is

$$\text{SNR} = \frac{2R^2 G P_s P_{LO}}{\sigma_T^2 + \sigma_s^2 + \sigma_{sp-sp}^2 + \sigma_{LO-sp}^2}, \tag{8.6.18}$$

where σ_T^2 is the thermal noise and the remaining three terms represent the contribution of shot noise, spontaneous-spontaneous beat noise, and local oscillator-
spontaneous beat noise. The shot noise σ_s^2 as well as σ_{LO-sp}^2 are dominated by
the local-oscillator power P_{LO}. Their explicit expressions are

$$\sigma_s^2 = 2q^2 \eta P_{LO} \Delta f / h\nu, \tag{8.6.19}$$
$$\sigma_{LO-sp}^2 = 2q^2 \eta^2 G F_n P_{LO} \Delta f / h\nu. \tag{8.6.20}$$

σ_{sp-sp}^2 is still given by Eq. (8.6.12), whereas σ_{s-sp}^2 has been neglected. It is
easy to see that shot noise σ_s^2 can still be neglected in comparison with σ_{LO-sp}^2,
since it is smaller by a factor $\eta G F_n$. Thermal noise σ_T^2 can also be neglected.
Therefore, by using $P_s = N_p h\nu B$ with $\Delta f = B/2$, the SNR is given by

$$\text{SNR} = \frac{2N_p}{F_n + h\nu G F_n^2 \Delta\nu_{\text{opt}}/(2P_{LO})}. \tag{8.6.21}$$

Equation (8.6.21) should be compared with Eq. (6.1.15), which represents
the shot-noise limit of heterodyne receivers. It shows that the SNR is degraded
by the amplifier noise, and the degradation is at least by a factor of F_n (if we
assume that $\eta = 1$). The second term in the denominator of Eq. (8.6.21) can
be made negligible by increasing the local-oscillator power in combination with
an optical filter. The SNR then reduces to

$$\text{SNR} = 2N_p/F_n. \tag{8.6.22}$$

For an ideal amplifier, $F_n = 2$ and SNR $= N_p$. Hence the SNR of a heterodyne
receiver is degraded by a factor of ηF_n when an optical amplifier is used to
preamplify the received signal. The SNR degradation is generally smaller in the
coherent case than in the case of direct detection. An important feature is that
an optical filter is not even necessary, as the second term in Eq. (8.6.21) can be
made negligible by increasing the local-oscillator power.

The receiver sensitivity can be obtained by using the results of Section 6.3.
It depends on the modulation formats (ASK, PSK, and FSK) and on the de-
modulation technique (synchronous versus asynchronous) used at the receiver.
Consider, as an example, the case of FSK asynchronous heterodyne receiver.
The BER is given by Eq. (6.4.25) or by

$$\text{BER} = \tfrac{1}{2} \exp(-\text{SNR}/4). \tag{8.6.23}$$

By using SNR from Eq. (8.6.21), Eq. (8.6.23) can be used to study how the
receiver sensitivity depends on the amplifier parameters. It is easy to verify
from Eq. (8.6.23) that SNR should exceed 80 (or 19 dB) to realize a BER of

$< 10^{-9}$. By using this value in Eq. (8.6.21), the receiver sensitivity in terms of the average number of photons/bit, $\bar{N}_p$, is given by

$$\bar{N}_p = N_p = 40[F_n + h\nu G F_n^2 \Delta\nu_{\text{opt}}/(2P_{LO})]. \qquad (8.6.24)$$

Since $\bar{N}_p = 40$ in the absence of an optical amplifier, the receiver sensitivity degrades by the factor in brackets in Eq. (8.6.24). In the limit of large P_{LO}, the sensitivity degrades by F_n. The main conclusion is that the amplifier noise figure F_n plays the crucial role in the performance of lightwave systems with optical amplifiers and should be minimized as much as possible.

8.6.2 Power Boosters

Improvements in the receiver sensitivity realized by using an EDFA can be used to increase the transmission distance of point-to-point fiber links used for intercity and interisland communications since intermediate repeaters can be eliminated. However, such lightwave systems often require transmitter powers as high as 100 mW when the transmission distance increases beyond 200 km. In an early experiment [119], a 1.8-Gb/s signal was transmitted over 250 km (without regeneration or in-line amplification) by using a combination of two EDFAs acting as a power amplifier at the transmitter and as a preamplifier at the receiver. The performance was improved in a 1992 transmission experiment in which a 2.5-Gb/s signal was transmitted over 318 km [120]. Bit rate was further increased to 5 Gb/s in another experiment that transmitted a signal over 226 km of conventional fibers with a total dispersion of 4100 ps/nm [121]. In this experiment, two EDFAs were used to boost the signal power from -8 dBm to 15.5 dBm (about 35 mW). This power level is large enough that SBS (see Section 8.4) becomes a problem. The experiment suppressed SBS by broadening the laser linewidth from 35 MHz to 160 MHz [see Eq. (8.4.5)]. Most experiments have used unidirectional transmission. In a novel scheme unrepeatered bidirectional transmission was realized by using two EDFAs on each side serving in a dual role [122]. Each EDFA acted as a power booster in the forward direction and as an optical preamplifier in the reverse direction. Even though the wavelength of two transmitters differed by 11 nm, the same EDFA can act as a power booster and preamplifier because of its wide gain bandwidth.

For transmission distances in excess of 300 km, it becomes necessary to employ in-line amplification. Several laboratory experiments have demonstrated repeaterless transmission at 2.5 Gb/s over more than 500 km by using two in-line amplifiers that were remotely pumped from the transmitter and receiver ends by using high-power pump lasers [123], [124]. In addition, an amplifier at the transmitter boosted the launched power close to 100 mW. SBS was suppressed through phase modulation of the optical carrier that broadened the carrier linewidth to above 200 MHz (under CW operation). The effect of GVD was reduced by using dispersion-compensating fibers (see Section 9.5). In a 1996 experiment. the 2.5-Gb/s signal was transmitted over 465 km by directly modulating a DFB laser [125]. The frequency chirping of the modulated signal broadened the spectrum enough that an external phase modulator was not required provided that the launched power was kept below 100 mW. The bit rate of repeaterless systems can be increased to 10 Gb/s by employing the same techniques used at 2.5 Gb/s. In a 1996 experiment [126], the 10-Gb/s signal was transmitted over 442 km using two remotely pumped in-line amplifiers. Two external modulators were used, one for SBS suppression and another for signal generation. These results clearly indicate the advances realized with the use of fiber amplifiers.

8.6.3   Local-Area-Network Amplifiers

EDFAs have been used to overcome distribution and other kinds of losses in LANs and broadcast networks [127]–[139]. An important application is for analog SCM multichannel video distribution networks whose power budget is inherently limited (see Section 7.5.1). The power budget in one experiment [128] was extended to 15 dB by amplifying the multichannel signal with an EDFA that provided 11.5-dB fiber-to-fiber gain. In another experiment [129], an EDFA amplified 100 FM video channels and six 622-Mb/s digital baseband channels simultaneously, covering a 34-nm spectral range. The amplified multichannel signal could be distributed to 4096 subscribers. The concept can be generalized to millions of subscribers by using multiple amplifiers. A broadcast network capable of serving 39.5 million subscribers was demonstrated [130] in 1990 by using two stages of fiber couplers. An EDFA, acting as a power amplifier, boosted the multichannel signal during each stage. The distribution network was capable of broadcasting 384 digital video channels over 27.7 km.

The application of EDFAs for analog video distribution has continued to attract attention during the 1990s because of its practical use by the telephone and CATV industry. In a 30-channel experiment [131], the power budget was as large as 45 dB. In a later experiment [132], the power budget was increased to 73 dB using just three EDFAs. The system was capable of transmitting the composite video signal to 0.52 million subscribers and is suitable for LANs such as fiber-to-the-home systems. In another experiment, the broadcast network was capable of transmitting 35 video channels to 4.2 million subscribers [133].

Experiments have also been performed to show that EDFAs can enhance the LAN performance when star couplers are used for establishing two-way communication between users. In one experiment [134], an EDFA was used as a preamplifier between the star coupler and the receiver. The 14-dB improvement in receiver sensitivity increases the number of users by a factor of 20. Another scheme makes use of star couplers with gain [135], [136]. Several star couplers and EDFAs are combined in such a way that the signal can be distributed without much loss. In addition, a single pump laser can be used to pump all EDFAs, making the scheme economically attractive. A photonic dual-bus architecture with optical amplifiers has also been proposed [138]. The use of fiber amplifiers permits multigigabit operation with 100 nodes over hundreds of kilometers.

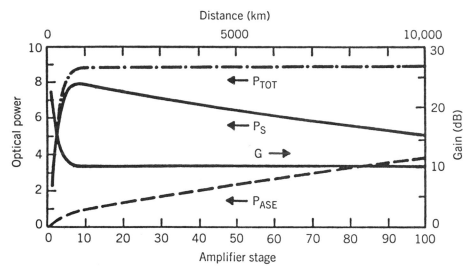

Figure 8.22 Variation of the signal power P_s and the ASE power P_{ASE} along a cascaded chain of optical amplifiers. The total power P_{TOT} becomes nearly constant after a few amplifiers. (After Ref. [141]. ©1991 IEEE. Reprinted with permission.)

8.6.4   Cascaded In-Line Amplifiers

Optical amplifiers are often cascaded to overcome the fiber loss in a long-haul (e.g., undersea) communication system. The design of such lightwave systems requires a consideration of many factors, the most important being amplifier noise, fiber dispersion, and fiber nonlinearity. This section considers some of the design issues.

Amplifier Noise

Amplifier noise affects the system performance in two ways. First, the *amplified spontaneous emission* (ASE) accumulates over many amplifiers. Second, as the level of ASE grows, it begins to saturate optical amplifiers and reduce the signal gain. The net result is that the signal level drops and the ASE level increases along the transmission line, degrading the SNR considerably at the receiver. Numerical simulations show that the system is self-regulating in the sense that the total power obtained by adding the signal and ASE powers remains relatively constant. Figure 8.22 shows this self-regulating behavior for a cascaded chain of 100 amplifiers with 100-km spacing and 35-dB small-signal gain. The power launched by the transmitter is 1 mW. The other parameters are $P_{out}^s = 8$ mW, $n_{sp} = 1.3$, and $G_0 \exp(-\alpha L_A) = 3$, where L_A is the amplifier spacing. The signal and ASE powers become comparable after 10,000 km, indicating the SNR problem at the receiver.

The *effective noise figure* for a cascaded chain of k amplifiers can be obtained

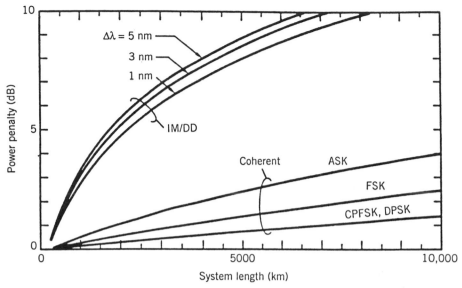

Figure 8.23 Power penalty as a function of the system length for a 2.5-Gb/s lightwave system with cascaded amplifiers spaced 100 km to compensate for the fiber loss. The top three curves show the performance of IM/DD systems for three values of the optical filter bandwidth $\Delta\lambda$. The bottom three curves show the performance of coherent systems with different modulation formats. (After Ref. [143]. ©1991 IEEE. Reprinted with permission.)

by considering the propagation of ASE and is given by [140]–[142]

$$F_n^{\text{eff}} = F_{n1} + \frac{F_{n2}}{G_1} + \frac{F_{n3}}{G_1 G_2} + \cdots + \frac{F_{nk}}{G_1 G_2 \cdots G_{k-1}}, \qquad (8.6.25)$$

where F_{nj} and G_j are the noise figure and the gain of the jth amplifier ($j = 1, \ldots, k$), respectively. This noise figure can be used to calculate the BER and the receiver sensitivity by following the analysis of Sections 8.6.1 and 8.6.2.

From the standpoint of system design, one is often interested in estimating the power penalty, defined as an increase in the optical power received in the presence of an amplifier chain. Figure 8.23 shows the power penalty calculated as a function of system length L_T for the case in which amplifiers are placed every 100 km to exactly balance the 20-dB fiber loss. The launched power is 1 mW, so that each amplifier receives 10 μW (-20 dBm) at its input. Gain saturation by ASE is neglected. The system operates at 2.5 Gb/s, and each amplifier is assumed to have a noise figure of 5 dB. In the direct-detection (IM/DD) case, the power penalty is quite large and exceeds 5 dB for $L_T > 2000$ km even when an optical filter is used to reduce the ASE. By contrast, the power penalty remains small for coherent lightwave systems.

Fiber Dispersion and Nonlinearity

Many experiments performed during the early 1990s demonstrated the potential benefits of in-line amplifiers for increasing the transmission distance of point-to-point fiber links. A 1990 experiment showed transmission over 904 km at 1.2 Gb/s using 12 EDFAs [144]. In a coherent transmission experiment [145], the 2.5-Gb/s signal was transmitted over 2223 km by using 25 EDFAs placed at approximately 80-km intervals, offering a total gain of more than 440 dB. The effective BL product for this experiment exceeds 5.5 (Tb/s)-km. As expected, fiber dispersion is the limiting factor for lightwave systems with in-line amplifiers. Indeed, the experiment was possible only because dispersion-shifted fibers were used throughout the link. Moreover, the zero-dispersion wavelength varied over different fiber segments in such a way that the total dispersion over the entire 2223-km length amounted to only −43 ps/nm at the operating wavelength of 1.55 μm. Even then, a 4.2-dB power penalty was observed in the experiment. By 1992, the total transmission distance could be increased beyond 10,000 km [146]–[149]. In one experiment [147], a 2.5-Gb/s signal was transmitted over 10,073 km using 199 EDFAs. The bit rate was extended to 10 Gb/s in a similar experiment, but the total distance was limited to 6000 km [148]. The performance can be improved further if care is taken to decrease the effect of fiber dispersion by operating very close to the zero-dispersion wavelength. Indeed, an effective transmission distance of 21,000 km at the bit rate of 2.5 Gb/s and of 14,300 km at 5 Gb/s was demonstrated in a laboratory experiment using the recirculating fiber-loop configuration [150].

In practice, it is difficult to realize a total transmission distance L_T of more than a few thousand kilometers. The reason is that both fiber dispersion and fiber nonlinearity limit L_T simply because their effects are cumulative [151]–[158]. A crude estimate of L_T can be obtained by neglecting the *nonlinear effects* during signal transmission. Since amplifiers compensate only for the fiber loss, dispersion limitations discussed in Section 5.2.2 and shown in Fig. 5.4 apply for lightwave systems with cascaded in-line amplifiers if L is replaced by L_T. From Eq. (5.2.3), dispersion limitation for systems making use of standard fibers ($\beta_2 \approx -20$ ps^2/km at 1.55 μm) is $B^2 L_T < 3000$ (Gb/s)^2-km but can be increased by a factor of 20 or more for lightwave systems with dispersion-shifted fibers. For example, a 10-Gb/s lightwave system is limited to $L_T < 6000$ km even with the use of dispersion-shifted fibers having an average dispersion such that $\beta_2 = -1$ ps^2/km.

The estimate above is crude since it does not include the nonlinear effects in optical fibers. Optical pulses propagate over thousands of kilometers in amplified lightwave systems. Even though power levels are relatively modest, the nonlinear effects can become important because of their accumulation over long distances [31]. For single-channel lightwave systems, the most dominant nonlinear phenomenon that limits the system performance is *self-phase modulation* (SPM). A rough estimate of the limitation imposed by the SPM can be obtained from Eq. (2.6.7) by replacing L_{eff} with the total link length L_T and P_{in} by the average power $\bar{P}$ over one amplification stage. The condition $\phi_{NL} \ll 1$ then

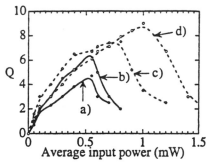

Figure 8.24 Q factor as a function of the average input power for a 9000-km fiber link: (a) 5-Gb/s operation with dispersion-shifted fibers; (b) improvement realized by inserting 150-GHz filters after every amplifier; (c) 6-Gb/s operation near the zero-dispersion wavelength; (d) 10-Gb/s operation with dispersion management. (After Ref. [157]. ©1996 IEEE. Reprinted with permission.)

limits the total link length to $L_T \ll L_{NL}$, where the *nonlinear length* is defined as $L_{NL} = (\bar{\gamma}\bar{P})^{-1}$. Typically, $\bar{\gamma} \approx 1$ W^{-1}/km, and the link length is limited to below 1000 km even for $\bar{P} = 1$ mW.

The estimate of the SPM-limited distance is too simplistic to be accurate since it completely ignores the role of fiber dispersion. In fact, since the dispersive and nonlinear effects act on the optical signal simultaneously, their mutual interplay becomes quite important. As discussed in Section 5.2.3, it is necessary to solve Eq. (5.2.4) numerically. A numerical approach was indeed adopted during the 1990s to quantify the impact of SPM on the performance of periodically amplified lightwave systems [151]–[158]. The effects of amplifier noise must also be included in numerical simulations for a realistic assessment of such lightwave systems. A general approach solves Eq. (5.2.4) in each fiber segment, amplifies the signal inside the EDFA, taking into account the noise spectral density given by Eq. (8.1.15), and repeats the procedure for the entire link. The system performance is quantified through the Q factor as defined in Eq. (4.5.10) since it is related directly to the BER through Eq. (4.5.9).

Solid curves in Fig. 8.24 show variation of the Q factor with the average input power for a NRZ, single-channel (IM/DD) lightwave system designed to operate at 5 Gb/s over 9000 km of dispersion-shifted fibers [$D = 1$ ps/(km-nm)] with 40-km amplifier spacing [157]. An optical filter of 150-GHz bandwidth, inserted after every amplifier, reduces the ASE for the curve (b) while no in-line filters are present in the case of curve (a). Since $Q < 6$ for all input powers, such a system cannot operate reliably in the absence of in-line filters ($Q > 6$ is required for a BER of $< 10^{-9}$). In the presence of optical filters, $Q > 6$ can be realized only at a specific value of the average input power (about 0.5 mW). This behavior can be understood by noting that as the input power increases, the system performance improves first because of a better SNR, but becomes worse at high input powers for which the nonlinear effects (SPM) begins to dominate.

The role of dispersion can be minimized (1) by operating close to the zero-dispersion wavelength of the fiber or (2) by dispersion management such that the fiber dispersion is alternated within each segments in such a way that the average dispersion is zero (see Chapter 9). In the first case, the GVD parameter β_2 fluctuates because of unintentional variations in the zero-dispersion wavelength of various fiber segments. The curve (c) is drawn for a 6-Gb/s system for the case of a Gaussian distribution of β_2 with a standard deviation of 0.3 ps^2/km. The filter bandwidth is taken to be 60 GHz [157]. The curve (d) shows the dispersion-managed case for a 10-Gb/s system with a filter bandwidth of 50 GHz. All other parameters remain the same. Clearly, system performance can improve considerably with dispersion management, although the input pump power needs to be optimized in each case.

It should be evident that the design of lightwave systems with a large number of in-line amplifiers is quite involved and requires control of many system parameters. One way to avoid the detrimental effects of fiber dispersion and nonlinearity is to make use of solitons. Solitons balance the two potentially damaging effects in such a way that the pulse shape is maintained as the optical signal propagates through the fiber link. Chapter 10 is devoted to a discussion of soliton communication systems.

WDM Lightwave Systems

The multichannel-amplification capability of EDFAs has been demonstrated in many system experiments [159]–[166]. In an early experiment [159], four 2.5-Gb/s channels with 2-nm channel spacing were transmitted over 459 km by using six EDFAs as in-line amplifiers. The crosstalk penalty was found to be negligible. A field experiment was also performed by transmitting four channels operating at 1.7 Gb/s over 70 km of fiber [160]. This experiment used coherent detection with a FSK asynchronous receiver. In a 1993 experiment, four channels were transmitted over 1500 km using 22 cascaded amplifiers [161]. By 1996, 55 channels, spaced apart by 0.8 nm and each operating at 20 Gb/s, were transmitted over 150 km by using two in-line amplifiers, resulting in a total bit rate of 1.1 Tb/s and the BL product of 165 (Tb/s)-km [162]. For long-haul applications, the aim is to transmit a large number of channels over a distance of more than 1000 km [163], [164]. In one test-bed experiment, a transmission distance of 6000 km at 20 Gb/s (8 channels at 2.5 Gb/s) has been realized with an amplifier spacing of 75 km [165]. On the high-bit-rate end, a 1996 experiment multiplexed sixteen 10-Gb/s channels to realize transmission at 160 Gb/s, but the link length was only 531 km [166]. In contrast, a fiber-loop experiment demonstrated that 100-Gb/s transmission (20 channels at 5 Gb/s) over a transoceanic distance of 9100 km is possible using the polarization-scrambling and forward-error-correction techniques [164].

The performance requirements of EDFAs are quite stringent for WDM applications. The reason is that small variations in the amplifier gain for individual channels grow exponentially over a chain of in-line amplifiers since the gain spectrum is the same for all amplifiers. Even a 0.2-dB gain difference grows to 20 dB over a chain of 100 in-line amplifiers, making channels powers vary by a factor of 100, an unacceptable variation range in practice. As a result, the number of channels is limited not only by the amplifier bandwidth but also by the spectral nonuniformities seen in Fig. 8.17. To amplify all channels by nearly the same amount, the double-peak nature of the EDFA gain spectrum forces one to pack all channels near one of the gain peaks. Such an approach limits the number of channels drastically.

The entire bandwidth of 30–40 nm can be used if the EDFA gain spectrum is flattened. Several *gain-flattening* techniques have been developed for this purpose [167]–[187]. In one approach, either a notch filter [167] or a Mach–Zehnder filter [168] is used to equalize the gain for various channels. Indeed, up to 128 channels have been transmitted over 480 km by using multiple EDFAs, whose gain was equalized through tunable filters [169], [170]. A combination of long-period fiber gratings can also be used for this purpose since it acts as an optical filter whose transmission spectrum can be tailored to counteract the EDFA gain nonuniformities. Indeed, an EDFA whose gain is flat to within 1 dB over a 40-nm bandwidth has been produced by using such a grating filter [171].

In a different approach, input powers of different channels are adjusted to reduce power variations at the receiver to an acceptable level [172], [173]. Another approach uses *inhomogeneous broadening* of the gain spectrum to equalize channel gains either by cooling each in-line amplifier [178] or by periodically inserting such inhomogeneously broadened amplifiers in a long chain of amplifiers [179]. In a similar approach, fiber amplifiers are designed to provide a flat gain spectrum either by codoping them with materials such as alumina [183] or by using *fluoride* fiber in place of silica fiber as the host medium in which erbium ions are doped [184]–[186]. Gain flatness to within 1.5 dB over a 75-nm bandwidth has been realized bu doping *tellurite* fiber with erbium ions [187]. The state of the art had advanced enough by 1996 that EDFAs with a relatively flat gain over 30-nm bandwidth were available commercially.

PROBLEMS

8.1 The Lorentzian gain profile of an optical amplifier has a FWHM of 1 THz. Calculate the amplifier bandwidths when it is operated to provide 20- and 30-dB gain. Neglect gain saturation.

8.2 An optical amplifier can amplify a 1-μW signal to the 1-mW level. What is the output power when a 1-mW signal is incident on the same amplifier? Assume that the saturation power is 10 mW.

8.3 Explain the concept of noise figure for an optical amplifier. Why does the SNR of the amplified signal degrade by 3 dB even for an ideal amplifier?

8.4 A 250-μm-long semiconductor laser is used as an FP amplifier by biasing it below threshold. Calculate the amplifier bandwidth by assuming 32% reflectivity for both facets and 30-dB peak gain. The group index $n_g = 4$. How much does the bandwidth change when both facets are coated to reduce the facet reflectivities to 1%?

8.5 Complete the derivation of Eq. (8.2.3) by using Eq. (8.2.1). What should be the facet reflectivities to ensure traveling-wave operation of a semiconductor laser amplifier designed to provide 20-dB gain. Assume that $R_1 = 2R_2$.

8.6 A semiconductor laser amplifier is used to amplify two channels separated by 1 GHz. Each channel can be amplified by 30 dB in isolation. What are the channel gains when both channels are amplified simultaneously? Assume that $P_{in}/P_s = 10^{-3}$, $\tau_c = 0.5$ ns, and $\beta_c = 5$.

8.7 Integrate Eq. (8.2.21) to obtain the time-dependent saturated gain given by Eq. (8.2.22). Use it to derive Eq. (8.2.24) for the energy gain.

8.8 Explain why semiconductor laser amplifiers impose a chirp on the pulse during amplification. Derive an expression for the imposed chirp when a Gaussian pulse is incident on the amplifier. Use Eq. (8.2.23) with $m = 1$ for the input pulse.

8.9 Discuss the origin of gain saturation in fiber Raman amplifiers. Solve Eqs. (8.3.2) and (8.3.3) with $\alpha_s = \alpha_p$ and derive Eq. (8.3.8) for the saturated gain.

8.10 Explain the differences between fiber amplifiers based on SRS and SBS. Why do the signal and pump have to counterpropagate for fiber Brillouin amplifiers?

8.11 Explain the gain mechanism in EDFAs. Use the three-level rate equations to derive an expression for the small-signal gain. You can assume a rapid transfer of the pumped population to the excited state.

8.12 Solve Eqs. (8.5.7) and (8.5.8) analytically, or numerically if an analytic solution is not possible, and plot the saturated amplifier gain as a function of the pump power for $\alpha_p L = 5$ and $\alpha_s L = 2$, where L is the amplifier length. You can negelect the fiber loss by setting $\alpha = \alpha' = 0$.

8.13 Starting from Eq. (8.6.10), derive Eq. (8.6.16) for the sensitivity of a direct-detection receiver when an EDFA is used as a preamplifier.

8.14 Calculate the receiver sensitivity at a BER of 10^{-9} and 10^{-12} by using Eq. (8.6.16). Assume that the receiver operates at 1.55 μm with 3-GHz bandwidth. The preamplifier has a noise figure of 4 dB, and a 1-nm optical filter is installed between the preamplifier and the detector.

8.15 Discuss the design issues for a 5000-km-long lightwave system operating at 2.5 Gb/s and using EDFAs every 100 km to overcome the fiber-cable loss of 0.2 dB/km. Can the system work with standard fibers with $D = 16$ ps/(km-nm)?

REFERENCES

[1] A. E. Siegman, *Lasers*, University Science Books, Mill Valley, CA, 1986.

[2] A. Yariv, *Opt. Lett.* **15**, 1064 (1990); H. Kogelnik and A. Yariv, *Proc. IEEE* **52**, 165 (1964).

[3] T. Mukai, Y. Yamamoto, and T. Kimura, *IEEE J. Quantum Electron.* **18**, 1560 (1982).

[4] T. Saitoh and T. Mukai, *IEEE J. Quantum Electron.* **23**, 1010 (1987).

[5] M. J. O'Mahony, *J. Lightwave Technol.* **6**, 531 (1988).

[6] N. A. Olsson, *J. Lightwave Technol.* **7**, 1071 (1989).

[7] T. Saitoh and T. Mukai, in *Coherence, Amplification, and Quantum Effects in Semiconductor Lasers*, Y. Yamamoto, Ed., Wiley, New York, 1991, Chap. 7.

[8] G. P. Agrawal, *Semiconductor Lasers*, Van Nostrand Reinhold, New York, 1993, Chap. 11.

[9] G.-H. Duan, in *Semiconductor Lasers: Past, Present, and Future*, G. P. Agrawal, Ed., AIP Press, Woodbury, NY, 1995, Chap. 10.

[10] C. E. Zah, J. S. Osinski, C. Caneau, S. G. Menocal, L. A. Reith, J. Salzman, F. K. Shokoohi, and T. P. Lee, *Electron. Lett.* **23**, 990 (1987).

[11] I. Cha, M. Kitamura, H. Honmou, and I. Mito, *Electron. Lett.* **25**, 1241 (1989).

[12] S. Cole, D. M. Cooper, W. J. Devlin, A. D. Ellis, D. J. Elton, J. J. Isaak, G. Sherlock, P. C. Spurdens, and W. A. Stallard, *Electron. Lett.* **25**, 314 (1989).

[13] G. Großkopf, R. Ludwig, R. G. Waarts, and H. G. Weber, *Electron. Lett.* **23**, 1387 (1987).

[14] N. A. Olsson, *Electron. Lett.* **24**, 1075 (1988).

[15] M. Sumida, *Electron. Lett.* **25**, 1913 (1989).

[16] M. Koga and T. Mutsumoto, *J. Lightwave Technol.* **9**, 284 (1991).

[17] G. P. Agrawal, *Opt. Lett.* **12**, 260 (1987).

[18] G. P. Agrawal, *Electron. Lett.* **23**, 1175 (1987).

[19] I. M. I. Habbab and G. P. Agrawal, *J. Lightwave Technol.* **7**, 1351 (1989).

[20] S. Ryu, K. Mochizuki, and H. Wakabayashi, *J. Lightwave Technol.* **7**, 1525 (1989).

[21] G. P. Agrawal and I. M. I. Habbab, *IEEE J. Quantum Electron.* **26**, 501 (1990).

[22] G. P. Agrawal, *Appl. Phys. Lett.* **51**, 302 (1987); *J. Opt. Soc. Am. B* **5**, 147 (1988).

[23] G. P. Agrawal and N. A. Olsson, *IEEE J. Quantum Electron.* **25**, 2297 (1989).

[24] G. P. Agrawal and N. A. Olsson, *Opt. Lett.* **14**, 500 (1989).

[25] N. A. Olsson, G. P. Agrawal, and K. W. Wecht, *Electron. Lett.* **25**, 603 (1989).

[26] N. A. Olsson, M. G. Öberg, L. A. Koszi, and G. J. Przybylek, *Electron. Lett.* **24**, 36 (1988).

[27] M. G. Öberg, N. A. Olsson, L. A. Koszi, and G. J. Przybylek, *Electron. Lett.* **24**, 38 (1988).

[28] K. T. Koai, R. Olshansky, and P. M. Hill, *IEEE Photon. Technol. Lett.* **2**, 926 (1990).

[29] R. H. Stolen, E. P. Ippen, and A. R. Tynes, *Appl. Phys. Lett.* **20**, 62 (1972).

[30] R. H. Stolen, *Proc. IEEE* **68**, 1232 (1980).

[31] G. P. Agrawal, *Nonlinear Fiber Optics*, 2nd ed., Academic Press, San Diego, CA, 1995.

[32] M. Ikeda, *Opt. Commun.* **39**, 148 (1981).

[33] M. Nakazawa, M. Tokuda, Y. Negishi, and N. Uchida, *J. Opt. Soc. Am. B* **1** 80 (1984).

[34] J. Hegarty, N. A. Olsson, and L. Goldner, *Electron. Lett.* **21**, 290 (1985).

[35] M. L. Dakss and P. Melman, *IEE Proc.* **135**, Pt. J, 95 (1988).

[36] T. Horiguchi, T. Sato, and Y. Koyamada, *IEEE Photon. Technol. Lett.* **4**, 64 (1992).

[37] T. Sato, T. Horiguchi, Y. Koyamada, and I. Sankawa, *IEEE Photon. Technol. Lett.* **4**, 923 (1992).

[38] S. G. Grubb, Paper SaA1, *Proc. Conf. on Optical Amplifiers and Applications*, Optical Society of America, Washington, DC, 1995.

[39] S. V. Chernikov, Y. Zhu, R. Kashyap, and J. R. Taylor, *Electron. Lett.* **31**, 472 (1995).

[40] E. M. Dianov, *Laser Phys.* **6**, 579 (1996).

[41] P. B.Hansen, A. J. Stentz, L. Eskilden, S. G. Grubb, T. A. Strasser, and J. R. Pedrazzani, *Electron. Lett.* **27**, 2164 (1996).

[42] P. B.Hansen, L. Eskilden, S. G. Grubb, A. J. Stentz, T. A. Strasser, J. Judkins, J. J. DeMarco, J. R. Pedrazzani, and D. J. DiGiovanni, *IEEE Photon. Technol. Lett.* **9**, 262 (1997).

[43] E. P. Ippen and R. H. Stolen, *Appl. Phys. Lett.* **21**, 539 (1972).

[44] D. Cotter, *Electron. Lett.* **18**, 495 (1982); *J. Opt. Commun.* **4**, 10 (1983).

[45] R. W. Tkach, A. R. Chraplyvy, and R. M. Derosier, *Electron. Lett.* **22**, 1011 (1986).

[46] E. Lichtman, A. A. Friesem, R. G. Waarts, and H. H. Yaffe, *J. Opt. Soc. Am. B* **4**, 1397 (1987); E. Lichtman and A. A. Friesem, *Opt. Commun.* **64**, 544 (1987).

[47] C. L. Tang, *J. Appl. Phys.* **37**, 2945 (1966).

[48] N. A. Olsson and J. P. van der Ziel, *Appl. Phys. Lett.* **48**, 1329 (1986); *Electron. Lett.* **22**, 488 (1986).

[49] N. A. Olsson and J. P. van der Ziel, *J. Lightwave Technol.* **5**, 147 (1987).

[50] J. A. Arnaud, *IEEE J. Quantum Electron.* **4**, 893 (1968).

[51] C. G. Atkins, D. Cotter, D. W. Smith, and R. Wyatt, *Electron. Lett.* **22**, 556 (1986).

[52] D. Cotter, D. W. Smith, C. G. Atkins, and R. Wyatt, *Electron. Lett.* **22**, 671 (1986).

[53] R. G. Waarts, A. A. Friesem, and Y. Hefetz, *Opt. Lett.* **13**, 152 (1988).

[54] A. R. Chraplyvy and R. W. Tkach, *Electron. Lett.* **22**, 1084 (1986).

[55] R. W. Tkach, A. R. Chraplyvy, and R. M. Derosier, *IEEE Photon. Technol. Lett.* **1**, 111 (1989).

[56] R. W. Tkach, A. R. Chraplyvy, R. M. Derosier, and H. T. Shang, *Electron. Lett.* **24**, 260 (1988).

[57] Y.-H. Lee, H.-W. Tsao, M.-S. Kao, and J. Wu, *J. Opt. Commun.* **13**, 99 (1992).

[58] Y. Sato, Y. Yamabayashi, and K. Aoyama, *J. Lightwave Technol.* **11**, 1652 (1993).

[59] C. J. Koester and E. Snitzer, *Appl. Opt.* **3**, 1182 (1964).

[60] S. B. Poole, D. N. Payne, R. J. Mears, M. E. Fermann, and R. E. Laming, *J. Lightwave Technol.* **4**, 870 (1986).

[61] M. J. F. Digonnet, Ed., *Rare-Earth Doped Fiber Lasers and Amplifiers*, Marcel Dekker, New York, 1993.

[62] A. Bjarklev, *Optical Fiber Amplifiers: Design and System Applications*, Artech House, Boston, 1993.

[63] E. Desurvire, *Erbium-Doped Fiber Amplifiers*, Wiley, New York, 1994.

[64] M. Shimizu, M. Yamada, H. Horiguchi, T. Takeshita, and M. Okayasu, *Electron. Lett.* **26**, 1641 (1990).

[65] M. Nakazawa, Y. Kimura, E. Yoshida, and K. Suzuki, *Electron. Lett.* **26**, 1936 (1990).

[66] B. Pederson, A. Bjarklev, H. Vendeltorp-Pommer, and J. H. Povlesen, *Opt. Commun.* **81**, 23 (1991).

[67] M. Horiguchi, K. Yoshino, M. Shimizu, and M. Yamada, *Electron. Lett.* **29**, 593 (1993).

[68] R. I. Laming, J. E. Townsend, D. N. Payne, F. Meli, G. Grasso, and E. J. Tarbox, *IEEE Photon. Technol. Lett.* **3**, 253 (1991).

[69] W. J. Miniscalco, *J. Lightwave Technol.* **9**, 234 (1991).

[70] J. L. Zyskind, E. Desurvire, J. W. Sulhoff, and D. J. DiGiovanni, *IEEE Photon. Technol. Lett.* **2**, 869 (1990).

[71] C. R. Giles and E. Desurvire, *J. Lightwave Technol.* **9**, 271 (1991).

[72] B. Pedersen, A. Bjarklev, O. Lumholt, and J. H. Povlsen, *IEEE Photon. Technol. Lett.* **3**, 548 (1991).

[73] K. Nakagawa, S. Nishi, K. Aida, and E. Yoneda, *J. Lightwave Technol.* **9**, 198 (1991).

[74] S. G. Grubb, W. H. Humer, R. S. Cannon, S. W. Nendetta, K. L. Sweeney, P. A. Leilabady, M. R. Keur, J. G. Kwasegroch, T. C. Munks, and P. W. Anthony, *Electron. Lett.* **28**, 1275 (1992).

[75] E. Delevaque, T. Georges, J. F. Bayon, M. Monerie, P. Niay, and P. Benarge, *Electron. Lett.* **29**, 1112 (1993).

[76] R. Olshansky, *Electron. Lett.* **24**, 1363 (1988).

[77] M. Yamada, M. Shimizu, M. Okayasu, T. Takeshita, M. Horiguchi, Y. Tachikawa, and E. Sugita, *IEEE Photon. Technol. Lett.* **2**, 205 (1990).

[78] R. I. Laming and D. N. Payne, *IEEE Photon. Technol. Lett.* **2**, 418 (1990).

[79] K. Kikuchi, *Electron. Lett.* **26**, 1851 (1990).

[80] R. I. Laming, M. N. Zervas, and D. N. Payne, *IEEE Photon. Technol. Lett.* **4**, 1345 (1992).

[81] K. Kikuchi, *IEEE Photon. Technol. Lett.* **5**, 221 (1993).

[82] M. Murakami and S. Saito, *IEEE Photon. Technol. Lett.* **4**, 1269 (1992).

[83] E. Desurvire, C. R. Giles, and J. R. Simpson, *J. Lightwave Technol.* **7**, 2095 (1989).

[84] K. Inoue, H. Toba, N. Shibata, K. Iwatsuki, A. Takada, and M. Shimizu, *Electron. Lett.* **25**, 594 (1989).

[85] C. R. Giles, E. Desurvire, and J. R. Simpson, *Opt. Lett.* **14**, 880 (1990).

[86] C. Y. Kuo, *J. Lightwave Technol.* **11**, 7 (1993).

[87] K. Kikushima, *J. Lightwave Technol.* **12**, 463 (1994).

[88] D. Lipka and D. Werner, *Electron. Lett.* **30**, 1940 (1994).

[89] A. Takada, K. Iwatsuki, and M. Saruwatari, *IEEE Photon. Technol. Lett.* **2**, 122 (1990).

[90] B. J. Ainslie, K. J. Blow, A. S. Gouveia-Neto, P. G. J. Wigley, A. S. B. Sombra, and J. R. Taylor, *Electron. Lett.* **26**, 186 (1990).

[91] I. Yu. Khrushchev, A. B. Grudinin, E. M. Dianov, D. V. Korobkin, V. A. Semenov, and A. M. Prokhorov, *Electron. Lett.* **26**, 456 (1990).

[92] A. B. Grudinin, E. M. Dianov, D. V. Korobkin, A. Yu. Makarenko, A. M. Prokhorov, and I. Yu. Khrushchev, *JETP Lett.* **51**, 135 (1990).

[93] M. Nakazawa, K. Kurokawa, H. Kubota, K. Suzuki, and Y. Kimura, *Appl. Phys. Lett.* **57**, 653 (1990).

[94] G. P. Agrawal, *Opt. Lett.* **16**, 226 (1991); *Phys. Rev. A* **44**, 7493 (1991).

[95] J. E. Pedersen, M. C. Brierley, S. F. Carter, and P. W. France, *Electron. Lett.* **26**, 329 (1990).

[96] Y. Durteste, M. Monerie, J. Y. Allain, and H. Poignant, *Electron. Lett.* **27**, 628 (1991).

[97] S. F. Carter, D. Szebesta, S. T. Davey, R. Wyatt, M. C. Brierley, and P. W. France, *Electron. Lett.* **27**, 628 (1991).

[98] Y. Miyajima, T. Sugawa, and Y. Fukasaku, *Electron. Lett.* **27**, 1706 (1991).

[99] Y. Ohishi, T. Kanamori, T. Nishi, and S. Takahashi, *IEEE Photon. Technol. Lett.* **3**, 715 (1991).

[100] T. Whitley, R. Wyatt, D. Szebesta, S. Davey, and J. R. Williams, *IEEE Photon. Technol. Lett.* **5**, 399 (1993).

[101] M. Shimizu, T. Kanamori, J. Temmyo, M. Wada, M. Yamada, Y. Terunuma, Y. Ohishi, and S. Sudo, *IEEE Photon. Technol. Lett.* **5**, 654 (1993).

[102] M. Yamada, M. Shimizu, T. Kanamori, Y. Ohishi, Y. Terunuma, K. Oikawa, K. Kitushima, Y. Miyamoto, and S. Sudo, *IEEE Photon. Technol. Lett.* **7**, 969 (1995).

[103] D. L. Williams, S. T. Davey, D. M. Spirit, and B. L. Ainslie, *Electron. Lett.* **26**, 517 (1990).

[104] G. R. Walker, D. M. Spirit, D. L. Williams, and S. T. Davey, *Electron. Lett.* **27**, 1390 (1991)

[105] D. N. Chen and E. Desurvire, *IEEE Photon. Technol. Lett.* **4**, 52 (1992).

[106] K. Rottwitt, J. H. Povlsen, A. Bjarklev, O. Lumholt, B. Pedersen, and T. Rasmussen, *IEEE Photon. Technol. Lett.* **4**, 714 (1992); **5**, 218 (1993).

[107] K. Rottwitt, J. H. Povlsen, and A. Bjarklev, *J. Lightwave Technol.* **11**, 2105 (1993).

[108] C. Lester, K. Bertilsson, K. Rottwitt, P. A. Andrekson, M. A. Newhouse, and A. J. Antos, *Electron. Lett.* **31**, 219 (1995).

[109] A. Altuncu, L. Noel, W. A. Pender, A. S. Siddiqui, T. Widdowson, A. D. Ellis, M. A. Newhouse, A. J. Antos, G. Kar, and P. W. Chu, *Electron. Lett.* **32**, 233 (1996).

[110] M. Nissov, H. N. Poulsen, R. J. Pedersen, B. F. Jørgensen, M. A. Newhouse, and A. J. Antos, *Electron. Lett.* **32**, 1905 (1996).

[111] P. P. Smyth, R. Wyatt, A. Fidler, P. Eardley, A. Sayles, and S. Graig-Ryan, *Electron. Lett.* **26**, 1604 (1990).

[112] R. C. Steele and G. R. Walker, *IEEE Photon. Technol. Lett.* **2**, 753 (1990).

[113] T. L. Blair and H. Nakano, *Electron. Lett.* **27**, 835 (1991).

[114] T. Saito, Y. Sunohara, K. Fukagai, S. Ishikawa, N. Henmi, S. Fujita, and Y. Aoki, *IEEE Photon. Technol. Lett.* **3**, 551 (1991).

[115] A. H. Gnauck and C. R. Giles, *IEEE Photon. Technol. Lett.* **4**, 80 (1992).

[116] F. F. Röhl and R. W. Ayre, *IEEE Photon. Technol. Lett.* **5**, 358 (1993).

[117] R. C. Steele, G. R. Walker, and N. G. Walker, *IEEE Photon. Technol. Lett.* **3**, 545 (1991).

[118] O. K. Tonguz and L. G. Kazovsky, *J. Lightwave Technol.* **9**, 174 (1991).

[119] K. Hagimoto, K. Iwatsuki, A. Takada, M. Nakazawa, M. Saruwatari, K. Aida, and K. Nakagawa, *Electron. Lett.* **25**, 662 (1989).

[120] Y. K. Park, S. W. Granlund, T. W. Cline, L. D. Tzeng, J. S. French, J.-M. P. Delavaux, R. E. Tench, S. K. Korotky, J. J. Veselka, and D. J. DiGiovanni, *IEEE Photon. Technol. Lett.* **4**, 179 (1992).

[121] Y. K. Park, O. Mizuhara, L. D. Tzeng, J.-M. P. Delavaux, T. V. Nguyen, M. L. Kao, P. D. Yates, and J. Stone, *IEEE Photon. Technol. Lett.* **5**, 79 (1993).

[122] E. Kannan and S. Frisken, *IEEE Photon. Technol. Lett.* **5**, 76 (1993).

[123] P. B. Hansen, L. Eskildsen, S. G. Grubb, A. M. Vengsarkar, S. K. Korotky, T. A. Strasser, J. E. J. Alphonsus, J. J. Veselka, D. J. DiGiovanni, D. W. Peckham, E. C. Beck, D. Truxal, W. Y. Cheung, S. G. Kosinski, D. Gasper, P. F. Wysocki, V. L. da Silva, and J. R. Simpson, *Electron. Lett.* **31**, 1460 (1995).

[124] S. S. Sian, O. Gautheron, M. S. Chaudhry, C. D. Stark, S. M. Webb, K. M. Guild, M. Mesic, J. M. Dryland, J. R. Chapman, A. R. Docker, E. Brandon, T. Barbier, P. Garabedian, and P. Bousselet, Paper PD26, *Proc. Optical Fiber Commun. Conf.*, Optical Society of America, Washington, DC, 1995.

[125] L. Eskildsen, P. B. Hansen, S. G. Grubb, A. M. Vengsarkar, T. A. Strasser, J. E. J. Alphonsus, D. J. DiGiovanni, D. W. Peckham, D. Truxal, and W. Y. Cheung, *IEEE Photon. Technol. Lett.* **8**, 724 (1996).

[126] P. B. Hansen, L. Eskildsen, S. G. Grubb, A. M. Vengsarkar, S. K. Korotky, T. A. Strasser, J. E. J. Alphonsus, J. J. Veselka, D. J. DiGiovanni, D. W. Peckham, and D. Truxal, *Electron. Lett.* **32**, 1018 (1996).

[127] H. Toba, K. Inoue, N. Shibata, K. Nosu, K. Iwatsuki, N. Takato, and M. Shimizu, *Electron. Lett.* **25**, 885 (1989).

[128] W. I. Way, M. M. Choy, A. Yi-Yan, M. Andrejco, M. Saifi, and C. Lin, *IEEE Photon. Technol. Lett.* **1**, 343 (1989).

[129] W. I. Way, M. W. Maeda, A. Yi-Yan, M. Andrejco, M. M. Choy, M. Saifi, and C. Lin, *Electron. Lett.* **26**, 139 (1990).

[130] A. M. Hill, R. Wyatt, J. F. Massicott, K. J. Blyth, D. S. Forrester, R. A. Lobbett, P. J. Smith, and D. B. Payne, *Electron. Lett.* **26**, 1882 (1990).

[131] P. M. Gabla, C. Bastide, Y. Cretin, P. Bousselet, A. Pitel, and J. P. Blondel, *IEEE Photon. Technol. Lett.* **4**, 510 (1992).

[132] H. Bülow, R. Fritschi, R. Heidemann, B. Junginger, H. G. Krimmel, and J. Otterbach, *IEEE Photon. Technol. Lett.* **4**, 1287 (1992).

[133] H. Bülow, R. Fritschi, R. Heidemann, B. Junginger, H. G. Krimmel, and J. Otterbach, *Electron. Lett.* **28**, 1836 (1992).

[134] A. E. Willner, E. Desurvire, H. M. Presby, C. A. Edwards, and J. R. Simpson, *IEEE Photon. Technol. Lett.* **2**, 669 (1990).

[135] A. E. Willner, A. A. M. Saleh, H. M. Presby, D. J. DiGiovanni, and C. A. Edwards, *IEEE Photon. Technol. Lett.* **3**, 250 (1991).

[136] Y. K. Chen, S. Chi, and J. W. Liaw, *IEEE Photon. Technol. Lett.* **5**, 230 (1993).

[137] W. Y. Guo and Y. K. Chen, *IEEE Photon. Technol. Lett.* **5**, 232 (1993).

[138] K. T. Koai and R. Olshansky, *IEEE Photon. Technol. Lett.* **5**, 482 (1993).

[139] A. Willner and S.-M. Hwang, *IEEE Photon. Technol. Lett.* **6**, 760 (1994).

[140] Y. Yamamoto and T. Mukai, *Opt. Quantum Electron.* **21**, S1 (1989).

[141] C. R. Giles and E. Desurvire, *J. Lightwave Technol.* **9**, 147 (1991).

[142] G. R. Walker, N. G. Walker, R. C. Steele, M. J. Creaner, and M. C. Brain, *J. Lightwave Technol.* **9**, 182 (1991).

[143] S. Ryu, S. Yamamoto, H. Taga, N. Edagawa, Y. Yoshida, and H. Wakabayashi, *J. Lightwave Technol.* **9**, 251 (1991).

[144] N. Edagawa, Y. Toshida, H. Taga, S. Yamamoto, K. Mochizuchi, and H. Wakabayashi, *Electron. Lett.* **26**, 66 (1990).

[145] S. Saito, T. Imai, and T. Ito, *J. Lightwave Technol.* **9**, 161 (1991).

[146] S. Saito, *J. Lightwave Technol.* **10**, 1117 (1992).

[147] T. Imai, M. Murakami, T. Fukuda, M. Aiki, and T. Ito, *Electron. Lett.* **28**, 1484 (1992).

[148] M. Murakami, T. Kataoka, T. Imai, K. Hagimoto, and M. Aiki, *Electron. Lett.* **28**, 2254 (1992).

[149] H. Taga, M. Suzuki, N. Edagawa, Y. Yoshida, S. Yamamoto, S. Akiba, and H. Wakabayashi, *Electron. Lett.* **28**, 2247 (1992).

[150] N. S. Bergano, J. Aspell, C. R. Davidson, P. R. Trischitta, B. M. Nyman, and F. W. Kerfoot, *Electron. Lett.* **27**, 1889 (1991).

[151] J. P. Hamaide, P. Emplit, and J. M. Gabriagues, *Electron. Lett.* **26**, 1451 (1990).

[152] D. Marcuse, *J. Lightwave Technol.* **9**, 356 (1991); **9**, 505 (1991); **9**, 1330 (1991).

[153] A. Naka and S. Saito, *Electron. Lett.* **28**, 2221 (1992).

[154] A. Mecozzi, *J. Opt. Soc. Am. B* **11**, 462 (1994).

[155] A. Naka and S. Saito, *J. Lightwave Technol.* **12**, 280 (1994).

[156] E. Lichtman, *J. Lightwave Technol.* **13**, 898 (1995).

[157] F. Matera and M. Settembre, *J. Lightwave Technol.* **14**, 1 (1996).

[158] N. Kikuchi and S. Sasaki, *Electron. Lett.* **32**, 570 (1996).

[159] H. Taga, Y. Yoshida, S. Yamamoto, and H. Wakabayashi, *Electron. Lett.* **26**, 500 (1990).

[160] P. M. Gabla, J. O. Frorud, E. Leclerc, S. Gauchard, and V. Havard, *IEEE Photon. Technol. Lett.* **4**, 717 (1992).

[161] H. Toga, N. Edagawa, Y. Yoshida, S. Yamamoto, and H. Wakabayashi, *Electron. Lett.* **29**, 485 (1993).

[162] H. Onaka, H. Miyata, G. Ishikawa, K. Otsuka, H. Ooi, Y. Kai, S. Kinoshita, M. Seino, H. Nishimoto, and T. Chikama, Paper PD19, *Proc. Optical Fiber Commun. Conf.*, Optical Society of America, Washington, DC, 1996.

[163] H. Taga, *J. Lightwave Technol.* **14**, 1287 (1996).

[164] N. S. Bergano and C. R. Davidson, *J. Lightwave Technol.* **14**, 1287 (1996).

[165] O. Gautheron, G. Bassier, V. Letellier, G. Grandpierre, and P. Bollaert, *Electron. Lett.* **32**, 1019 (1996).

[166] S. Artigaud, M. Chbat, P. Nouchi, F. Chiquet, D. Bayart, L. Hamon, A. Pitel, F. Goudeseune, P. Bousselet, and J.-L. Beylat, *Electron. Lett.* **32**, 1389 (1996).

[167] M. Tachibana, R. I. Laming, P. R. Morkel, and D. N. Payne, *IEEE Photon. Technol. Lett.* **3**, 118 (1991).

[168] K. Inoue, K. Tominato, and H. Toba, *IEEE Photon. Technol. Lett.* **3**, 718 (1991)

[169] H. Toba, K. Nakanishi, K. Oda, K. Inoue, and T. Kominato, *IEEE Photon. Technol. Lett.* **5**, 248 (1993).

[170] K. Oda, M. Fukutoku, H. Toba, and T. Kominato, *Electron. Lett.* **30**, 982 (1994).

[171] P. F. Wysocki, J. Judkins, R. Espindola, M. Andrejco, A. Vengsarkar, and K. Walker, Paper PD2, *Proc. Optical Fiber Commun. Conf.*, Optical Society of America, Washington, DC, 1997.

[172] A. R. Chraplyvy, J. A. Nagel, and R. W. Tkach, *IEEE Photon. Technol. Lett.* **4**, 920 (1992).

[173] A. R. Chraplyvy, R. W. Tkach, K. C. Reichmann, P. D. Magill, and J. A. Nagel, *IEEE Photon. Technol. Lett.* **5**, 428 (1993).

[174] M. Wilkinson, A. Bebbington, S. A. Cassidy, and P. McKee, *Electron. Lett.* **28**, 131 (1992).

[175] R. Kashyap, R. Wyatt, and R. J. Campbell, *Electron. Lett.* **29**, 154 (1993).

[176] R. Kashyap, R. Wyatt, and P. F. McKee, *Electron. Lett.* **29**, 1025 (1993).

[177] S. F. Su, R. Olshansky, D. A. Smith, and J. E. Baran, *Electron. Lett.* **29**, 477 (1993).

[178] E. L. Goldstein, V. da Silva, L. Eskildsen, M. Andrejco, and Y. Silberberg, *IEEE Photon. Technol. Lett.* **5**, 543 (1993).

[179] L. Eskildsen, E. L. Goldstein, V. da Silva, M. Andrejco, and Y. Silberberg, *IEEE Photon. Technol. Lett.* **5**, 1188 (1993).

[180] A. Willner and S.-M. Hwang, *IEEE Photon. Technol. Lett.* **5**, 1023 (1993).

[181] J. F. Massicott, S. D. Wilson, R. Wyatt, R. R. Armitage, R. Kashyap, D. Williams, and R. A. Lobbett, *Electron. Lett.* **30**, 962 (1994).

[182] M. Semenkoff, M. Guibert, J. F. Kerdiles, and Y. Sorel, *Electron. Lett.* **30**, 1411 (1994).

[183] S. Yoshida, S. Kuwano, and K. Iwashita, *Electron. Lett.* **31**, 1765 (1995).

[184] B. Clesca, D. Ronarch, D. Bayart, Y. Sorel, L. Hamon, M. Giubert, J.-L. Beylat, J. F. Kerdiles, and M. Semenkoff, *IEEE Photon. Technol. Lett.* **6**, 509 (1994).

[185] D. Bayart, B. Clesca, L. Hamon, and J.-L. Beylat, *IEEE Photon. Technol. Lett.* **6**, 613 (1994).

[186] M. Yamada, T. Kanamori, Y. Terunuma, K. Oikawa, M. Shimizu, S. Sudo, and K. Sagawa, *IEEE Photon. Technol. Lett.* **8**, 882 (1996).

[187] A. Mori, Y. Ohishi, M. Yamada, H. Ono, Y. Nishida, K. Oikawa, and S. Sudo, Paper PD1, *Proc. Optical Fiber Commun. Conf.*, Optical Society of America, Washington, DC, 1997.

Chapter 9

DISPERSION COMPENSATION

It should be clear from Chapter 8 that with the advent of the optical amplifier, fiber loss is no longer a major limiting factor for optical communication systems. Indeed, most state-of-the-art lightwave systems are limited by fiber dispersion rather than fiber loss. In some sense, optical amplifiers solve the loss problem but, at the same time, worsen the dispersion problem since, in contrast with electronic regenerators, an optical amplifier does not restore the amplified signal to its original state. As a result, dispersion-induced degradation of the transmitted signal accumulates over multiple amplifiers. During the 1990s several dispersion-compensation schemes have been developed to address the dispersion problem [1]. In this chapter we review these schemes with an emphasis on the underlying physics and the improvement realized in practice. In Section 9.1 we explain why dispersion compensation is needed. Sections 9.2 and 9.3 are devoted to the electronic methods used at the receiver or transmitter for dispersion compensation. In Sections 9.4–9.6 we consider the use of high-dispersion optical elements along the fiber link, such as a two-mode fiber or a fiber grating. The technique of optical phase conjugation, also known as midspan spectral inversion, is discussed in Section 9.7. Section 9.8 is devoted to the broadband dispersion compensation needed when optical TDM or WDM is used for increasing the system capacity. In the final section we consider dispersion compensation in long-haul lightwave systems.

9.1 DISPERSION LIMITATIONS

In Section 2.4 we discussed the limitations imposed on system performance by dispersion-induced pulse broadening. As shown by the dashed line in Fig. 2.13, the group-velocity dispersion (GVD) effects can be minimized using a narrow-linewidth laser and operating close to the zero-dispersion wavelength λ_{ZD} of the fiber. However, it is not always practical to match the operating wavelength

λ with λ_{ZD}. An example is provided by third-generation lightwave systems operating near 1.55 μm and using optical transmitters containing a DFB laser. Such systems generally use the existing worldwide network, consisting of more than 5×10^7 km of the "standard" single-mode fiber with $\lambda_{\text{ZD}} \approx 1.31$ μm. Since the dispersion parameter $D \approx 16$ ps/(km-nm) in the 1.55-μm region of such fibers, the GVD severely limits system performance when the bit rate exceeds 2 Gb/s (see Fig. 2.13). For a directly modulated DFB laser, one can use Eq. (2.4.26) for estimating the limiting transmission distance and obtain

$$L < (4B|D|\sigma_\lambda)^{-1}, \tag{9.1.1}$$

where σ_λ is the RMS spectral width with a typical value of 0.15 nm because of spectral broadening induced by frequency chirping (see Section 3.3.7). Using $D = 16$ ps/(km-nm) in Eq. (9.1.1), lightwave systems operating at 2.5 Gb/s are limited to $L \approx 42$ km. Indeed, such systems use electronic regenerators, spaced apart by about 40 km, and cannot benefit from the availability of optical amplifiers. Furthermore, their bit rate cannot be increased beyond 2.5 Gb/s because the regenerator spacing becomes too small to be feasible economically.

System performance can be improved considerably by using an external modulator and thus avoiding spectral broadening induced by frequency chirping. This option has become practical with the commercialization of transmitters containing DFB lasers with a monolithically integrated modulator. The $\sigma_\lambda = 0$ line in Fig. 2.13 provides the dispersion limit when such transmitters are used with standard telecommunication fibers. The limiting transmission distance can be estimated from Eq. (2.4.31) and is given by

$$L < (16|\beta_2|B^2)^{-1}, \tag{9.1.2}$$

where β_2 is the GVD coefficient related to D by Eq. (2.3.5). By using a typical value $\beta_2 = -20$ ps^2/km at 1.55 μm, $L < 500$ km at 2.5 Gb/s. Although improved considerably compared with the case of directly modulated DFB lasers, this dispersion limit becomes of concern when in-line amplifiers are used for loss compensation. Moreover, if the bit rate is increased to 10 Gb/s, the GVD-limited transmission distance drops to 30 km, a value so low that optical amplifiers cannot be used in designing such lightwave systems. It is evident from Eq. (9.1.2) that the relatively large GVD of standard single-mode fibers severely limits the performance of 1.55-μm systems designed to use the existing telecommunication network at a bit rate of 10 Gb/s or more.

The recent development of dispersion-compensation schemes is aimed at solving this practical problem. The basic idea behind all such schemes is quite simple and can be understood by using the pulse-propagation equation, derived in Section 2.4.1 and written as

$$\frac{\partial A}{\partial z} + \frac{i}{2}\beta_2 \frac{\partial^2 A}{\partial t^2} - \frac{1}{6}\beta_3 \frac{\partial^3 A}{\partial t^3} = 0, \tag{9.1.3}$$

where A is the amplitude of the pulse envelope. The effect of third-order dispersion is included by the β_3 term. In practice, this term can be neglected when

$|\beta_2|$ exceeds 1 ps²/km. Equation (9.1.3) has been solved in Section 2.4.2, and the solution is given as Eq. (2.4.15). In the specific case of $\beta_3 = 0$ the solution is given by

$$A(z,t) = \frac{1}{2\pi} \int_{-\infty}^{\infty} \tilde{A}(0,\omega) \exp\left(\frac{i}{2}\beta_2 z\omega^2 - i\omega t\right) d\omega, \qquad (9.1.4)$$

where $\tilde{A}(0,\omega)$ is the Fourier transform of $A(0,t)$.

Dispersion-induced degradation of the optical signal is caused by the phase factor $\exp(i\beta_2 z\omega^2/2)$, acquired by spectral components of the pulse during its propagation in the fiber. All dispersion-compensation schemes attempt to cancel this phase factor so that the input signal can be restored. Actual implementation can be carried out at the receiver, at the transmitter, or along the fiber link. In the following sections we consider various implementations separately.

9.2   POSTCOMPENSATION TECHNIQUES

Electronic techniques can be used for compensation of GVD within the receiver. The philosophy behind this approach is that even though the optical signal has been degraded by GVD, one may be able to equalize the effects of dispersion electronically if the fiber acts as a *linear system*. It is relatively easy to compensate for dispersion if a coherent heterodyne receiver is used for signal detection. The reason can be understood from the discussion in Section 6.1. A heterodyne receiver first converts the optical signal into a microwave signal at the intermediate frequency ω_{IF} while preserving both the amplitude and phase information A microwave bandpass filter whose impulse response is governed by the transfer function

$$H(\omega) = \exp[-i(\omega - \omega_{IF})^2\beta_2 L/2], \qquad (9.2.1)$$

where L is the fiber length, should restore to its original form the signal received. This conclusion follows from the standard theory of linear systems (see Section 4.3.2) by using Eq. (9.1.4) with $z = L$. This technique is most practical for dispersion compensation in coherent lightwave systems [2]. In a synchronous PSK transmission experiment [3], a 31.5-cm-long *microstrip line* was used for dispersion equalization. Its use made it possible to transmit the 8-Gb/s signal over 188 km of standard fiber having a dispersion of 18.5 ps/(km-nm). In a 1993 experiment, the technique was extended to homodyne detection using *single-sideband transmission* [4], and the 6-Gb/s signal could be recovered at the receiver after propagating over 270 km of standard fiber. Microstrip lines can be designed to compensate for GVD acquired over fiber lengths as long as 4900 km for a lightwave system operating at a bit rate of 2.5 Gb/s [5].

As discussed in Chapter 6, use of a coherent receiver is often not practical. An electronic dispersion equalizer for a direct-detection receiver is much more practical. A linear electronic circuit cannot compensate GVD in this case. The problem lies in the fact that all phase information is lost during direct detection since a photodetector responds to optical intensity only (see Chapter 4). As a result, no linear equalization technique can recover a signal that has spread outside its allocated bit slot. Nevertheless, several nonlinear equalization techniques have been developed that permit recovery of the degraded signal [6]–[9]. In one method, the decision threshold, normally kept fixed at the center of the eye diagram (see Section 4.3.3), is varied depending on the preceding bits. In another, the decision about a given bit is made after examining the analog waveform over a multiple-bit interval surrounding the bit in question [6]. The main difficulty with all such techniques is that they require electronic logic circuits, which must operate at the bit rate and whose complexity increases exponentially with the number of bits over which an optical pulse has spread because of GVD-induced pulse broadening. Consequently, electronic equalization is generally limited to low bit rates and to transmission distances of only a few dispersion lengths.

An optoelectronic equalization technique based on a *transversal filter* has also been proposed [10]. In this technique, a power splitter at the receiver splits the received optical signal into several branches. Fiber-optic delay lines introduce variable delays in different branches. The optical signal in each branch is converted into photocurrent by using variable-sensitivity photodetectors, and the summed photocurrent is used by the decision circuit. The technique can extend the transmission distance by about a factor of 3 for a lightwave system operating at 5 Gb/s.

9.3   PRECOMPENSATION TECHNIQUES

This approach to dispersion compensation modifies the characteristics of input pulses at the transmitter before they are launched into the fiber link. The underlying idea can be understood from Eq. (9.1.4). It consists of changing the spectral amplitude $\tilde{A}(0, \omega)$ of the input pulse in such a way that GVD-induced degradation is eliminated, or at least reduced, substantially. Clearly, if the spectral amplitude is changed as

$$\tilde{A}(0, \omega) \to \tilde{A}(0, \omega) \exp(-i\omega^2 \beta_2 L/2), \tag{9.3.1}$$

where L is the fiber length, GVD will be compensated exactly, and the pulse will retain its shape at the fiber output. Unfortunately, it is not easy to implement Eq. (9.3.1) in practice. In a simple approach, the input pulse is chirped suitably to minimize the GVD-induced pulse broadening. Since the frequency chirp is applied at the transmitter before propagation of the pulse, this technique is referred to as pre-chirping.

9.3.1   Pre-Chirping

A simple way to understand the role of pre-chirping is based on the theory presented in Section 2.4, where propagation of chirped Gaussian pulses in optical fibers is discussed. The input amplitude is taken to be

$$A(0, t) = A_0 \exp\left[-\frac{1 + iC}{2} \left(\frac{t}{T_0} \right)^2 \right], \tag{9.3.2}$$

where C is the chirp parameter. As seen in Fig. 2.12, for values of C such that $\beta_2 C < 0$, the input pulse initially compresses in a dispersive fiber. Thus, a suitably chirped pulse can propagate over longer distances before it broadens outside its allocated bit slot. As a rough estimate of the improvement, consider the case in which pulse broadening by a factor of up to $\sqrt{2}$ is acceptable. By using Eq. (2.4.17) with $T_1/T_0 = \sqrt{2}$, the transmission distance is given by

$$L = \frac{C + \sqrt{1 + 2C^2}}{1 + C^2} L_D, \qquad (9.3.3)$$

where $L_D = T_0^2/|\beta_2|$ is the dispersion length. For unchirped Gaussian pulses, $C = 0$ and $L = L_D$. However, L increases by 36% for $C = 1$. Note also that $L < L_D$ for large values of C. In fact, the maximum improvement by a factor of $\sqrt{6}$ occurs for $C = 1/\sqrt{2}$. These features clearly illustrate that the pre-chirp technique requires careful optimization. Even though the pulse shape is rarely Gaussian in practice, the pre-chirp technique can increase the transmission distance by a factor of about 2 when used with care. As early as 1986, a super-Gaussian model [11] suitable for NRZ transmission predicted such an improvement, a feature also evident in Fig. 2.14, which shows the results of numerical simulations for chirped super-Gaussian pulses.

The pre-chirp technique was considered during the 1980s in the context of directly modulated semiconductor lasers [11]–[14]. Such lasers chirp the pulse automatically through the carrier-induced index changes governed by the linewidth enhancement factor β_c (see Section 3.3.7). Unfortunately, the chirp parameter C is negative ($C = -\beta_c$) for directly modulated semiconductor lasers. Since β_2 in the 1.55-μm wavelength region is also negative for standard fibers, the condition $\beta_2 C < 0$ is not satisfied. In fact, as seen in Fig. 2.12, the chirp induced through direct modulation increases GVD-induced pulse broadening, thereby reducing the transmission distance drastically. Several schemes during the 1980s considered the possibility of shaping the current pulse appropriately in such a way that the transmission distance improved over that realized without current-pulse shaping [12]–[14].

In the case of external modulation, optical pulses are nearly chirp-free. The pre-chirp technique in this case imposes a frequency chirp with a positive value of the chirp parameter C so that the condition $\beta_2 C < 0$ is satisfied. Several schemes have been proposed for this purpose [15]–[22]. In a simple approach shown schematically in Fig. 9.1, the frequency of the DFB laser is first modulated (FM) before the laser output is passed to an external modulator for amplitude modulation (AM). The resulting optical signal exhibits simultaneous AM and FM. From a practical standpoint, FM of the optical carrier can be realized by modulating the injection current of the DFB laser by a small amount (~ 1 mA). Although such a direct modulation of the DFB laser also modulates the optical power sinusoidally, the magnitude is small enough that it does not interfere with the detection process.

It is clear from Fig. 9.1 that FM of the optical carrier, followed by external AM, generates a signal that consists of chirped pulses. The amount of chirp

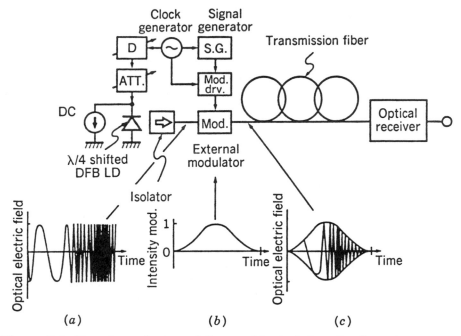

Figure 9.1 Experimental setup of the pre-chirp technique for dispersion compensation. The three traces at bottom show (a) FM output of the DFB laser, (b) pulse shape produced by external modulator, and (c) pre-chirped pulse used for signal transmission. (After Ref. [19]. ©1994 IEEE. Reprinted with permission.)

can be determined as follows. Assuming that the pulse shape is Gaussian, the optical signal can be written as

$$E(0,t) = A_0 \exp(-t^2/T_0^2) \exp[-i\omega_0(1 + \delta \sin \omega_m t)t], \qquad (9.3.4)$$

where the carrier frequency ω_0 of the pulse is modulated sinusoidally at the frequency ω_m with a modulation depth δ. Near the pulse center, $\sin(\omega_m t) \approx \omega_m t$, and Eq. (9.3.4) becomes

$$E(0,t) = A_0 \exp\left[-\frac{1+iC}{2}\left(\frac{t}{T_0}\right)^2\right] \exp(-i\omega_0 t), \qquad (9.3.5)$$

where the chirp parameter C is given by

$$C = 2\delta\omega_m\omega_0 T_0^2. \qquad (9.3.6)$$

The sign and magnitude of the chirp parameter C can be controlled by changing the FM parameters δ and ω_m.

Phase modulation of the optical carrier also leads to a positive chirp, as can be verified by replacing Eq. (9.3.4) with

$$E(0,t) = A_0 \exp(-t^2/T_0^2) \exp[-i\omega_0 t + i\delta \cos(\omega_m t)] \qquad (9.3.7)$$

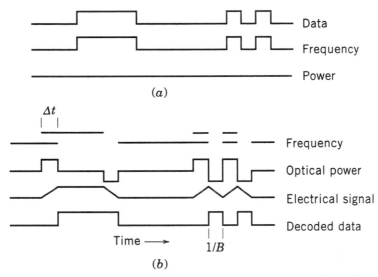

Figure 9.2 Schematic illustration of dispersion compensation through pure FSK modulation: (a) Three traces showing the NRZ electrical signal and the frequency and power variations of the optical signal at transmitter; (b) two traces showing the frequency and power variations of the optical signal and two traces showing the electrical signal and the reconstructed data stream at the receiver. (After Ref. [26]. ©1994 IEEE. Reprinted with permission.)

and using $\cos x \approx 1 - x^2/2$. An advantage of the phase-modulation technique is that the external modulator itself can modulate the carrier phase. The simplest solution is to employ an external modulator whose refractive index can be changed electronically in such a way that it imposes a frequency chirp with $C > 0$ [15]. As early as 1991, a 5-Gb/s signal was transmitted over 256 km [17] using a LiNbO$_3$ modulator such that values of C were in the range 0.6–0.8. These experimental values are in agreement with the Gaussian-pulse theory on which Eq. (9.3.3) is based. Other types of semiconductor modulators, such as an electroabsorption modulator [18] or a Mach–Zehnder (MZ) modulator [20], can also chirp the optical pulse with $C > 0$, and have indeed been used to demonstrate transmission beyond the dispersion limit [21]. With the development of DFB lasers containing a monolithically integrated electroabsorption modulator, the implementation of the pre-chirp technique has become quite practical. In a 1996 experiment, a 10-Gb/s NRZ signal was transmitted over 100 km of standard fiber using such a transmitter [22].

9.3.2 Pure Frequency Modulation

Simultaneous AM and FM of the optical signal is not essential for dispersion compensation. In a different approach [23]–[29], often referred to as *dispersion-supported transmission* [26], pure FM is used for signal transmission. The FSK format, discussed in Section 6.2.3, is used to generate the optical signal such that the laser wavelength is switched by a constant amount $\Delta\lambda$ between 1 and 0 bits. During propagation inside the dispersive fiber, the two wavelengths travel at slightly different speeds. The time delay between the 1 and 0 bits is determined by the wavelength shift $\Delta\lambda$ and is given by $\Delta T = DL\Delta\lambda$, as shown in Eq. (2.3.4). The wavelength shift $\Delta\lambda$ is chosen such that $\Delta T = 1/B$. Figure 9.2 shows schematically how such a one-bit delay produces a three-level optical signal at the receiver. In essence, because of fiber dispersion, the pure FSK signal is converted into a signal whose amplitude is modulated. The signal can be decoded at the receiver by using an electrical integrator in conjunction with a decision circuit [26].

Several transmission experiments have demonstrated the usefulness of the FSK scheme [24]–[27]. All of these experiments were concerned with increasing the transmission distance of a 1.55-μm lightwave system operating at 10 Gb/s over standard fibers. By 1994, transmission of the 10-Gb/s signal over 253 km of standard fiber had been realized [26]. In a 1995 experiment, even at a bit rate of 20 Gb/s, the signal could be transmitted over 53 km [27]. These values should be compared with the prediction of Eq. (9.1.2). Clearly, the transmission distance can be improved by a large factor by using the FSK technique.

9.3.3 Duobinary Coding

Another approach to increasing the transmission distance consists of generating an optical signal at the transmitter whose bandwidth at a given bit rate is reduced compared with the conventional on–off keying. One scheme makes use of *duobinary coding*, which can reduce the signal bandwidth by 50% [30]. In the simplest duobinary scheme, the two successive bits in the digital bit stream are summed, forming a three-level duobinary code at half the bit rate. Since the GVD-induced degradation depends on the signal bandwidth, the transmission distance should improve for a reduced-bandwidth signal.

In an experimental comparison between the binary and duobinary schemes, the 10-Gb/s signal could be transmitted over 30- to 40-km-longer fiber lengths with duobinary coding [31]. The duobinary scheme can be combined with the pre-chirping technique. Indeed, transmission of a 10-Gb/s signal over 160 km of the standard fiber has been realized by combining duobinary coding with an external modulator capable of producing a frequency chirp with $C > 0$ [31]. Since chirping increases the signal bandwidth, it is hard to understand why it would help. It appears that phase reversals occurring in practice when a duobinary signal is generated are responsible primarily for improvement realized with duobinary coding [32]. In fact, a new dispersion-compensation scheme, called phase-shaped binary transmission, has been proposed to take advantage of phase reversals [33]. A MZ modulator can be used to induce phase reversals in the electric field associated with a conventional binary signal [34]. The use of duobinary transmission increases signal-to-noise requirements and requires decoding at the receiver. Despite these shortcomings, it may find applications at high bit rates for dispersion-limited lightwave systems when the objective is to increase the transmission distance by 30–40 km.

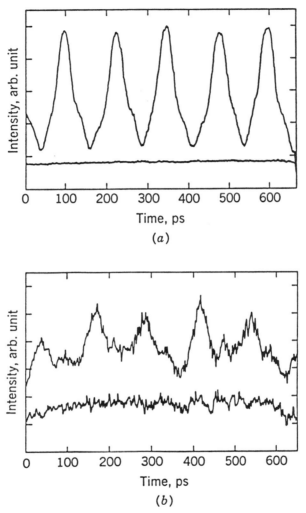

Figure 9.3 Streak-camera traces of the 16-Gb/s signal transmitted over 70 km of standard fiber (a) with and (b) without compensation of dispersion resulting from amplifier-induced chirp. Bottom trace shows the background level in both cases. (After Ref. [36]. ©1989 IEE. Reprinted with permission.)

9.3.4 Amplifier-Induced Chirp

A simple pre-chirp technique, proposed and demonstrated in 1989, amplifies the modulator output in a semiconductor laser amplifier operating in the gain-saturation regime [35]–[39]. Gain saturation leads to time-dependent variations in the carrier density, which, in turn, chirp the amplified pulse through carrier-induced variations in the refractive index (see Section 8.2.4). The amount of chirp is given by Eq. (8.2.26) and depends on the input pulse shape. As seen in Fig. 8.8, the chirp is nearly linear over most of the pulse. The semiconductor laser amplifier not only amplifies the pulse but also imposes a frequency chirp such that the chirp parameter $C > 0$. Because of this chirp, the input pulse can be compressed in a fiber with $\beta_2 < 0$. Such a compression was observed in an experiment [35] in which 40-ps input pulses were compressed to 23 ps when they were propagated over 18 km of standard fiber.

The potential of this technique for dispersion compensation was demonstrated in a 1989 experiment by transmitting a 16-Gb/s signal, obtained from a mode-locked external-cavity semiconductor laser, over 70 km of fiber [36]. Figure 9.3 compares the streak-camera traces of the signal obtained with and without dispersion compensation. From Eq. (9.1.2), in the absence of amplifier-induced chirp, the transmission distance at 16 Gb/s is limited by GVD to about 14 km for a fiber with $D = 15$ ps/(km-nm). The use of the amplifier in the gain-saturation regime increased the transmission distance fivefold, a feature that makes this approach to dispersion compensation quite attractive. It has an added benefit that it can compensate for the coupling and insertion losses that invariably occur in a transmitter by amplifying the signal before it is launched into the optical fiber. Moreover, this technique can be used for simultaneous compensation of fiber loss and GVD [39] if semiconductor laser amplifiers are used periodically along the fiber link in place of erbium-doped fiber amplifiers.

9.3.5   Fiber-Induced Chirp

As discussed in Section 2.6.2, the intensity-dependent refractive index of a non-linear pulse can chirp the pulse through the phenomenon of self-phase modulation (SPM). Thus, a simple pre-chirp technique consists of passing the modulator output through a nonlinear medium before launching it into the transmission fiber. By using Eq. (2.6.7), the optical signal at the fiber input is given by

$$A(0, t) = \sqrt{P(t)} \exp[i\bar{\gamma} L_m P(t)], \qquad (9.3.8)$$

where $P(t)$ is the power of the pulse, L_m is the length of the nonlinear medium, and $\bar{\gamma}$ is the nonlinear parameter. In the case of Gaussian pulses for which $P(t) = P_0 \exp(-t^2/T_0^2)$, the chirp is nearly linear, and Eq. (9.3.8) can be approximated by

$$A(0, t) = \sqrt{P_0} \exp\left[-\frac{1 + iC}{2}\left(\frac{t}{T_0}\right)^2\right] \exp(-i\bar{\gamma} L_m P_0), \qquad (9.3.9)$$

where the chirp parameter is given by $C = 2\bar{\gamma} L_m P_0$. For $\bar{\gamma} > 0$, the chirp parameter C is positive, and is thus suitable for dispersion compensation.

Since $\bar{\gamma} > 0$ for silica fibers, the transmission fiber itself can be used for chirping the pulse. This approach was suggested [40] in a 1986 study. It takes advantage of higher-order solitons (see Chapter 10) which go through a stage of initial compression. Figure 9.4 shows the GVD-limited transmission distance as a function of the average launch power for 4- and 8-Gb/s lightwave systems. It indicates the possibility of doubling the transmission distance by optimizing the average power of the input signal to about 3 mW.

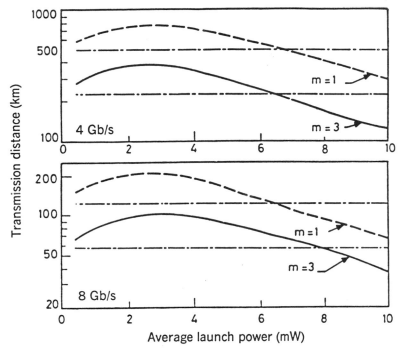

Figure 9.4 Dispersion-limited transmission distance as a function of the average launch power for Gaussian ($m = 1$) and super-Gaussian ($m = 3$) input pulses at bit rates of 4 and 8 Gb/s. Horizontal lines correspond to the linear case. The power-dependent improvement is due to the fiber-induced chirp. (After Ref. [40]. ©1986 IEE. Reprinted with permission.)

9.4   HIGH-DISPERSION FIBERS

Several optical techniques take advantage of the linear nature of Eq. (9.1.3) and are capable of compensating the fiber GVD completely if the average optical power is low enough that nonlinear effects in optical fibers are negligible. In the simplest approach, the optical signal is propagated over multiple fiber segments with different dispersion characteristics. The basic idea can be understood by considering just two segments and writing Eq. (9.1.4) as

$$A(L, t) = \frac{1}{2\pi} \int_{-\infty}^{\infty} \tilde{A}(0, \omega) \exp\left(\frac{i}{2}\omega^2(\beta_{21}L_1 + \beta_{22}L_2) - i\omega t\right) d\omega, \qquad (9.4.1)$$

where $L = L_1 + L_2$, and β_{2j} is the GVD parameter of the fiber segment of length L_j ($j = 1, 2$). By using $D_j = -(2\pi c/\lambda^2)\beta_{2j}$, the condition for dispersion compensation becomes

$$D_1 L_1 + D_2 L_2 = 0. \qquad (9.4.2)$$

Since $A(L, t) = A(0, t)$ when Eq. (9.4.2) is satisfied, the pulse shape is restored to its input form.

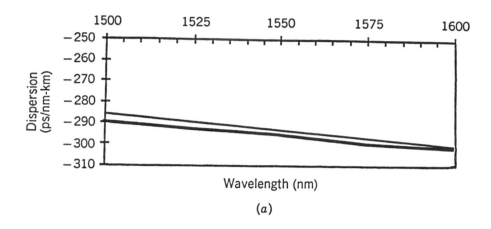

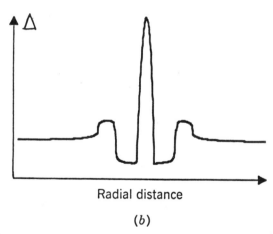

Figure 9.5 (a) Dispersion characteristics of two DCFs with a high figure of merit. (b) index profile of a DCF. (After Ref. [45]. ©1996 OSA. Reprinted with permission.)

Equation (9.4.2) shows that the *dispersion-compensating fiber* (DCF) must have normal GVD at 1.55 μm ($D_2 < 0$) since $D_1 > 0$ for standard telecommunication fibers. Moreover, its length should should be chosen to satisfy

$$L_2 = -(D_1/D_2)L_1. \tag{9.4.3}$$

For practical reasons, L_2 should be as small as possible. This is possible only if the DCF has a large negative value of D_2.

The idea of using a DCF has been around since 1980 [41]. However, it was only after the advent of optical amplifiers in 1990 that the development of DCFs accelerated in pace. There are two basic approaches to designing DCFs. In one approach, the DCF supports a single mode, but it is designed with a relatively small value of the fiber parameter V defined by Eq. (2.2.38). As discussed in Section 2.2.3, the fundamental mode is weakly confined for $V \approx 1$ (see Fig. 2.7). Since a large fraction of the mode propagates inside the cladding layer where the refractive index is smaller, the waveguide contribution to the GVD is quite different for such fibers, resulting in $D \sim -100$ ps/(km-nm). A depressed-cladding design is often used in practice [42]–[45]. Unfortunately, DCFs also have a relatively high loss because of an increase in the bending losses ($\alpha \approx 0.4$–1.0 dB/km). The ratio $|D|/\alpha$ is often used as a figure of merit M for characterizing various DCFs [44]. By 1994, DCFs with $M \approx 150$ ps/(nm-dB) were available commercially. Recent advances have produced DCFs whose dispersion exceeds -200 ps/(km-nm) with $M > 400$ ps/(nm-dB). Figure 9.5 shows the refractive-index profile and the measured dispersion characteristics of two such DCFs.

Single-mode DCFs suffer from several problems. First, 1 km of DCF compensates dispersion for only 10–12 km of standard fiber. Second, their loss is relatively high at the 1.55-μm operating wavelength ($\alpha \sim 0.5$ dB/km) because of increased bending losses. Third, because of a relatively small mode diameter, the optical intensity is larger at a given input power, resulting in enhanced nonlinear effects. Considerable work is being done to address these issues and improve the performance of single-mode DCFs.

Most of the problems associated with a single-mode DCF can be solved to some extent by using a *two-mode fiber* designed with values of V such that the higher-order mode is near cutoff ($V \approx 2.5$). Such fibers have almost the same loss as the single-mode fiber but can be designed such that the dispersion parameter D for the higher-order mode has large negative values [46]–[48]. Indeed, values of D as large as -770 ps/(km-nm) has been measured for elliptical-core fibers [48]. A 1-km length of such a DCF can compensate the GVD for a 40-km-long fiber link, adding relatively little to the total link loss. An added advantage of the two-mode DCF is that it allows for broadband dispersion compensation [46]. This aspect is discussed later in the chapter.

The use of a two-mode DCF requires a mode-conversion device capable of converting the energy from the fundamental mode to the higher-order mode supported by the DCF. Several such all-fiber devices have been developed [49]–[51]. The all-fiber nature of the mode-conversion device is important from the standpoint of compatibility with the fiber network. Moreover, such an approach reduces the insertion loss. Additional requirements on a mode converter are that it should be polarization insensitive and should operate over a broad bandwidth. Almost all practical mode-conversion devices use a two-mode fiber with a built-in grating that provides coupling between the two modes. The grating period Λ is chosen to match the mode-index difference $\delta\bar{n}$ of the two modes ($\Lambda = \lambda/\delta\bar{n}$) and is typically ~ 100 μm. Such gratings have been made by using several different mechanisms, such as a periodic stress [49], microbending [50], [51], and photosensitivity [52], [53]. Insertion losses are typically below 1 dB, with coupling efficiencies as high as 99%.

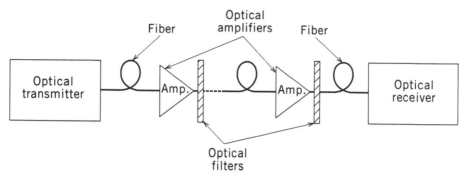

Figure 9.6 Schematic illustration of dispersion compensation in a long-haul fiber link by using an optical filter after each optical amplifier. Optical filters serve a dual purpose, as they also reduce the amplifier noise by filtering out the amplified spontaneous emission.

9.5 OPTICAL EQUALIZING FILTERS

A shortcoming of DCFs is that relatively long lengths ($\sim$ 1 km) are required to compensate for the GVD acquired over 10 km of standard fiber. This adds considerably to the link loss, especially in the case of long-haul applications. For this reason, several other optical schemes for GVD compensation have been developed. Most of them can be classified under the category of *optical equalizing filters*. In this section we consider interferometric filters, and the next section is devoted to fiber gratings.

The function of optical equalizing filters is easily understood from Eq. (9.1.4). Since the GVD affects the optical signal though the spectral phase $\exp(i\beta_2 z\omega^2/2)$, it is evident that an optical filter whose transfer function cancels this phase will restore the signal. Unfortunately, no optical filter, except for an optical fiber itself, has a transfer function capable of compensating the GVD exactly. Nevertheless, several optical filters have provided partial GVD compensation by mimicking the ideal transfer function. Consider an optical filter with the transfer function $H(\omega)$. If this filter is placed after a fiber of length L, the filtered optical signal from Eq. (9.1.4) can be written as

$$A(L,t) = \frac{1}{2\pi} \int_{-\infty}^{\infty} \tilde{A}(0,\omega)H(\omega)\exp\left(\frac{i}{2}\beta_2 L\omega^2 - i\omega t\right) d\omega, \qquad (9.5.1)$$

By expanding the phase of $H(\omega)$ in a Taylor series and retaining up to the quadratic term,

$$H(\omega) = |H(\omega)|\exp[i\phi(\omega)] \approx |H(\omega)|\exp[i(\phi_0 + \phi_1\omega + \tfrac{1}{2}\phi_2\omega^2)], \qquad (9.5.2)$$

where $\phi_m = d^m\phi/d\omega^m (m = 0, 1, \dots)$ is evaluated at the optical carrier frequency ω_0. The constant phase ϕ_0 and the time delay ϕ_1 do not affect the pulse shape and can be ignored. The spectral phase introduced by the fiber can be compensated by choosing an optical filter such that $\phi_2 = -\beta_2 L$. The pulse can be recovered perfectly only if $|H(\omega)| = 1$ and the cubic and higher-order terms in the Taylor expansion in Eq. (9.5.2) are negligible. Figure 9.6 shows schematically how such an optical filter can be placed after each amplifier so that both the fiber loss and GVD can be compensated along a long-haul fiber link. If the transfer function $H(\omega) = 0$ outside the signal bandwidth, the optical filter can also reduce the amplifier noise by filtering out the spontaneous emission.

9.5.1 Fabry–Perot Interferometers

An interferometer, by its nature, is sensitive to the frequency of the input light and acts as an optical filter because of its frequency-dependent transmission characteristics. A simple example is provided by the Fabry–Perot (FP) interferometer, discussed in Sections 7.2.1 and 8.2.1 in different contexts. A reflective FP interferometer, often called the *Gires–Tournois interferometer*, is designed with a partially reflective front mirror and a 100% reflective back mirror. Its transfer function is given by [54]

$$H_{\rm FP}(\omega) = H_0 \frac{1 + r\exp(i\omega T)}{1 + r\exp(-i\omega T)}, \qquad (9.5.3)$$

where H_0 is a constant taking into account all losses, $|r|^2$ is the front-mirror reflectivity, and T is the round-trip time within the FP cavity. Since $|H_{\rm FP}(\omega)|$ is frequency independent, only the spectral phase is modified by the FP filter. However, the phase $\phi(\omega)$ of $H_{\rm FP}(\omega)$ is far from ideal. It is a periodic function that peaks at the FP resonances (longitudinal-mode frequencies discussed in Section 3.3.2). In the vicinity of each peak, a spectral region exists in which the phase variation is nearly quadratic. In fact, by expanding $\phi(\omega)$ in a Taylor series, ϕ_2 is given by

$$\phi_2 = 2T^2 r(1 - r)/(1 + r)^3. \qquad (9.5.4)$$

As an example, for a 1-cm-long FP cavity with $r = 0.8$, $\phi_2 \approx 550$ ps^2. With a proper design, such a device can compensate for GVD acquired over several hundred kilometers of standard fiber by adjusting the cavity length suitably [54]. In a 1991 experiment [55], such an all-fiber device was used to transmit the 8-Gb/s signal over 130 km of standard fiber. The relatively high insertion loss of 8 dB was compensated by using an optical amplifier. A loss of 6 dB was attributed to a 3-dB fiber coupler used to separate the reflected signal from the incident signal. A relatively high loss and a narrow bandwidth of such filters limit the use of FP filters in practical lightwave systems.

9.5.2 Mach–Zehnder Interferometers

As discussed in Section 7.2.1, a MZ interferometer acts as an optical filter. An optical equalizer based on a combination of several MZ interferometers was proposed in 1992 [56] and has been fabricated in the form of a *planar lightwave circuit* by using silica waveguides [57]. Figure 9.7(a) shows the device schematically. The device is 52×71 mm^2 in size and exhibits a chip loss of 8 dB. It

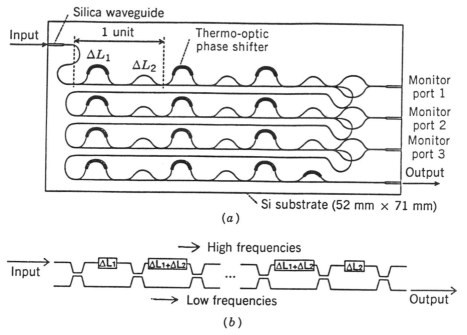

Figure 9.7 (a) planar lightwave circuit for dispersion compensation making use of a chain of Mach–Zehnder interferometers; (b) unfolded view of the device. (After Ref. [57]. ©1996 IEEE. Reprinted with permission.)

consists of 12 couplers with asymmetric arm lengths that are cascaded in series. A chromium heater is deposited on one arm to provide thermo-optic control of the optical phase by changing the arm length. The main advantage of such a device is that its dispersion-equalization characteristics can be controlled by changing the arm lengths and the number of MZ interferometers.

The operation of the MZ device can be understood from the unfolded view shown in Fig. 9.7(b). The device is designed such that the higher-frequency components propagate in the longer arm of the MZ interferometers. As a result, they experience more delay than the lower-frequency components taking the shorter route. The relative delay introduced by such a device is just the opposite of that introduced by an optical fiber in the anomalous-dispersion regime. The transfer function $H(\omega)$ can be obtained analytically and is used to optimize the device design and performance [58]. In a 1994 implementation [59], a planar lightwave circuit with only five MZ interferometers provided a relative delay of 836 ps/nm. Such a device is only a few centimeters long, but it is capable of compensating for 50 km of fiber dispersion. Its main limitations are a relatively narrow bandwidth ($\sim$ 10 GHz) and sensitivity to input polarization. However, it acts as a programmable optical filter whose GVD as well as the operating wavelength can be adjusted. In one device, the GVD could be varied over the range -681 to 786 ps/nm [60].

9.6 FIBER BRAGG GRATINGS

A fiber Bragg grating acts as an optical filter because of the existence of a *stop band*, the frequency region in which most of the incident light is reflected back [61]. The stop band is centered at the *Bragg wavelength* $\lambda_B = 2\bar{n}\Lambda$, where Λ is the grating period and $\bar{n}$ is the modal index. The periodic nature of index variations couples the forward- and backward-propagating waves at wavelengths close to the Bragg wavelength and, as a result, provides frequency-dependent reflectivity to the incident signal over a bandwidth determined by the grating strength. In essence, a fiber grating acts as a reflection filter. Although the use of such gratings for dispersion compensation was proposed in the 1980s [62], [63], it is only during the 1990s that fabrication technology based on the *photosensitivity* of optical fibers advanced enough to make their use practical. This section covers the basic theory and the experimental progress realized in recent years. Further details can be found in several review articles [64]–[67].

9.6.1 Uniform-Period Gratings

Bragg gratings are generally analyzed using the *coupled-mode equations* that describe the coupling between the forward- and backward-propagating waves at a given frequency ω and are written as [61]

$$
\begin{aligned}
dA_f/dz &= i\delta A_f + i\kappa A_b, & (9.6.1) \\
dA_b/dz &= -i\delta A_b - i\kappa A_f, & (9.6.2)
\end{aligned}
$$

where A_f and A_b are the spectral amplitudes of the two waves,

$$
\delta = \beta - \beta_B = \frac{2\pi}{\lambda} - \frac{2\pi}{\lambda_B} \quad \text{and} \quad \kappa = \frac{\pi n_g \Gamma}{\lambda_B}. \qquad (9.6.3)
$$

Here δ is the detuning from the Bragg wavelength and the *coupling coefficient* κ corresponds to a fiber grating whose refractive index varies as $n(z) = \bar{n} + n_g \cos(2\pi z/\Lambda)$. The fiber mode is characterized by the mode index $\bar{n}$ and the confinement factor Γ defined as in Eq. (2.2.50).

The coupled-mode equations can be solved analytically owing to their linear nature. The transfer function of the grating, acting as a reflective filter, is found to be [61]

$$
H(\omega) = r(\omega) = \frac{A_b(0)}{A_f(0)} = \frac{i\kappa \sin(qL_g)}{q\cos(qL_g) - i\delta\sin(qL_g)}, \qquad (9.6.4)
$$

where $q = \sqrt{\delta^2 - \kappa^2}$ and L_g is the grating length. Figure 9.8 shows the reflectivity $|H(\omega)|^2$ and the phase of $H(\omega)$ for $\kappa L_g = 2$ and 3. The grating reflectivity becomes nearly 100% within the stop band for $\kappa L_g = 3$. However, since the phase is nearly linear in that region, grating dispersion governed by the second derivative of phase, ϕ_2 in Eq. (9.5.2), is relatively small. Quite large values of the GVD are possible close to but outside the stop band, but rapid variations of both the amplitude $|H(\omega)|$ and ϕ_2 make the use of such uniform gratings for dispersion compensation far from being practical.

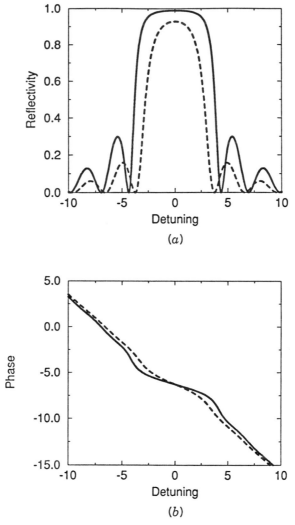

Figure 9.8 (a) Magnitude and (b) phase of the reflectivity, plotted as a function of detuning δL_g, for a uniform fiber grating for $\kappa L_g = 2$ (dashed curve) and $\kappa L_g = 3$ (solid curve).

The problem can be solved by using the *apodization technique* in which the index change n_g is made nonuniform across the grating, resulting in z-dependent κ. In practice, such an apodization occurs naturally when an ultraviolet Gaussian beam is used to write the grating holographically [64]–[67]. For such gratings, κ peaks in the center and tapers down to zero at both ends. A better approach consists of making a grating such that κ varies linearly over the entire length of the fiber grating. In a 1996 experiment [68], such a 11-cm-long grating was used to compensate the GVD for a 10-Gb/s signal transmitted over

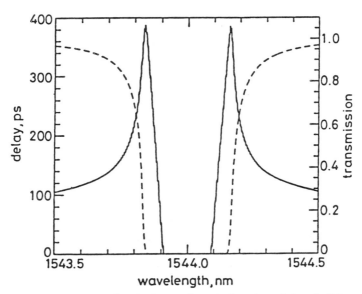

Figure 9.9 Transmittivity (dashed curve) and the time delay (solid curve) as a function of wavelength for a uniform-pitch grating for which $\kappa(z)$ varies linearly from 0 to 6 cm^{-1} over the 11-cm length. (After Ref. [68]. ©1996 IEE. Reprinted with permission.)

100 km of standard fiber. The coupling coefficient $\kappa(z)$ varied smoothly from 0 to 6 cm^{-1} over the grating length. Figure 9.9 shows the transmission characteristics of such a grating calculated by solving the coupled-mode equations numerically. The solid curve shows the group delay related to the phase derivative $d\phi/d\omega$. In a 0.1-nm-wide wavelength region near 1544.2 nm, the group delay varies almost linearly at a rate of about 2000 ps/nm, indicating that the grating can compensate for the GVD acquired over 100 km of standard fiber while providing more than 50% transmission to the incident light. Indeed, such a grating compensated GVD over 106 km for a 10-Gb/s signal with only a 2-dB power penalty at a BER of 10^{-9} [68]. In the absence the grating, the penalty was infinitely large because of the existence of a BER floor.

The tapering of the coupling coefficient along the grating length can also be used for dispersion compensation when the signal wavelength lies within the stop band and the grating acts as a reflection filter. Numerical solutions of the coupled-mode equations for a uniform-period grating for which $\kappa(z)$ varies linearly from 0 to 12 cm^{-1} over the 12-cm length show that the V-shaped group-delay profile, centered at the Bragg wavelength, can be used for dispersion compensation if the wavelength of the incident signal is offset from the center of the stop band such that the signal spectrum sees a linear variation of the group delay. Such a 8.1-cm-long grating was capable of compensating the GVD acquired over 257 km of standard fiber by a 10-Gb/s signal [69].

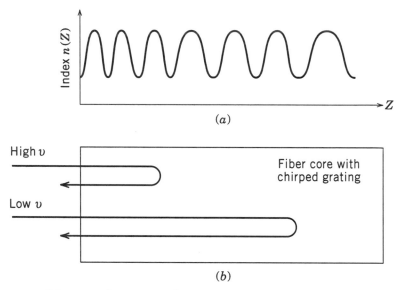

Figure 9.10 Schematic illustration of dispersion compensation by a linearly chirped fiber grating: (a) index profile $n(z)$ along the grating length; (b) reflection of low and high frequencies at different locations within the grating because of variations in the Bragg wavelength.

9.6.2   Chirped Fiber Gratings

Gratings with a uniform period have a relatively narrow stop band (typically < 0.1 nm) whereas a broadband grating is often required in practice. A solution is provided by the *chirped fiber grating* [63] in which the optical period, $\bar{n}\Lambda$, of the grating varies linearly over its length. Since the Bragg wavelength $\lambda_B = 2\bar{n}\Lambda$ also varies along the grating length, different frequency components of an incident optical pulse are reflected at different points, depending on where the Bragg condition is satisfied locally.

It is easy to understand the operation of such a chirped fiber grating from Fig. 9.10. For the compensation of GVD, the optical period of the grating, $\bar{n}\Lambda$, should increase along its length to provide "normal" GVD ($\beta_2 > 0$). In a standard fiber with anomalous dispersion at the 1.55-μm wavelength, the high-frequency components of an optical pulse propagate faster than the low-frequency components. Since the Bragg wavelength increases along the grating length, the low-frequency components travel further into the grating before being reflected and experience more delay than the high-frequency components. As a result, the relative delay induced by the grating is just the opposite of the fiber and can compensate the fiber dispersion. From this simple picture, the dispersion parameter D_g of a chirped grating of length L_g can be determined by using the relation $T_R = D_g L_g \Delta\lambda$, where T_R is the round-trip time inside the grating and $\Delta\lambda$ is the difference in the Bragg wavelengths at the two ends of the grating. Since $T_R = 2\bar{n}L_g/c$, the grating dispersion is given by a remarkably simple expression,

$$D_g = 2\bar{n}/c\Delta\lambda. \qquad (9.6.5)$$

As an example, $D_g \approx 5 \times 10^7$ ps/(km-nm) for a grating bandwidth $\Delta\lambda = 0.2$ nm. Because of such large values of D_g, a 10-cm-long chirped grating can compensate for the GVD acquired over 300 km of standard fiber.

Chirped fiber gratings have been fabricated by using several different methods [64]–[67]. It is important to note that it is the optical period $\bar{n}\Lambda$ that needs to be varied along the grating (z axis), and thus chirping can be induced either by varying the physical grating period Λ or by changing the effective mode index $\bar{n}$ along z. In the commonly used *dual-beam holographic technique*, the fringe spacing of the interference pattern is made nonuniform by using dissimilar curvatures for the interfering wavefronts [70], resulting in Λ variations. In practice, cylindrical lenses are used in one or both arms of the interferometer. In a *double-exposure technique* [71], a moving mask is used to vary $\bar{n}$ along z during the first exposure. A uniform-period grating is then written over the same section of the fiber by using the *phase-mask technique*. Many other variations are possible. For example, chirped fiber gratings have been fabricated by tilting or stretching the fiber, by using strain or temperature gradients, and by stitching together multiple uniform sections.

The potential of chirped fiber gratings for dispersion compensation has been demonstrated in several transmission experiments [72]–[78]. In 1994, GVD compensation over 160 km of standard fiber at 10 and 20 Gb/s was realized [73]. In 1995, a 12-cm-long chirped grating was used to compensate GVD over 270 km of fiber at 10 Gb/s [74]. Later, the transmission distance was increased to 400 km using a 10-cm-long apodized chirped fiber grating [75]. This is a remarkable performance by an optical filter that is only 10 cm long and represents improvement by a factor of 20 if we note from Eq. (9.1.2) that the distance is limited to about 20 km in the absence of dispersion compensation.

Figure 9.11 shows the measured reflectivity and the group delay (related to the phase derivative $d\phi/d\omega$) as a function of the wavelength for the 10-cm-long grating with a bandwidth $\Delta\lambda = 0.12$ nm chosen to ensure that the 10-Gb/s signal fits within the stop band of the grating. For such a grating, the period Λ changes by only 0.008% over its entire length. Perfect dispersion compensation occurs over the spectral range for which $d\phi/d\omega$ varies linearly. The slope of the group delay (about 5000 ps/nm) is a measure of the dispersion-compensation capability of the grating. Such a grating can recover the 10-Gb/s signal by compensating the GVD acquired over 400 km of the standard fiber. The chirped grating should be apodized in such a way that the coupling coefficient peaks in the middle but vanishes at the grating ends. The apodization is essential to remove the ripples that occur for gratings with a constant κ.

It is clear from Eq. (9.6.5) that D_g of a chirped grating is ultimately limited by the bandwidth $\Delta\lambda$ over which GVD compensation is required, which in turn is determined by the bit rate B. Further increase in the transmission distance at a given bit rate is possible only if the signal bandwidth is reduced or a pre-chirp technique is used at the transmitter. In a 1996 system trial [77], pre-

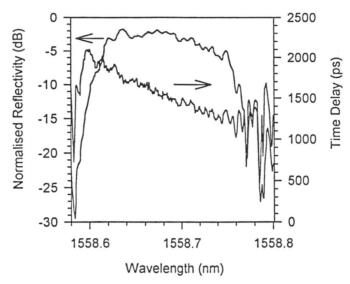

Figure 9.11 Measured reflectivity and time delay for a 10-cm-long, apodized and linearly chirped, fiber grating having a bandwidth of 0.12 nm. The slope of the linear time delay is a measure of the dispersion-compensation capability of the grating. (After Ref. [78]. ©1996 IEEE. Reprinted with permission.)

chirping of the 10-Gb/s optical signal was combined with the two chirped fiber gratings, cascaded in series, to increase the transmission distance to 537 km. The bandwidth-reduction technique can also be combined with the grating. As discussed in Section 9.3.2, a duobinary coding scheme can reduce the bandwidth by up to 50%. In a 1996 experiment, the transmission distance of a 10-Gb/s signal was extended to 700 km by using a 10-cm-long chirped grating in combination with a phase-alternating duobinary scheme [78]. Figure 9.11 shows the measured reflectivity and the time delay for the apodized chirped grating used in the experiment. The grating bandwidth was reduced to 0.073 nm, too narrow for the 10-Gb/s signal but wide enough for the reduced-bandwidth duobinary signal.

Chirped fiber gratings appear to be quite useful for dispersion compensation and are likely to find commercial applications. The main limitation of such gratings is that they work as a reflection filter. A 3-dB fiber coupler is often used to separate the reflected signal from the incident one. However, its use imposes a 6-dB loss that adds to other insertion losses. An *optical circulator*, a three-port device in which the reflected signal is directed toward the third port, can reduce the insertion loss to below 1 dB. Another approach makes use of a grating-based transmission filter for dispersion compensation. Two or more fiber gratings can be combined to form a transmission filter, and such filters can be used for dispersion compensation with a relatively low insertion loss [79]. A single grating can be converted into a transmission filter by introducing a phase shift in the middle of the grating [80]. A Moiré grating, formed by superimposing

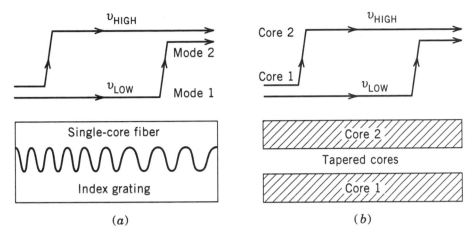

Figure 9.12 Schematic illustration of dispersion compensation by two fiber-based transmission filters: (a) chirped dual-mode coupler; (b) tapered dual-core fiber. (After Ref. [82]. ©1994 IEEE. Reprinted with permission.)

two chirped gratings formed on the same piece of fiber, also has a transmission peak within its stop band [81]. The bandwidth of such transmission filters is relatively small, although it can be made large enough for passing a 10-Gb/s signal. Two other techniques, discussed next, can act effectively as a transmission filter suitable for dispersion compensation.

9.6.3 Chirped Mode Couplers

A chirped mode coupler is an all-fiber device designed on the concept of chirped distributed resonant coupling [82]. Figure 9.12 shows the operation of two such devices schematically. The basic idea behind a chirped mode coupler [83] is quite simple. Rather than coupling the forward and backward propagating waves of the same mode (as is done in a fiber grating), the chirped grating couples the two spatial modes of a dual-mode fiber. Such a device is similar to the mode converter discussed in Section 9.4 in the context of a DCF except that the grating period is varied linearly over the fiber length. The signal is transferred from the fundamental mode to a higher-order mode by the grating, but different frequency components travel different lengths before being transferred because of the chirped nature of the grating that couples the two modes. If the grating period increases along the coupler length, the coupler can compensate for the fiber GVD. The signal remains propagating in the forward direction, but ends up in a higher-order mode of the coupler. A uniform-grating mode converter can be used to reconvert the signal back into the fundamental mode if necessary.

A variant of the same idea uses the coupling between the fundamental modes of a dual-core fiber with dissimilar cores [84]. If the two cores are close enough, evanescent-wave coupling between the modes leads to a transfer of energy from one core to another, similar to the case of a directional coupler. When the spacing between the cores is linearly tapered, such a transfer takes place at different points along the fiber, depending on the frequency of the propagating signal. Thus, a dual-core fiber with the linearly tapered core spacing can compensate for fiber GVD. Such a device keeps the signal propagating in the forward direction, although it is physically transferred to the neighboring core. This scheme can also be implemented in the form of a compact device by using semiconductor waveguides since the supermodes of two coupled waveguides exhibit a large amount of GVD that is also tunable [85].

9.7   OPTICAL PHASE CONJUGATION

The use of *optical phase conjugation* (OPC) for dispersion compensation was proposed [86] in 1979. However, it was only in 1993 that the OPC technique was implemented experimentally; it has attracted considerable attention since then [87]–[105]. In contrast with other optical schemes discussed in this chapter, the OPC is a nonlinear optical technique. This section describes the principle behind it and discusses its implementation in practical lightwave systems.

9.7.1   Principle of Operation

The simplest way to understand how OPC can compensate for GVD is to take the complex conjugate of Eq. (9.1.3) and obtain

$$\frac{\partial A^*}{\partial z} - \frac{i}{2}\beta_2 \frac{\partial^2 A^*}{\partial t^2} - \frac{1}{6}\beta_3 \frac{\partial^3 A^*}{\partial t^3} = 0. \qquad (9.7.1)$$

A comparison of Eqs. (9.1.3) and (9.7.1) shows that the propagation of phase-conjugated field A^* is equivalent to changing the sign of the GVD parameter β_2. This observation suggests immediately that if the optical field is phase-conjugated in the middle of the fiber link, the dispersion acquired over the first half will be exactly compensated in the second-half section of the link. Since the β_3 term does not change sign on phase conjugation, OPC cannot compensate for the third-order dispersion. In fact, it is easy to show, by keeping the higher-order terms in the Taylor expansion in Eq. (2.4.4), that OPC compensates for all even-order dispersion terms while leaving the odd-order terms unaffected.

The effectiveness of midspan OPC for dispersion compensation can also be verified by using Eq. (9.1.4). The optical field just before OPC is obtained by using $z = L/2$ in this equation. The propagation of the phase-conjugated field A^* in the second-half section then yields

$$A^*(L,t) = \frac{1}{2\pi} \int_{-\infty}^{\infty} \tilde{A}^*(L/2,\omega) \exp\left(\frac{i}{4}\beta_2 L\omega^2 - i\omega t\right) d\omega, \qquad (9.7.2)$$

where $\tilde{A}^*(L/2,\omega)$ is the Fourier transform of $A^*(L/2,t)$ and is given by

$$\tilde{A}^*(L/2,\omega) = \tilde{A}^*(0,-\omega) \exp(-i\omega^2\beta_2 L/4). \qquad (9.7.3)$$

By substituting Eq. (9.7.3) in Eq. (9.7.2), one finds that $A(L,t) = A^*(0,t)$. Thus, except for a phase reversal induced by the OPC, the input field is completely recovered, and the pulse shape is restored to its input form. Since the signal spectrum after OPC becomes the mirror image of the input spectrum, the OPC technique is also referred to as *midspan spectral inversion*.

9.7.2 Compensation of Self-Phase Modulation

As discussed in Section 2.6.2, the nonlinear phenomenon of SPM leads to fiber-induced chirping of the transmitted signal. Section 9.3.5 indicated that this chirp can be used to advantage with a proper design. Optical solitons (see Chapter 10) also use the SPM to their advantage. However, in most non-soliton systems, the SPM-induced nonlinear effects degrade the signal quality, especially when the signal is propagated over long distances by using multiple optical amplifiers (see Section 8.6.4).

The OPC technique differs from all other dispersion-compensation schemes in one important way: Under certain conditions, it can compensate simultaneously for both the GVD and SPM. This feature of OPC was noted in the early 1980s [106] and has been studied extensively after 1993 [104]. It is easy to show that both the GVD and SPM are compensated perfectly in the absence of fiber loss. Pulse propagation in a lossy fiber is governed by Eq. (5.2.4) or by

$$\frac{\partial A}{\partial z} + \frac{i}{2}\beta_2 \frac{\partial^2 A}{\partial t^2} = i\bar{\gamma}|A|^2 A - \frac{1}{2}\alpha A. \qquad (9.7.4)$$

where $\bar{\gamma}$ governs the SPM and α accounts for the fiber loss. This equation is studied in Chapter 10 in the context of optical solitons. When $\alpha = 0$, A^* satisfies the same equation when one takes the complex conjugate of Eq. (9.7.4) and changes z to $-z$. As a result, midspan OPC can compensate for SPM and GVD simultaneously.

The presence of fiber loss destroys this important property of midspan OPC. The reason is intuitively obvious if we note that that the SPM-induced phase shift is power dependent. As a result, much larger phase shifts are induced in the first-half of the link than the second half, and OPC cannot compensate for the nonlinear effects. Equation (9.7.4) can be used to study the impact of fiber loss. By making the substitution

$$A(z,t) = B(z,t)\exp(-\alpha z/2), \qquad (9.7.5)$$

Eq. (9.7.4) can be written as

$$\frac{\partial B}{\partial z} + \frac{i}{2}\beta_2 \frac{\partial^2 B}{\partial t^2} = i\gamma(z)|B|^2 B, \qquad (9.7.6)$$

where $\gamma(z) = \bar{\gamma}\exp(-\alpha z)$. The effect of fiber loss is mathematically equivalent to the loss-free case but with a z-dependent nonlinear parameter. By taking the complex conjugate of Eq. (9.7.6) and changing z to $-z$, it is easy to see that perfect SPM compensation can occur only if $\gamma(z) = \gamma(L - z)$. This condition cannot be satisfied for $\alpha \neq 0$.

One may think that the problem can be solved by amplifying the signal after midspan OPC such that the signal power becomes equal to the input power before the signal is launched in the second-half section of the fiber link. Although such an approach can reduce the impact of SPM, it does not lead to perfect compensation of the SPM. The reason can be understood by noting that propagation of a phase-conjugated signal is equivalent to propagating a *time-reversed* signal [107]. Thus, perfect SPM compensation can occur only if the power variations are symmetric around the midspan point where the OPC is performed so that $\gamma(z) = \gamma(L - z)$ in Eq. (9.7.6). In practice, signal transmission does not satisfy this property. One can come close to SPM compensation if the signal is amplified often enough that the power does not vary by a large amount during each amplification stage. This approach is, however, not practical since it requires closely spaced amplifiers.

Perfect compensation of both GVD and SPM can be realized by using *dispersion-decreasing fibers*, a topic covered in Section 10.4.4. To see how such a scheme can be implemented, assume that β_2 in Eq. (9.7.6) is a function of z. By making the transformation,

$$\xi = \int_0^z \gamma(z)\, dz, \qquad\qquad (9.7.7)$$

Eq. (9.7.6) can be written as [104]

$$\frac{\partial B}{\partial \xi} + \frac{i}{2} B(\xi) \frac{\partial^2 B}{\partial t^2} = |B|^2 B, \qquad\qquad (9.7.8)$$

where $B(\xi) = \beta_2(\xi)/\gamma(\xi)$. Both GVD and SPM are compensated if $B(\xi) = B(\xi_L - \xi)$, where ξ_L is the value of ξ at $z = L$. A simple solution is provided by the case in which the dispersion is tailored in exactly the same way as $\gamma(z)$. Since fiber loss makes $\gamma(z)$ to vary exponentially, both GVD and SPM can be compensated exactly in a dispersion-decreasing fiber whose GVD decreases exponentially. This approach is quite general and applies even when in-line amplifiers are used.

9.7.3   Generation of Phase-Conjugated Signal

The implementation of the midspan OPC technique requires a nonlinear optical element that generates the phase-conjugated signal. The most commonly used method makes use of *four-wave mixing* (FWM) in a nonlinear medium. Since the optical fiber itself is a nonlinear medium, a simple approach is to use a few-kilometer-long fiber especially designed to maximize the FWM efficiency.

The FWM phenomenon in optical fibers has been studied extensively [61]. Its use requires injection of a pump signal at a frequency ω_p that is shifted from the signal frequency ω_s by a small amount (~ 0.5 THz). The fiber nonlinearity generates the phase-conjugated signal at the frequency $\omega_c = 2\omega_p - \omega_s$ provided

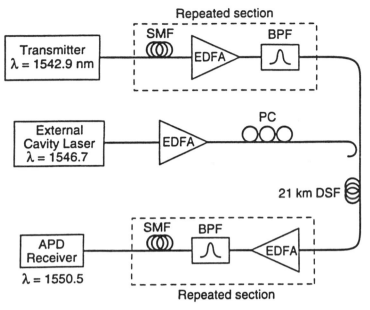

Figure 9.13 Experimental setup for dispersion compensation through midspan spectral inversion in a 21-km-long dispersion-shifted fiber (DSF). (After Ref. [88]. ©1993 IEEE. Reprinted with permission.)

that the *phase-matching condition* $k_c = 2k_p - k_s$ is approximately satisfied, where $k_j = n(\omega_j)\omega_c/c$ is the wave number for the optical field of frequency ω_j. In single-mode fibers, the phase-matching condition can be approximately satisfied if the zero-dispersion wavelength of the fiber is chosen to coincide with the pump wavelength. This was the approach adopted in the 1993 experiments in which the potential of OPC for dispersion compensation was first demonstrated. In one experiment [87], the 1546-nm signal was phase conjugated by using FWM in a 23-km fiber with pumping at 1549 nm. The 6-Gb/s signal was transmitted over 152 km of standard fiber in a coherent transmission experiment employing the FSK format. In another experiment [88], the 10-Gb/s signal (IM/DD format) was transmitted over 360 km. The midspan OPC was performed in a 21-km fiber by using a pump laser whose wavelength was tuned exactly to the zero-dispersion wavelength of the fiber. The pump and signal wavelengths differed by 3.8 nm. Figure 9.13 shows the experimental setup. A bandpass filter (BPF) is used to separate the phase-conjugated signal from the pump.

Several factors need to be considered while implementing the midspan OPC technique in practice. First, since the signal wavelength changes from ω_s to $\omega_c = 2\omega_p - \omega_s$ at the phase conjugator, the GVD parameter β_2 becomes different in the the second-half section. As a result, perfect compensation occurs only if the phase conjugator is slightly offset from the midpoint of the fiber link. The exact location L_p can be determined by using the condition $\beta_2(\omega_s)L_p = \beta_2(\omega_c)(L - L_p)$, where L is the total link length. By expanding $\beta_2(\omega_c)$ in a

Taylor series around the signal frequency ω_s, L_p is found to be

$$\frac{L_p}{L} = \frac{\beta_2 + \delta\beta_3}{2\beta_2 + \delta\beta_3},\tag{9.7.9}$$

where $\delta = \omega_c - \omega_s$ is the frequency shift of the signal induced by the OPC technique. For a typical wavelength shift of 6 nm, the phase-conjugator location changes by about 1%. The effect of residual dispersion and SPM in the phase-conjugation fiber itself can also affect the placement of phase conjugator [100].

A second factor that need to be addressed is that the FWM process in optical fibers is polarization sensitive. Since signal polarization is not controlled in telecommunication fibers, it varies randomly (see Section 6.5.3). Such random variations affect the FWM efficiency, making the standard OPC technique unsuitable for practical purposes. Fortunately, the FWM scheme can be modified to make it polarization insensitive. In one approach, two orthogonally polarized pump beams at different wavelengths, located symmetrically on the opposite sides of the zero-dispersion wavelength λ_{ZD} of the fiber, are used [90]. This scheme has another advantage: the phase-conjugate wave can be generated at the same frequency as the signal by choosing λ_{ZD} such that it coincides with the signal frequency. This feature can readily be verified by noting that $\omega_c = \omega_{p1} + \omega_{p2} - \omega_s$, when pump frequencies $\omega_{p1} \neq \omega_{p2}$. Polarization insensitivity of OPC can also be realized by using a single pump in combination with a fiber grating and an *orthoconjugate mirror* [96], but the device works in the reflective mode and requires the separation of the conjugate wave from the signal by using a 3-dB coupler or an optical circulator.

The relatively low efficiency of the OPC process in optical fibers is also of some concern. Typically, the conversion efficiency η_c is below 1%, making it necessary to amplify the phase-conjugated signal [88]. Effectively, the insertion loss of the phase conjugator exceeds 20 dB. However, the FWM process is not inherently a low-efficiency process and, in principle, it can even provide net gain [61]. Indeed, the analysis of the FWM equations shows that η_c increases considerably by increasing the pump power while decreasing the signal power; it can even exceed 100% by optimizing the power levels and the pump-signal wavelength difference [98]. High pump powers are often avoided because of the onset of stimulated Brillouin scattering (SBS) occurring near 10 mW. However, SBS can be suppressed by modulating the pump at a frequency ~ 100 MHz. In a 1994 experiment [108], 35% conversion efficiency was realized by using this technique.

The FWM process in a semiconductor laser amplifier has also been used to generate the phase-conjugated signal for dispersion compensation. This approach was first used in a 1993 experiment [89] to demonstrate transmission of a 2.5-Gb/s signal, obtained through direct modulation of a semiconductor laser, over 100 km of standard fiber. Later, in a 1995 experiment the same approach was used to demonstrate the transmission of a 40-Gb/s signal over 200 km of standard fiber [99]. Highly nondegenerate FWM in semiconductor laser amplifiers was proposed [110] as early as 1987 and has been studied extensively in the context of wavelength conversion [111]. Its main advantage is that the phase-conjugated signal can be generated in a device of length 1 mm or less. The conversion efficiency is also typically higher than that of FWM in an optical fiber because of amplification, although this advantage is offset by the relatively large coupling losses resulting from the need to couple the signal back into the fiber. By a proper choice of the pump-signal detuning, conversion efficiencies of more than 100% (net gain for the phase-conjugated signal) have been realized for FWM in semiconductor laser amplifiers [112]. Such a performance makes this approach quite attractive for dispersion compensation.

9.7.4 Effect of Periodic Amplification

Most of the experimental work on dispersion compensation has considered transmission distances of several hundred kilometers. For long-haul applications, one may ask whether the OPC technique can compensate for GVD acquired over several thousand kilometers in fiber links which use amplifiers periodically for loss compensation. This question has been studied mainly through numerical simulations.

In one set of simulations, the NRZ-format data stream is modeled by considering a pseudo-random sequence of 128 bits with a super-Gaussian shape for optical pulses [101]. Fiber loss is compensated by ideal, noise-less amplifiers, and GVD is compensated by an ideal OPC process. The system performance is judged by the amount of eye closure. The results show that a 10-Gb/s signal can be transmitted over 6000 km if the average launch power is kept below 3 mW to reduce the effects of fiber nonlinearity. As discussed in Section 9.7.2, OPC compensates for GVD and SPM only partially when the nonlinear effects become important. In another set of computer simulations [102], the amplifier spacing was found to play an important role. The signal could be propagated over distances as long as 9000 km, by keeping the amplifiers only 40 km apart. The choice of the operating wavelength with respect to the zero-dispersion wavelength was also critical. In the anomalous-dispersion region ($\beta_2 < 0$), the periodic variation of the signal power along the fiber link can lead to the generation of additional sidebands through the phenomenon of *modulation instability* [109]. This instability can be avoided if the dispersion parameter is relatively large [$D > 10$ ps/(km-nm)]. This is the case for standard fibers near 1.55 μm. It should be remarked that the maximum transmission distance depends critically on many factors, such as the FWM efficiency, the input power, and the amplifier spacing, and may decrease to below 3000 km, depending on the operating parameters [103].

The use of OPC for long-haul lightwave systems requires periodic use of optical amplifiers and phase conjugators. These two optical elements can be combined into one by using *parametric amplifiers* which not only generate the phase-conjugated signal through the FWM process but also amplify it. The analysis of such a long-haul system shows that 20- to 30-ps input pulses can travel over thousands of kilometers despite a high fiber dispersion, and the total transmission distance can exceed 15,000 km for dispersion-shifted fibers with $\beta_2 = -2$ ps^2/km near 1.55 μm [113].

9.8   BROADBAND COMPENSATION

Although most experiments on dispersion compensation have focused on 10-Gb/s lightwave systems, considerable effort is under way to extend the system capacity to beyond 100 Gb/s by using channel-multiplexing techniques, such as TDM and WDM, discussed in Chapter 7. At such high bit rates, a dispersion-compensation method should be compatible with the broad bandwidth occupied by the multichannel signal. In this section we discuss how broadband operation can be realized in practice by considering TDM and WDM systems separately.

9.8.1   Time-Division Multiplexing

A TDM optical signal uses ultrashort pulses (width ~ 1 ps) since the bit slot at a bit rate of 100 Gb/s is only 10 ps wide. For such short optical pulses, the pulse spectrum becomes broad enough ($\Delta \nu > 100$ GHz) that it is difficult to compensate GVD over the entire bandwidth of the pulse because of the frequency dependence of β_2. Equation (2.4.34) can be used to estimate the maximum transmission distance L, limited by the third-order dispersion β_3, when only second-order dispersion is compensated. The result is

$$L \leq 0.034(|\beta_3|B^3)^{-1}. \qquad (9.8.1)$$

This limitation is shown in Fig. 2.13 by the dashed line. At 200 Gb/s, L is limited to about 50 km and drops to only 3.4 km at 500 Gb/s if we use a typical value $\beta_3 = 0.08$ ps^3/km. Clearly, it is essential to use techniques that compensate for both the second- and third-order dispersions simultaneously when the bit rate exceeds 100 Gb/s for a TDM system.

The simplest solution to broadband dispersion compensation is provided by high-dispersion fibers designed to have a negative dispersion slope so that both β_2 and β_3 have opposite signs, in comparison with the standard telecommunication fiber. The necessary conditions for designing such fibers can be obtained by solving Eq. (9.1.3) by using the Fourier method. For a fiber link containing two different fibers of lengths L_1 and L_2, the conditions for broadband dispersion compensation are given by

$$\beta_{21}L_1 + \beta_{22}L_2 = 0 \qquad \text{and} \qquad \beta_{31}L_1 + \beta_{32}L_2 = 0, \qquad (9.8.2)$$

where β_{2j} and β_{3j} are second- and third-order dispersion parameters for the fiber of length L_j. The first condition is the same as Eq. (9.4.2). By using Eq. (9.4.3), the second condition can be used to find the third-order dispersion parameter for the DCF:

$$\beta_{32} = (D_2/D_1)\beta_{31} = -(L_1/L_2)\beta_{31}. \qquad (9.8.3)$$

It is generally difficult to satisfy Eq. (9.8.2) over a wide wavelength range, although considerable success has been realized in recent years. For TDM systems, the signal bandwidth is typically ~ 100 GHz, and it is sufficient to satisfy

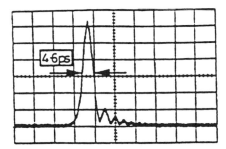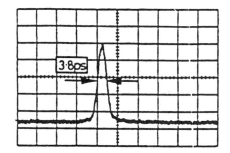

Figure 9.14 Experimentally observed pulse shapes after a 2.6-ps input pulse has propagated over 300 km of dispersion-shifted fiber ($\beta_2 = 0$). The left and right traces compare the improvement realized by compensating the third-order dispersion β_3 through a planar lightwave circuit. (After Ref. [117]. ©1996 IEE. Reprinted with permission.)

Eq. (9.8.2) over a 2- to 3-nm bandwidth. This requirement is easily met for two-mode DCFs discussed in Section 9.4, although it requires precise control of the ratio L_2/L_1 [114]. Indeed, numerical simulations show that for a 1-ps pulse, L_1 can exceed 100 km by using optimized lengths of such a two-mode DCF. Assuming that a 2-ps bit slot is required for a RZ system making use of 1-ps pulse, transmission over 100 km at bit rates as high as 500 Gb/s appears to be feasible with the use of dispersion compensation.

Several TDM experiments have been performed to demonstrate the potential of dispersion-compensation techniques. In one experiment [115], the 100-Gb/s TDM signal was transmitted over 560 km with 80-km amplifier spacing by managing the fiber dispersion. In another experiment, the bit rate was extended to 400 Gb/s by using 0.98-ps optical pulses within a 2.5-ps time slot [116]. Without compensation of the third-order dispersion, the pulse broadened to 2.3 ps after 40 km and exhibited a long oscillatory tail extending over 5–6 ps, a characteristic feature of the third-order dispersion [61]. With partial compensation of the third-order dispersion, the oscillatory tail disappeared and the pulse width reduced to 1.6 ps, making it possible to recover the 400-Gb/s data with high accuracy.

Optical filters can also be designed to compensate for the third-order dispersion. The planar lightwave circuits based on multiple MZ interferometric filters (see Section 9.5.2) have proved quite successful because of the programmable nature of such filters. In one experiment [117], such a filter was designed to have a dispersion slope of -15.8 ps/nm^2 over a 170-GHz bandwidth. It was used to compensate the third-order dispersion over 300 km of a dispersion-shifted fiber with $\beta_3 \approx 0.05$ ps/(km-nm^2) at the operating wavelength. Figure 9.14 compares the pulse shapes at the fiber output observed with and without β_3 compensation when a 2.6-ps pulse was transmitted over 300 km of such a fiber. The equalizer eliminates the long oscillatory tail and reduces the width of the main peak from 4.6 to 3.8 ps. The increase in the pulse width from its input value of 2.6 ps is attributed to polarization-mode dispersion (PMD). A phase-diversity detection scheme has been proposed for PMD compensation [118] and should be useful whenever system performance is limited by PMD.

The third-order dispersion can also be compensated by using chirped fiber gratings. The discussion in Section 9.6.2 was limited to linearly chirped gratings which are designed to compensate for the effects of second-order dispersion only. Such gratings work quite well for dispersion compensation in 10-Gb/s systems but would be unsuitable at high bit rates ∼ 100 Gb/s. A quadratically chirped fiber Bragg grating can be designed to compensate simultaneously for both second- and third-order dispersions. Numerical simulations show that such gratings should be able to compensate for dispersion over 800 km of standard fiber over a bandwidth of 1 nm [119], and thus allow 100-Gb/s transmission over the existing telecommunication network.

9.8.2 Wavelength-Division Multiplexing

As discussed in Section 7.2, WDM lightwave systems transmit multiple channels by modulating several optical carriers with a wavelength spacing ∼ 1 nm. The entire signal may thus occupy a bandwidth of 20–30 nm, although it is bunched in spectral packets of bandwidth ∼ 0.1 nm, depending on the bit rate of individual channels. The main difference compared with the TDM case is that, since relatively wide (> 50 ps) optical pulses are used for individual channels, the third-order dispersion plays little role during pulse transmission. However, because of the wavelength dependence of β_2, second-order dispersion must be compensated over a wide spectral range covered by the WDM signal. Two distinct approaches have been used for broadband dispersion compensation in WDM systems. In one approach, a DCF fiber, capable of compensating GVD over the entire signal bandwidth, is employed. In the second approach, advantage is taken of the periodic nature of the signal spectrum, and an optical filter with the periodic dispersion characteristics is used.

Consider first the case of optical filters with periodic dispersion characteristics. A FP filter has multiple transmission peaks spaced apart periodically by the free spectral range given by Eq. (7.3.1). The use of such filters for channel selection has been discussed in Section 7.3.3. They can also provide GVD compensation by detuning the FP transmission peak from the central wavelength of each channel. However, it is difficult to design FP filters with a large amount of dispersion. Fiber gratings provide large GVD but are generally suitable only for single-channel systems. However, a new kind of fiber grating, referred to as *sampled fiber grating* [120], has been made such that it has multiple stop bands. Its fabrication is quite simple. Rather than making a single long grating, multiple short-length gratings are written with uniform spacing among them. (Each short section is a sample, hence the name "sampled" grating.) Such a grating has multiple reflectivity peaks whose wavelength spacing is determined by the sample period and is thus controllable during the fabrication process. Moreover, if each sample is chirped, the dispersion characteristics of each reflectivity peak are governed by the amount of chirp introduced. Such a grating was used to demonstrate simultaneous compensation of fiber dispersion over 240 km for two 10-Gb/s channels [120]. In principle, such gratings should work for WDM systems with 10–15 channels, although it becomes difficult to match the grating and fiber dispersions for multiple channels when the number of channels increases.

High-dispersion fibers discussed in Section 9.4 offer an alternative. Elliptical-core two-mode DCFs [48] are quite suitable for broadband dispersion compensation in WDM systems since the dispersion parameter D has not only large negative values (~ 500 ps/(km-nm), but $|D|$ also increases with increasing wavelength, thereby counteracting the corresponding increase in standard fibers. A similar trend is observed in single-mode DCFs, although $|D|$ is smaller by about a factor of 2 [45]. Many experiments have demonstrated the potential of DCFs. In a 1993 experiment [121], four 2.5-Gb/s channels were transmitted over 67 km of fiber by using dispersion compensation over 20-nm bandwidth. In a 1995 field demonstration, two 10-Gb/s channels were transmitted over 360 km of installed fiber [122], thereby establishing the use of DCFs under realistic operating conditions. In a 1996 experiment [123], a 40-Gb/s WDM signal (16 channels at 2.5 Gb/s) was transmitted over 427 km of fiber by combining dispersion compensation with forward-error correction.

Considerable effort is under way to design WDM systems with a total capacity in excess of 100 Gb/s. In one experiment [124], 8 channels, 1.6 nm apart, each operating at 20 Gb/s, were transmitted over 232 km of standard fiber by using multiple DCFs. The residual dispersion for each channel was relatively small (~ 100 ps/nm for the entire span) since all channels were compensated simultaneously by the DCFs. This approach was extended in 1996 to transmit a 1.1-Tb/s WDM signal (55 channels, each operating at 20 Gb/s) over 150 km of standard fiber [125]. Since channel spacing was 0.6 nm, the same DCF compensated fiber dispersion over the 32-nm bandwidth, indicating the broadband nature of dispersion compensation in this experiment. A small amount of residual dispersion (~ 100 ps/nm for the entire span) even helped in suppressing the FWM-induced interchannel crosstalk. This experiment clearly demonstrates the ultimate potential of dispersion compensation for WDM systems.

9.9   LONG-HAUL LIGHTWAVE SYSTEMS

This chapter has so far focused on lightwave systems in which dispersion compensation is achieved over fiber lengths of a few hundred kilometers. The question one may ask is which dispersion-compensation technique is best from a system-design standpoint if the transmission distance is several thousand kilometers. If the optical signal is regenerated electronically every 100–200 km, all techniques discussed in this chapter should work well since various kinds of degradations do not accumulate over long lengths. By contrast, if the signal is maintained in the optical domain over the entire link by using periodic amplification, nonlinear effects such as SPM and FWM [61] may affect the system differently for different dispersion-compensation schemes. This issue has been studied extensively [126]–[136]. The design aspects related to the OPC technique have been covered in Section 9.7.4. In this section we consider long-haul

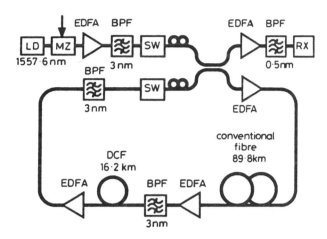

Figure 9.15 Recirculating fiber loop used to demonstrate transmission of a 10-Gb/s signal over 2245 km of standard fiber using a DCF periodically. Components used include laser diode (LD), modulator (MZ), bandpass filter (BPF), optical switch (SW), fiber amplifier (EDFA), and optical receiver (RX). (After Ref. [128]. ©1996 IEE. Reprinted with permission.)

lightwave systems making use of dispersion-compensating elements such as a DCF or a fiber grating.

Because of cost considerations, most laboratory experiments use a fiber loop in which the optical signal is forced to recirculate many times to simulate a long-haul lightwave system. Figure 9.15 shows a schematic of such a recirculating fiber loop used to demonstrate transmission of a 10-Gb/s signal over 2245 km of standard fiber [128]. It uses 16.2 km of DCF to compensate the dispersion over 89.8 km of fiber. Two bandpass filters are inserted inside the loop to reduce the effects of amplifier noise (see Chapter 8). The major nonlinear phenomenon affecting the performance of a single-channel system is SPM. Indeed, it was found that the transmission distance depended on the signal power at the DCF input and was maximum for an optimum value of about 0.57 mW. Clearly, SPM occurring inside the DCF plays an important role since the nonlinear effects are enhanced within the DCF because of its reduced effective core area. The bit rate was extended in a later experiment to 80 Gb/s by multiplexing eight 10-Gb/s channels (0.8-nm channel spacing) over the recirculating fiber loop [129]. Despite the use of gain-equalizing filters, the total transmission distance was reduced to 1171 km because of various nonlinear effects such as FWM. In another experiment [130], the bit rate was extended to 160 Gb/s by multiplexing sixteen 10-Gb/s channels, but FWM limited the distance to 1000 km.

The GVD and the power variations along a dispersion-compensated fiber link depend on the relative locations of the DCFs and optical amplifiers. The question thus arises whether one can extend the system performance by optimizing the GVD profile along the link. The answer is affirmative. In fact, this technique has been used and is referred to as *dispersion management*. Its use is most beneficial if the system designer has the liberty of choosing fibers with different dispersion characteristics. The underlying idea is quite simple and consists of mixing fibers with positive and negative GVDs such that not only is the total dispersion reduced for all channels but also the nonlinear effects are minimized simultaneously. The simplest scheme consists of GVD reversal from amplifier to amplifier and was used to transmit eight 20-Gb/s channels over 300 km [137]. In a 1995 experiment, a 340-Gb/s WDM signal (17 channels at 20 Gb/s) was transmitted over 150 km using dispersion management [138]. In another experiment [139], dispersion management was used to transmit a 100-Gb/s WDM signal over 560 km with 80-km amplifier spacing.

The optimization of system performance for various GVD profiles along the fiber link has also been studied. In a 1994 experiment [127], a 1000-km-long fiber loop containing 31 fiber amplifiers was used to study three different dispersion profiles. A maximum transmission distance of 12,000 km was realized for the case in which short sections of normal-GVD fibers were used to compensate for the anomalous GVD of long sections. Numerical simulations have also been used to analyze the effectiveness of various dispersion profiles [131]–[135]. In general, local GVD should be kept relatively large to suppress the nonlinear effects while minimizing the average dispersion for all channels. In one study, the effects of FWM were minimized by placing the dispersion compensator at the end of fiber link rather than at the end of each amplifier segment [136]. The main conclusion is that the optimization of a WDM system employing dispersion compensation depends on many design parameters, such as the launch power, amplifier spacing, and the location of DCFs.

PROBLEMS

9.1 What is the dispersion-limited transmission distance for a 1.55-μm lightwave system making use of direct modulation at 10 Gb/s? Assume that frequency chirping broadens the Gaussian-shape pulse spectrum by a factor of 6 from its transform-limited width. Use $D = 17$ ps/(km-nm) for fiber dispersion.

9.2 How much improvement in the dispersion-limited transmission distance is expected if an external modulator is used in place of direct modulation for the lightwave system of Problem 9.1?

9.3 Solve Eq. (9.1.3) by using the Fourier transform method. Use the solution to find an analytic expression for the pulse shape after a Gaussian input pulse has propagated to $z = L$ in a fiber with $\beta_2 = 0$.

9.4 Use the result obtained in Problem 9.3 to plot the pulse shape after a 1-ps (FWHM) Gaussian pulse is transmitted over 20 km of dispersion-shifted fiber with $\beta_2 = 0$ and $\beta_3 = 0.08$ ps^3/km. How would the pulse shape change if the sign of β_3 is inverted?

9.5 Use Eqs. (9.1.4) and (9.3.2) to plot the pulse shapes for $C = -1, 0$, and 1 when 50-ps (FWHM) chirped Gaussian pulses are transmitted over 100 km of standard fiber with $D = 16$ ps/(km-nm). Compare the three cases and comment on their relative merits.

9.6 The pre-chirp technique is used for dispersion compensation in a 10-Gb/s lightwave system operating at 1.55 μm and transmitting the 1 bits as chirped Gaussian pulses of 40 ps width (FWHM). Pulse broadening by up to 50% can be tolerated. What is the optimum value of the chirp parameter C, and how far the signal can be transmitted for this optimum value?

9.7 The pre-chirp technique in Problem 9.6 is implemented through frequency modulation of the optical carrier. Determine the modulation frequency for a maximum change of 10% from the average value.

9.8 Repeat Problem 9.7 for the case in which the pre-chirp technique is implemented through sinusoidal modulation of the carrier phase.

9.9 The transfer function of an optical filter is given by

$$H(\omega) = \exp[-(1 + ib)\omega^2/\omega_f^2].$$

What is the impulse response of this filter? Use Eq. (9.5.1) to find the pulse shape at the filter output when a Gaussian pulse is launched at the fiber input. How would you optimize the filter to minimize the effect of fiber dispersion?

9.10 Use the result obtained in Problem 9.9 to compare the pulse shapes before and after the filter when 30-ps (FWHM) Gaussian pulses are propagated over 100 km of fiber with $\beta_2 = -20$ ps^2/km. Assume that the filter bandwidth is the same as the pulse spectral width (FWHM in both cases) and the filter parameter b is optimized to minimize the effect of fiber dispersion. What is the optimum value of b?

9.11 Derive Eq. (9.5.3) by considering multiple round trips inside a FP filter with one 100%-reflectivity mirror.

9.12 Solve Eqs. (9.6.1) and (9.6.2) and show that the grating reflectivity is given by Eq. (9.6.4).

9.13 Write a computer program to solve Eqs. (9.6.1) and (9.6.2) for chirped fiber gratings such that both δ and κ vary with z. Use it to plot the amplitude and phase of the reflectivity of a grating in which the period varies linearly by 0.01% over the 10-cm length. Assume that $\kappa L = 4$ and a Bragg wavelength of 1.55 μm at the input end of the grating.

9.14 Explain how midspan OPC compensates for fiber dispersion. Show that the OPC process inverts the signal spectrum.

9.15 Prove that both SPM and GVD can be compensated through midspan OPC only if the fiber loss $\alpha = 0$. Show also that simultaneous compensation of SPM and GVD can occur when $\alpha \neq 0$ if GVD decreases along the fiber length. What is the optimum GVD profile of such a fiber?

REFERENCES

[1] R. Jopson and A. Gnauck, *IEEE Commun. Mag.* **33** (6), 336 (1995).

[2] K. Iwashita and N. Takachio, *J. Lightwave Technol.* **8**, 367 (1990).

[3] N. Takachio, S. Norimatsu, and K. Iwashita, *IEEE Photon. Technol. Lett.* **4**, 278 (1992).

[4] K. Yonenaga and N. Takachio, *IEEE Photon. Technol. Lett.* **5**, 949 (1993).

[5] S. Yamazaki, T. Ono, and T. Ogata, *J. Lightwave Technol.* **11**, 603 (1993).

[6] J. H. Winters and R. D. Gitlin, *IEEE Trans. Commun.* **38**, 1439 (1990).

[7] J. H. Winters, *J. Lightwave Technol.* **8**, 1487 (1990).

[8] J. C. Cartledge, R. G. McKay, and M. C. Nowell, *J. Lightwave Technol.* **10**, 1105 (1992).

[9] J. H. Winters, *Proc. SPIE* **1787**, 346 (1992).

[10] R. I. MacDonald, *IEEE Photon. Technol. Lett.* **6**, 565 (1994).

[11] G. P. Agrawal and M. J. Potasek, *Opt. Lett.* **11**, 318 (1986).

[12] R. Olshansky and D. Fye, *Electron. Lett.* **20**, 80 (1984).

[13] L. Bickers and L. D. Westbrook, *Electron. Lett.* **21**, 103 (1985).

[14] T. L. Koch and R. C. Alferness, *J. Lightwave Technol.* **3**, 800 (1985).

[15] F. Koyoma and K. Iga, *J. Lightwave Technol.* **6**, 87 (1988).

[16] T. Saito, N. Henmi, S. Fujita, M. Yamaguchi, and M. Shikada, *IEEE Photon. Technol. Lett.* **3**, 74 (1991).

[17] A. H. Gnauck, S. K. Korotky, J. J. Veselka, J. Nagel, C. T. Kemmerer, W. J. Minford, and D. T. Moser, *IEEE Photon. Technol. Lett.* **3**, 916 (1991).

[18] E. Devaux, Y. Sorel, and J. F. Kerdiles, *J. Lightwave Technol.* **11**, 1937 (1993).

[19] N. Henmi, T. Saito, and T. Ishida, *J. Lightwave Technol.* **12**, 1706 (1994).

[20] J. C. Cartledge, H. Debrégeas, and C. Rolland, *IEEE Photon. Technol. Lett.* **7**, 224 (1995).

[21] J. A. J. Fells, M. A. Gibbon, I. H. White, G. H. B. Thompson, R. V. Penty, C. J. Armistead, E. M. Kinber, D. J. Moule, and E. J. Thrush, *Electron. Lett.* **30**, 1168 (1994).

[22] K. Morito, R. Sahara, K. Sato, and Y. Kotaki, *IEEE Photon. Technol. Lett.* **8**, 431 (1996).

[23] B. Wedding, *Electron. Lett.* **28**, 1298 (1992).

[24] B. Wedding and B. Franz, *Electron. Lett.* **29**, 402 (1993).

[25] Y. Sorel, J. F. Kerdiles, C. Kazmierski, M. Blez, D. Mathoorasing, and A. Ougazzaden, *Electron. Lett.* **29**, 973 (1993).

[26] B. Wedding, B. Franz, and B. Junginger, *J. Lightwave Technol.* **12**, 1720 (1994).

[27] B. Wedding, K. Koffers, B. Franz, D. Mathoorasing, C. Kazmierski, P. P. Monteiro, and J. N. Matos, *Electron. Lett.* **31**, 566 (1995).

[28] J. Binder and U. Kohn, *IEEE Photon. Technol. Lett.* **6**, 558 (1994).

[29] C. Bungarzeanu, *IEEE Photon. Technol. Lett.* **6**, 858 (1994).

[30] M. Schwartz, *Information, Transmission, Modulation, and Noise*, 4th ed., McGraw-Hill, New York, 1990, Sec. 3.10.

[31] G. May, A. Solheim, and J. Conradi, *IEEE Photon. Technol. Lett.* **6**, 648 (1994).

[32] D. Penninckx, L. Pierre, J.-P. Thiery, B. Clesca, M. Chbat, and J.-L. Beylat, *Electron. Lett.* **32**, 1023 (1996).

[33] D. Penninckx, M. Chbat, L. Pierre, and J.-P. Thiery, *IEEE Photon. Technol. Lett.* **9**, 259 (1997).

[34] A. J. Price and N. Le Mercier, *Electron. Lett.* **6**, 58 (1995).

[35] G. P. Agrawal and N. A. Olsson, *Opt. Lett.* **14**, 500 (1989).

[36] N. A. Olsson, G. P. Agrawal, and K. W. Wecht, *Electron. Lett.* **25**, 603 (1989).

[37] N. A. Olsson and G. P. Agrawal, *Appl. Phys. Lett.* **55**, 13 (1989).

[38] G. P. Agrawal and N. A. Olsson, *IEEE J. Quantum Electron.* **25**, 2297 (1989).

[39] G. P. Agrawal and N. A. Olsson, U.S. Patent 4,979,234 (1990).

[40] M. J. Potasek and G. P. Agrawal, *Electron. Lett.* **22**, 759 (1986).

[41] C. Lin, H. Kogelnik, and L. G. Cohen, *Opt. Lett.* **5**, 476 (1980).

[42] D. S. Larner and V. A. Bhagavatula, *Electron. Lett.* **21**, 1171 (1985).

[43] A. M. Vengsarkar and W. A. Reed, *Opt. Lett.* **18**, 924 (1993).

[44] A. J. Antos and D. K. Smith, *J. Lightwave Technol.* **12**, 1739 (1994).

[45] D. W. Hawtof, G. E. Berkey, and A. J. Antos, Paper PD6, *Proc. Optical Fiber Commun. Conf.*, Optical Society of America, Washington, DC, 1996.

[46] C. D. Poole, J. M. Wiesenfeld, A. R. McCormick, and K. T. Nelson, *Opt. Lett.* **17**, 985 (1992).

[47] C. D. Poole, J. M. Wiesenfeld, and D. J. DiGiovanni, *IEEE Photon. Technol. Lett.* **5**, 194 (1993).

[48] C. D. Poole, J. M. Wiesenfeld, D. J. DiGiovanni, and A. M. Vengsarkar, *J. Lightwave Technol.* **12** 1746 (1994).

[49] R. C. Youngquist, J. L. Brooks, and H. J. Shaw, *Opt. Lett.* **9**, 177 (1984).

[50] J. N. Blake, B. Y. Kim, and H. J. Shaw, *Opt. Lett.* **11**, 177 (1986).

[51] C. D. Poole, C. D. Townsend, and K. T. Nelson, *J. Lightwave Technol.* **9**, 598 (1991).

[52] H. G. Park and B. Y. Kim, *Electron. Lett.* **25**. 797 (1989).

[53] F. Bilodeau, K. O. Hill, B. Malo, D. C. Johnson, and I. M. Skinner, *Electron. Lett.* **27**, 682(1991).

[54] L. J. Cimini, L. J. Greenstein, and A. A. M. Saleh, *J. Lightwave Technol.* **8**, 649 (1990).

[55] A. H. Gnauck, C. R. Giles, L. J. Cimini, J. Stone, L. W. Stulz, S. K. Korotoky, and J. J. Veselka, *IEEE Photon. Technol. Lett.* **3**, 1147 (1991).

[56] T. Ozeki, *Opt. Lett.* **17**, 375 (1992).

[57] K. Takiguchi, K. Okamoto, S. Suzuki, and Y. Ohmori, *IEEE Photon. Technol. Lett.* **6**, 86 (1994).

[58] M. Sharma, H. Ibe, and T. Ozeki, *J. Lightwave Technol.* **12**, 1759 (1994).

[59] K. Takiguchi, K. Okamoto, and K. Moriwaki, *IEEE Photon. Technol. Lett.* **6**, 561 (1994).

[60] K. Takiguchi, K. Jinguji, K. Okamoto, and Y. Ohmori, *Electron. Lett.* **31**, 2192 (1995).

[61] G. P. Agrawal, *Nonlinear Fiber Optics*, 2nd ed., Academic Press, San Diego, CA, 1995.

[62] D. K. W. Lam, B. K. Garside, and K. O. Hill, *Opt. Lett.* **7**, 291 (1982).

[63] F. Ouellette, *Opt. Lett.* **12**, 622, 1987.

[64] K. O. Hill, B. Malo, F. Bilodeau, and D. C. Johnson, *Annu. Rev. Mater. Sci.* **23**, 125 (1993).

[65] R. J. Campbell and R. Kashyap, *Int. J. Optoelectron.* **9**, 33 (1994).

[66] R. Kashyap, *Opt. Fiber Technol.* **1**, 17 (1994).

[67] I. Bennion, J. A. R. Williams, L. Zhang, and K. Sugden, *Opt. Quantum Electron.* **28**, 93 (1996).

[68] B. J. Eggleton, T. Stephens, P. A. Krug, G. Dhosi, Z. Brodzeli, and F. Ouellette, *Electron. Lett.* **32**, 1610 (1996).

[69] T. Stephens, P. A. Krug, Z. Brodzeli, G. Dhosi, F. Ouellette, and L. Poladian, *Electron. Lett.* **32**, 1599 (1996).

[70] M. C. Farries, K. Sugden, D. C. J. Reid, I. Bennion, A. Molony, and M. J. Goodwin, *Electron. Lett.* **30**, 891 (1994).

[71] K. O. Hill, F. Bilodeau, B. Malo, T. Kitagawa, S. Thériault, D. C. Johnson, and J. Albert, *Opt. Lett.* **19**, 1314 (1994).

[72] K. O. Hill, S. Thériault, B. Malo, F. Bilodeau, T. Kitagawa, D. C. Johnson, J. Albert, K. Takiguchi, T. Kataoka, and K. Hagimoto, *Electron. Lett.* **30**, 1755 (1994).

[73] D. Garthe, R. E. Epworth, W. S. Lee, A. Hadjifotiou, C. P. Chew, T. Bricheno, A. Fielding, H. N. Rourke, S. R. Baker, K. C. Byron, R. S. Baulcomb, S. M. Ohja, and S. Clements, *Electron. Lett.* **30**, 2159 (1994).

[74] P. A. Krug, T. Stephens, G. Yoffe, F. Ouellette, P. Hill, and G. Dhosi, *Electron. Lett.* **31**, 1091 (1995).

[75] W. H. Loh, R. I. Laming, X. Gu, M. N. Zervas, M. J. Cole, T. Widdowson, and A. D. Ellis, *Electron. Lett.* **31**, 2203 (1995).

[76] R. I. Laming, N. Robinson, P. L. Schrivner, M. N. Zervas, S. Barcelos, L. Reekie, and J. A. Tucknott, *IEEE Photon. Technol. Lett.* **8**, 428 (1996).

[77] W. H. Loh, R. I. Laming, N. Robinson, A. Cavaciuti, F. Vaninetti, C. J. Anderson, M. N. Zervas, and M. J. Cole, *IEEE Photon. Technol. Lett.* **8**, 944 (1996).

[78] W. H. Loh, R. I. Laming, A. D. Ellis, and D. Atkinson, *IEEE Photon. Technol. Lett.* **8**, 1258 (1996).

[79] S. V. Chernikov, J. R. Taylor, and R. Kashyap, *Opt. Lett.* **20**, 1586 (1995).

[80] G. P. Agrawal and S. Radic, *IEEE Photon. Technol. Lett.* **6**, 995 (1994).

[81] L. Zhang, K. Sugden, I. Bennion, and A. Molony, *Electron. Lett.* **31**, 477 (1995).

[82] F. Ouellette, J.-F. Cliche, and S. Gagnon, *J. Lightwave Technol.* **12**, 1278 (1994).

[83] F. Ouellette, *Opt. Lett.* **16**, 303 (1991).

[84] F. Ouellette, *Electron. Lett.* **27**, 1668 (1991).

[85] U. Peschel, T. Peschel, and F. Lederer, *Appl. Phys. Lett.* **67**, 2111 (1995).

[86] A. Yariv, D. Fekete, and D. M. Pepper, *Opt. Lett.* **4**, 52 (1979).

[87] S. Watanabe, N. Saito, and T. Chikama, *IEEE Photon. Technol. Lett.* **5**, 92 (1993).

[88] R. M. Jopson, A. H. Gnauck, and R. M. Derosier, *IEEE Photon. Technol. Lett.* **5**, 663 (1993).

[89] M. C. Tatham, G. Sherlock, and L. D. Westbrook, *Electron. Lett.* **29**, 1851 (1995).

[90] R. M. Jopson, and R. E. Tench, *Electron. Lett.* **29**, 2216 (1993).

[91] S. Watanabe, T. Chikama, G. Ishikawa, T. Terahara, and H. Kuwahara, *IEEE Photon. Technol. Lett.* **5**, 1241 (1993).

[92] K. Kikuchi and C. Lorattanasane, *IEEE Photon. Technol. Lett.* **6**, 104 (1994).

[93] W. Pieper, C. Kurtzke, R. Schnable, D. Breuer, R. Ludwig, K. Petermann, and H. G. Weber, *Electron. Lett.* **6**, 724 (1994).

[94] S. Watanabe, G. Ishikawa, T. Naito, and T. Chikama, *J. Lightwave Technol.* **12**, 2139 (1994).

[95] W. Wu, P. Yeh and S. Chi, *IEEE Photon. Technol. Lett.* **6**, 1448 (1994).

[96] C. R. Giles, V. Mizrahi, and T. Erdogan, *IEEE Photon. Technol. Lett.* **7**, 126 (1995).

[97] M. E. Marhic, N. Kagi, T.-K. Chiang, and L. G. Kazovsky, *Opt. Lett.* **20**, 863 (1995).

[98] S. Wabnitz, *IEEE Photon. Technol. Lett.* **7**, 652 (1995).

[99] A. D. Ellis, M. C. Tatham, D. A. O. Davies, D. Nesser, D. G. Moodie, and G. Sherlock, *Electron. Lett.* **31**, 299 (1995).

[100] M. Yu, G. P. Agrawal, and C. J. McKinstrie, *IEEE Photon. Technol. Lett.* **7**, 932 (1995).

[101] X. Zhang, F. Ebskamp, and B. F. Jorgensen, *IEEE Photon. Technol. Lett.* **7**, 819 (1995).

[102] C. Lorattanasane and K. Kikuchi, *IEEE Photon. Technol. Lett.* **7**, 1375 (1995).

[103] X. Zhang and B. F. Jorgensen, *Electron. Lett.* **32**, 753 (1996).

[104] S. Watanabe and M. Shirasaki, *J. Lightwave Technol.* **14**, 243 (1996).

[105] A. Røyset, S. Y. Set, I. Goncharenko, and R. I. Laming, *IEEE Photon. Technol. Lett.* **8**, 449 (1996).

[106] R. A. Fisher, B. R. Suydam, and D. Yevick, *Opt. Lett.* **8**, 611 (1983).

[107] R. A. Fisher, Ed., *Optical Phase Conjugation*, Academic Press, San Diego, CA, 1983.

[108] S. Watanabe and T. Chikama, *Electron. Lett.* **30**, 163 (1994).

[109] F. Matera, A. Mecozzi, M. Romagnoli, and M. Settembre, *Opt. Lett.* **18**, 1499 (1993).

[110] G. P. Agrawal, *Appl. Phys. Lett.* **51**, 302 (1987).

[111] G. P. Agrawal, in *Semiconductor Lasers: Past, Present, Future*, G. P. Agrawal, Ed., AIP Press, Woodbury, NY, 1995, Chap. 8.

[112] A. D'Ottavi, F. Martelli, P. Spano, A. Mecozzi, and S. Scotti, *Appl. Phys. Lett.* **68**, 2186 (1996).

[113] R.-D. Li, P. Kumar, W. L. Kath, and J. N. Kutz, *IEEE Photon. Technol. Lett.* **5**, 669 (1993).

[114] C.-C. Chang and A. M. Weiner, *IEEE Photon. Technol. Lett.* **6**, 1392 (1994).

[115] S. Kawanishi, H. Takara, O. Kamatani, T. Morioka, and M. Saruwatari, *Electron. Lett.* **32**, 470 (1996).

[116] S. Kawanishi, H. Takara,, T. Morioka, O. Kamatani, K. Takiguchi, T. Kitoh, and M. Saruwatari, *Electron. Lett.* **32**, 916 (1996).

[117] K. Takiguchi, S. Kawanishi, H. Takara, K. Okamoto, and Y. Ohmori, *Electron. Lett.* **32**, 755 (1996).

[118] B. W. Hakki, *IEEE Photon. Technol. Lett.* **9**, 121 (1997).

[119] J. A. R. Williams, I. Bennion, and N. J. Doran, *Opt. Commun.* **116**, 62 (1995).

[120] F. Ouellette, P. A. Krug, T. Stephens, G. Dhosi, and B. Eggleton, *Electron. Lett.* **31**, 899 (1995).

[121] H. Izadpanah, E. Goldstein, and C. Lin, *Electron. Lett.* **29**, 364 (1993).

[122] Y. K. Park, P. D. Yeates, J.-M. P. Delavaux, O. Mizuhara, T. V. Nguyen, L. D. Tzeng, R. E. Tench, B. W. Hakki, C. D. Chen, R. J. Nuyts, and K. Ogawa, *IEEE Photon. Technol. Lett.* **7**, 816 (1995).

[123] S. Sian, S. M. Webb, K. M. Guild, and D. R. Terrence, *Electron. Lett.* **32**, 50 (1996).

[124] R. W. Tkach, R. M. Derosier, A. H. Gnauck, A. M. Vengsarkar, D. W. Peckham, J. J. Zyskind, J. W. Sulhoff, and A. R. Chraplyvy, *IEEE Photon. Technol. Lett.* **7**, 1369 (1995).

[125] H. Onaka, H. Miyata, G. Ishikawa, K. Otsuka, H. Ooi, Y. Kai, S. Kinoshita, M. Seino, H. Nishimoto, and T. Chikama, Paper PD19, *Proc. Optical Fiber Commun. Conf.*, Optical Society of America, Washington, DC, 1996.

[126] D. Marcuse, A. R. Chraplyvy, and T. W. Tkach, *J. Lightwave Technol.* **12**, 885 (1994).

[127] H. Taga, S. Yamamoto, N. Edagawa, Y. Yoshida, S. Akiba, and H. Wakabayashi, *J. Lightwave Technol.* **12**, 1616 (1994).

[128] N. Kikuchi, S. Sasaki, and K. Sekine, *Electron. Lett.* **31**, 375 (1995).

[129] S. Sekine, N. Kikuchi, S. Sasaki, and Y. Uchida, *Electron. Lett.* **31**, 1080 (1995).

[130] K. Oda, M. Fukutoku, M. Fukui, T. Kitoh, and H. Toba, *Proc. Optical Fiber Commun. Conf.*, Optical Society of America, Washington, DC, 1995.

[131] F. Kuppers, A. Mattheus, and R. Ries, *Pure Appl. Opt.* **4**, 459 (1995).

[132] A. Naka and S. Saito, *J. Lightwave Technol.* **13**, 862 (1995).

[133] N. Kikuchi and S. Sasaki, *J. Lightwave Technol.* **13**, 868 (1995).

[134] J. Nakagawa and K. Hotate, *J. Opt. Commun.* **16**, 202 (1995).

[135] F. Matera and M. Settembre, *J. Lightwave Technol.* **14**, 1 (1996).

[136] M. E. Marhic, N. Kagi, T.-K. Chiang, and L. G. Kazovsky, *IEEE Photon. Technol. Lett.* **8**, 145 (1996).

[137] A. H. Gnauck, A. R. Chraplyvy, R. W. Tkach, and R. M. Derosier, *Electron. Lett.* **30**, 1241 (1994).

[138] A. R. Chraplyvy, A. H. Gnauck, R. W. Tkach, R. M. Derosier, C. R. Giles, B. M. Nyman, G. A. Ferguson, J. W. Sulhoff, and J. L. Zyskind, *IEEE Photon. Technol. Lett.* **7**, 98 (1995).

[139] S. Kawanishi, H. Takara, O. Kamatani, T. Morioka, and M. Saruwatari, *Electron. Lett.* **32**, 470 (1996).

Chapter 10

SOLITON COMMUNICATION SYSTEMS

The word *soliton* was coined [1] in 1965 to describe the particlelike properties of pulse envelopes in dispersive nonlinear media: Under certain conditions, the pulse envelope not only propagates undistorted but also survives collisions just as particles do. The existence of solitons in optical fibers and their use for optical communications were suggested [2] in 1973, and by 1980 solitons were observed experimentally [3]. The potential of solitons for long-haul optical communication was demonstrated in 1988 in an experiment in which fiber loss was compensated using the technique of Raman amplification [4]. Remarkable progress made during the decade of 1990s [5]–[7] has converted optical solitons into a practical candidate for the next generation of lightwave communication systems. Although soliton lightwave systems were not commercially available in 1996, several field trials making use of solitons were in the planning stage.

In this chapter we describe soliton communication systems with emphasis on the physics and design of such systems. The basic concepts behind fiber solitons are introduced in Section 10.1, where we also discuss the properties of such solitons. Section 10.2 covers how fiber solitons can be used for optical communications and how the design of such lightwave systems differs from that of conventional systems. The design and performance issues for soliton lightwave systems operating in the average-soliton regime are considered in Section 10.3 with emphasis on the various physical mechanisms that limit the system performance. In Section 10.4 we consider several new techniques that can be used for increasing the bit rate of single-channel soliton systems. The use of solitons for multichannel lightwave systems is discussed in Section 10.5.

10.1   FIBER SOLITONS

The existence of fiber solitons is the result of a balance between *group-velocity dispersion* (GVD) and *self-phase modulation* (SPM), both of which, as discussed in Sections 2.4 and 5.2, limit the performance of fiber-optic communication systems when acting independently on optical pulses propagating inside the fiber. One can develop an intuitive understanding of how such a balance is possible by following the analysis of Section 2.4. As shown there, the GVD broadens optical pulses during their propagation inside the fiber except when the pulse is initially chirped in the right way (see Fig. 2.12). More specifically, a chirped pulse can be compressed during the early stage of propagation whenever the GVD parameter β_2 and the chirp parameter C happen to have opposite signs, so that $\beta_2 C$ is negative. SPM, resulting from the intensity dependence of the refractive index, imposes a chirp on the optical pulse such that $C > 0$. Since $\beta_2 < 0$ in the 1.55-μm wavelength region, the condition $\beta_2 C < 0$ is readily satisfied. Moreover, since the SPM-induced chirp is power dependent, it is not difficult to imagine that under certain conditions, SPM and GVD may cooperate in such a way that the SPM-induced chirp is just right to cancel the GVD-induced broadening of the pulse. The optical pulse would then propagate undistorted in the form of a soliton.

10.1.1   Nonlinear Schrödinger Equation

The mathematical description of fiber solitons requires solution of the wave equation in a dispersive nonlinear medium. A simple approach is to start with the propagation equation [Eq. (2.4.7)], satisfied by the slowly varying pulse envelope $A(z,t)$ in the presence of GVD, and modify it to include the effects of fiber nonlinearity responsible for SPM. Equation (2.6.5) provides a clue for such a modification. Since silica fibers are only weakly nonlinear (the intensity-dependent change in the refractive index is typically $< 10^{-9}$), the effect of SPM can be included by adding a nonlinear term on the right side of Eq. (2.4.7) so that this equation takes the form [6]

$$\frac{\partial A}{\partial z} + \beta_1 \frac{\partial A}{\partial t} + \frac{i}{2}\beta_2 \frac{\partial^2 A}{\partial t^2} - \frac{1}{6}\beta_3 \frac{\partial^3 A}{\partial t^3} = i\gamma |A|^2 A - \frac{\alpha}{2}A, \qquad (10.1.1)$$

where the fiber loss is included through α, $\beta_1 = v_g^{-1}$ with v_g representing the group velocity, β_2 and β_3 account for fiber dispersion, and γ is the nonlinearity parameter defined as

$$\gamma = 2\pi n_2 / \lambda A_{\text{eff}}. \qquad (10.1.2)$$

Here n_2 is the nonlinear-index coefficient, λ is the optical wavelength, and A_{eff} is the *effective core area* introduced in Section 2.6.2 (the bars over γ and n_2 appearing there are dropped for notational simplicity). The parameters β_2 and γ govern the effects of GVD and SPM, respectively. Equation (10.1.1) is quite accurate for describing evolution of optical pulses as short as 5 ps. For pulses shorter than that, several higher-order nonlinear effects, discussed later in Section 10.4, need to be included [6]. Equation (10.1.1) can be used in most cases of practical interest.

To discuss soliton solutions of Eq. (10.1.1), we first set $\alpha = 0$ and $\beta_3 = 0$. The fiber loss is included later in Section 10.3, while the effects of third-order dispersion (TOD) are considered in Section 10.4. It is useful to write Eq. (10.1.1) in a normalized form by introducing

$$\tau = \frac{t - \beta_1 z}{T_0}, \qquad \xi = \frac{z}{L_D}, \qquad U = \frac{A}{\sqrt{P_0}}, \qquad (10.1.3)$$

where T_0 is a measure of the pulse width, P_0 is the peak power of the pulse, and the dispersion length L_D is defined as

$$L_D = T_0^2/|\beta_2|. \qquad (10.1.4)$$

Equation (10.1.1) then takes the form

$$i\frac{\partial U}{\partial \xi} - \text{sgn}(\beta_2)\frac{1}{2}\frac{\partial^2 U}{\partial \tau^2} + N^2|U|^2 U = 0, \qquad (10.1.5)$$

where $\text{sgn}(\beta_2) = +1$ or -1, depending on whether β_2 is positive (normal GVD) or negative (anomalous GVD). The parameter N is defined by

$$N^2 = \gamma P_0 L_D = \gamma P_0 T_0^2/|\beta_2|. \qquad (10.1.6)$$

It represents a dimensionless combination of the pulse and fiber parameters. It will be seen later that N has a physical significance. Equation (10.1.5) is known in the soliton literature [8]–[11] as the *nonlinear Schrödinger equation* (NSE).

10.1.2 Fundamental and Higher-Order Solitons

The NSE belongs to a special class of nonlinear partial differential equations that can be solved exactly by using a mathematical technique known as the *inverse scattering method* [8]–[10]. Although the NSE supports solitons for both normal and anomalous GVD, pulselike solitons are found only for the case of anomalous dispersion [12]. In the case of normal dispersion ($\beta_2 > 0$), the solutions occur in the form of a dip in a constant background. Such solutions, referred to as dark solitons, are discussed in Section 10.1.3. This chapter focuses on pulselike (bright) solitons since they are used almost exclusively for optical communications.

It is common to introduce $u = NU$ as a renormalized amplitude and write Eq. (10.1.5) in its canonical form as

$$i\frac{\partial u}{\partial \xi} + \frac{1}{2}\frac{\partial^2 u}{\partial \tau^2} + |u|^2 u = 0, \qquad (10.1.7)$$

where $\beta_2 < 0$ was assumed. This equation has been solved by the inverse scattering method [12]. Details of this method are available in several books devoted entirely to solitons [8]–[10]. The main results can be summarized as follows. When an input pulse having an initial amplitude

$$u(0, \tau) = N \text{sech}(\tau) \tag{10.1.8}$$

is launched into the fiber, its shape remains unchanged during propagation when $N = 1$ but follows a periodic pattern for integer values of $N > 1$ such that the input shape is recovered at $\xi = m\pi/2$, where m is an integer. The optical pulse corresponding to $N = 1$ is called the *fundamental soliton*. Pulses corresponding to other integer values of N are called *higher-order solitons*. The parameter N represents the order of the soliton. By noting that $\xi = z/L_D$, the soliton period z_0, defined as the distance over which higher-order solitons recover their original shape, is given by

$$z_0 = \frac{\pi}{2} L_D = \frac{\pi}{2} \frac{T_0^2}{|\beta_2|}. \tag{10.1.9}$$

The *soliton period* z_0 and *soliton order* N play an important role in the theory of optical solitons. Figure 10.1 shows the pulse evolution for the first-order ($N = 1$) and third-order ($N = 3$) solitons over one soliton period by plotting the pulse shape $|u(\xi, \tau)|^2$ (top row) and the frequency chirp (bottom row), defined as the time derivative of the soliton phase. Only the fundamental soliton remains chirp-free during propagation while maintaining its shape.

The solution corresponding to the fundamental soliton can be obtained by solving Eq. (10.1.7) directly without recourse to the inverse scattering method. The approach consists of assuming that a solution of the form

$$u(\xi, \tau) = V(\tau) \exp[i\phi(\xi, \tau)] \tag{10.1.10}$$

exists, where V must be independent of ξ for Eq. (10.1.10) to represent a fundamental soliton that maintains its shape during propagation. The phase ϕ can depend on both ξ and τ. If Eq. (10.1.10) is substituted in Eq. (10.1.7) and the real and imaginary parts are separated, one obtains two real equations for V and ϕ. The phase equation shows that ϕ should be of the form $\phi(\xi, \tau) = K\xi$, where K is a constant. The function $V(\tau)$ is then found to satisfy the following second-order nonlinear differential equation:

$$\frac{d^2 V}{d\tau^2} = 2V(K - V^2). \tag{10.1.11}$$

This equation can be solved by multiplying it by $2(dV/d\tau)$ and integrating over τ. The result is

$$(dV/d\tau)^2 = 2KV^2 - V^4 + C, \tag{10.1.12}$$

where C is a constant of integration. By using the boundary condition that both V and $dV/d\tau$ should vanish at $|\tau| = \infty$, C is found to be 0. The constant K is determined by using the other boundary condition that $V = 1$ and $dV/d\tau = 0$ at the soliton peak, assumed to occur at $\tau = 0$. Its use provides $K = 1/2$, and

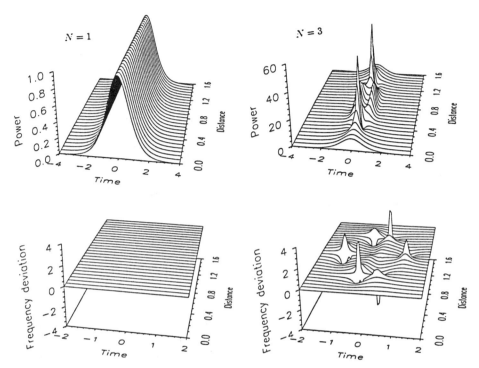

Figure 10.1 Evolution of the first-order (left column) and third-order (right column) solitons over one soliton period. Top and bottom rows show the pulse shape and chirp profile, respectively.

hence $\phi = \xi/2$. Equation (10.1.12) is easily integrated to obtain $V(\tau) = \mathrm{sech}(\tau)$. We have thus obtained the well-known "sech" solution [6]–[10],

$$u(\xi, \tau) = \mathrm{sech}(\tau) \exp(i\xi/2), \qquad (10.1.13)$$

for the fundamental soliton by direct integration of the NSE. It shows that the input pulse acquires a phase shift $\xi/2$ as it propagates inside the fiber, but its amplitude remains unchanged. It is this property of the fundamental soliton that makes it an ideal candidate for optical communications. In essence, the effects of fiber dispersion are exactly compensated by the fiber nonlinearity when (1) the input pulse has a "sech" shape and (2) its width and peak power are related by Eq. (10.1.6) in such a way that $N = 1$.

An important property of optical solitons is that they are remarkably stable against perturbations. Thus, even though the fundamental soliton requires a specific shape and a certain peak power corresponding to $N = 1$ in Eq. (10.1.6), it can be generated even when the pulse shape and the peak power deviate from the ideal conditions. Figure 10.2 shows the numerically simulated evolution of a Gaussian input pulse for which $N = 1$ but $u(0, \tau) = \exp(-\tau^2/2)$. As seen there, the pulse adjusts its shape and width in an attempt to become a

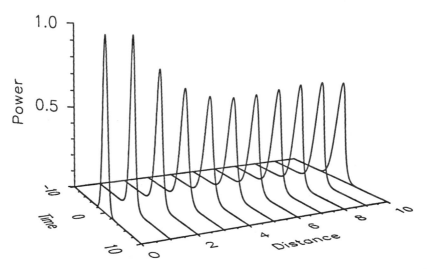

Figure 10.2 Evolution of a Gaussian pulse with $N = 1$ over the range $\xi = 0\text{–}10$. The pulse evolves toward the fundamental soliton by changing its shape, width, and peak power.

fundamental soliton and attains a "sech" profile for $\xi \gg 1$. A similar behavior is observed when N deviates from 1. It turns out that the Nth-order soliton can be formed when the input value of N is in the range $N - 1/2$ to $N + 1/2$ [13]. In particular, the fundamental soliton can be excited for values of N in the range 0.5–1.5. Figure 10.3 shows the evolution for $N = 1.2$ over the range $\xi = 0\text{–}10$ by solving the NSE numerically with the initial condition $u(0, \tau) = 1.2 \operatorname{sech}(\tau)$. The pulse width and the peak power oscillate initially but eventually become constant after the input pulse has adjusted itself to satisfy the condition $N = 1$ in Eq. (10.1.6).

In general, small deviations from the ideal conditions are not detrimental for soliton propagation since the input pulse is able to adjust its parameters to form a fundamental soliton. Some pulse energy is lost during the dynamic adaptation phase in the form of *dispersive waves*. It will be seen later that such dispersive waves affect the system performance and should be minimized in practice by matching the input conditions as close to the ideal requirements as possible. When solitons adapt to perturbations adiabatically, a perturbation theory developed specifically for solitons can be used to study how the soliton amplitude, width, frequency, speed, and phase evolve along the fiber. Such a perturbation theory is quite useful in the design of soliton communication systems as discussed in Section 10.3.

10.1.3 Dark Solitons

The NSE can be solved by the inverse scattering method even in the case of normal dispersion. The intensity profile of the resulting solutions exhibits a dip in a uniform background, and it is the dip that remains unchanged during

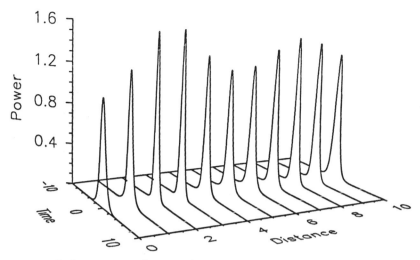

Figure 10.3 Pulse evolution for a "sech" pulse with $N = 1.2$ over the range $\xi = 0\text{--}10$. The pulse evolves toward the fundamental soliton ($N = 1$) by adjusting its width and peak power.

propagation inside the fiber. For this reason, such solutions of the NSE are called *dark solitons*. Even though the dark solitons were discovered in the 1970s [14], [15], it is only recently that they have been studied thoroughly [16]–[26]. This section describes the properties of dark solitons with emphasis on their differences from bright solitons.

The NSE describing dark solitons is obtained from Eq. (10.1.7) by changing the sign of the second term. This equation can be solved by postulating a solution of the form given by Eq. (10.1.10) and following the procedure outlined in Section 10.1.2. The general solution can be written as [23]

$$u_d(\xi, \tau) = (\eta \tanh\zeta - i\kappa) \exp(iu_0^2\xi), \qquad (10.1.14)$$

where

$$\zeta = \eta(\tau - \kappa\xi), \quad \eta = u_0 \cos\phi, \quad \kappa = u_0 \sin\phi, \qquad (10.1.15)$$

u_0 is the amplitude of the CW background, ϕ is an internal phase angle in the range $0 < \phi < \pi/2$), and η and κ are the amplitude and velocity of the dark soliton, respectively.

One important difference between the bright and dark solitons is that the velocity κ of a dark soliton depends on its amplitude η through the internal phase angle ϕ. For $\phi = 0$, Eq. (10.1.14) reduces to $u_d(\xi, \tau) = u_0 \tanh(u_0\tau) \exp(iu_0^2\xi)$, a form that shows that the soliton power drops to zero at the center of the dip. Such a soliton is referred as the *black* soliton. When $\phi \neq 0$, the intensity does not drop to zero at the dip center; such solitons are called *gray* solitons. It is common to introduce a blackness parameter B [19] related to the internal angle ϕ as $B = \cos\phi$. Another interesting feature of dark solitons is their phase profile. In contrast with bright solitons which have a constant phase, the phase

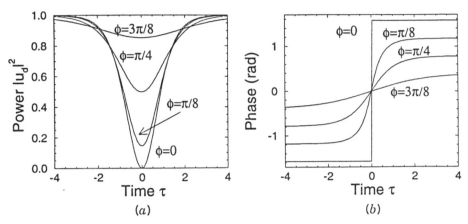

Figure 10.4 (a) Intensity and (b) phase profiles of dark solitons for several values of the internal phase ϕ.

of a dark soliton changes across its width. Figure 10.4 shows the intensity and phase profiles for several values of ϕ. For a black soliton ($\phi = 0$), a phase shift of π occurs exactly at the center of the dip. For other values of ϕ, the phase changes by an amount $\pi - 2\phi$ in a more gradual fashion.

Dark solitons have been observed experimentally by using relatively broad optical pulses with a narrow dip at the pulse center [16]–[18]. Numerical simulations show [19] that the central dip can propagate as a dark soliton despite the nonuniform background as long as the background intensity is uniform in the vicinity of the dip. Higher-order dark solitons do not follow a periodic evolution pattern similar to that shown in Fig. 10.1 for the third-order bright soliton. The numerical results obtained by solving the NSE with the initial condition $u(0, \tau) = \tanh(\tau)$ show that for $N > 1$, the input pulse forms a fundamental dark soliton by narrowing its width while ejecting several dark-soliton pairs in the process [20].

Several techniques can be used to generate dark solitons, including electric modulation in one arm of a Mach–Zehnder interferometer [22], nonlinear conversion of a beat signal in a dispersion-decreasing fiber [24], and conversion of a NRZ signal into a RZ signal and then into dark solitons by using a balanced Mach–Zehnder interferometer [25]. In a 1995 experiment [26], a 10-Gb/s signal was transmitted over 1200 km by using dark solitons. The transmission distance was limited by the asymmetry of the dark soliton originating from the time response of the electronic circuit used to generate them. Improvements are likely to occur with the development of sources capable of generating a dark-soliton bit stream with little amplitude and width fluctuations.

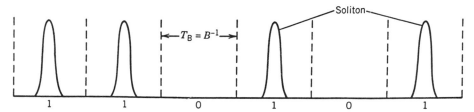

Figure 10.5 Soliton bit stream in RZ format. Each soliton occupies a small fraction of the bit slot so that neighboring soliton are spaced far apart.

10.2   SOLITON-BASED COMMUNICATIONS

Solitons are attractive for optical communications because they are able to main-tain their width even in the presence of fiber dispersion. However, their use requires substantial changes in system design compared with conventional non-soliton systems. In this section we discuss several such issues.

10.2.1   Information Transmission with Solitons

As discussed in Section 1.2.3, two distinct modulation formats can be used to generate a digital bit stream (see Fig. 1.6). The NRZ format is commonly used because its signal bandwidth is about 50% smaller than with the RZ format. However, the NRZ format cannot be used when solitons are used as information bits. The reason is easily understood by noting that the soliton width must be a small fraction of the bit slot to ensure that the neighboring solitons are well separated. Mathematically, the soliton solution (10.1.13) of the NSE is valid only when it occupies the entire time window extending to $|\tau| \to \infty$. It remains approximately valid for a train of solitons only when individual solitons are well isolated. This requirement can be used to relate the soliton width T_0 to the bit rate B as

$$B = \frac{1}{T_B} = \frac{1}{2q_0 T_0}, \qquad (10.2.1)$$

where T_B is the duration of the bit slot and $2q_0 = T_B/T_0$ is the separation between neighboring solitons in normalized units. Figure 10.5 shows a soliton bit stream in the RZ format.

The input pulse characteristics needed to excite the fundamental soliton can be obtained by setting $\xi = 0$ in Eq. (10.1.13). In physical units, the amplitude of the pulse is given by

$$A(0,t) = \sqrt{P_0}\,\text{sech}(t/T_0). \qquad (10.2.2)$$

The peak power P_0 is obtained from Eq. (10.1.6) by setting $N = 1$ and is related to the pulse width T_0 and the fiber parameters as

$$P_0 = |\beta_2|/\gamma T_0^2. \qquad (10.2.3)$$

ne width parameter T_0 used for normalization is related to the full width at
half maximum (FWHM) of the soliton as

$$T_s = 2T_0 \ln(1 + \sqrt{2}) \simeq 1.763T_0. \tag{10.2.4}$$

The pulse energy for the fundamental soliton is obtained by using

$$E_s = \int_{-\infty}^{\infty} |A(0,t)|^2 \, dt = 2P_0T_0. \tag{10.2.5}$$

Assuming that 1 and 0 bits are equally likely to occur, the average power of the
RZ signal becomes $\bar{P}_s = E_s(B/2) = P_0/2q_0$. As a simple example, $T_0 = 10$ ps
for a 10-Gb/s soliton system if we choose $q_0 = 5$. The FWHM of the soliton is
about 17.6 ps when $T_0 = 10$ ps. The peak power of the input pulse is 5 mW
by using $\beta_2 = -1$ ps^2/km and $\gamma = 2$ W^{-1}/km as typical values for dispersion-
shifted fibers. This value of the peak power corresponds to a pulse energy of
0.1 pJ and an average power level of only 0.5 mW.

10.2.2 Soliton Interaction

An important design parameter of soliton lightwave systems is the pulse width
T_s. As discussed earlier, each soliton pulse occupies only a fraction of the bit
slot. For practical reasons, one would like to pack solitons as tightly as possible. However, the presence of pulses in the neighboring bits perturbs the
soliton simply because the combined optical field is not a solution of the NSE.
This phenomenon, referred to as *soliton interaction*, has been studied extensively [27]–[31].

One can understand the implications of soliton interaction by solving the
NSE numerically with the input amplitude consisting of a soliton pair so that

$$u(0,\tau) = \text{sech}(\tau - q_0) + r \, \text{sech}[r(\tau + q_0)] \exp(i\theta), \tag{10.2.6}$$

where r is the relative amplitude of the two solitons, θ is the relative phase, and
$2q_0$ is the initial (normalized) separation. Figure 10.6 shows the evolution of a
soliton pair with a separation $q_0 = 3.5$ for several values of the parameters r
and θ. Clearly, soliton interaction depends strongly both on the relative phase
θ and the amplitude ratio r.

Consider first the case of equal-amplitude solitons ($r = 1$). The two solitons
attract each other in the in-phase case ($\theta = 0$) such that they collide periodically along the fiber length. However, for $\theta = \pi/4$, the solitons separate from
each other after an initial attraction stage. For $\theta = \pi/2$, the solitons repel each
other even more strongly, and their spacing increases with distance. From the
standpoint of system design, such behavior is not acceptable. It would lead
to jitter in the arrival time of solitons, since the relative phase of neighboring
solitons is not likely to remain well controlled. One way to avoid soliton interaction is to increase q_0, since the extent of interaction depends on soliton spacing.
For sufficiently large q_0, deviations in the soliton position are expected to be

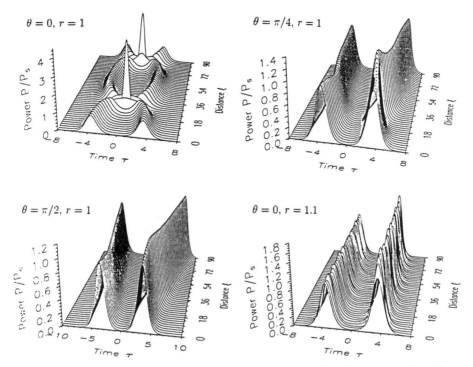

Figure 10.6 Evolution of a soliton pair over 90 dispersion lengths showing the effects of soliton interaction for four different choices of amplitude ratio r and relative phase θ. Initial spacing $q_0 = 3.5$ in all four cases.

small enough that the soliton remains at its initial position within the bit slot throughout the transmission distance.

The dependence of soliton separation on q_0 can be studied analytically by using the inverse scattering method [27], [28]. A perturbative approach can be used for $q_0 \gg 1$. In the specific case of $r = 1$ and $\theta = 0$, the soliton separation $2q_s$ at any distance ξ is given by [28]

$$2\exp[2(q_s - q_0)] = 1 + \cos[4\xi \exp(-q_0)]. \tag{10.2.7}$$

This relation shows that $q_s(\xi)$ varies periodically along the fiber with the oscillation period

$$\xi_p = (\pi/2)\exp(q_0). \tag{10.2.8}$$

This result is valid for $q_0 \gg 1$. A more accurate expression, valid for arbitrary values of q_0, is given by [30]

$$\xi_p = \frac{\pi \sinh(2q_0) \cosh(q_0)}{2q_0 + \sinh(2q_0)}. \tag{10.2.9}$$

Equation (10.2.8) is quite accurate for $q_0 > 3$. Its predictions are in agreement with Fig. 10.6 drawn for the case $q_0 = 3.5$. It can be used for system design as follows. If $\xi_p L_D$ is much greater than the total transmission distance L_T, soliton interaction can be neglected since soliton spacing would deviate little from its initial value. For $q_0 = 6$, $\xi_p \approx 634$. The dispersion length typically exceeds 100 km. Hence $L_T \ll \xi_p L_D$ can be realized even for $L_T = 10,000$ km. By using $L_D = T_0^2/|\beta_2|$ and $T_0 = (2Bq_0)^{-1}$ from Eq. (10.2.1), the condition $L_T \ll \xi_p L_D$ can be written in the form of the following design criterion:

$$B^2 L_T \ll \frac{\pi \exp(q_0)}{8q_0^2 |\beta_2|}. \qquad (10.2.10)$$

For the purpose of illustration, let us choose $\beta_2 = -1$ ps^2/km, since most soliton communication systems use dispersion-shifted fibers. Equation (10.2.10) then implies that $B^2 L_T \ll 4.4$ (Tb/s)^2-km if we use $q_0 = 6$ to avoid soliton interaction. The pulse width at a given bit rate B can be determined from Eq. (10.2.1). For example, $T_s = 14.7$ ps at $B = 10$ Gb/s when $q_0 = 6$.

A relatively large soliton spacing, necessary to avoid soliton interaction, limits the bit rate of soliton communication systems. The spacing can be reduced by up to a factor of 2 by using unequal amplitudes for the neighboring solitons. As seen in Fig. 10.6, the separation for two in-phase solitons does not change by more than 10% for an initial soliton spacing as small as $q_0 = 3.5$ if their initial amplitudes differ by 10% ($r = 1.1$). Note that the peak power deviates by only 1% from its ideal value corresponding to $N = 1$. Since such small changes in the peak power are not detrimental for the soliton nature of pulse propagation, this scheme is feasible in practice and can be useful for increasing the system capacity. The design of such systems would, however, require attention to many details. Soliton interaction can also be modified by other factors, such as the initial frequency chirp imposed on the input pulse.

10.2.3   Frequency Chirp

To propagate as a fundamental soliton inside the optical fiber, the input pulse should not only have a "sech" profile but also be chirp-free. Many sources of short optical pulses have a frequency chirp imposed on them. The initial chirp can be detrimental to soliton propagation simply because it disturbs the exact balance between the GVD and SPM. This subsection considers how the frequency chirp affects the soliton nature of the optical pulse [32]–[35].

The effect of an initial frequency chirp can be studied by solving Eq. (10.1.7) numerically with an input amplitude

$$u(0, \tau) = \mathrm{sech}(\tau) \exp(-iC\tau^2/2), \qquad (10.2.11)$$

where C is the chirp parameter introduced in Section 2.4.2. The quadratic form of phase variation corresponds to a linear frequency chirp such that the optical frequency increases with time (up-chirp) for positive values of C. Figure 10.7 shows pulse evolution for the case $N = 1$ and $C = 0.5$. The pulse shape changes considerably even for $C = 0.5$. The pulse is initially compressed mainly because of the positive chirp; initial compression occurs even in the absence of nonlinear

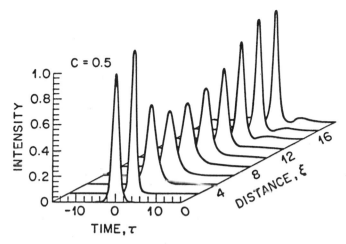

Figure 10.7 Evolution of a chirped optical pulse for the case $N = 1$ and $C = 0.5$. For $C = 0$ the pulse shape does not change, since the pulse propagates as a fundamental soliton.

effects (see Section 2.4.2). The pulse then broadens but is eventually compressed a second time with the tails gradually separating from the main peak. The main peak evolves into a soliton over a propagation distance $\xi > 15$. A similar behavior occurs for negative values of C, although the initial compression does not occur in that case. The formation of a soliton is expected for small values of $|C|$ since solitons are generally stable under weak perturbations. But the input pulse does not evolve toward a soliton when $|C|$ exceeds a critical valve C_{crit}. For the case $N = 1$, the soliton seen in Fig. 10.7 does not form if C is increased from 0.5 to 2.

The critical value C_{crit} of the chirp parameter can be obtained by using the inverse scattering method [32]–[34]. It depends on N and is found to be $C_{\text{crit}} = 1.64$ for $N = 1$. It also depends on the form of the phase factor [34] in Eq. (10.2.11). From the standpoint of system design, the initial chirp should be minimized as much as possible. This is necessary because even if the chirp is not detrimental for $|C| < C_{\text{crit}}$, a part of the pulse energy is shed as a dispersive waves during the press of soliton formation [32]. For instance, only 83% of the input energy is converted into a soliton for the case $C = 0.5$ shown in Fig. 10.7, and this fraction reduces to 62% for $C = 0.8$.

10.2.4 Soliton Transmitters

Soliton communication systems require an optical source capable of producing chirp-free picosecond pulses at a high repetition rate with a shape as close to the "sech" shape as possible. The source should operate in the wavelength region near 1.55 μm, where the fiber loss is minimum and where EDFAs can be used to compensate for the residual fiber loss. A semiconductor laser, commonly used

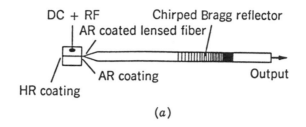

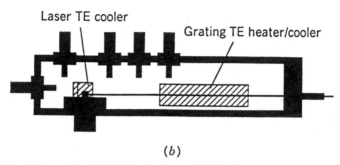

Figure 10.8 Schematic of (a) the device and (b) the package for a hybrid soliton pulse source. (After Ref. [39]. ©1995 IEEE. Reprinted with permission.)

for nonsoliton lightwave systems, remains the laser of choice even for soliton systems.

Early experiments on soliton transmission used the technique of gain switching [36]–[38] for generating picosecond optical pulses of duration 20–40 ps by biasing the laser below threshold and pumping it high above threshold periodically by applying current pulses. The repetition rate is determined by the frequency of current modulation and is typically $\sim$ 1–10 GHz. A problem with the gain-switching technique is that the emitted pulses are chirped because of the carrier-induced index changes governed by the linewidth enhancement factor β_c (see Section 3.3.7). However, the pulse can be made nearly chirp-free by passing it through an optical fiber with normal GVD ($\beta_2 > 0$), which compresses the pulse at the same time. The compression mechanism can be understood by using the analysis of Section 2.4.2 and noting that gain switching produces pulses with a frequency chirp such that the chirp parameter C is negative. In a demonstration of this technique [37], 14-ps optical pulses were obtained at the 3-GHz repetition rate by passing the gain-switched pulse through a polarization-preserving, dispersion-shifted, 3.7-km-long fiber with $\beta_2 = 23$ ps^2/km at the 1.55-μm operating wavelength. An EDFA was used to amplify the optical pulse to the power level required for launching fundamental solitons. In another experiment [38], gain-switched pulses were simultaneously amplified and compressed in an EDFA after having first been passed through a narrowband optical filter. It was possible to generate nearly transform-limited, 17-ps-wide optical pulses at repetition rates of 6–24 GHz.

Mode-locked semiconductor lasers are also suitable for soliton communications and are often preferred since the mode-locked pulse train emitted from such lasers is nearly chirp-free. Active mode locking is generally used by modulating the laser current at a frequency equal to the frequency difference between the two neighboring longitudinal modes (see Section 3.3.2). However, a solitary semiconductor laser has a relatively short cavity length (typically 0.5 mm or less), resulting in a modulation frequency of more than 50 GHz. An external-cavity configuration is often used to increase the cavity length and reduce the modulation frequency. In a practical approach, the pigtail that is invariably attached to an optical transmitter is used to form the external cavity by etching a chirped fiber grating. Figure 10.8 shows the design of such a hybrid soliton-pulse source. The use of a chirped fiber grating (see Section 9.6.2) provides a wavelength stability to within 0.1 nm while offering a self-tuning mechanism that allows to mode-lock the laser over a wide range of modulation frequencies [39]. A thermoelectric heater can be used to tune the operating wavelength over a range of 7 nm by changing the grating pitch. Such a source produces soliton-like pulses of widths $\sim$ 20 ps at a repetition rate as large 10 GHz and has been used in many transmission experiments.

The main drawback of external-cavity semiconductor lasers stems from their hybrid nature. A monolithic source of picosecond pulses is preferred in practice. Several approaches have been used to produce such a source. Monolithic semiconductor lasers with a cavity length of about 4 mm can be actively mode-locked to produce a 10-GHz pulse train. Passive mode locking of a monolithic DBR laser has produced 3.5-ps pulses at a repetition rate of 40 GHz [40]. An electroabsorption modulator, integrated with a semiconductor laser, offers another alternative. Such transmitters are commonly used for nonsoliton lightwave systems (see Section 3.4). They can also be used to produce a pulse train by using the nonlinear nature of the absorption response of the modulator. Chirp-free pulses of 10- to 20-ps duration at a repetition rate as high as 20 GHz have been produced by this technique [41]. By 1996, the repetition rate of such modulator-integrated lasers has been increased to 50 GHz [42]. The *quantum-confinement Stark effect* in a multiquantum-well modulator can also be used to produce a pulse train suitable for soliton transmission [43].

Mode-locked fiber lasers provide an alternative to semiconductor sources although such lasers still need a semiconductor laser for pumping. An EDFA is placed within a Fabry–Perot (FP) or ring cavity to make a fiber laser [6]. Both active and passive mode locking techniques have been used to produce short optical pulses. Active mode locking requires modulation at a high-order harmonic of the longitudinal-mode spacing because of relatively long cavity lengths ($>$ 1 m) that are typically used for fiber lasers. Such harmonically mode-locked fiber lasers use an intracavity $LiNbO_3$ modulator and have been used in soliton transmission experiments [44]. A semiconductor laser amplifier can also be used for active mode locking, producing pulses shorter than 10 ps at a repetition rate as high as 20 GHz [45]. Passively mode-locked fiber lasers use either a multiquantum-well device that acts as a fast saturable absorber or fiber nonlinearity to produce phase shifts that lead to an effective saturable absorber [6], [46]. The repetition rate of such fiber lasers is relatively low and difficult to control, making them often unsuitable for soliton systems.

In a different approach, nonlinear pulse shaping in a dispersion-tailored fiber (see Section 10.4.4) is used to produce a train of ultrashort pulses. The basic idea consists of injecting a CW beam, with weak sinusoidal modulation imposed on it, into such a fiber. The combined effect of GVD, SPM, and decreasing dispersion is to convert the sinusoidally modulated signal into a train of ultrashort solitons [47]. The repetition rate of pulses is governed by the frequency of the initial sinusoidal modulation, often generated by beating two optical signals. Two DFB semiconductor lasers or a two-mode fiber laser can be used for this purpose. By 1993, this technique led to the development of an integrated fiber source capable of producing a soliton pulse train at a high repetition rate by using a *comb-like* dispersion profile produced by splicing pieces of low- and high-dispersion fibers [47]. A *dual-frequency* fiber laser was used to generate the beat signal and to produce a 2.2-ps soliton train at the 59-GHz repetition rate. In another experiment [48], a 40-GHz soliton train of 3-ps pulses was generated by using a single DFB laser whose output was modulated by using a Mach–Zehnder modulator before launching it into a dispersion-tailored fiber with a comblike GVD profile.

A simple method of pulse-train generation modulates the phase of the output obtained from a DFB semiconductor laser, followed by an optical bandpass filter [49]. Phase modulation generates FM sidebands on both sides of the carrier frequency, and the optical filter selects the sidebands on one side of the carrier. Such a device generates a stable pulse train of widths ~ 20 ps at a repetition rate that is controlled by the phase modulator. It can also be used as a dual-wavelength source by filtering sidebands on both sides of the carrier frequency, with a typical channel spacing of about 0.8 nm at the 1.55-μm wavelength. Another simple technique uses a single Mach–Zehnder modulator, driven by an electrical data stream in the NRZ format, to convert the CW output of a DFB laser into an optical bit stream in the RZ format [50]. Although optical pulses launched from such transmitters typically do not have the "sech" shape of a soliton, they can be used for soliton systems because of the soliton-formation capability of the fiber discussed in Section 10.1.2.

10.2.5   Loss-Induced Soliton Broadening

As discussed in Section 10.1.2, solitons use fiber nonlinearity to maintain their width even in the presence of fiber dispersion. However, this property holds only if the fiber loss is negligible. It is not difficult to see that a decrease in the soliton energy because of fiber loss would lead to soliton broadening simply because the reduced peak power weakens the nonlinear effect necessary to counteract the GVD. This section considers the loss-induced broadening of a soliton.

The fiber loss in included by the last term in Eq. (10.1.1). In normalized

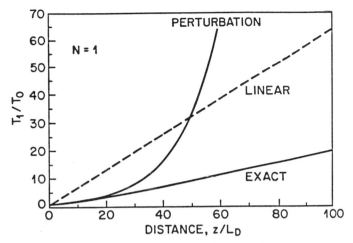

Figure 10.9 Soliton broadening in a lossy fiber ($\Gamma = 0.07$) for a fundamental soliton. Perturbation theory agrees with numerical results (curve marked "exact") for $\Gamma\xi < 1$. Dashed curve shows the behavior expected in the absence of nonlinear effects. (After Ref. [52]. ©1985 Elsevier. Reprinted with permission.)

units, the NSE becomes [see Eq. (10.1.7)]

$$i\frac{\partial u}{\partial \xi} + \frac{1}{2}\frac{\partial^2 u}{\partial \tau^2} + |u|^2 u = -\frac{i}{2}\Gamma u, \tag{10.2.12}$$

where

$$\Gamma = \alpha L_D = \alpha T_0^2/|\beta_2| \tag{10.2.13}$$

represents the fiber loss over one dispersion length. For $\Gamma \ll 1$, the last term can be treated as a small perturbation, resulting in the approximate solution [51]

$$u(\xi, \tau) \approx \text{sech}[\exp(-\Gamma\xi)\tau]\exp\{(i/4\Gamma)[1 - \exp(-2\Gamma\xi)]\}. \tag{10.2.14}$$

In terms of the input width T_0, the soliton width increases exponentially as

$$T_1(\xi) = T_0 \exp(\Gamma\xi) = T_0 \exp(\alpha z). \tag{10.2.15}$$

Such an exponential increase of the soliton width cannot be expected to continue for arbitrarily long distances. Numerical solutions of Eq. (10.2.12) indeed show a slower increase for $\xi \gg 1$ [52]. Figure 10.9 shows the broadening factor T_1/T_0 as a function of ξ when a fundamental soliton is launched into a fiber with $\Gamma = 0.07$. The perturbative result is also shown for comparison; it is reasonably accurate up to $\Gamma\xi = 1$. The dashed line in Fig. 10.9 shows the expected pulse broadening in the absence of nonlinear effects. The important point to note is that soliton broadening is much less compared with the linear case. Thus, the nonlinear effects can be beneficial for optical communication systems even when solitons cannot be maintained perfectly because of fiber loss. In one study [53] an increase in the repeater spacing by more than a factor of

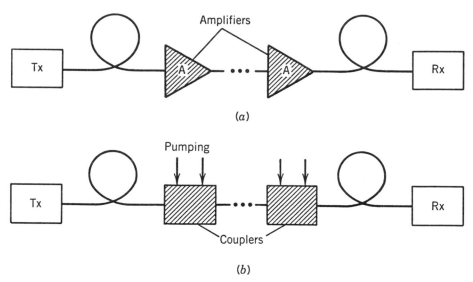

Figure 10.10 (a) Lumped and (b) distributed amplification schemes for compensation of fiber loss in soliton communication systems.

2 was predicted by using higher-order solitons. The required peak power was relatively modest (about 3 mW) at a bit rate of 8 Gb/s.

In long-haul application, solitons are transmitted over long fiber lengths (up to 10,000 km or more) without using electronic repeaters. To overcome the effect of fiber loss, solitons need to be amplified periodically to recover their original width, peak power, and energy [54], [55]. In the following section we consider two amplification schemes.

10.2.6   Soliton Amplification

The simplest scheme for *soliton amplification*, shown in Fig. 10.10(a), is the same as that used for nonsoliton communication systems. An optical amplifier is placed periodically along the fiber link, and its gain is adjusted such that the fiber loss between two amplifiers is exactly compensated by the amplifier gain. An important design parameter is the spacing L_A between amplifiers—it should be as large as possible to minimize the overall cost. For nonsoliton lightwave systems, L_A is typically 80–100 km. For soliton communication systems, L_A is restricted to much smaller values [54]. The reason is that optical amplifiers boost the soliton energy to the input level locally over a relatively short distance without allowing for gradual recovery of the fundamental soliton. The amplified soliton adjusts its width dynamically in the fiber section following the optical amplifier. However, it also sheds a part of its energy as dispersive waves during this adjustment phase. The dispersive part can accumulate to significant levels over a large number of amplification stages and must be avoided. One way to reduce the dispersive part is to reduce the amplifier spacing L_A such that the soliton is not perturbed much over this length. Numerical simulations [54] show that this is the case when L_A is a fraction of the dispersion length ($L_A \ll L_D$). The dispersion length L_D depends on both the pulse width T_0 and the GVD parameter β_2, and can vary from 10 to 1000 km, depending on their values. Typically, L_D is less than 50 km for 10-Gb/s systems operating at 1.55 μm by using standard fibers. The use of lumped amplifiers then requires $L_A < 10$ km, a rather small value from the standpoint of system design. However, L_D can exceed 200 km when $|\beta_2| < 1$ ps^2/km at the operating wavelength. Amplifier spacings of 30–50 km then become feasible.

An alternative solution is offered by the technique of *distributed amplification* in which solitons are amplified all along the same fiber link that is used for data transmission [55]. Stimulated Raman scattering (SRS) was used as early as 1985 [56] for this purpose. Distributed amplification can also be achieved by doping the transmission fiber lightly with a rare-earth element such as erbium (see Section 8.5.8). Figure 10.10(b) shows the distributed-amplification scheme in which the pump lasers periodically inject CW light in both directions by using WDM fiber couplers. The pump wavelength is chosen such that it provides the gain at the signal wavelength. Since the gain is distributed over the entire fiber length, solitons can be amplified adiabatically in such a way that N is maintained near its input value of $N = 1$ despite the fiber loss. Indeed, if the gain were exactly equal to the fiber loss at each point inside the fiber, N would remain 1 and the soliton would maintain itself over arbitrarily long distances. This condition is not satisfied in practice because the pump power does not remain constant along the fiber (the pump suffers loss along with the soliton). The pump-station spacing L_A depends on the fiber loss at the pump wavelength and on the extent that the soliton energy is allowed to deviate from its input value. Typically, $L_A = 40$–50 km if at most 20% deviations in the soliton energy are tolerable [57]. However, L_A can exceed L_D by a large margin, in contrast with the *lumped-amplification scheme* for which $L_A \ll L_D$.

Early experiments on soliton amplification concentrated on the *Raman-amplification scheme*. An experiment in 1985 demonstrated that fiber loss over 10 km can be compensated by the Raman gain while maintaining the soliton width [56]. Two color-center lasers were used in this experiment. One laser produced 10-ps pulses at 1.56 μm, which were propagated as fundamental solitons in a 10-km-long fiber. The other laser was operated continuously at 1.46 μm and acted as a pump for amplifying 1.56-μm solitons. In the absence of the Raman gain, the soliton broadened by about 50% because of loss-induced broadening. This amount of broadening was in agreement with Eq. (10.2.14), which predicts $T_1/T_0 = 1.51$ for $z = 10$ km and $\alpha = 0.18$ dB/km, the values used in the experiment. When the pump power was about 125 mW, the 1.8-dB Raman gain compensated the fiber loss and the output pulse was nearly identical with the input pulse.

A latter experiment [4] demonstrated soliton transmission over 4000 km by using the Raman-amplification scheme. This experiment used a 42-km fiber loop whose loss was exactly compensated by injecting the CW pump light from a 1.46-μm color-center laser. The solitons were allowed to circulate many times along the fiber loop, and their width was monitored after each round trip. The 55-ps solitons could be circulated along the loop up to 96 times without a significant increase in their pulse width, indicating soliton recovery over 4000 km. The distance could be increased to 6000 km with further optimization. This experiment demonstrated in 1988 that soliton communication over transoceanic distances was feasible in principle. The main drawback was that Raman amplification required pump lasers emitting more than 500 mW of CW power near 1.46 μm. It is difficult to obtain such high powers from semiconductor lasers, and the color-center lasers used in the experimental demonstration were too bulky to be useful for practical lightwave systems.

The situation changed considerably in 1989 with the availability of EDFAs. Since then, many experiments have used EDFAs to amplify solitons. The results show that under certain conditions, solitons can be maintained over long distances despite the lumped nature of the amplification process. The next section is devoted to the design issues associated with such lightwave systems. Distributed amplification is considered in Section 10.4.

10.3 SOLITON SYSTEM DESIGN

Consider a long-haul soliton link in which the fiber loss is compensated periodically by using the lumped-amplification scheme. The way a soliton reacts to the energy loss strongly depends on the loss per dispersion length, αL_D, and the amplifier spacing L_A. If $\alpha L_D \ll 1$ and $L_A \gg L_D$, the soliton can adjust to the energy loss adiabatically. This regime is referred to as the quasi-adiabatic regime and is discussed in Section 10.4.1. On the other hand, if the amplifier spacing L_A is much smaller than the dispersion length L_D, the soliton shape is not distorted significantly despite the energy loss. In such a system, solitons can be amplified hundreds of time while preserving their shape. Since soliton evolution is then governed by the average soliton energy over one amplifier spacing, this mode of operation is referred to as the *average-soliton regime*. In this section we discuss the issues that need attention when the average-soliton regime is used for the design of soliton communication systems.

10.3.1 Average-Soliton Regime

The periodic amplification of solitons can be accounted for by adding a gain term to Eq. (10.2.12) and writing it as [58]

$$i\frac{\partial u}{\partial \xi} + \frac{1}{2}\frac{\partial^2 u}{\partial \tau^2} + |u|^2 u = -\frac{i}{2}\Gamma u + i(\sqrt{G} - 1)\sum_{m=1}^{N_A} \delta(\xi - m\xi_A)u, \qquad (10.3.1)$$

where N_A is the total number of amplifiers, $\xi_A = L_A/L_D$, $\Gamma = \alpha L_D$, and $G = \exp(\Gamma \xi_A)$ is the amplifier gain needed to compensate for the fiber loss. The delta function accounts for the lumped nature of amplification at locations $\xi = m\xi_A$. The factor $\sqrt{G} - 1$ represents the change in the soliton amplitude during amplification.

Because of the rapid variations in the soliton energy introduced by the lumped-amplification scheme, it is useful to make the transformation

$$u(\xi, \tau) = a(\xi)\, v(\xi, \tau), \qquad (10.3.2)$$

where $a(\xi)$ contains rapid variations and $v(\xi, \tau)$ is a slowly varying function of ξ. By substituting Eq. (10.3.2) in Eq. (10.3.1), $v(\xi, \tau)$ is found to satisfy

$$i\frac{\partial v}{\partial \xi} + \frac{1}{2}\frac{\partial^2 v}{\partial \tau^2} + a^2(\xi)|u|^2 u = 0, \qquad (10.3.3)$$

where $a(\xi)$ is obtained by solving

$$\frac{\partial a}{\partial \xi} = -\frac{1}{2}\Gamma a + (\sqrt{G} - 1)\sum_{m=1}^{N_A} \delta(\xi - m\xi_A)a. \qquad (10.3.4)$$

By noting that the last term in Eq. (10.3.4) is periodic and contributes only at $\xi = m\xi_A$, $a(\xi)$ is found to be a periodic function of ξ. In each period, $a(\xi)$ decreases exponentially with a jump to its initial value at the end of the period.

The concept of average soliton makes use of the fact that $a^2(\xi)$ in Eq. (10.3.3) varies rapidly with a period $\xi_A \ll 1$. Since solitons evolve little over a short distance ξ_A, one can replace $a^2(\xi)$ by its average value over one period. This approximation can be justified by assuming a solution of Eq. (10.3.3) in the form $v = \bar{v} + \delta v$, where $\bar{v}$ is the *average soliton* satisfying the standard NSE

$$i\frac{\partial \bar{v}}{\partial \xi} + \frac{1}{2}\frac{\partial^2 \bar{v}}{\partial \tau^2} + \langle a^2(\xi)\rangle |\bar{v}|^2 \bar{v} = 0, \qquad (10.3.5)$$

and δv is a perturbation. The practical importance of the average-soliton concept stems from the fact that the perturbation δv turns out to be relatively small when $\xi_A \ll 1$ since the leading-order correction varies as ξ_A^2 rather than ξ_A [58]. As a result, the average-soliton description is quite accurate even for $\xi_A = 0.2$. The input peak power P_{in} of the average soliton is chosen such that $\langle a^2(\xi)\rangle = 1$ in Eq. (10.3.5). By using $G = \exp(\Gamma\xi_A)$, it is given by

$$P_{\text{in}} = \frac{G\ln G}{G - 1}\, P_0, \qquad (10.3.6)$$

where P_0 is the peak power in lossless fibers. As an example, $G = 10$ and $P_{\text{in}} \approx 2.56\, P_0$ for 50-km amplifier spacing and a fiber loss of 0.2 dB/km.

Figure 10.11 shows the soliton evolution in the average-soliton regime over a distance of 10,000 km for the case in which solitons are amplified every 50 km. When soliton width corresponds to a dispersion length of 200 km, soliton is preserved quite well even after 200 amplification stages since the condition $\xi_A \ll 1$ is reasonably well satisfied. However, if the dispersion length reduces to 25 km, the soliton is destroyed because it no longer propagates in the average-soliton regime.

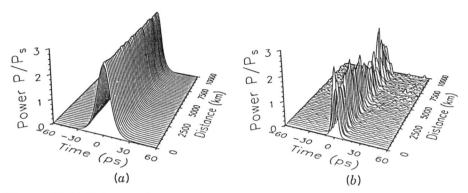

Figure 10.11 Soliton evolution in the average-soliton regime over a distance of 10,000 km with $L_A = 50$ km, $\alpha = 0.22$ dB/km, and $\beta_2 = 0.5$ ps^2/km for (a) $L_D = 200$ km and (b) $L_D = 25$ km. The soliton is destroyed when amplifier spacing exceeds the dispersion length.

The condition $\xi_A \ll 1$ or $L_A \ll L_D$, required to operate within the average-soliton regime, can be related to the width T_0 by using $L_D = T_0^2/|\beta_2|$. The resulting condition is

$$T_0 \gg \sqrt{|\beta_2| L_A}. \qquad (10.3.7)$$

Since the bit rate B is related to T_0 through Eq. (10.2.1), the condition (10.3.7) can be written in the form of the following design criterion:

$$B^2 L_A \ll (4q_0^2|\beta_2|)^{-1}. \qquad (10.3.8)$$

By choosing typical values $\beta_2 = -0.5$ ps^2/km, $L_A = 50$ km, and $q_0 = 5$, we obtain $T_0 \gg 5$ ps and $B \ll 20$ GHz. Clearly, use of the average-soliton regime imposes a severe limitation on both the bit rate and amplifier spacing of soliton communication systems.

10.3.2   Amplifier Noise

The use of in-line optical amplifiers affects the soliton evolution considerably. The reason is that amplifiers, needed to restore the soliton energy, also add noise originating from *amplified spontaneous emission* (ASE). As discussed in Section 8.1.3, the spectral density of ASE depends on the amplifier gain G itself and is given by Eq. (8.1.15).

To understand how ASE effects soliton evolution, we should consider the most general form of the fundamental soliton [5],

$$u_s(\xi, \tau) = \eta \, \mathrm{sech}[\eta(\tau + \Omega\xi - q)] \exp[-i\Omega\tau + i(\eta^2 - \Omega^2)\xi/2 + i\phi], \qquad (10.3.9)$$

where the parameters η, q, Ω, and ϕ represent the amplitude, position, frequency and phase of the input pulse at $\xi = 0$, respectively. This form of soliton shows clearly how the soliton amplitude and width are inversely related. It also shows that a frequency shift Ω changes the soliton speed directly in proportion to the amount of shift simply because the group velocity depends on the carrier frequency of the soliton. Equation (10.3.9) reduces to the soliton solution (10.1.13) when $\eta = 1$, $q = 0$, $\Omega = 0$, and $\phi = 0$. However, these values appropriate for an ideal soliton can change in the presence of amplifier noise and other perturbations.

The effect of ASE is to change randomly the values of the four soliton parameters η, q, Ω, and ϕ in Eq. (10.3.9) at the output of each amplifier. The variances of such fluctuations for the four soliton parameters can be calculated by treating ASE as a perturbation. According to the *adiabatic perturbation theory* developed for solitons [59], the evolution of the four soliton parameters because of a perturbation $\epsilon(u_s)$ in governed by [60]

$$\frac{d\eta}{d\xi} = \text{Im} \int_{-\infty}^{\infty} \epsilon(u_s) u_s^* \, d\tau \tag{10.3.10}$$

$$\frac{d\Omega}{d\xi} = \text{Re} \int_{-\infty}^{\infty} \epsilon(u_s) \tanh[\eta(\tau - q)] u_s^* \, d\tau \tag{10.3.11}$$

$$\frac{dq}{d\xi} = -\Omega + \frac{1}{\eta} \text{Im} \int_{-\infty}^{\infty} \epsilon(u_s)(\tau - q) u_s^* \, d\tau \tag{10.3.12}$$

$$\frac{d\phi}{d\xi} = q \frac{d\Omega}{d\xi} + \frac{\eta^2 - \Omega^2}{2}$$
$$\quad - \frac{1}{\eta} \text{Re} \int_{-\infty}^{\infty} \epsilon(u_s)\{1 - \eta(\tau - q) \tanh[\eta(\tau - q)]\} u_s^* \, d\tau, \tag{10.3.13}$$

where Re and Im stand for the real and imaginary parts, respectively. These evolution equations play an important role in the design of soliton systems since they can be used for an arbitrary form of perturbation.

In the case of ASE, the perturbation term is given by

$$\epsilon(u_s) = i\, n(\xi, \tau) \exp(-i\Omega\tau + i\phi), \tag{10.3.14}$$

where $n(\xi, \tau)$ is the random noise, assumed to be complex to include both the amplitude and phase fluctuations. It vanishes on average and its variance is related to the spectral noise density S_{ASE} given by Eq. (8.1.15). By using Eqs. (10.3.10)–(10.3.13), the variances of fluctuations are given by [61], [62]

$$\sigma_\eta^2 = 2n_{sp}F_G/N_s, \tag{10.3.15}$$

$$\sigma_\Omega^2 = 2n_{sp}F_G/(3N_s), \tag{10.3.16}$$

$$\sigma_q^2 = \pi^2 n_{sp}F_G/(6N_s), \tag{10.3.17}$$

$$\sigma_\phi^2 = (\pi^2/6 + 2)n_{sp}F_G/(3N_s), \tag{10.3.18}$$

where n_{sp} is the *spontaneous-emission factor* as defined in Eq. (8.1.16), $F_G = (G - 1)^2/(G \ln G)$, and N_s is the number of photons contained in a fundamental soliton of energy E_s given by Eq. (10.2.5), (i.e., $N_s = 2P_0T_0/\hbar\omega_0$). The functional form of F_G results from the enhanced peak power needed in the average-soliton regime [see Eq. (10.3.6)].

Amplitude fluctuations, as one might expect, lead to a degradation of the signal-to-noise ratio (SNR) of the soliton bit stream. Although not immediately obvious, frequency fluctuations induce timing jitter that affects the performance of soliton communication systems and limits the total transmission distance. The timing-jitter issue is covered in the next subsection.

10.3.3 Timing Jitter

If optical amplifiers compensate for the fiber loss, one may ask what limits the total transmission distance of a soliton link. The answer is provided by the timing jitter. A soliton communication system can operate reliably only if all solitons arrive at the receiver within their assigned bit slot. Several physical mechanisms induce deviations in the soliton position from its original location at the bit center. In this section we consider several such mechanisms.

Gordon–Haus Jitter

As seen in the preceding section, optical amplifiers add noise which affects both the amplitude and frequency of the amplified soliton. Amplitude fluctuations degrade the SNR of the soliton bit stream. The SNR degradation, although undesirable, is not a major limiting factor. In fact, frequency fluctuations affect system performance much more drastically by inducing the timing jitter. The origin of timing jitter can be understood by noting from Eq. (10.3.9) that a change in the soliton frequency by Ω affects the group velocity or the speed at which the soliton propagates through the fiber. Since Ω fluctuates because of amplifier noise, soliton transit time through the fiber link also becomes random. Fluctuations in the arrival time of a soliton at the receiver are referred to as the *Gordon–Haus timing jitter* [63], [64].

It is relatively easy to calculate the variance of the timing jitter by using the soliton perturbation theory. In the absence of other perturbations within the fiber, one can set $\epsilon(u_s) = 0$ in Eqs. (10.3.10)–(10.3.13). Equation (10.3.12) governing the soliton position is easily integrated to yield $q(\xi) = -\Omega\xi$. At the end of one amplifier spacing, jitter becomes $q = -\Omega\xi_A$. Since the random frequency shifts accumulate from amplifier to amplifier, the total timing jitter for a series of N_A amplifiers is obtained by adding all contributions and becomes

$$q = -\xi_A \sum_{p=1}^{N_A} \sum_{i=1}^{p} \Omega_i, \qquad (10.3.19)$$

where Ω_i is the frequency shift induced by the ith amplifier. In the limit of large N_A, the sum over N_A can be replaced by an integral, and the variance of timing jitter is found to be

$$\sigma_{\text{GH}}^2 = \langle q^2 \rangle - \langle q \rangle^2 = N_A^3 \xi_A^2 \sigma_\Omega^2 / 3, \qquad (10.3.20)$$

where σ_Ω^2 is given by Eq. (10.3.16).

Since a soliton should arrive within its allocated bit slot for its correct identification at the receiver, the arrival-time jitter should be a small fraction of the bit
slot. This requirement can be written as $\sigma_{GH}T_0 < f_j/B$, where f_j is the fraction
of the bit slot by which soliton can move without affecting system performance
adversely. By using $L_T = N_A \xi_A L_D$, $B = (2q_0 T_0)^{-1}$, $N_s = 2P_0 T_0/(\hbar\omega_0)$, and
$P_0 = (\gamma L_D)^{-1}$ in Eq. (10.3.20), the bit rate–distance product BL_T is found to
be limited by (in physical units)

$$
BL_T < \left(\frac{9\pi f_j^2 L_A}{n_{sp} F_G q_0 \lambda h \gamma D} \right)^{1/3},
\tag{10.3.21}
$$

where $D = -2\pi c \beta_2/\lambda^2$ is the dispersion parameter introduced in Section 2.3.
The tolerable value of f_j depends on the acceptable BER and on details of the
receiver design; typically, $f_j \sim 0.1$. To see how amplifier noise limits the total
transmission distance, consider a specific soliton communication system operating at 1.55 μm. By using typical parameter values, $q_0 = 6$, $\alpha = 0.2$ dB/km,
$\gamma = 3$ W^{-1}/km, $D = 1$ ps/(km-nm), $n_{sp} = 2$, $L_A = 50$ km, $F_G = 3.518$, and
$f_j = 0.1$, BL_T must be below 48 (Tb/s)-km. For a 10-Gb/s system, the transmission distance is limited to 4800 km. It will be seen later that the jitter can
be reduced substantially using optical filters (see Section 10.3.4).

Acoustic Jitter

A timing-jitter mechanism that limits the total transmission distance has its
origin in the simple phenomenon of acoustic-wave generation [65]. Confinement
of the optical field within the fiber core creates a field gradient in the radial
direction of the fiber. This gradient of electric field leads to the generation
of acoustic waves through *electrostriction*, a phenomenon that creates density
variations in response to variations in the electric field. Since the refractive index
of fused silica is related to the material density, one can associate a change
in refractive index (and hence in the soliton velocity) with the generation of
acoustic wave. Such index changes last for about 2 ns, roughly the time required
by the acoustic wave to traverse the fiber core. Since solitons follow one another
on a much shorter time scale (0.1 ns for $B = 10$ Gb/s), the acoustic wave
generated by a single soliton affects tens or even hundreds of the following
solitons. Such an acoustic-wave-assisted interaction among solitons has been
observed experimentally [66].

If a bit stream were composed of only 1 bits such that a soliton occupied
each bit slot, all solitons would be shifted in time by the same amount through
emission of acoustic waves, creating a uniform shift of the soliton train with no
impact on the timing jitter. However, since an information-coded bit stream
consists of a random string of 1 and 0 bits, the change in the group velocity
of a given soliton depends on the presence or absence of solitons in the preceding tens of bit slots. As a result, different solitons acquire slightly different
velocities, resulting in timing jitter. Thus, acoustic jitter has a deterministic
origin, in contrast with Gordon–Haus jitter, which is stochastic in nature. The

deterministic nature of *acoustic jitter* makes it possible to reduce its impact in practice by moving the detection window at the receiver through an automatic tracking circuit [67] or by using a suitable coding scheme [68].

Attempts have been made to evaluate the extent of timing jitter by using simple analytical models [65]. For a fiber with a core area of 35 μm^2 and at bit rates above 2 Gb/s, the acoustic jitter can be approximated by

$$\sigma_{\text{acou}} = 1.38 \times 10^{-8} (1 - 1.18/B)^{1/2} D^2 B^{3/2} L^2, \tag{10.3.22}$$

where σ_{acou} is expressed in ps, D in ps/(km-nm), L in km, and B in Gb/s. As an example, $\sigma_{\text{acou}} = 6.56$ ps when $D = 0.4$ ps/(km-nm), $L = 10,000$ km, and $B = 10$ Gb/s. Although generally smaller than Gordon–Haus jitter, the acoustic jitter can contribute significantly to the total timing jitter of a soliton communication system, especially if fiber dispersion is large. Similar to the Gordon–Haus case, acoustic jitter can also be reduced considerably by using optical filters (see Section 10.3.4).

Polarization-Mode Dispersion

In soliton communication systems, all solitons are launched with the same state of linear polarization at the input of a fiber link. However, as solitons are periodically amplified, their state of polarization becomes random because the ASE added at every amplifier is of random polarization. Such polarization fluctuations lead to timing jitter in the arrival time of individual solitons through fiber birefringence since the two orthogonally polarized components travel with slightly different group velocities. The phenomenon of *polarization-mode dispersion* (PMD) has been discussed in Section 2.3.5, and its effects are quantified through the PMD parameter D_p. The timing jitter introduced by the combination of ASE and PMD can be written as [69]

$$\sigma_{\text{pol}}^2 = \frac{\pi n_{sp} F_G}{16 N_s} \frac{D_p^2 L^2}{L_A}. \tag{10.3.23}$$

Note that σ_{pol} increases linearly with transmission distance L. As an estimate, $\sigma_{\text{pol}} = 0.38$ ps for a long-haul communication system having $\alpha = 0.2$ dB/km, $L_A = 50$ km, $n_{sp} = 2$, $N_s = 5 \times 10^5$, $D_p = 0.1$ ps/$\sqrt{\text{km}}$, and $L = 10,000$ km. Such a low value of σ_{pol} is unlikely to affect 10-Gb/s soliton communication systems for which the bit slot is 100 ps wide. However, for fibers having larger values of the PMD parameter ($D_p > 1$ ps/$\sqrt{\text{km}}$), the PMD-induced timing jitter becomes important enough that its impact should be considered together with other sources of the timing jitter.

Soliton Interaction

In deriving Eq. (10.3.21) for the timing jitter, solitons were assumed to be sufficiently far apart that their interaction is negligible. However, to maximize the bit rate, solitons are sometimes packed closely together. In the absence of amplifier noise, solitons shift their position in a deterministic manner because of the attractive or repulsive forces experienced by them (see Section 10.2.2). Since the interaction force between two solitons is strongly dependent on their separation and relative phase, both of which fluctuate due to amplifier noise, soliton interaction considerably modifies the Gordon–Haus timing jitter. By considering noise-induced fluctuations of the relative phase of neighboring solitons, timing jitter of interacting solitons is generally found to be enhanced by amplifier noise [70]. However, for a large input phase difference close to π between neighboring solitons, phase randomization leads to a reduction of the timing jitter.

An important consequence of soliton interaction is that the statistics of the timing jitter deviates considerably from the Gaussian statistics expected for Gordon–Haus jitter in the absence of soliton interaction [71]. Such non-Gaussian corrections can occur even when soliton interaction is relatively weak ($q_0 > 5$). They manifest through an enhancement of the bit-error rate and must be accounted for an accurate estimate of the system performance. When solitons are packed so tightly that soliton interaction becomes quite significant, the probability density function of the timing jitter develops a five-peak structure [72]. Equation (10.3.20) cannot be used to evaluate timing jitter in that case. The use of numerical simulations is essential to study the impact of timing jitter on a bit stream composed of interacting solitons.

10.3.4   Timing-Jitter Control

It should be clear from the preceding discussion that the timing jitter ultimately limits the performance of soliton communication systems. It is essential to find a solution to the timing-jitter problem before the use of solitons can become practical. Several techniques were developed during the 1990s for controlling the timing jitter [73], [74].

Optical Bandpass Filters

Optical filters such as a FP étalon (see Section 7.2.1) have been used since 1991 [75] to realize soliton transmission beyond the Gordon–Haus limit [76], [77]. This approach makes use of the fact that the ASE occurs over the entire amplifier bandwidth, whereas the soliton spectrum is a small fraction of it. The bandwidth of optical filters is chosen such that they let the soliton pass but block most of the ASE. Unfortunately, the transmission distance shows only a modest improvement since the timing jitter is reduced by less than 50% when all in-line optical filters have the same center frequency.

The filter technique can be improved dramatically by allowing the center frequency of the successive filters to increase (or decrease) along the link. Such *sliding-frequency filters* avoid the accumulation of ASE occurring when fixed-frequency filters are used [78], [79]. Sliding-frequency filters also reduce the growth of dispersive waves [80], [81] that are generated by strongly perturbed solitons. There remains the choice of up-sliding or down-sliding, depending on whether the center frequency of filters increases or decreases along the link—up-sliding is found to provide better performance [78], [82]. The reason can be understood by considering the third-order dispersion associated with filters. The use of sliding-frequency filters may be difficult to implement in practice because of the need to maintain a precise frequency control. In one scheme, the optical filter is designed such that it automatically offsets its peak-transmission frequency from the carrier frequency of the incident soliton [83]. In an alternative approach, significant reduction of the timing jitter has been realized [84], [85] by periodically sliding the signal frequency while using fixed-frequency filters.

The soliton perturbation theory can be used to study how optical filters benefit a soliton communication system. The effect of a bandpass filter is to modify the soliton spectrum such that

$$\tilde{u}(\xi_f, \omega) \rightarrow T_f(\omega)\tilde{u}(\xi_f, \omega), \tag{10.3.24}$$

where $\tilde{u}(\xi_f, \omega)$ is the soliton spectral amplitude obtained by taking the Fourier transform of Eq. (10.3.9) and $T_f(\omega)$ is the transmission coefficient of the optical filter located at ξ_f. If the action of a filter is assumed to be distributed over the distance ξ_f and the filter spectrum is approximated by a parabola over a range covering the soliton spectrum, the perturbation term for the soliton perturbation theory becomes [80]–[82]

$$\epsilon(u_s) = i \left[\delta g - C_f \left(i \frac{\partial}{\partial \tau} - \omega_f \right)^2 \right] u_s, \tag{10.3.25}$$

where δg is the excess gain required to compensate for the loss introduced by each filter and C_f is related to the curvature of the filter spectrum $T_f(\omega)$. For a FP filter, C_f is given by [60]

$$C_f = \frac{1}{2T_0^2 \Delta \nu_f^2 \xi_f} \frac{R}{(1-R)^2}, \tag{10.3.26}$$

where $\Delta \nu_f$ is the free-spectral range and R is the mirror reflectivity of the FP filter. For a sliding-frequency filter, the center frequency ω_f becomes ξ dependent and should be written as $\omega_f = \omega_f' \xi$, where ω_f' is the linear sliding rate of the center frequency of filters. The use of $\epsilon(u_s)$ in Eqs. (10.3.10)–(10.3.13) then predicts how the four soliton parameters evolve in the presence of optical filters. The results show that the soliton frequency slides with the filters, keeping the soliton train intact, but the ASE accumulated over multiple amplifiers is filtered out later when the soliton spectrum has shifted by more than its own width. As a result, the timing jitter is considerably reduced with the use of sliding-frequency filters.

Figure 10.12 shows the predicted reduction by plotting the timing jitter with and without filters as a function of distance at several bit rates. The bit-rate dependence is solely due to the acoustic jitter; the $B = 0$ curves show the contribution of the Gordon–Haus jitter alone. Optical filters help in reducing both types of timing jitter and permit transmission of 10-Gb/s solitons over

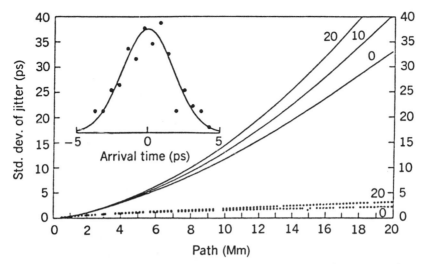

Figure 10.12 Timing jitter with (dotted curves) and without (solid curves) sliding-frequency filters as a function of distance at several bit rates. The bit-rate dependence is due to the inclusion of acoustic jitter. The inset shows a Gaussian fit to the numerically simulated Gordon–Haus jitter at 10,000 km for a 10-Gb/s system with sliding-frequency filters. (After Ref. [78]. ©1992 OSA. Reprinted with permission.)

more than 20,000 km. In the absence of filters, the timing jitter becomes so large that a 10-Gb/s soliton system cannot be operated for distances beyond 8000 km. The inset in Fig. 10.12 shows a Gaussian fit to the Gordon–Haus jitter of 10-Gb/s solitons at a distance of 10,000 km obtained by solving the NSE numerically after including the effects of both the ASE and sliding-frequency filters [78]. The timing-jitter distribution is approximately Gaussian with a standard deviation of about 1.76 ps. In the absence of filters, the jitter exceeds 10 ps under the same conditions.

Optical filters benefit a soliton communication system in several ways. Their use reduces not only timing jitter but also soliton interaction [86], [87], making it possible to pack solitons closer. The physical mechanism behind the reduction of soliton interaction is related to the change in the soliton phase at each filter. A rapid variation of the relative phase between neighboring solitons, occurring as a result of filtering, averages out the soliton interaction by alternating the nature of the interaction force from attractive to repulsive. The reduction of soliton interaction is even more effective if the filter frequency is alternately shifted up and down in a zigzag pattern [88].

Even though FP filters are most common, other types of filters have also been used [89]. In particular, optical filters having a relatively flat transmission passband (tophat-like) are of interest since they minimize energy loss. The Butterworth filters [91] can reduce the accumulation of ASE without requiring sliding of their central frequency along the fiber link. Moreover, they can be more effective in reducing the soliton interaction [92].

Synchronous Modulators

Solitons can also be controlled in the time domain. A 1991 experiment demonstrated [93] transmission of a soliton train over 1 million kilometers using the technique of *synchronous intensity modulation*, implemented using a LiNbO$_3$ modulator. The technique works by introducing additional losses for solitons that have shifted from their original position (center of the bit slot). The modulator forces solitons to move toward its transmission peak where the loss is minimum. Mathematically, the action of modulator is to change the soliton amplitude as

$$u_s(\xi_m, \tau) \to T_m(\tau - \tau_m) u_s(\xi_m, \tau), \qquad (10.3.27)$$

where $T_m(\tau)$ is the transmission coefficient of the modulator located at $\xi = \xi_m$. For the case of sinusoidal modulation, commonly used in practice, $T_m(\tau)$ can be approximated by a parabola in the vicinity of $\tau = \tau_m$. The soliton perturbation theory can again be used in a manner similar to the case of optical filters. The results show that synchronous modulators force the soliton to move toward their transmission peak, and such forcing reduces the timing jitter considerably.

Synchronous modulation can be combined with optical filters to control solitons simultaneously in both the time and frequency domains. A numerical study [94], followed by an experimental realization [95], indicated the possibility of achieving arbitrarily long transmission distances by this combination. Synchronous intensity modulation also permits a relatively large amplifier spacing [96] because it can reduce the impact of dispersive waves. This property of modulators has been exploited to transmit a 20-Gb/s soliton train over 150,000 km with an amplifier spacing of 105 km [97]. In another experiment, a single synchronous modulator, inserted just after the transmitter, allowed transmission of a 20-Gb/s signal over 3000 km [98], well beyond the 2300-km Gordon–Haus limit of the system without modulation. In this experiment, the clock signal used to generate the soliton train was also used to drive the modulator. In contrast, when in-line synchronous modulators are used, the signal clock must be regenerated electronically.

Synchronous modulation can also be implemented by using a phase modulator [99], [100]. One can understand the effect of periodic phase modulation by recalling that a frequency shift $\delta\omega = -\partial\phi(t)/\partial t$ is associated with any phase variation $\phi(t)$. Since a change in soliton frequency is equivalent to a change in the group velocity, phase modulation induces a temporal displacement. Synchronous phase modulation is implemented in such a way that the soliton experiences a frequency shift only if it moves away from the center of the bit slot, thereby confining it to its original position despite the timing jitter induced by ASE and other sources. Intensity and phase modulations can be combined together [101] to further improve the system performance. Similar to the case of sliding-frequency optical filters, synchronous modulators help a soliton communication system in several other ways. Among other things, they reduce soliton interaction, clamp the level of amplifier noise, and inhibit the growth of dispersive waves [102].

Other Techniques

Numerous ways through which solitons interact with each other and with other optical fields lend themselves to many diverse techniques for soliton control. In one approach, helpful in reducing soliton interaction, the amplitude of neighboring solitons is alternated between two values differing typically by 10%. Such a difference in the amplitude η results in different rates of phase accumulation with increasing ξ [see Eq. (10.3.9)] for the two types of solitons. As a result, the phase difference θ between neighboring solitons changes with propagation, resulting in an averaging of soliton interaction. Recall that the interaction force depends on the phase difference between the neighboring solitons and changes from attractive to repulsive with changes in θ (see Section 10.2.2). Such a technique has been used successfully [103] to transmit solitons at a bit rate of 20 Gb/s over 11,500 km, a distance larger than the distance over which two solitons would collide in the absence of amplitude alternation. In a variation of this technique, the soliton frequency is altered to reduce interaction among the neighboring solitons [90]. Relatively small frequency shifts ($\sim$ 100 MHz), imposed on specific solitons in a bit stream, can double the transmission distance because of the reduced soliton interaction.

Another approach to soliton control consists of inserting a *fast saturable absorber* (FSA) periodically along the fiber link. The FSA absorbs low-intensity light such as dispersive waves but leaves a soliton intact as it becomes transparent at high intensities. To be effective, it should respond at a time scale shorter than the soliton width. It is difficult to find an absorber that can respond at such short time scales. However, nonlinear phase effects in fibers can be used to make an interferometer that acts like a FSA. For example, a nonlinear optical-loop mirror [6] or a nonlinear amplifying-loop mirror can act as a FSA and reduce considerably the timing jitter of solitons [104], [105]. The same device can also stabilize the soliton amplitude simultaneously [106].

Re-timing of a soliton train can also be realized [107] by taking advantage of the nonlinear phenomenon of *cross-phase modulation* (XPM) in optical fibers [6]. The technique overlaps the soliton data stream and another pulse train composed of only 1 bits (generated through clock recovery, for example) in a few-kilometer fiber where XPM induces a phase shift on the soliton data stream whose magnitude can be controlled. Such a phase modulation of the soliton translates into a net frequency shift only when the soliton does not lie in the middle of the bit slot. Similar to the case of synchronous phase modulation, the direction of the frequency shift is such that the soliton is confined to the center of the bit slot.

Other nonlinear effects occurring in optical fibers [6] can also be exploited to control the soliton parameters. In one study [108], stimulated Raman scattering was used for this purpose. If a pump beam, modulated at the signal bit rate and upshifted in frequency by the Raman shift (about 13 THz), is copropagated with the soliton bit stream, it simultaneously provides gain (through Raman amplification) and phase modulation (through XPM) to each soliton. Such a technique results in both phase and intensity modulations of the soliton stream

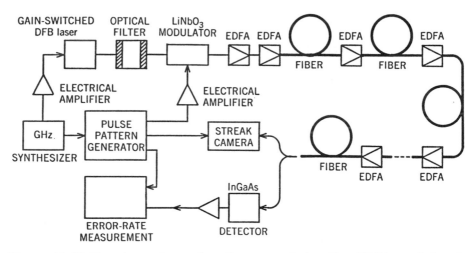

Figure 10.13 Experimental setup for soliton transmission. Two EDFAs used after the LiNbO$_3$ modulator act as power boosters. (After Ref. [110]. ©1990 IEEE. Reprinted with permission.)

and can reduce the timing jitter. Another approach [109] makes use of four-wave mixing (FWM) for soliton reshaping and for controlling the soliton parameters.

10.3.5 Experimental Progress

The first experiment [4] that demonstrated the possibility of soliton transmission over transoceanic distances was performed in 1988 by using a recirculating fiber loop whose loss was compensated through the Raman-amplification scheme. The main drawback from a practical standpoint was that the experiment used two color-center lasers for soliton generation and amplification. The availability of diode-pumped EDFAs by 1989 provided an opportunity to use them as in-line amplifiers for compensating the fiber loss. Many experimental demonstrations of soliton communications were carried out worldwide starting in 1990.

The soliton experiments can be divided into two categories, depending on whether a linear fiber link or a recirculating fiber loop is used in the experiment. The experiments using fiber link are more realistic as they mimic the actual field conditions. Several 1990 experiments [110]–[112] demonstrated soliton transmission over fiber lengths $\sim$ 100 km at bit rates of up to 5 Gb/s. Input pulses were generated by using gain-switched or mode-locked semiconductor lasers and amplified by EDFAs to increase their peak power up to levels required for launching fundamental solitons into a dispersion-shifted fiber. Figure 10.13 shows the experiment setup schematically. A LiNbO$_3$ intensity modulator is used to impose the signal on the pulse train. The soliton-bit stream is transmitted through several fiber sections with losses of each section compensated by using an EDFA. The amplifier spacing is chosen to satisfy the criterion $L_A \ll L_D$ and is typically

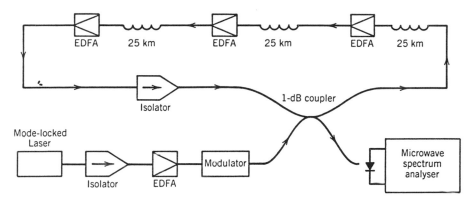

Figure 10.14 Recirculating-loop configuration used to demonstrate soliton transmission over 12,000 km at 2.5 Gb/s. (After Ref. [114]. ©1991 IEE. Reprinted with permission.)

in the range 25–40 km. In a 1991 experiment [113], solitons were transmitted over 1000 km at 10 Gb/s. The 45-ps-wide solitons permitted an amplifier spacing of 50 km in the average-soliton regime.

Since 1991, most soliton transmission experiments have used a recirculating fiber-loop configuration because of cost considerations. In a 1991 experiment [114], 2.5-Gb/s solitons were transmitted over 12,000 km by using a 75-km fiber loop containing three EDFAs, spaced apart by 25 km (see Fig. 10.14). In this experiment, the bit rate–distance product, $BL = 30$ (Tb/s)-km, was limited mainly by the amplifier-induced timing jitter as discussed in Section 10.3.3. The timing-jitter problem was later solved by using optical filters. In a 1991 experiment [75], the 2.5-Gb/s signal could be transmitted over 14,000 km when fixed-frequency filters were used inside the recirculating fiber loop. Soon afterward, the use of sliding-frequency filters [78] resulted in transmission over 15,000 km of a 5-Gb/s signal [115]. Moreover, when the bit rate was doubled by using the WDM technique (discussed later in Section 10.5), the resulting 10-Gb/s signal could still be transmitted over 11,000 km. In a 1993 experiment, the use of sliding-frequency optical filters provided soliton transmission over 20,000 km at 10 Gb/s in a single channel, and over 13,000 km at 20 Gb/s, in a two-channel experiment [79]. Further improvements have resulted in soliton transmission over 35,000 km at 10 Gb/s and over 24,000 km at 15 Gb/s [116]. The BL product of 360 (Tb/s)-km realized in these experiments clearly demonstrates the potential of soliton communication systems for long-haul and undersea applications.

In the time-domain approach for the soliton control, the timing-jitter problem is solved by using a synchronous modulator within the fiber loop. In a 1991 experiment, solitons at 10 Gb/s could be maintained over 1 million kilometers when a LiNbO$_3$ modulator was used within the 510-km loop incorporating ED-FAs with 50-km spacing [93]. In a later experiment, solitons were controlled in both the time and frequency domains by using both modulators and optical filters inside the fiber loop. No performance degradation was observed even after

1 million kilometers, suggesting that such an approach can maintain solitons over unlimited distances [95].

The terrestrial use of soliton systems in the 1.3-μm wavelength region is also attractive for transmission distances $\sim$ 1000 km. The motivation behind the development of such systems stems from the need to update the existing terrestrial fiber links to 10 Gb/s and beyond while making use of the 50–60 million kilometers of the installed fiber base. This worldwide network of standard telecommunication fibers has a relatively high dispersion at the 1.55-μm wavelength [$\beta_2 \approx -21$ ps/(km-nm)], resulting in a dispersion length $L_D \sim 10$ km for a 10-Gb/s soliton system. Since the required amplifier spacing is 30–40 km, it is not possible to satisfy the condition $L_A \ll L_D$ near the 1.55-μm wavelength regime. However, if the operating wavelength is chosen near 1.3 μm, L_D exceeds 200 km, making it easy to satisfy the condition $L_A \ll L_D$. The penalty to be paid is that the fiber loss is higher at 1.3 μm, and thus a larger amplifier gain in needed to compensate for it.

Since practical fiber amplifiers operating at 1.3 μm are not readily available, semiconductor laser amplifiers [117] provide an alternative, especially with the progress realized in reducing their polarization sensitivity. The main drawback of semiconductor laser amplifiers (see Section 8.2) stems from the fact that the amplified pulse is heavily chirped because of the dynamic index changes that occur with gain saturation [118]. Another drawback is that the carrier density does not recover quickly after the amplification of a single soliton, leaving less gain for the following solitons and producing pattern-dependent power fluctuations. Since soliton amplification is accompanied with pulse-energy-dependent frequency shifts, chirping, and unequal gains, it results in variations in the solitons amplitude and frequency which degrade the system performance considerably. Sliding-frequency filters [119] or a combination of fixed-frequency filters and acousto-optic modulators [120] can solve most of these problems. The design of 1.3-μm soliton communication systems is under extensive study, and several European field trials are expected to take place by 1998.

10.4 HIGH-CAPACITY SOLITON SYSTEMS

The average-soliton regime discussed in Section 10.3 has been used almost exclusively for the design of soliton lightwave systems. However, its use limits the bit rate to below 20 Gb/s for practical amplifier spacings, as dictated by the condition in Eq. (10.3.8). Several attempts have been made to increase the bit rate of soliton communication systems beyond this limit. In a 1992 experiment [121], the 32-Gb/s solitons were transmitted over 90 km by using 16-ps pulses obtained from a mode-locked semiconductor laser. Since the bit slot is only about 31 ps wide at the bit rate of 32 Gb/s, neighboring 16-ps solitons were so close to each other [the parameter $q_0 \approx 1$ from Eq. (10.2.1)] that soliton interaction limited the transmission distance to 90 km. Nonetheless, in a 1993 experiment [122], the bit rate was extended to 80 Gb/s using polarization-division multiplexing (discussed later) such that the neighboring bit slots carried orthogonally polarized soliton pulses. The 80-Gb/s signal was transmitted over 80 km by this technique. The same technique was later used to extend the bit rate to 160 Gb/s [123] by multiplexing two 80-Gb/s channels. At such high bit rates, the bit slot is so small (10 ps wide at 100 Gb/s) that typically the soliton width is 3 ps or less. For such ultrashort solitons several higher-order nonlinear effects become quite important. This sections is devoted to the design of such high-speed soliton communication systems.

10.4.1   Quasi-adiabatic Regime

In Section 10.3 we discussed a regime where the amplifier spacing is short compared with the dispersion length ($L_A \ll L_D$). In this average-soliton regime, large loss-induced energy variations may occur over one amplifier spacing with little distortion of the soliton. As the soliton width is decreased in order to increase the bit rate, the condition $L_A \ll L_D$ no longer holds, resulting in shape distortion and considerable emission of dispersive waves when $L_A \sim L_D$. A further decrease of the soliton width leads to a new regime in which $L_D \ll L_A$ and considerable dynamic evolution of the soliton occurs over one amplifier spacing. The fate of a soliton in such a regime depends strongly on the loss per dispersion length and is governed by the dimensionless parameter $\Gamma = \alpha L_D$. If $\Gamma \gg 1$, solitons are strongly perturbed and cannot survive over long lengths. On the other hand, if $\Gamma \ll 1$, each soliton can adapt to losses adiabatically by increasing its width and decreasing its peak power while preserving its soliton nature. For this reason, this regime is referred to as the *quasi-adiabatic regime*.

The condition $\Gamma \ll 1$ can be related to the bit rate by using $\Gamma = \alpha L_D = \alpha T_0^2/|\beta_2|$ and Eq. (10.2.1) to become

$$B \gg \frac{1}{2q_0}\sqrt{\frac{\alpha}{|\beta_2|}}. \tag{10.4.1}$$

For $\alpha = 0.2$ dB/km, $q_0 = 5$, and $\beta_2 = -1$ ps^2/km, $B \gg 21$ Gb/s, or the bit rate should exceed 40 Gb/s. Clearly, the quasi-adiabatic regime is most suitable for high-speed soliton communication systems.

The soliton perturbation theory can be used to study how the soliton width increases in the quasi-adiabatic regime because of fiber losses. From Eq. (10.2.12), the perturbation $\epsilon(u) = -i\Gamma/2$. The use of this perturbation in Eq. (10.3.10) then shows that the soliton amplitude decreases as $\exp(-\Gamma\xi)$. Since soliton width varies inversely with the soliton amplitude, it increases exponentially as

$$T_s(z) = T_s(0)\,\exp(\Gamma\xi) = T_s(0)\,\exp(\alpha z). \tag{10.4.2}$$

For periodically amplified solitons with amplifier spacing L_A, the width increases by a factor of $\exp(\alpha L_A)$ before the soliton reaches the amplifier. Since each soliton also sheds a part of its energy in the form of dispersive waves, whose emission should be minimized as much as possible, the soliton width should not increase by a large factor. A compromise consists of choosing L_A such that $\alpha L_A \leq 1$, so that the soliton width does not increase by more than a factor

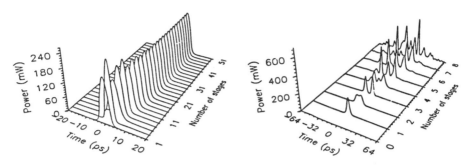

Figure 10.15 Stabilization of a 4.5-ps soliton by FSAs in the quasi-adiabatic regime with 20-km amplifier spacing. For comparison, soliton evolution without FSAs is shown on the right.

of 2.7. For $\alpha = 0.2$ dB/km, the amplifier spacing is then limited to about 22 km. The main point to note is that the soliton character of the optical pulse is maintained, despite an increase in soliton width, because of the adiabatic nature of soliton evolution.

According to Eq. (10.4.1), operation in the adiabatic regime requires the use of ultrashort solitons ($T_s < 5$ ps). As the soliton width decreases, higher-order dispersive and nonlinear effects become important. These effects are included by adding three terms to the right side of Eq. (10.2.12), resulting in a generalized NSE [6]:

$$i\frac{\partial u}{\partial \xi} + \frac{1}{2}\frac{\partial^2 u}{\partial \tau^2} + |u|^2 u = -\frac{i}{2}\Gamma u + i\delta\frac{\partial^3 u}{\partial \tau^3} - is\frac{\partial}{\partial \tau}(|u|^2 u) + \tau_R u\frac{\partial |u|^2}{\partial \tau}, \quad (10.4.3)$$

where the three dimensionless parameters are defined as

$$\delta = \frac{\beta_3}{6|\beta_2|T_0}, \qquad s = \frac{1}{\omega_0 T_0}, \qquad \tau_R = \frac{T_R}{T_0}. \quad (10.4.4)$$

The terms involving δ, s, and τ_R take into account, respectively, the effects of TOD, self-steepening, and Raman amplification. Self-steepening becomes important only for solitons shorter than 100 fs and can safely be neglected for bit rates below 1 Tb/s. The Raman effect leads to a continuous downshift of the carrier frequency, an effect known as the *soliton self-frequency shift* (SSFS) [124], [125]. Its origin can be understood by noting that for ultrashort solitons, the pulse spectrum becomes so broad that the high-frequency components of the pulse can transfer energy through Raman amplification to the low-frequency components of the same pulse. SSFS is negligible for $T_s > 10$ ps but becomes of considerable importance for short solitons ($T_s < 5$ ps) since it scales as T_s^{-4}. Since $T_R \sim 5$ fs, the Raman term remains a small perturbation even for $T_s = 1$ ps [6]. As an example of the smallness of the three higher-order terms, $\delta \sim 0.01$, $s \sim 0.001$, and $\tau_R \sim 0.005$ for $T_0 = 1$ ps and typical values of fiber parameters. The TOD term becomes quite important when β_2 is below 0.1 ps^2/km since δ can exceed 0.1 in that case.

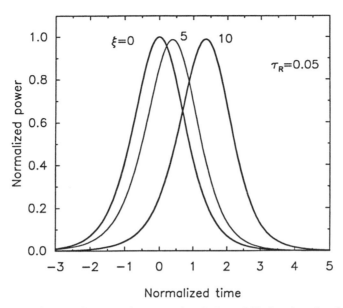

Figure 10.16 Pulse envelopes at distances $\xi = 0$, 5, and 10 showing the slowing down of the fundamental soliton because of the self-frequency shift induced by stimulated Raman scattering.

The operation of a soliton communication system in the quasi-adiabatic regime requires removal of dispersive waves. Numerical simulations based on Eq. (10.4.3) show that dispersive waves, accumulated over multiple amplification stages, eventually destroy the integrity of the soliton bit stream because of its interaction with dispersive waves. Figure 10.15 shows the destruction of a 4.5-ps soliton only after seven amplifiers because of this interaction. A FSA inserted before each amplifier can solve the interaction problem since it blocks the low-power dispersive waves while letting the soliton pass through it. Several kinds of FSAs have been considered for this purpose, including multiquantum-well structures [126], twin-core fibers [127], nonlinear amplifying-loop mirrors [128], [129], and devices based on the nonlinear polarization evolution [130]. Figure 10.15 shows how a 4.5-ps soliton can propagate over 50 amplifiers when FSAs are inserted before each amplifier. Numerical simulations show that the amplifier spacing is limited to 20–25 km, in agreement with the qualitative estimate of 22 km obtained by using the criterion $\alpha L_A = 1$.

The SSFS becomes the most limiting factor in the quasi-adiabatic regime when FSAs are used to control the growth of dispersive waves. Its effect on soliton communication systems can be understood by noting that a shift in the carrier frequency translates into a change in the group velocity. In particular, since the carrier frequency shifts toward longer wavelengths, the soliton slows down. Figure 10.16 shows such a slowing down of the fundamental soliton by solving Eq. (10.4.3) numerically with $\Gamma = 0$, $\delta = 0$, $s = 0$, and $\tau_R = 0.05$. By the time the soliton has propagated over 10 dispersion lengths, it has been

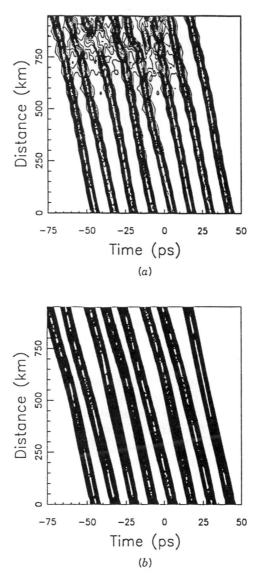

Figure 10.17 (a) Destruction of a 80-Gb/s soliton train in the quasi-adiabatic regime because of dispersive-wave accumulation; (b) its restoration with the use of fast saturable absorbers. Tilt of a soliton trajectory is due to the effects of third-order dispersion. (After Ref. [135]. ©1995 IEE. Reprinted with permission.)

delayed by a significant fraction of its own width. Because of SSFS, solitons do not arrive at their expected position, although the change in the arrival time is deterministic in nature. If the fiber loss is neglected by assuming that it is compensated periodically through in-line amplifiers, the delay in the arrival time of the soliton (in ps) is approximately given by [131]

$$t_0 = 6.8 \times 10^{-3} (|\beta_2| L_T / T_s^2)^2. \tag{10.4.5}$$

The delay becomes significant when the total transmission distance exceeds the dispersion length by a large factor.

A simple shift in the arrival time of the soliton is not of any concern as long as all bits are delayed by the same amount. However, the soliton width can fluctuate from bit to bit because of energy fluctuations induced by the amplifier noise. Such width fluctuations would be converted into timing jitter by SSFS. The Raman-induced timing jitter is discussed later in this section. Optical filters can be used to control the accumulation of the SSFS [132]. Because the SSFS must stay a small fraction (typically 10%) of the filter bandwidth, the width of the soliton is limited to values above 1 ps. By considering simultaneously the effects of lumped gain with filtering, the SSFS, and the FSA, a simple model has been used to predict the width of the steady-state soliton for a given amplifier spacing [133]. The results show that the amplifier spacing needed for stable transmission of solitons drops from 20 km to 5 km as the soliton width is reduced from 5 ps to 2 ps. Since the bit slot is only 10 ps at 100 Gb/s, one must use short solitons to avoid soliton interaction. However, such a system requires an amplifier spacing too small to be practical.

The transmission of ultrashort solitons in the quasi-adiabatic regime has been demonstrated [134] by using 3-ps solitons with $L_A = 25$ km and 3-nm bandwidth filters in dispersion-shifted fibers having $\beta_2 = -0.25$ ps^2/km and $\beta_3 = 0.1$ ps^3/km. A close packing of solitons ($q_0 \approx 3$) allowed transmission at a bit rate of 80 Gb/s, but the transmission distance was limited to 500 km. Since FSAs were not used in this experiment, accumulation of dispersive waves is expected to occur. Indeed, numerical simulations, performed by using Eq. (10.4.3) with the experimental values of the system parameters, show that the transmission distance is limited by the interaction of a soliton with dispersive wave and other solitons [135]. The distance can be increased by inserting periodically along the fiber link either FSAs or synchronous modulators, both of which control the growth of dispersive waves. Figure 10.17 shows the results of numerical simulations of the 80-Gb/s experiment together with the improvement occurring with the insertion of FSAs before every amplifier. To increase the bit rate further, the use of polarization-division multiplexing has been considered. This topic is covered next.

10.4.2 Polarization-Division Multiplexing

Since a single-mode optical fiber supports two orthogonal states of polarization for the same fundamental mode, a new kind of multiplexing, known as the *polarization-division multiplexing* (PDM), can be used to double the capacity of fiber-optic communication systems. In PDM, two channels at the same wavelength are transmitted through the fiber such that their pulse trains are orthogonally polarized at the fiber input. At first glance, such a scheme should not work unless polarization-preserving fibers are used since the polarization state changes randomly in conventional fibers because of birefringence fluctuations. However, even though the polarization states of each channel does change at the end of the fiber link in an unpredictable manner, their orthogonal nature is preserved, making it possible to isolate each channel through simple optical techniques.

While implementing PDM for soliton bit streams, a nonlinear phenomenon, known as *soliton self-trapping* and mediated by cross-phase modulation [6], tends to destroy the orthogonal nature of the two bit streams. For this reason, PDM is typically implemented [136] by interleaving the two soliton bit streams in the time domain (TDM) such that the two neighboring solitons have orthogonal states of polarization. Since soliton interaction is much weaker for orthogonally polarized solitons, the main advantage of PDM lies in the reduction of soliton-soliton interaction, which can be reduced further [137] using sliding-frequency filters. The effective bit rate increases simply because solitons can be packed more tightly when the PDM technique is used [138].

An important factor limiting the performance of PDM soliton systems is the fiber birefringence. Even the best optical fibers exhibit residual birefringence that varies along the fiber (typically on a scale smaller than 1 km) because of stress and core-diameter variations. Associated with fiber birefringence is the polarization-mode dispersion (PMD), which indicates a relative delay between the two polarization components of a PDM signal. In fact, PMD seriously limits the use of PDM for linear systems making use of the NRZ format. The limitation for linear systems arises because of the frequency dependence of the PMD, which leads to pulse depolarization (different parts of the pulse have different polarizations). However, the situation is different for solitons. The natural tendency of a soliton to preserve its integrity under various perturbations also holds for perturbations affecting its state of polarization. Unlike linear pulses, the state of polarization is kept constant across the entire soliton (no depolarization across the pulse), and the effect of polarization perturbations is to induce a small change in the state of polarization of the entire soliton (a manifestation of its particlelike nature). Such resistance of solitons to PMD, however breaks down for large amounts of PMD. The breakdown limit has been estimated to be [139]

$$D_p \leq 0.3\, D^{1/2}, \qquad (10.4.6)$$

where D_p is the PMD parameter (see Section 2.3.5) expressed in ps/$\sqrt{\text{km}}$ and D is the dispersion parameter in units of ps/(nm-km). Since typically $D_p <$ 0.1 ps/$\sqrt{\text{km}}$ for high-quality optical fibers, D must exceed 0.06 ps/(nm-km). In practice, D is larger than 0.1 ps/(nm-km), and PMD is a minor problem for most soliton communication systems.

The potential of PDM for high-capacity soliton systems has been demonstrated. In one experiment performed in the quasi-adiabatic regime [123], the reduced soliton-soliton interaction allowed transmission at 160 Gb/s over 200 km. The soliton width (FWHM) was 1.6 ps in this experiment, as the bit slot is only 6.4 ps wide at a bit rate of 160 Gb/s. The transmission distance was limited by the interaction of dispersive waves with the soliton bit stream. As discussed earlier, it should be possible to increase the transmission distance with the use of FSAs or synchronous modulators.

Two additional mechanisms that can generate timing jitter through fiber birefringence [69] should be considered while evaluating the performance of soliton lightwave systems making use of PDM. The first is due to accumulation of the delay between two orthogonally polarized solitons in a birefringent fiber. Fortunately, because of random birefringence variations, such a timing jitter tends to average out over long transmission distances. However, orthogonally polarized solitons can experience significant temporal shifts locally during their propagation in fibers with random birefringence fluctuations. Such local temporal shifts may bring a soliton pair close enough to enhance soliton-soliton interaction and thus prevent recovery of the polarization state. For this reason it is important to ensure that solitons do not deviate too much from the center of the bit slot at any point along the fiber link. The second source of PMD-induced jitter originates from amplifier noise and has already been discussed in Section 10.3.3. Equation (10.3.23) shows that the resulting timing jitter is relatively small compared with other sources of timing jitter.

An important consideration in the design of PDM systems is related to *polarization-dependent loss* or *polarization-dependent gain*. If a communication system contains multiple elements that amplify or attenuate the two polarization components of a soliton differently, the polarization state is easily altered. In fact, for the worst situation in which the soliton polarization is oriented at 45° from the low-loss (or high-gain) direction, the state of polarization rotates by 45° and gets aligned with the low-loss direction only after 30–40 amplifiers, for a gain or loss anisotropy of < 0.2 dB [140]. Even though the axes of polarization-dependent gain or loss are likely to be evenly distributed along the soliton link, such effects may still become an important source of timing jitter.

An extension of the PDM technique, called *polarization-multilevel coding*, has also been suggested [141]. In this technique, the information coded in each bit is contained in the angle that the soliton state of polarization makes with one of the principal birefringence axes. This technique is also limited by random variations of fiber birefringence and by randomization of the polarization angle by the amplifier noise and has not yet been implemented because of its complexity.

10.4.3   Distributed Amplification

As discussed earlier, the distributed-amplification scheme is inherently superior to lumped amplification since its use provides a nearly lossless fiber by locally compensating the loss at every point along the fiber link. In fact, this scheme was used as early as 1985 to demonstrate soliton transmission over a 10 km of fiber [56], and than later extended in 1988 to a 4000-km length by using a recirculating fiber loop [4]. In these experiments, the gain was provided through Raman amplification by pumping the transmission fiber at a wavelength of about 1.46 μm from a color-center laser. With the advent of the EDFA, lumped amplification became more common because of the low pump powers needed by such amplifiers. However, with the adoption of lumped amplification came the limitations imposed by the average-soliton regime. Distributed amplification avoids these limitations and is thus still being pursued. In one scheme, the transmission fiber itself is doped lightly and pumped periodically to provide enough gain that fiber loss is locally compensated (see Section 8.5.8). Several experiments have demonstrated that solitons can be propagated in such active fibers [142]–[146].

Ideally, distributed amplification requires a constant gain g_s per unit length to match the constant loss rate α of fibers. However, since the pump power decreases because of fiber loss and its absorption by dopants, the gain is nonuniform along the fiber length and cannot compensate precisely for fiber loss everywhere. A bidirectional pumping scheme is used in practice to reduce the gain nonuniformities [see Fig. 10.10(b)]. Residual variations in the soliton energy E_s can be studied by solving [57]

$$\frac{dE_s}{dz} = [-\alpha + g_s(z)]E_s. \tag{10.4.7}$$

The gain coefficient $g_s(z)$ is calculated by using the analysis of Section 8.5.3 [6]. Assuming that the fiber is equally pumped at both ends and neglecting the gain saturation, $g_s(z)$ can be approximated by

$$g_s(z) = g_0\{\exp(-\alpha_p z) + \exp[-\alpha_p(L_A - z)]\}, \tag{10.4.8}$$

where α_p is the fiber loss at the pump wavelength and L_A is the *pump-station spacing*. The gain constant g_0 is related to the pump power injected at both ends. Variations of soliton energy along the fiber length are obtained by integrating Eq. (10.4.7) and are given by

$$\ln\left(\frac{E_s(z)}{E_{\text{in}}}\right) = \alpha L_A\left(\frac{\sinh[\alpha_p(z - L_A/2)] + \sinh(\alpha_p L_A/2)}{2\sinh(\alpha_p L_A/2)} - \frac{z}{L_A}\right), \tag{10.4.9}$$

where E_{in} is the soliton energy at the fiber input. The range of energy variations increases with L_A because of more severe depletion of the pump for longer fibers. Nevertheless, this range can be made much smaller than that occurring for the lumped-amplification case.

The effect of *energy excursion* on solitons depends on the ratio $\xi_A = L_A/L_D$, where L_A now stands for the pump-station spacing. When $\xi_A \ll 1$, little soliton reshaping occurs even with large power excursion, and stable transmission over long distances can be achieved. This regime is similar to the average-soliton regime for passive fibers. For $\xi_A \gg 1$, solitons evolve adiabatically with little emission of dispersive waves (the quasi-adiabatic regime). For intermediate values of ξ_A, a more complicated behavior occurs. In particular, dispersive waves and solitons are resonantly amplified when $\xi_A \simeq 4\pi$. The periodic amplification of solitons and dispersive waves and their mutual interaction lead to unstable and chaotic behavior in this regime [57].

Even though the qualitative description of the distributed and lumped amplification seems similar, two important differences should be noted. First, the soliton-energy excursion is much smaller in active fibers than in passive fibers. This feature makes it possible to push the average-soliton limit to $\xi_A \approx 0.7$ [142]–[144]. Second, for an active fiber pumped to transparency, the energy of the output soliton in each section of a chain of active fibers is identical to its value at the input. Consequently, in the quasi-adiabatic regime of distributed amplification, the width of the output soliton coincides with its value at the input end, in contrast with the lumped-amplification case, where considerable soliton broadening occurs.

Modeling of soliton communication systems making use of distributed amplification requires the addition of a gain term to the generalized NSE given by Eq. (10.4.3). However, it is important to include the effects of a finite gain bandwidth. In a simple approach, the equation governing soliton evolution is written as [6]

$$i\frac{\partial u}{\partial \xi} + \frac{1}{2}\frac{\partial^2 u}{\partial \tau^2} + |u|^2 u = -\frac{i}{2}\Gamma u + \frac{i}{2}g_s\left(u + \tau_2^2\frac{\partial^2 u}{\partial \tau^2}\right) + i\delta\frac{\partial^3 u}{\partial \tau^3} + \tau_R u\frac{\partial |u|^2}{\partial \tau}, \quad (10.4.10)$$

where g_s represents the distributed gain and τ_2 is inversely related to the gain bandwidth. The shock term has been neglected because of its negligible effects on soliton propagation in most cases of practical interest.

Numerical simulations based on Eq. (10.4.10) show the expected benefit of the distributed-amplification scheme for high-capacity soliton communication systems. However, for soliton widths below 5 ps, the SSFS leads to considerable changes in the soliton evolution since it changes the gain and dispersion experienced by solitons. Fortunately, the finite gain bandwidth also reduces the amount of SSFS, stabilizing the soliton frequency close to the gain peak [145]. Under certain conditions, the SSFS can become so large that it cannot be compensated, and the soliton just moves out of the gain window, loosing all its energy. It should be stressed that Eq. (10.4.10) approximates the gain spectrum by a parabola in the vicinity of the gain peak. Its use is justified for solitons whose spectrum is much narrower than the gain bandwidth but become questionable for femtosecond solitons. The SSFS affects the evolution of femtosecond solitons drastically, as observed experimentally [146].

10.4.4   Dispersion-Decreasing Fibers

An interesting scheme for stable propagation of solitons restores the balance between GVD and SPM in a lossy fiber by changing the dispersion properties of the transmission fiber [147]. Such fibers are called *dispersion-decreasing fibers* (DDFs) because their GVD must decrease in such a way that it compensates for the reduced SPM experienced by the soliton weakened from fiber loss. Since the soliton peak power decreases exponentially, the requirement that $N = 1$ in Eq. (10.1.6) can be maintained if the GVD decreases exponentially as

$$|\beta_2(z)| = |\beta_2(0)| \exp(-\alpha z). \quad (10.4.11)$$

For such a dispersion profile, the soliton should keep its width constant even in a lossy fiber. DDFs are also beneficial for WDM application. This topic is treated in Section 10.5.

Fibers with a nearly exponential GVD profile have been fabricated. A practical technique for making such DDFs consists of reducing the core diameter along the fiber length in a controlled manner during the fiber-drawing process. Variations in the fiber diameter change the waveguide contribution to β_2 and reduce its magnitude. Typically, GVD varies by a factor of 10 over a length of 20–40 km. The accuracy realized by the use of this technique is estimated to be better than 0.1 ps^2/km [148], [149]. Since fibers with a continuously varying dispersion are not always readily available, a simple approach consists of approximating the exponential profile with a *staircase* by splicing together several constant-dispersion fibers with different β_2 values [150], [151].

Propagation of short solitons in DDFs has been demonstrated in two experiments by using a 40-km DDF and soliton widths down to 2 ps [149], [152]. In both cases, the 40-km DDF resulted from splicing two 20-km sections of DDFs having their GVD matched at the splice. The soliton preserved its width in spite of an energy loss of more than 8 dB. DDFs have also been used for data transmission. In a recirculating-loop experiment [153], a 6.5-ps soliton train at 10 Gb/s was transmitted over a distance of 300 km. The acoustic timing jitter is believed to be the limiting factor because of a relatively high value of the average dispersion for the DDF used in the experiment. The effects of SSFS and TOD would also limit the performance of such systems, especially through the timing jitter, as discussed next.

Timing Jitter

The propagation of ultrashort solitons through DDFs can be studied by using Eq (10.4.3) after including variations of GVD along the fiber length. The resulting equation becomes

$$i\frac{\partial u}{\partial \xi} + \frac{1}{2}p(\xi)\frac{\partial^2 u}{\partial \tau^2} + |u|^2 u = -\frac{i}{2}\Gamma u + i\delta\frac{\partial^3 u}{\partial \tau^3} + \tau_R u\frac{\partial |u|^2}{\partial \tau}, \qquad (10.4.12)$$

where $p(\xi) = |\beta_2(\xi)/\beta_2(0)|$ is the normalized GVD at ξ and the shock term has been neglected by setting $s = 0$. The distance ξ is normalized to the dispersion length $L_D = T_0^2/|\beta_2(0)|$, defined by using the GVD value at the fiber input.

Because of the ξ dependence of the second term, Eq. (10.4.12) is no longer a standard NSE. However, it can be transformed into a perturbed NSE by using

$$v = p^{-1/2}u, \qquad \xi' = \int_0^\xi p(\xi)\,d\xi. \qquad (10.4.13)$$

These transformations renormalize the soliton amplitude and the distance scale to the local GVD. In terms of v and ξ', Eq. (10.4.12) becomes

$$i\frac{\partial v}{\partial \xi'} + \frac{1}{2}\frac{\partial^2 v}{\partial \tau^2} + |v|^2 v = -\left(\frac{i\Gamma}{2p} + \frac{1}{2p}\frac{dp}{d\xi'}\right)v + \frac{i\delta}{p}\frac{\partial^3 v}{\partial \tau^3} + \tau_R v\frac{\partial |v|^2}{\partial \tau}. \qquad (10.4.14)$$

If the GVD profile is chosen such that $dp/d\xi' = -\Gamma$, or $p(\xi) = \exp(-\Gamma\xi)$, fiber loss has no effect on soliton propagation. This is the same profile as in Eq. (10.4.11) obtained on physical grounds.

Since the left side of Eq. (10.4.14) corresponds to the standard NSE, one can apply the soliton perturbation theory to study the effects of SSFS and TOD on the timing jitter. By using Eq. (10.3.10)–(10.3.12) with the last two terms of Eq. (10.4.14) acting as perturbation, the soliton displacement at the end of a single DDF of length ξ_A is found to be given by [154]

$$q(\xi_A) = -d_{GH}\Omega_0 + \tau_R d_R \eta_0^4 + \delta\xi_A \eta_0^2 + 3\delta\xi_A \Omega_0^2 + q_0, \qquad (10.4.15)$$

where two cross terms resulting from the combined effect of SSFS and TOD have been neglected since they are proportional to the product of δ and τ_R. The parameters d_{GH} and d_R depend on the amplifier spacing and the fiber loss and are defined as

$$d_{GH} = [1 - \exp(-\Gamma\xi_A)]/\Gamma, \qquad (10.4.16)$$

$$d_R = \frac{8}{15\Gamma^2}\left(\frac{1}{2} - \exp(-\Gamma\xi_A) + \exp(-2\Gamma\xi_A)\right). \qquad (10.4.17)$$

The parameters η_0, Ω_0, and q_0 in Eq. (10.4.15) represent the soliton amplitude, frequency, and position at $\xi = 0$. Because of the amplifier noise, η_0, Ω_0, and q_0 fluctuate around their average values of 1, 0, and 0, respectively. By considering such fluctuations and adding the contribution of N_A amplifiers in a long amplifier chain ($N_A \gg 1$), timing jitter can be calculated by following a procedure similar to that of Section 10.3.3. The result is given by [155]

$$\sigma_{\text{DDF}}^2 = \left(\frac{4}{5}N_A^5\tau_R^2 d_R^2 + 2N_A^4\tau_R\delta d_R\xi_A + \frac{4}{3}N_A^3\delta^2\xi_A^2\right)\sigma_\eta^2 + \frac{1}{3}N_A^3 d_{GH}^2\sigma_\Omega^2 + N_A\sigma_q^2,$$

$$(10.4.18)$$

where σ_η^2, σ_Ω^2, and σ_q^2 are given by Eqs. (10.3.15)–(10.3.17). The term with d_{GH} represents the Gordon–Haus jitter and is a generalization of Eq. (10.3.20) for DDFs. The term $N_A\sigma_q^2$ comes from the direct effect of the ASE on soliton position and is negligible in practice. The two terms involving d_R are related to the *Raman-induced timing jitter*, while the terms containing δ are due to the TOD. The Raman-induced timing jitter originates from the SSFS. Its origin can be understood as follows. Fluctuations in the soliton amplitude introduced by amplifier noise result in width fluctuations, which are converted to fluctuations in the soliton frequency by the SSFS, which are in turn translated into position fluctuations by the GVD. The first Raman term (proportional to N^5) generally dominates for solitons shorter than 5 ps. The TOD contribution to timing jitter becomes important if the minimum dispersion of DDFs falls below 0.1 ps/(km-nm).

Figure 10.18 shows the individual contributions of amplitude, frequency, and position fluctuations together with the total timing jitter as a function of transmission distance for soliton widths in the range $T_s = 1$–40 ps by choosing $\beta_2^{\min} = -0.1$ ps^2/km and $L_A = 80$ km. For $T_s > 10$ ps, timing jitter originates

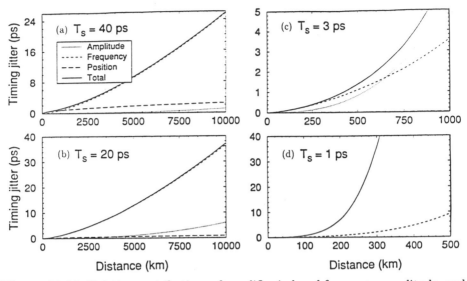

Figure 10.18 Relative contributions of amplifier-induced frequency, amplitude, and position fluctuations to the timing jitter as a function of propagation distance in dispersion-decreasing fibers for several soliton widths. Total timing jitter is shown by a solid line.

mostly from frequency fluctuations (Gordon–Haus jitter) since the contributions of Raman and TOD effects are small for such relatively broad solitons. When shorter solitons are used, the contribution of higher-order effects, especially the Raman effect, increases rapidly with transmission distance. For 3-ps or shorter solitons, the contribution of amplitude fluctuations to the timing jitter (mediated through the Raman effect) becomes so important that the total transmission distance is limited to only a few hundred kilometers in the absence of a soliton-control mechanism. Since the effect of amplitude fluctuations on the timing jitter increases more rapidly than that of frequency fluctuations [N^5 versus N^3 dependences in Eq. (10.4.18)], the former will dominate for long distances. For a transoceanic distance of 10,000 km, amplitude fluctuations dominate for soliton widths below 7 ps. For 1-ps solitons, amplitude fluctuations totally dominate the timing jitter at all distances. These results clearly indicate that timing jitter becomes the limiting factor when the bit rate of a soliton lightwave system exceeds 50 Gb/s, requiring solitons shorter than 5 ps.

Control of Timing Jitter

The increase in the timing jitter brought by the Raman and TOD effects and a shorter bit slot at higher bit rates ($\sim$ 10 ps at B = 100 Gb/s) make the control of timing jitter essential before such systems can become practical. As discussed in Section 10.3.4, both optical filters and synchronous modulators should help in reducing the timing jitter. The technique of *optical phase conjugation* (OPC),

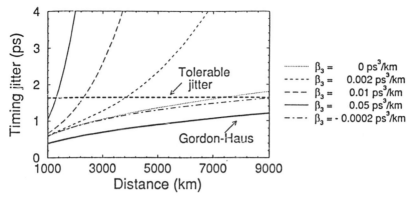

Figure 10.19 Effect of third-order dispersion on timing jitter in a periodically amplified DDF-based soliton communication system making use of phase conjugation. Thick solid curve shows the Gordon–Haus jitter.

discussed in Section 9.7 in the context of dispersion compensation, is also quite effective in reducing the timing jitter of average solitons. The same technique turns out to be beneficial to DDF-based systems. The implementation of OPC requires either parametric amplifiers [6] in place of EDFAs or insertion of a nonlinear optical device before each amplifier that changes the soliton amplitude from u_s to u_s^* while preserving all other features of the bit stream. Such a change is equivalent to inverting the soliton spectrum around the wavelength of the pump laser used for the FWM process. As discussed in Section 9.7, a few-kilometer-long fiber with suitable dispersion characteristics can be used for spectral inversion. The timing jitter changes considerably because of OPC. The soliton perturbation theory can again be used, although the details are quite cumbersome. Following the same procedure as outlined above, the timing jitter is found to be [156]

$$\sigma_{\text{OPC}}^2 = \left(8 N_A \tau_R^2 d_R^2 + 4 N_A^2 \tau_R d_R \delta \xi_A + \frac{4}{3} N_A^3 \delta^2 \xi_A^2 \right) \sigma_\eta^2 + \frac{1}{2} N_A d_{GH}^2 \sigma_\Omega^2 + N_A \sigma_q^2.$$

(10.4.19)

This equation should be compared with Eq. (10.4.18) obtained without OPC. The N_A dependence of the Gordon–Haus contribution has changed from N_A^3 to N_A. The Raman-induced jitter is reduced even more dramatically—from N_A^5 to N_A. These reductions come from the OPC-induced spectral inversion, which provides compensation for the effects of both the GVD and SSFS. However, OPC does not compensate for the TOD. As a result, the term involving δ^2 is the same, while the cross term with $d_R \delta$ is reduced, as it involves the Raman effect.

The effects of TOD are shown in Fig. 10.19 by plotting the timing jitter of 2-ps solitons undergoing periodic OPC. The dashed horizontal line represents the tolerable timing jitter. For comparison, the timing jitter obtained by considering only the Gordon–Haus term in Eq. (10.4.19) is shown by the thick solid line. Other curves correspond to different values of the TOD parameter β_3. For

$\beta_3 = 0.05$ ps^3/km, a typical value for dispersion-shifted fibers, the transmission distance is limited by TOD to below 1500 km. However, considerable improvement occurs when β_3 is reduced. Transmission over 7500 km is possible for $\beta_3 = 0$ and the distance can be increased further for slightly negative values of β_3. Such values of β_3 are possible in suitably designed dispersion-compensating fibers. It should be stressed that a complete description of timing jitter requires inclusion of other effects, such as dispersive waves, acoustic waves, and PMD. Dispersive waves are likely to become important for low average dispersion while the acoustic effect may take over for large values of average dispersion.

10.4.5 Dispersion Management

Since DDFs are not readily available, a simple approach makes use of *dispersion management*, a technique that consists of using multiple sections of constant-dispersion fibers whose lengths and GVDs are judiciously chosen. Generally speaking, operation at a wavelength where the GVD is low should improve system performance, as it lowers the required average power, reduces the timing jitter, and lowers the magnitude of the SSFS. However, as the operating wavelength approaches the zero-dispersion wavelength, several factors affect system performance [157]. For single-channel systems, the effects of TOD can induce severe pulse distortion and emission of dispersive waves when $|\beta_2| < 0.1$ ps^2/km. In WDM systems, as discussed in Section 7.3, the FWM can lead to considerable crosstalk. Dispersion management provides a solution since it can lower the average GVD of the entire link while keeping the GVD of each section large enough that the TOD effects remain negligible. Variants of this scheme are also referred to as *partial soliton communication* [158] and *dispersion allocation* [159].

In its simplest form, a relatively short segment of dispersion-compensating fiber (DCF) is added to the transmission fiber, the same technique as that used extensively for nonsoliton systems (see Section 9.4). However, Eq. (9.4.16) cannot be used to find the DCF length since its derivation neglects the nonlinear effects completely. When the nonlinear effects are included, the required DCF length is about 90% of that predicted by Eq. (9.4.16) [160]–[162]. In one experiment, the compensation of GVD by 90% through DCFs enabled transmission of a 20-Gb/s soliton bit stream over 5520 km of a linear fiber link containing amplifiers at 40-km intervals [162].

The use of the DCFs in soliton links forces each soliton to propagate in the normal-dispersion regime with a large positive value of β_2. At first sight, such a scheme should not even work, since DCFs cannot support solitons and would lead to considerable broadening and chirping of the pulse. The reason solitons are not destroyed in DCFs is again attributed to a relatively short length of the DCF section compared with the dispersion length in the average-soliton regime. In fact, because of dispersion compensation, not only the peak power but also the width and shape of the average soliton oscillates periodically—such pulses are referred to as breathing solitons [163]. Moreover, numerical simulations show that the pulse shape is no longer "sech" but close to a Gaussian profile [164]. This behavior can be understood from the theory of FM mode-locked lasers and by noting that a periodically amplified fiber link is similar to a laser cavity as far as pulse evolution is concerned. More specifically, the DCF generates frequency chirp (FM) through SPM while the anomalous GVD in the dispersion-shifted fiber compresses the chirped pulse. Soliton propagation is also affected by the position of the DCF. For example, placing the DCF just after the soliton is amplified degrades system performance by introducing an excess of SPM [165]. Increasing the number of fiber segments permits a longer amplifier spacing [159] and improves transmission fidelity [166].

A general method capable of determining the optimum design of a dispersion-managed system with many fiber segments is difficult to establish because of the multiple effects which should be taken into account simultaneously. In one approach [166], fiber segments of the same length but having a Gaussian distribution of GVD are grouped in pairs. Each pair is composed of two segments of nearly opposite GVD values such that the average GVD has a small negative value for each pair. In a field experiment on the Tokyo network [167], the concept of dispersion allocation has been applied successfully to demonstrate soliton transmission at 10 Gb/s over the 2000 km, despite the fact that the installed fibers were not originally intended to support solitons.

Lowering of average GVD through dispersion compensation not only reduces pulse distortion but also lowers the timing jitter [158]. The reason is easy to understand from Eq. (10.3.21). Indeed, in a 20-Gb/s experiment [160], the timing jitter became low enough when the average dispersion over each fiber segment was reduced to a value near -0.025 ps^2/km that a 20-Gb/s signal could be transmitted over transoceanic distances. In another 20-Gb/s experiment [168], solitons were transmitted over 9000 km without using any in-line optical filters since the periodic use of DCFs reduced the timing jitter by more than a factor of 3. It turns out that periodic dispersion compensation reduces the Gordon–Haus jitter even more than that predicted by the analysis of Section 10.3.3 for the average value of GVD [169]. The reason is that the peak power (or the energy) of the average soliton is enhanced in a dispersion-managed system compared with that predicted by Eq. (10.3.6). Dispersion compensation can also be used to increase the amplifier spacing. In one set of computer simulations, the use of a chirped fiber grating (see Section 9.6.2) allowed an amplifier spacing of 100 km for 40-Gb/s solitons transmitted over 7500 km [170]. The pulse shape just after the amplifier was nearly Gaussian.

The use of dispersion compensation makes it possible to design soliton systems in which the average value of β_2 becomes almost zero. An interesting question one may ask is whether solitons exist in such a system. Pulse propagation at the zero-dispersion wavelength was studied as early as 1986 [171]. In general, the pulse spectrum shifts slightly into the anomalous-dispersion regime, and a soliton is formed. Similar behavior occurs in the case of a dispersion-managed system with $\beta_2 = 0$ on average. A solitonlike but chirped pulse propagates through the fiber in a manner analogous to the case of a mode-locked laser discussed earlier [172].

Dispersion management can also be used to approximate the exponential dispersion profile of a DDF by a *staircase-like profile*. How should one select the length and GVD of each fiber segment for emulating a DDF? The answer is not obvious, and several methods have been proposed [173]–[175]. In one approach, power deviations are minimized in each section [173]. In another approach, M sections of different GVD D_m and different lengths L_m ($m = 1$–M) are chosen such that the product $D_m L_m$ is the same for each section [174]. In a third approach, D_m and L_m are selected to minimize the emission of dispersive waves [175]. The dispersion-management issue for solitons remains an active area of research in 1997 [176]–[180].

10.5 WDM SOLITON SYSTEMS

As discussed in Chapter 7, the capacity of a lightwave system can be increased considerably by using the WDM technique. A WDM soliton system transmits over the same fiber several soliton bit streams, distinguishable through their different carrier frequencies. In this section we focus on main issues involved in the design of such systems.

10.5.1 Soliton Collisions

The new feature that becomes important for WDM systems is the possibility of collisions among solitons belonging to different channels because of their different group velocities. To understand the importance of such collisions, let us consider the NSE once again. Since dispersion management plays an important role in the design of WDM systems, it is better to use Eq. (10.4.12) but neglect the higher-orders terms by setting $\tau_R = \delta = 0$ by assuming that the system is designed to minimize their influence. The resulting equation becomes

$$i\frac{\partial u}{\partial \xi} + \frac{1}{2}p(\xi)\frac{\partial^2 u}{\partial \tau^2} + |u|^2 u = -\frac{i}{2}\Gamma u, \qquad (10.5.1)$$

where $p(\xi)$ is the normalized GVD profile. Similar to the case of a DDF, one can transform Eq. (10.5.1) into a perturbed NSE. However, instead of the transformations in Eq. (10.4.13), it is more appropriate to use $u' = u\exp(-\Gamma\xi/2)$ and $\xi' = \int_0^\xi p(\xi)\,d\xi$ [181]. In the transformed variables, Eq. (10.5.1) becomes

$$i\frac{\partial u'}{\partial \xi'} + \frac{1}{2}\frac{\partial^2 u'}{\partial \tau^2} + b|u'|^2 u' = 0, \qquad (10.5.2)$$

where $b(\xi) = \exp(-\Gamma\xi)/p(\xi)$. In what follows, primes on u' and ξ' are dropped for notational convenience.

The effects of *interchannel collisions* on the performance of WDM systems can be extracted by considering the simplest case of two WDM channels. Consider a collision between two solitons separated in frequency by $\Omega_{\rm ch}$, related to the channel spacing $f_{\rm ch}$ as $\Omega_{\rm ch} = 2\pi T_0 f_{\rm ch}$. The actual carrier frequencies of the two channels are $\omega_0 \pm \Omega_{\rm ch}/2T_0$. By replacing u by $u_1 + u_2$, noting that the carrier frequencies for u_1 and u_2 are different, and neglecting the FWM terms, we obtain the following two coupled equations:

$$i\frac{\partial u_1}{\partial \xi} + \frac{1}{2}\frac{\partial^2 u_1}{\partial \tau^2} + b(|u_1|^2 + 2|u_2|^2)u_1 = 0, \tag{10.5.3}$$

$$i\frac{\partial u_2}{\partial \xi} + \frac{1}{2}\frac{\partial^2 u_2}{\partial \tau^2} + b(|u_2|^2 + 2|u_1|^2)u_2 = 0. \tag{10.5.4}$$

These equations are identical with the coupled NSEs [6] obtained for two co-propagating pulses interacting through cross-phase modulation (XPM). The last nonlinear term is responsible for the XPM and is the origin of perturbations of solitons in a WDM system. This term is important only when two solitons overlap temporally during a collision.

It is useful to define a *collision length* L_{coll} as the distance over which the two solitons overlap before the faster-moving soliton overtakes the slower one. It is difficult to be precise of the instant at which a collision begins or ends. A common convention uses $2T_s$ for the duration of the collision by assuming that a collision begins and ends when the solitons overlap at their half-power points [181]. Since the relative speed of the two solitons is $\Delta V = (|\beta_2|\Omega_{ch}/T_0)^{-1}$, the collision length is given by $L_{coll} = (\Delta V)(2T_s)$ or

$$L_{coll} = \frac{2T_sT_0}{|\beta_2|\Omega_{ch}} = \frac{0.28}{q_0|\beta_2|Bf_{ch}}, \tag{10.5.5}$$

where the relations $T_s = 1.763T_0$ and $B = (2q_0T_0)^{-1}$ were used. As an example, for $B = 10$ Gb/s, $q_0 = 5$, and $\beta_2 = -0.5$ ps^2/km, the collision length $L_{coll} \sim 100$ km for a channel spacing of 100 GHz, but reduces to below 10 km for channels spaced apart by more than 1 THz.

10.5.2   Collision-Induced Frequency Shifts

Since XPM induces a time-dependent phase shift on each soliton, it leads to a shift in the soliton frequency during a collision. The soliton perturbation theory can be used to calculate the frequency shift by using the XPM term as a perturbation. The results show that the carrier frequencies of the two solitons change during collision by the same amount but in opposite directions. The *collision-induced frequency shift* Ω_c for the slow-moving soliton is obtained from [181]

$$\frac{d\Omega_c}{d\xi} = \frac{b(\xi)}{\Omega_{ch}}\frac{d}{d\xi}\left(\int_{-\infty}^{\infty} \operatorname{sech}^2(\tau - \Omega_{ch}\xi/2)\operatorname{sech}^2(\tau + \Omega_{ch}\xi/2)\,d\tau\right). \tag{10.5.6}$$

Consider first the case of a constant-dispersion fiber with negligible losses so that $b = 1$ in Eq. (10.5.6). The integration can then be performed analytically, and the frequency shift becomes

$$\Omega_c(\xi) = \frac{4[\Omega_{ch}\xi\cosh(\Omega_{ch}\xi) - \sinh(\Omega_{ch}\xi)]}{\Omega_{ch}[\sinh(\Omega_{ch}\xi)]^3}. \tag{10.5.7}$$

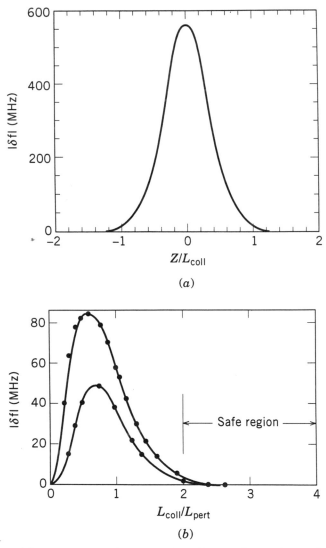

Figure 10.20 (a) Frequency shift during the collision of two 50-ps solitons with 75-GHz channel spacing in a lossless fiber; (b) residual frequency shift after a collision because of periodic perturbation of solitons; $L_{\text{pert}} = 20$ and 40 km for lower and upper curves, respectively. (After Ref. [181]. ©1991 IEEE. Reprinted with permission.)

What is the impact of frequency shifts occurring during a soliton collision? Since the speed of a soliton changes with its frequency, collisions speed up or slow down a soliton, depending on whether the soliton frequency increases or decreases. Figure 10.20(a) shows how soliton frequency changes for the slower-moving soliton during a collision of two 50-ps solitons with a channel spacing of 75 GHz. The frequency shift δf builds up over one collision length as the two solitons approach each other, reaches a peak value of about 0.6 GHz at the point of maximum overlap, and then decreases back to zero as the two solitons separate. The maximum frequency shift can be obtained by setting $\xi = 0$ in Eq. (10.5.7)] and is given by $\Omega_c^{max} = 4/(3\Omega_{ch})$. In physical units, it becomes

$$\delta f_{max} = (3\pi^2 T_0^2 f_{ch})^{-1}. \tag{10.5.8}$$

The main consequence of the frequency shift and the associated changes in the soliton speed is that the soliton position shifts from its location at the bit center after the collision is over. At the end of the collision, each soliton recovers the frequency and velocity it had before the collision, but its position and phase are altered [12]. The temporal shift $\Delta\tau_c$ is calculated by integrating Eq. (10.5.7) over ξ. The final shift at the end of the collision is found to be

$$\Delta\tau_c = 4/\Omega_{ch}^2. \tag{10.5.9}$$

If all bit slots were occupied, *collision-induced temporal shift* will be of no consequence since all solitons of a channel will be shifted by the same amount. However, the 1 and 0 bits occur randomly in a real bit stream. Since different solitons of a channel shift by different amounts, soliton collisions induce a timing jitter even in lossless fibers.

10.5.3   Collision-Induced Limitations

The situation is worse in practical soliton systems operating in the average-soliton regime, in which fiber loss is compensated periodically through lumped amplifiers. The reason is that soliton collisions are adversely affected by the loss-amplification cycle. Physically, large energy variations occurring over a collision length destroy the symmetric nature of the collision. Mathematically, the ξ dependence of $b(\xi)$ in Eq. (10.5.6) changes the frequency shift. As a result, solitons may not recover their original frequency and velocity after the collision is over. Equation (10.5.6) or numerical simulations can be used to calculate such *residual frequency shifts* for a given form of $b(\xi)$. Figure 10.20(b) shows the residual frequency shift after a complete collision of initially well-separated solitons as a function of the ratio L_{coll}/L_{pert}, where L_{pert} is either the amplifier spacing L_A or the period of GVD perturbations if that period is longer than L_A [181]. The simulations assume 50-ps solitons, an average dispersion of 1 ps/(km-nm), $L_A = 20$ km, and a channel spacing of 75 GHz. Consider the case when $L_{pert} = L_A$. The residual frequency shift increases rapidly as L_{coll} approaches L_A and can become ~ 0.1 GHz. Such shifts are not acceptable since they accumulate over multiple solitons and produce velocity changes large enough to move the soliton out of the bit slot.

As seen in Fig. 10.20(b), when L_{coll} is large enough that a collision lasts over several amplifier spacings, the effects of loss-amplification cycles begin to average out, and the residual frequency shift decreases. It virtually vanishes for $L_{coll} > 3L_A$. Since the collision length L_{coll} is inversely related to the channel spacing Ω_{ch}, the condition $L_{coll} > 3L_A$ sets a limit on the maximum separation between the two outermost channels of a WDM system. For a WDM system with equal channel spacing, the shortest collision length is obtained by replacing Ω_{ch} in Eq. (10.5.5) with $N_{ch}\Omega_{ch}$, and the number of WDM channels is limited to

$$N_{ch} < \frac{1.763}{\Omega_{ch}}\left(\frac{2L_D}{3L_A}\right). \tag{10.5.10}$$

One may think that the number of channels can be increased by reducing the channel spacing Ω_{ch}. However, as the channels become closely spaced, the overlap of soliton spectra results in interchannel crosstalk for $\Omega_{ch} < 4\Delta\omega_s$, where $\Delta\omega_s$ is the spectral width (FWHM) of solitons [182], [183]. Another constraint on channel spacing is imposed by optical filters, which typically require that $\Omega_{ch} \geq 5\Delta\omega_s$ to minimize interchannel crosstalk. By using $\Omega_{ch} = 5\Delta\omega_s$, Eq. (10.5.10) can be written as the following criterion:

$$N_{ch} < L_D/5L_A. \tag{10.5.11}$$

By using $L_D = T_0^2/|\beta_2|$ and $B = (2q_0T_0)^{-1}$, this condition can also be written as a simple design rule:

$$N_{ch}B^2L_A < (20q_0^2|\beta_2|)^{-1}. \tag{10.5.12}$$

For typical values, $q_0 = 5$, $|\beta_2| = -0.5$ ps^2/km, and $L_A = 40$ km, the condition becomes $B\sqrt{N_{ch}} < 10$ Gb/s. The number of channels can be as large as 16 at a relatively low bit rate of 2.5 Gb/s but reduces to four at $B = 5$ Gb/s, and only a single channel is allowed at 10 Gb/s. Clearly, soliton collisions limit the usefulness of WDM severely.

10.5.4 Timing Jitter

In addition to the sources of timing jitter discussed in Section 10.3.3 for a single isolated channel, several other sources of jitter become important for WDM systems. First, each collision of solitons generates a temporal shift [see Eq. (10.5.9)] of the same magnitude for both solitons but in opposite directions. Even though the temporal shift $\Delta\tau_c$ scales as Ω_{ch}^{-2} and decreases rapidly with increasing Ω_{ch}, the number of collisions increases linearly with Ω_{ch}. As a result, the total time shift after transmission scales as Ω_{ch}^{-1}. Second, the number of collisions that two neighboring solitons in a given channel undergo is slightly different. This difference arises because adjacent solitons in a given channel interact with two different soliton groups shifted by one bit period. Thus, a relative time shift appears among solitons of the same channel that becomes a source of timing jitter in WDM systems because of its dependence on the bit patterns of the copropagating channels [184]. Third, collisions involving more than two solitons can occur and should be considered. In the limit of a large channel spacing (negligible overlap of soliton spectra), multisoliton interactions are well described by pairwise collisions [185]. This property of solitons allows the calculation of timing jitter by considering only pairwise interactions.

Two other mechanisms of timing jitter should be considered for realistic WDM systems operating in the average-soliton regime. As discussed earlier, energy variations due to loss-amplification cycles make collisions asymmetric when L_{coll} becomes shorter than or comparable to the amplifier spacing L_A. Asymmetric collisions leave residual frequency shifts which affect the solitons all along the fiber link because of a change in the group velocity. This mechanism can be made ineffective by ensuring that L_{coll} exceeds $3L_A$. The second mechanism produces a residual frequency shift when solitons from different channels overlap at the input of the transmission link, resulting in an incomplete collision. For instance, two solitons from different channels, injected synchronously into the fiber link, will acquire a net frequency shift of $4/(3\Omega_{ch})$ since the first half of the collision is absent. Such residual frequency shifts are generated only over the first few amplification stages, but pertain over the whole transmission length and become an important source of timing jitter [186], [187]. Their magnitude can be reduced by appropriately delaying each channel to minimize temporal overlaps at the injection point.

Similar to the case of single-channel systems, sliding-frequency optical filters can be used to reduce the timing jitter in WDM systems [188], [189]. Typically, FP filters are used since their periodic transmission windows allow a filtering of all channels simultaneously. For best operation, the mirrors reflectivities are kept low (below 25%) to reduce the finesse, resulting in a low contrast. Such low-contrast filters remove less energy from solitons but are as effective as filters with higher contrast. Their use allows the channel spacing to be as little as five times the spectral width of the soliton [190]. The physical mechanism remains the same as discussed in Section 10.3.4 for single-channel systems. More specifically, the collision-induced frequency shift is altered because the filter forces the soliton frequency toward its transmission peak. The net result is to reduce considerably the temporal shift occurring in the absence of filters [189]. Residual frequency shifts due to incomplete collisions are also damped by optical filters, reducing their effect on the timing jitter. Filtering can also relax the condition in Eq. (10.5.10) by allowing L_{coll} to approach L_A [191] and thus helps to increase the number of channels in a WDM system. As an additional benefit, optical filters also reduce the channel-to-channel variations in the average power, occurring, for example, because of a spectrally nonuniform amplifier gain.

The technique of synchronous modulation can be applied to WDM systems for controlling the timing jitter [192]. In a WDM experiment involving four channels, each operating at 10 Gb/s, transmission over transoceanic distances has been achieved by using synchronous modulators every 500 km [193]. When the synchronous modulators were inserted every 250 km, three channels, each operating at 20 Gb/s, could be transmitted over transoceanic distances [194]. To implement this scheme, demultiplexing was necessary to isolate each channel. Theses experiments also required dispersion management, as described next.

10.5.5   Dispersion Management

The discussion so far assumes that the GVD of the fiber link is constant. As discussed in Section 10.4.5, the performance of a single high-speed channel can improve significantly if the DDF or some other dispersion-management technique is used. One may ask whether WDM systems can also benefit from dispersion management, and the answer is yes! In fact, dispersion management is essential if a WDM soliton system is designed to transmit more than two or three channels.

Similar to the case of single-channel soliton systems discussed in Section 10.4.5, the simplest scheme consists of using two fiber segments with different GVDs between each pair of optical amplifiers. The optimization of such WDM systems often requires extensive numerical simulations. In one study, the amplifier spacing could be doubled with an optimized dispersion-allocation scheme [195]. In another set of numerical simulations [196], it was found that the condition $L_{coll} > 3L_A$ can be relaxed considerably. In fact, L_{coll} could approach L_A, indicating that dispersion compensation can be used to increase the number of channels in a WDM soliton system.

It is intuitively clear that DDFs with a continuously varying GVD profile should help a WDM system even more than the DCFs. We can use Eq. (10.5.2) to find the optimum GVD profile. By tailoring the fiber dispersion as $p(\xi) = \exp(-\Gamma\xi)$, the same exponential profile encountered in Section 10.4.4, the parameter $b = 1$, resulting in an unperturbed NSE. As a result, soliton collisions become symmetric despite the fiber loss and irrespective of the ratio L_{coll}/L_A. Consequently, no residual frequency shifts occur after a soliton collision in WDM systems making use of DDFs with an exponentially decreasing GVD.

As a practical alternative to DDFs, the *staircase approximation* can be used for the exponential profile, making it possible to use multiple constant-dispersion fibers [197]. The number of fiber segments used to approximate an exponential profile is not unique, and several schemes have been evaluated numerically [198]. An experiment in 1996 achieved transmission of seven 10-Gb/s channels over 9400 km using only four fiber segments in a recirculating fiber loop [199]. In a 1997 experiment, five 20-Gb/s channels were transmitted over 10,000 km by using the same four-segment approach in a 250-km recirculating fiber loop in combination with synchronous modulators [200]. The lower limit on collision length in such systems is approximatively given by $L_{coll} > L_A/3M$, where M is the number of segments.

The use of DDFs is also essential for avoiding the FWM process in soliton WDM systems. Equations (10.5.3) and (10.5.4) neglect the FWM by assuming that the FWM process is not phase-matched because the solitons typically do not propagate very close to the zero-dispersion wavelength. However, this assumption breaks down for average solitons because of the periodic nature of variations in the soliton energy resulting from the loss-amplification cycles. Such periodic variations of the peak power create a *nonlinear-index grating* that can nearly phase-match the FWM process in WDM soliton systems [174]. The

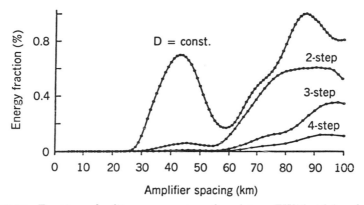

Figure 10.21 Fraction of soliton energy transferred to a FWM sideband during a single collision as a function of amplifier spacing when the exponential GVD profile between two neighboring amplifiers is approximated by a staircase with 2, 3, and 4 steps. The case of constant-dispersion fibers is also shown for comparison. The average dispersion is 0.5 ps/(km-nm) in all cases. (After Ref. [174]. ©1996 OSA. Reprinted with permission.)

phase-matching condition [6] can be written as

$$|\beta_2|(\Omega_{\rm ch}/T_0)^2 = 2\pi p/L_A, \qquad (10.5.13)$$

where p is an integer and the amplifier spacing L_A is the period of the index grating. As a result of *pseudo phase matching*, a few percent of the soliton energy can be transferred to the FWM sidebands in constant-dispersion fibers [174]. Moreover, FWM occurring during the simultaneous collision of three solitons leads to permanent frequency shifts for the slowest- and fastest-moving solitons as well as in energy exchange among all three channels [201].

The FWM can be avoided by using DDFs with an exponential GVD profile. The reason is the same as discussed earlier. The use of DDFs restores the symmetric nature of soliton collisions. As a result, the energy transferred to the FWM sidebands during the first half of a collision is returned back to the soliton during the second half of the same collision. Thus, the spectral sidebands generated through FWM do not grow with propagation of solitons. In practice, the staircase approximation is often used for the exponential profile since its use permits the use of multiple constant-dispersion fibers between two amplifiers. Figure 10.21 shows the residual energy remaining in a FWM sideband as a function of amplifier length L_A when the exponential GVD profile is approximated by using $m = 2$, 3, and 4 fiber sections chosen such that the product $D_m L_m$ is the same for all m [174]. Here D_m is the dispersion parameter of the mth section of length L_m. The case of constant-dispersion fibers is also shown for comparison. The double-peak nature of the curve in this case is due to the phase-matching condition in Eq. (10.5.13), which can be satisfied for different values of the integer p since a peak occurs whenever $L_A = 2\pi p L_D/\Omega_{\rm ch}^2$. The numerical simulations consider 20-ps solitons in two channels spaced 75 GHz apart and assume an average dispersion of 0.5 ps/(km-nm). Clearly, FWM can be nearly suppressed, with as few as three fiber sections for an amplifier spacing below 60 km. The use of a DDF, or its staircase equivalent, appears to be essential for WDM soliton systems for controlling both the timing jitter and FWM.

PROBLEMS

10.1 A soliton communication system is operating at 1.55 μm by using fibers with $D = 2$ ps/(km-nm). The effective core area of the fiber is 50 μm^2. Calculate the peak power and the pulse energy required for fundamental solitons of 30-ps width (FWHM).

10.2 What is the soliton period for the communication system of Problem 10.1?

10.3 Verify by direct substitution that the soliton solution given by Eq. (10.1.13) satisfies Eq. (10.1.7).

10.4 Write a computer program to solve the nonlinear Schrödinger equation numerically by using the split-step Fourier method (see Section 2.4 of Ref. [6]). You can use any programming language, including software packages such as Mathematica and Matlab.

10.5 Verify numerically by propagating a fundamental soliton over 10 dispersion lengths that the shape of the soliton does not change on propagation. Repeat your simulation for a Gaussian input pulse shape with the same peak power and explain your results.

10.6 A 10-Gb/s soliton lightwave system is designed with $q_0 = 5$ to ensure well-separated solitons in the RZ bit stream. Calculate the pulse width, peak power, energy of the pulse, and average power of the RZ signal assuming that $\beta_2 = -1$ ps^2/km and $\gamma = 2$ W^{-1}/km for the dispersion-shifted fiber used for soliton transmission.

10.7 A soliton communication system is designed to transmit data over 5000 km at $B = 5$ Gb/s. What should be the pulse width (FWHM) to ensure that the neighboring solitons do not interact during transmission? The dispersion parameter $D = 2$ ps/(km-nm) at the operating wavelength. Assume that soliton interaction is negligible when $B^2 L_T$ in Eq. (10.2.10) is 10% of its maximum allowed value.

10.8 Complete the derivation of Eq. (10.3.6) by performing the average over one amplifier spacing for the quantity $a^2(\xi)$, where $a(\xi)$ is the solution of Eq. (10.3.4).

10.9 A 10-Gb/s soliton communication system is designed with an amplifier spacing of 50 km. What should be the peak power of the input pulse to ensure that a fundamental soliton is maintained in an average sense in a fiber with 0.2 dB/km loss? Assume that $T_s = 20$ ps, $\beta_2 = -0.5$ ps^2/km and $\gamma = 2$ W^{-1}/km. What is the average launched power for such a system?

10.10 Calculate the maximum bit rate for a soliton lightwave system designed with $q_0 = 5$, $\beta_2 = -1$ ps^2/km, and $L_A = 50$ km. Assume that the condition (10.3.8) is satisfied when $B^2 L_A$ is at the 20% level. What is the soliton width at the maximum bit rate?

10.11 Complete the derivation of Eq. (10.3.20) for Gordon–Haus jitter using Eq. (10.3.19) as the starting point. Show all the steps clearly.

10.12 Use the soliton perturbation theory and Eq. (10.3.25) to compare the reduction in Gordon–Haus jitter for fixed- and sliding-frequency optical filters. How sensitive are your results to the sliding rate?

10.13 Explain in physical terms how sliding-frequency optical filters and synchronous modulators help to reduce the timing jitter in soliton communication systems.

10.14 Use the soliton perturbation theory to show that the soliton width increases exponentially in the quasi-adiabatic regime because of fiber loss.

10.15 Complete the derivation of Eq. (10.4.9) by integrating Eq. (10.4.7). Plot the energy ratio $E_s(z)/E_{\text{in}}$ for $L_A = 20$, 40, 60, and 80 km by taking $\alpha = 0.2$ dB/km and $\alpha_p = 0.25$ dB/km.

10.16 Use Eq. (10.4.9) to determine the pump-station spacing L_A for which the soliton energy deviates at most 20% from its input value.

10.17 Consider soliton evolution in a dispersion-decreasing fiber by using Eq. (10.4.12) without higher-order terms. Use the transformations given in Eq. (10.4.13) to prove that soliton remains unperturbed for an exponential dispersion profile $p(\xi) = \exp(-\Gamma\xi)$.

10.18 Prove by integrating Eq. (10.5.5) that collision-induced frequency shift is given by Eq. (10.5.6) when $b = 1$.

10.19 Explain how soliton collisions limit the number of channels in a WDM soliton system. Define the collision length and use this definition to derive an expression for the maximum number of channels by using the condition $L_{\text{coll}} > L_A$, where L_A is the amplifier spacing.

10.20 Explain how the periodic amplification of solitons helps to satisfy the phase-matching condition for four-wave mixing in a WDM soliton system. Derive an expression for the amplifier spacing in terms of the channel spacing for which the phase-matching condition is satisfied.

REFERENCES

[1] N. Zabusky and M. D. Kruskal, *Phys. Rev. Lett.* **15**, 240 (1965).

[2] A. Hasegawa and F. Tappert, *Appl. Phys. Lett.* **23**, 142 (1973).

[3] L. F. Mollenauer, R. H. Stolen, and J. P. Gordon, *Phys. Rev. Lett.* **45**, 1095 (1980).

[4] L. F. Mollenauer and K. Smith, *Opt. Lett.* **13**, 675 (1988).

[5] Y. Kodama and A. Hasegawa, *Progress in Optics*, Vol. 30, E. Wolf, Ed., North-Holland, Amsterdam, 1992, Chap. 4.

[6] G. P. Agrawal, *Nonlinear Fiber Optics*, 2nd ed., Academic Press, San Diego, CA, 1995, Chap. 5.

[7] A. Hasegawa and Y. Kodama, *Solitons in Optical Communications*, Clarendon Press, Oxford, 1995.

[8] R. K. Bullough and P. J. Caudrey, Eds., *Solitons*, Springer-Verlag, Berlin, 1980.

[9] M. J. Ablowitz and H. Segur, *Solitons and the Inverse Scattering Transform*, Society for Industrial and Applied Mathematics, Philadelphia, 1981.

[10] R. K. Dodd, J. C. Eilbeck, J. D. Gibbon, and H. C. Morris, *Solitons and Nonlinear Wave Equations*, Academic Press, San Diego, CA, 1984.

[11] G. P. Agrawal, in *Contemporary Nonlinear Optics*, G. P. Agrawal and R. W. Boyd, Eds., Academic Press, San Diego, CA, 1992, Chap. 2.

[12] V. E. Zakharov and A. B. Shabat, *Sov. Phys. JETP* **34**, 62 (1972).

[13] J. Satsuma and N. Yajima, *Prog. Theor. Phys.* **55**, 284 (1974).

[14] V. E. Zakharov and A. B. Shabat, *Sov. Phys. JETP* **37**, 823 (1973).

[15] A. Hasegawa and F. Tappert, *Appl. Phys. Lett.* **23**, 171 (1973).

[16] P. Emplit, J. P. Hamaide, F. Reynaud, C. Froehley, and A. Barthelemy, *Opt. Commun.* **62**, 374 (1987).

[17] D. Krökel, N. J. Halas, G. Giuliani, and D. Grischowsky, *Phys. Rev. Lett.* **60**, 29 (1988).

[18] A. M. Weiner, J. P. Heritage, R. J. Hawkins, R. N. Thurston, E. M. Kirschner, D. E. Leaird, and W. J. Tomlinson, *Phys. Rev. Lett.* **61**, 2445 (1988).

[19] W. J. Tomlinson, R. J. Hawkins, A. M. Weiner, J. P. Heritage, and R. N. Thurston, *J. Opt. Soc. Am. B* **6**, 329 (1989).

[20] W. Zhao and E. Bourkoff, *Opt. Lett.* **14**, 703 (1989); *Opt. Lett.* **14**, 808 (1989); *Opt. Lett.* **15**, 405 (1990).

[21] Y. S. Kivshar, *Phys. Rev. A* **42**, 1757 (1990); *IEEE J. Quantum Electron.* **29**, 250 (1993).

[22] W. Zhao and E. Bourkoff, *J. Opt. Soc. Am. B* **9**, 1134 (1992).

[23] Y. S. Kivshar and X. Yang, *Phys. Rev. E* **49**, 1657 (1994).

[24] D. J. Richardson, R. P. Chamberlin, L. Dong, and D. N. Payne, *Electron. Lett.* **30**, 1326 (1994).

[25] M. Nakazawa and K. Suzuki, *Electron. Lett.* **31**, 1084 (1995).

[26] M. Nakazawa and K. Suzuki, *Electron. Lett.* **31**, 1076 (1995).

[27] V. I. Karpman and V. V. Solovev, *Physica* **3D**, 487 (1981).

[28] J. P. Gordon, *Opt. Lett.* **8**, 596 (1983).

[29] F. M. Mitschke and L. F. Mollenauer, *Opt. Lett.* **12**, 355 (1987).

[30] C. Desem and P. L. Chu, *IEE Proc.* **134**, 145 (1987).

[31] Y. Kodama and K. Nozaki, *Opt. Lett.* **12**, 1038 (1987).

[32] C. Desem and P. L. Chu, *Opt. Lett.* **11**, 248 (1986).

[33] K. J. Blow and D. Wood, *Opt. Commun.* **58**, 349 (1986).

[34] A. I. Maimistov and Y. M. Sklyarov, *Sov. J. Quantum Electron.* **17**, 500 (1987).

[35] A. S. Gouveia-Neto, A. S. L. Gomes, and J. R. Taylor, *Opt. Commun.* **64**, 383 (1987).

[36] K. Iwatsuki, A. Takada, and M. Saruwatari, *Electron. Lett.* **24**, 1572 (1988).

[37] K. Iwatsuki, A. Takada, S. Nishi, and M. Saruwatari, *Electron. Lett.* **25**, 1003 (1989).

[38] M. Nakazawa, K. Suzuki, and Y. Kimura, *Opt. Lett.* **15**, 588 (1990).

[39] P. A. Morton, V. Mizrahi, G. T. Harvey, L. F. Mollenauer, T. Tanbun-Ek, R. A. Logan, H. M. Presby, T. Erdogan, A. M. Sergent, and K. W. Wecht, *IEEE Photon. Technol. Lett.* **7**, 111 (1995).

[40] S. Arahira, Y. Matsui, T. Kunii, S. Oshiba, and Y. Ogawa, *IEEE Photon. Technol. Lett.* **5**, 1362 (1993).

[41] M. Suzuki, H. Tanaka, N. Edagawa, Y. Matsushima, and H. Wakabayashi, *Fiber Integ. Opt.* **12**, 358 (1993).

[42] K. Sato, I. Kotaka, Y. Kondo, and M. Yamamoto, *Appl. Phys. Lett.* **69**, 2626 (1996).

[43] K. Wakita and I. Kotaka, *Microwave Opt. Tech. Lett.* **7**, 120 (1995).

[44] G. T. Harvey and L. F. Mollenauer, *Opt. Lett.* **18**, 107 (1993).

[45] D. M. Patrick, *Electron. Lett.* **30**, 43 (1994).

[46] I. N. Duling, Ed., *Compact Sources of Ultrashort Pulses*, Cambridge University Press, New York, 1995.

[47] S. V. Chernikov, J. R. Taylor, and R. Kashyap, *Electron. Lett.* **29**, 1788 (1993); *Opt. Lett.* **19**, 539 (1994).

[48] E. A. Swanson and S. R. Chinn, *IEEE Photon. Technol. Lett.* **7**, 114 (1995).

[49] P. V. Mamyshev, *Opt. Lett.* **19**, 2074 (1994).

[50] J. J. Veselka, S. K. Korotky, P. V. Mamyshev, A. H. Gnauck, G. Raybon, and N. M. Froberg, *IEEE Photon. Technol. Lett.* **8**, 950 (1996).

[51] A. Hasegawa and Y. Kodama, *Proc. IEEE* **69**, 1145 (1981); *Opt. Lett.* **7**, 285 (1982).

[52] K. J. Blow and N. J . Doran, *Opt. Commun.* **52**, 367 (1985).

[53] M. J. Potasek and G. P. Agrawal, *Electron. Lett.* **22**, 759 (1986).

[54] Y. Kodama and A. Hasegawa, *Opt. Lett.* **7**, 339 (1982); **8**, 342 (1983).

[55] A. Hasegawa, *Opt. Lett.* **8**, 650 (1983); *Appl. Opt.* **23**, 3302 (1984).

[56] L. F. Mollenauer, R. H. Stolen, and M. N. Islam, *Opt. Lett.* **10**, 229 (1985).

[57] L. F. Mollenauer, J. P. Gordon, and M. N. Islam, *IEEE J. Quantum Electron.* **22**, 157 (1986).

[58] A. Hasegawa and Y. Kodama, *Phys. Rev. Lett.* **66**, 161 (1991).

[59] V. I. Karpman and E. M. Maslov, *Sov. Phys. JETP* **46**, 281 (1977).

[60] T. Georges, *Opt. Fiber Technol.* **1**, 97 (1995).

[61] T. Georges and F. Favre, *J. Opt. Soc. Am. B* **10**, 1880 (1993).

[62] H. A. Haus and Y. Lai, *J. Opt. Soc. Am. B* **7**, 386 (1990).

[63] J. P. Gordon and H. A. Haus, *Opt. Lett.* **11**, 665 (1986).

[64] D. Marcuse, *J. Lightwave Technol.* **10**, 273 (1992).

[65] E. M. Dianov, A. V. Luchnikov, A. N. Pilipetskii, and A. M. Prokhorov, *Sov. Lightwave Commun.* **1**, 235 (1991); *Appl. Phys.* **54**, 175 (1992).

[66] K. Smith and L. F. Mollenauer, *Opt. Lett.* **14**, 1284 (1989).

[67] L. F. Mollenauer, *Opt. Lett.* **21**, 384 (1996).

[68] A. N. Pilipetskii and C. R. Menyuk, *Opt. Lett.* **22**, 28 (1997).

[69] L. F. Mollenauer and J. P. Gordon, *Opt. Lett.* **19**, 375 (1994).

[70] T. Georges and F. Fabre, *Opt. Lett.* **16**, 1656 (1991).

[71] C. R. Menyuk, *Opt. Lett.* **20**, 285 (1995).

[72] A. N. Pinto, G. P. Agrawal, and J. F. da Rocha, Paper WA3, *Proc. OSA Annual Meeting*, Optical Society of America, Washington, DC, 1996.

[73] S. Wabnitz, Y. Kodama, and A. B. Aceves, *Opt. Fiber Technol.* **1**, 187 (1995).

[74] N. J. Smith and N. J. Doran, *Opt. Fiber Technol.* **1**, 218 (1995).

[75] L. F. Mollenauer, M. J. Neubelt, M. Haner, E. Lichtman, S. G. Evangelides, and B. M. Nyman, *Electron. Lett.* **27**, 2055 (1991).

[76] A. Mecozzi, J. D. Moores, H. A. Haus, and Y. Lai, *Opt. Lett.* **16**, 1841 (1991).

[77] Y. Kodama and A. Hasegawa, *Opt. Lett.* **17**, 31 (1992).

[78] L. F. Mollenauer, J. P. Gordon, and S. G. Evangelides, *Opt. Lett.* **17**, 1575 (1992).

[79] L. F. Mollenauer, E. Lichtman, M. J. Neubelt, and G. T. Harvey, *Electron. Lett.* **29**, 910 (1993).

[80] Y. Kodama and S. Wabnitz, *Opt. Lett.* **19**, 162 (1994).

[81] M. Romagnoli, S. Wabnitz, and M. Midrio, *Opt. Commun.* **104**, 293 (1994).

[82] E. A. Golovchenko, A. N. Pilipetskii, C. R. Menyuk, J. P. Gordon, and L. F. Mollenauer, *Opt. Lett.* **20**, 539 (1995).

[83] H. Kim, J. H. Jang, and Y. C. Chung, *IEEE Photon. Technol. Lett.* **8**, 1193 (1996).

[84] G. Aubin, T. Montalant, J. Moulu, B. Nortier, F. Pirio, and J.-B. Thomine, *Electron. Lett.* **31**, 52 (1995).

[85] H. Toda, H. Yamagishi, and A. Hasegawa, *Opt. Lett.* **20**, 1002 (1995).

[86] Y. Kodama and S. Wabnitz, *Electron. Lett.* **27**, 1931 (1991).

[87] V. V. Afanasjev, *Opt. Lett.* **18**, 790 (1993).

[88] J.-C. Dung, S. Chi, and S. Wen, *Opt. Lett.* **20**, 1862 (1995).

[89] M. Suzuki, N. Edagawa, H. Taga, H. Tanaka, S. Yamamoto, and S. Akiba, *Electron. Lett.* **30**, 1083 (1994).

[90] A. L. J. Teixeira, G. P. Agrawal, and J. R. F. da Rocha, *Electron. Lett.* **32**, 1995 (1996).

[91] A. Mecozzi, *Opt. Lett.* **20**, 1859 (1995).

[92] S. Wabnitz and E. Westin, *Opt. Lett.* **21**, 1235 (1996).

[93] M. Nakazawa, E. Yamada, H. Kubota, and K. Suzuki, *Electron. Lett.* **27**, 1270 (1991).

[94] M. Nakazawa, H. Kubota, E. Yamada, and K. Suzuki, *Electron. Lett.* **28**, 1099 (1992).

[95] M. Nakazawa, K. Suzuki, E. Yamada, H. Kubota, Y. Kimura, and M. Takaya, *Electron. Lett.* **29**, 729 (1993).

[96] H. Kubota and M. Nakazawa, *Electron. Lett.* **29**, 1780 (1993).

[97] G. Aubin, E. Jeanny, T. Montalant, J. Moulu, F. Pirio, J.-B. Thomine, and F. Devaux, *Electron. Lett.* **31**, 1079 (1995).

[98] M. Nakazawa, K. Suzuki, H. Kubota, E. Yamada, and Y. Kimura, *Electron. Lett.* **30**, 1331 (1994).

[99] S. Wabnitz, *Electron. Lett.* **29**, 1711 (1993).

[100] N. J. Smith, W. J. Firth, K. J. Blow, and K. Smith, *Opt. Lett.* **19**, 16 (1994).

[101] S. Bigo, O. Audouin, and E. Desurvire, *Electron. Lett.* **31**, 2191 (1995).

[102] N. J. Smith, K. J. Blow, W. J. Firth, and K. Smith, *Opt. Commun.* **102**, 324 (1993).

[103] M. Suzuki, N. Edagawa, H. Taga, H. Tanaka, S. Yamamoto, and S. Akiba, *Electron. Lett.* **30**, 1083 (1994).

[104] M. Matsumoto, H. Ikeda, and A. Hasegawa, *Opt. Lett.* **19**, 183 (1994).

[105] E. Yamada and M. Nakazawa, *IEEE J. Quantum Electron.* **30**, 1842 (1994).

[106] A. Takada and W. Imajuku, *Electron. Lett.* **32**, 677 (1996).

[107] T. Widdowson, D. J. Malyon, A. D. Ellis, K. Smith, and K. J. Blow, *Electron. Lett.* **30**, 990 (1994).

[108] S. Kumar and A. Hasegawa, *Opt. Lett.* **20**, 1856 (1995).

[109] V. S. Grigoryan, A. Hasegawa, and A. Maruta, *Opt. Lett.* **20**, 857 (1995).

[110] M. Nakazawa, K. Suzuki, and Y. Kimura, *IEEE Photon. Technol. Lett.* **2**, 216 (1990).

[111] N. A. Olsson, P. A. Andrekson, P. C. Becker, J. R. Simpson, T. Tanbun-Ek, R. A. Logan, H. Presby, and K. Wecht, *IEEE Photon. Technol. Lett.* **2**, 358 (1990).

[112] K. Iwatsuki, S. Nishi, and K. Nakagawa, *IEEE Photon. Technol. Lett.* **2**, 355 (1990).

[113] E. Yamada, K. Suzuki, and M. Nakazawa, *Electron. Lett.* **27**, 1289 (1991).

[114] L. F. Mollenauer, B. M. Nyman, M. J. Neubelt, G. Raybon, and S. G. Evangelides, *Electron. Lett.* **27**, 178 (1991).

[115] L. F. Mollenauer, E. Lichtman, G. T. Harvey, M. J. Neubelt, and B. M. Nyman, *Electron. Lett.* **28**, 792 (1992).

[116] L. F. Mollenauer, P. V. Mamyshev, and M. J. Neubelt, *Opt. Lett.* **19**, 704 (1994).

[117] G. P. Agrawal, in *Semiconductor Lasers: Past, Present, and Future*, G. P. Agrawal, Ed., AIP Press, Woodbury, NY, 1995, Chap. 8.

[118] G. P. Agrawal and N. A. Olsson, *IEEE J. Quantum Electron.* **25**, 2297 (1989).

[119] A. Mecozzi, *Opt. Lett.* **20**, 1616 (1995).

[120] S. Wabnitz, *Opt. Lett.* **20**, 1979 (1995).

[121] P. A. Andrekson, N. A. Olsson, M. Haner, J. R. Simpson, T. Tanbun-Ek, R. A. Logan, D. Coblentz, H. M. Presby, and K. W. Wecht, *IEEE Photon. Technol. Lett.* **4**, 76 (1992).

[122] K. Iwatsuki, K. Suzuki, S. Nishi, and M. Saruwatari, *IEEE Photon. Technol. Lett.* **5**, 245 (1993).

[123] M. Nakazawa, K. Suzuki, E. Yoshida, E. Yamada, T. Kitoh, and M. Kawachi, *Electron. Lett.* **31**, 565 (1995).

[124] J. P. Gordon, *Opt. Lett.* **11**, 662 (1986).

[125] F. M. Mitschke and L. F. Mollenauer, *Opt. Lett.* **11**, 659 (1986).

[126] D. Atkinson, W. H. Loh, V. V. Afanasjev, A. B. Grudinin, A. J. Seeds, and D. N. Payne, *Opt. Lett.* **19**, 1514 (1994).

[127] R. Vallée and R.-J. Essiambre, *Opt. Lett.* **19**, 2095 (1994).

[128] M. Matsumoto, H. Ikeda, T. Uda, and A. Hasegawa, *J. Lightwave Technol.* **13**, 658 (1995).

[129] N. J. Smith and N. J. Doran, *J. Opt. Soc. Am. B* **12**, 1117 (1995).

[130] F. M. Knox, P. Harper, P. N. Kean, N. J. Doran, and I. Bennion, *Electron. Lett.* **31**, 1467 (1995).

[131] D. Wood, *J. Lightwave Technol.* **8**, 1097 (1990).

[132] B. A. Malomed, *J. Opt. Soc. Am. B* **11**, 1261 (1994).

[133] R.-J. Essiambre and G. P. Agrawal, *J. Opt. Soc. Am. B* **12**, 2420 (1995).

[134] M. Nakazawa, E. Yoshida, E. Yamada, K. Suzuki, T. Kitoh, and M. Kawachi, *Electron. Lett.* **30**, 1777 (1994).

[135] R.-J. Essiambre and G. P. Agrawal, *Electron. Lett.* **31**, 1461 (1995).

[136] S. G. Evangelides, L. F. Mollenauer, J. P. Gordon, and N. S. Bergano, *J. Lightwave Technol.* **10**, 28 (1992).

[137] S. Wabnitz, *Opt. Lett.* **20**, 261 (1995).

[138] C. De Angelis, S. Wabnitz, and M. Haelterman, *Electron. Lett.* **29**, 1568 (1993).

[139] L. F. Mollenauer, K. Smith, J. P. Gordon, and C. R. Menyuk, *Opt. Lett.* **14**, 1219 (1989).

[140] T. Widdowson, A. Lord, and D. J. Malyon, *Electron. Lett.* **30**, 879 (1994).

[141] M. Midrio, P. Franco, M. Crivellari, M. Romagnoli, and F. Matera, *Electron. Lett.* **31**, 1473 (1995).

[142] D. M. Spirit, I. W. Marshall, P. D. Constantine, D. L. Williams, S. T. Davey, and B. J. Ainslie, *Electron. Lett.* **27**, 222 (1991).

[143] K. Rottwitt, J. H. Povlsen, S. Gundersen, and A. Bjarklev, *Opt. Lett.* **18**, 867 (1993).

[144] C. Lester, K. Bertilsson, K. Rottwitt, P. A. Andrekson, M. A. Newhouse, and A. J. Antos, *Electron. Lett.* **31**, 219 (1995).

[145] M. Nakazawa, H. Kubota, K. Kurakawa, and E. Yamada, *J. Opt. Soc. Am. B* **8**, 1811 (1991).

[146] K. Kurokawa and M. Nakazawa, *IEEE J. Quantum Electron.* **28**, 1922 (1992).

[147] K. Tajima, *Opt. Lett.* **12**, 54 (1987).

[148] V. A. Bogatyrjov, M. M. Bubnov, E. M. Dianov, and A. A. Sysoliatin, *Pure Appl. Opt.* **4**, 345 (1995).

[149] D. J. Richardson, R. P. Chamberlin, L. Dong, and D. N. Payne, *Electron. Lett.* **31**, 1681 (1995).

[150] Q. Ren and H. Hsu, *IEEE J. Quantum Electron.* **24**, 2059 (1988).

[151] S. Chi and M.-C. Lin, *Electron. Lett.* **27**, 237 (1991).

[152] A. J. Stentz, R. Boyd, and A. F. Evans, *Opt. Lett.* **20**, 1770 (1995).

[153] D. J. Richardson, L. Dong, R. P. Chamberlin, A. D. Ellis, T. Widdowson, and W. A. Pender, *Electron. Lett.* **32**, 373 (1996).

[154] R.-J. Essiambre and G. P. Agrawal, *Opt. Commun.* **131**, 274 (1996).

[155] R.-J. Essiambre and G. P. Agrawal, *J. Opt. Soc. Am. B* **14**, 314 (1997).

[156] R.-J. Essiambre and G. P. Agrawal, *J. Opt. Soc. Am. B* **14**, 323 (1997).

[157] A. D. Ellis, T. Widdowson, and X. Shan, *Electron. Lett.* **32**, 381 (1996).

[158] H. Kubota and M. Nakazawa, *Opt. Commun.* **87**, 15 (1992).

[159] M. Nakazawa and H. Kubota, *Electron. Lett.* **31**, 216 (1995).

[160] M. Suzuki, I. Morita, N. Edagawa, S. Yamamoto, H. Taga, and S. Akiba, *Electron. Lett.* **31**, 2027 (1995).

[161] S. Wabnitz, I. Uzunov, and F. Lederer, *IEEE Photon. Technol. Lett.* **8**, 1091 (1996).

[162] A. Naka, T. Matsuda, and S. Saito, *Electron. Lett.* **32**, 1694 (1996).

[163] I. R. Gabitov and S. K. Turitsyn, *Opt. Lett.* **21**, 327 (1996).

[164] M. Nakazawa, H. Kubota, A. Sahara, and K. Tamura, *IEEE Photon. Technol. Lett.* **8**, 1088 (1996).

[165] F. M. Knox, W. Forysiak, and N. J. Doran, *J. Lightwave Technol.* **13**, 1955 (1995).

[166] R. Ohhira, A. Hasegawa, and Y. Kodama, *Opt. Lett.* **20**, 701 (1995).

[167] M. Nakazawa, Y. Kimura, K. Suzuki, H. Kubota, T. Komukai, E. Yamada, T. Sugawa, E. Yoshida, T. Yamamoto, T. Imai, A. Sahara, H. Nakazawa, O. Yamauchi, and M. Umezawa, *Electron. Lett.* **31**, 992 (1995).

[168] I. Morita, M. Suzuki, N. Edagawa, S. Yamamoto, H. Taga, and S. Akiba, *IEEE Photon. Technol. Lett.* **8**, 1573 (1996).

[169] N. J. Smith, W. Forysiak, and N. J. Doran, *Electron. Lett.* **32**, 2085 (1996).

[170] A. B. Grudinin and I. A. Goncharenko, *Electron. Lett.* **32**, 1602 (1996).

[171] G. P. Agrawal and M. J. Potasek, *Phys. Rev. A* **33**, 1765 (1986).

[172] M. Nakazawa, H, Kubota, and K. Tamura, *IEEE Photon. Technol. Lett.* **8**, 452 (1996).

[173] W. Forysiak, F. M. Knox, and N. J. Doran, *Opt. Lett.* **19**, 174 (1994).

[174] P. V. Mamyshev and L. F. Mollenauer, *Opt. Lett.* **21**, 396 (1996).

[175] T. Georges and B. Charbonnier, *Opt. Lett.* **21**, 1232 (1996).

[176] T. Georges and B. Charbonnier, *IEEE Photon. Technol. Lett.* **9**, 127 (1997).

[177] I. R. Gabitov, E. G. Shapiro, and S. K. Turitsyn, *Opt. Commun.* **134**, 317 (1997); *Phys. Rev. E* **55**, 3624 (1997).

[178] J. M. Jacob, E. A. Golovchenko, A. N. Pilipetskii, G. M. Carter, and C. R. Menyuk, *IEEE Photon. Technol. Lett.* **9**, 130 (1997).

[179] E. A. Golovchenko, J. M. Jacob, A. N. Pilipetskii, C. R. Menyuk, and G. M. Carter, *Opt. Lett.* **22**, 289 (1997).

[180] S. Kumar and A. Hasegawa, *Opt. Lett.* **22**, 372 (1997).

[181] L. F. Mollenauer, S. G. Evangelides, and J. P. Gordon, *J. Lightwave Technol.* **9**, 362 (1991).

[182] A. F. Benner, J. R. Sauer, and M. J. Ablowitz, *J. Opt. Soc. Am. B* **10**, 2331 (1993).

[183] P. K. A. Wai, C. R. Menyuk, and B. Raghavan, *J. Lightwave Technol.* **14**, 1449 (1996).

[184] R. B. Jenkins, J. R. Sauer, S. Chakravarty, and M. J. Ablowitz, *Opt. Lett.* **20**, 1964 (1995).

[185] S. Chakravarty, M. J. Ablowitz, J. R. Sauer, and R. B. Jenkins, *Opt. Lett.* **20**, 136 (1995).

[186] Y. Kodama and A. Hasegawa, *Opt. Lett.* **16**, 208 (1991).

[187] T. Aakjer, J. H. Povlsen, and K. Rottwitt, *Opt. Lett.* **18**, 1908 (1993).

[188] L. F. Mollenauer, E. Lichtman, G. T. Harvey, M. J. Neubelt, and B. M. Nyman, *Electron. Lett.* **28**, 792 (1992).

[189] A. Mecozzi and H. A. Haus, *Opt. Lett.* **17**, 988 (1992).

[190] E. A. Golovchenko, A. N. Pilipetskii, and C. R. Menyuk, *Opt. Lett.* **21**, 195 (1996).

[191] M. Midrio, P. Franco, F. Matera, M. Romagnoli, and M. Settembre, *Opt. Commun.* **112**, 283 (1994).

[192] E. Desurvire, O. Leclerc, and O. Audouin, *Opt. Lett.* **21**, 1026 (1996).

[193] M. Nakazawa, K. Suzuki, H, Kubota, Y. Kimura, E. Yamada, K. Tamura, T. Komukai, and T. Imai, *Electron. Lett.* **32**, 828 (1996).

[194] M. Nakazawa, K. Suzuki, H. Kubota, and E. Yamada, *Electron. Lett.* **32**, 1686 (1996).

[195] E. Kolltveit, J.-P. Hamaide, and O. Audouin, *Electron. Lett.* **32**, 1858 (1996).

[196] S. Wabnitz, *Opt. Lett.* **21**, 638 (1996).

[197] A. Hasegawa and S. Kumar, *Opt. Lett.* **21**, 39 (1996).

[198] S. Cardinal, E. Desurvire, J.-P. Hamaide, and O. Audouin, *Electron. Lett.* **33**, 77 (1997).

[199] L. F. Mollenauer, P. V. Mamyshev, and M. J. Neubelt, *Electron. Lett.* **32**, 471 (1996).

[200] M. Nakazawa, K. Suzuki, H. Kubota, A. Sahara, and E. Yamada, Paper PD21, *Proc. Optical Fiber Commun. Conf.*, Optical Society of America, Washington, DC, 1997.

[201] S. G. Evangelides and J. P. Gordon, *J. Lightwave Technol.* **14**, 1639 (1996).

Appendix A

International System of Units

The international system of units (known as the SI, short for *Système International*) is used in this book. The following table lists the main physical quantities together with their units, symbols, and dimensions.

Quantity	Unit	Symbol	Dimensions
length	meter	m	—
mass	kilogram	kg	—
time	second	s	—
current	ampere	A	—
temperature	kelvin	K	—
frequency	hertz	Hz	s^{-1}
force	newton	N	$(kg\text{-}m)/s^2$
energy	joule	J	N-m
power	watt	W	J/s
pressure	pascal	Pa	N/m^2
electric charge	coulomb	C	A-s
potential	volt	V	J/C
conductance	siemens	S	A/V
resistance	ohm	W	V/A
capacitance	farad	F	C/V
magnetic flux	weber	Wb	V-s
magnetic induction	tesla	T	Wb/m^2
inductance	henry	H	Wb/A

Appendix B

Decibel Units

In the design of optical communication systems the optical power can vary over several orders of magnitude as the signal travels from the transmitter to the receiver. Such large variations are handled most conveniently using decibel units, abbreviated dB, commonly used by engineers in many different fields. Any ratio R can be converted into decibels by using the general definition

$$R \text{ (in dB)} = 10 \log_{10} R. \tag{B.1}$$

The logarithmic nature of the decibel allows a large ratio to be expressed as a much smaller number. For example, 10^9 and 10^{-9} correspond to 90 dB and -90 dB, respectively. Since $R = 1$ corresponds to 0 dB, ratios smaller than 1 are negative in the decibel system. Furthermore, negative ratios cannot be written using decibel units.

The most common use of the decibel scale occurs for power ratios. For instance, the signal-to-noise ratio (SNR) of an optical or electrical signal is given by

$$\text{SNR} = 10 \log_{10}(P_S/P_N), \tag{B.2}$$

where P_S and P_N are the signal and noise powers, respectively. The fiber loss can also be expressed in decibel units by noting that the loss corresponds to a decrease in the optical power during transmission and thus can be expressed as a power ratio. For example, if a 1-mW signal reduces to 1 μW after transmission over 100 km of fiber, the 30-dB loss over the entire fiber span translates into a loss of 0.3 dB/km. The same technique can be used to define the insertion loss of any component. For instance, a 1-dB loss of a fiber connector implies that the optical power is reduced by 1 dB (by about 20%) when the signal passes through the connector. The bandwidth of an optical filter is defined at the 3-dB point, corresponding to 50% reduction in the signal power. The modulation bandwidth of LEDs in Section 3.2.3 and of semiconductor lasers in Section 3.3.7 is also defined at the 3-dB point, at which the modulated powers drops by 50%.

Since the losses of all components in a fiber-optic communication systems are expressed in dB, it is useful to express the transmitted and received powers

535

also by using a decibel scale. This is achieved by using a derived unit, denoted as dBm and defined as

$$\text{power (in dBm)} = 10 \log_{10}\left(\frac{\text{power}}{1 \text{ mW}}\right), \tag{B.3}$$

where the reference level of 1 mW is chosen simply because typical values of the transmitted power are in that range (the letter m in dBm is a reminder of the 1-mW reference level). In this decibel scale for the absolute power, 1 mW corresponds to 0 dBm, whereas powers below 1 mW are expressed as negative numbers. For example, a 10-μW power corresponds to -20 dBm. The advantage of decibel units becomes clear when the power budget, discussed in Chapter 5 is considered. Because of the logarithmic nature of the decibel scale, the power budget can be made simply by subtracting various losses from the transmitter power expressed in dBm units.

Appendix C

Acronyms

Each scientific field has its own jargon, and the field of optical communications is not an exception. Although an attempt was made to avoid extensive use of acronyms, many still appear throughout the book. Each acronym is defined the first time it appears in a chapter so that the reader does not have to search the entire text to find its meaning. As a further help, we list all acronyms here, in alphabetical order.

ac	alternating current
AM	amplitude modulation
AON	all-optical network
APD	avalanche photodiode
ASE	amplified spontaneous emission
ASK	amplitude-shift keying
ATM	asynchronous transfer mode
BER	bit-error rate
BH	buried heterostructure
BPF	bandpass filter
C^3	cleaved-coupled cavity
CATV	common-antenna (cable) television
CDM	code-division multiplexing
CNR	carrier-to-noise ratio
CPFSK	continuous-phase frequency-shift keying
CSMA	carrier-sense multiple access
CSO	composite second-order
CVD	chemical vapor deposition
CW	continuous wave
CTB	composite triple beat
DBR	distributed Bragg reflector
dc	direct current
DCF	dispersion-compensating fiber
DFB	distributed feedback

DIP	dual in-line package
DPSK	differential phase-shift keying
EDFA	erbium-doped fiber amplifier
FDDI	fiber distributed data interface
FDM	frequency-division multiplexing
FET	field-effect transistor
FM	frequency modulation
FP	Fabry–Perot
FSK	frequency-shift keying
FWHM	full width at half maximum
FWM	four-wave mixing
GVD	group-velocity dispersion
HBT	heterojunction-bipolar transistor
HDTV	high-definition television
HEMT	high-electron-mobility transistor
HFC	hybrid fiber-coaxial
HIPPI	high-performance parallel interface
IC	integrated circuit
IF	intermediate frequency
IMD	intermodulation distortion
IM/DD	intensity modulation with direct detection
IMP	intermodulation product
ISDN	integrated services digital network
ISI	intersymbol interference
LAN	local-area network
LED	light-emitting diode
L-I	light-current
LO	local oscillator
LPE	liquid-phase epitaxy
LPF	low-pass filter
MAN	metropolitan-area network
MBE	molecular-beam epitaxy
MCVD	modified chemical vapor deposition
MOCVD	metal-organic chemical vapor deposition
MONET	multiwavelength optical network
MPEG	motion-picture entertainment group
MPN	mode-partition noise
MQW	multiquantum well
MSK	minimum-shift keying
MSM	metal–semiconductor–metal
MSR	mode-suppression ratio
MTTF	mean time to failure
MZ	Mach–Zehnder
NA	numerical aperture
NEP	noise-equivalent power
NOLM	nonlinear optical-loop mirror

NRZ	non-return to zero
NSE	nonlinear Schrödinger equation
OC	optical carrier
OEIC	opto-electronic integrated circuit
OOK	on-off keying
OPC	optical phase conjugation
OTDM	optical time-division multiplexing
OVD	outside-vapor deposition
OXC	optical cross-connect
PCM	pulse-code modulation
PDF	probability density function
PDM	polarization-division multiplexing
PIC	photonic integrated circuit
PM	phase modulation
PMD	polarization-mode dispersion
PON	passive optical network
PSK	phase-shift keying
RIN	relative intensity noise
RMS	root-mean-square
RZ	return to zero
SAGCM	separate absorption, grading, charge, and multiplication
SAGM	separate absorption, grading, and multiplication
SAM	separate absorption and multiplication
SBS	stimulated Brillouin scattering
SCM	subcarrier multiplexing
SDH	synchronous digital hierarchy
SI	système international
SLA	semiconductor laser amplifier
SLM	single longitudinal mode
SNR	signal-to-noise ratio
SONET	synchronized optical network
SPM	self-phase modulation
SRS	stimulated Raman scattering
STM	synchronous transport module
STS	synchronous transport signal
TCP/IP	transmission control protocol/internet protocol
TDM	time-division multiplexing
TE	transverse electric
TM	transverse magnetic
TW	traveling wave
VAD	vapor-axial deposition
VCSEL	vertical-cavity surface-emitting laser
VPE	vapor-phase epitaxy
VSB	vestigial sideband
WAN	wide-area network
WDM	wavelength-division multiplexing

WDMA	wavelength-division multiple access
WGR	waveguide-grating router
XPM	cross-phase modulation
YIG	yttrium iron garnet

Index

absorption coefficient, 139, 400
absorption rate, 79
accelerated aging, 128
acoustic jitter, *see* timing jitter
acoustic phonon, 385, 390
acoustic waves, 248, 301, 385, 491
activation energy, 129
add/drop multiplexer, 306
adiabatic amplification, 485
amplification factor, 362, 370, 375, 382, 388
amplified spontaneous emission, 393, 398, 410, 488
amplifier spacing, 410, 414, 484, 487, 498, 503
amplifiers
    applications of, 366
    bandwidth of, 362
    Brillouin, 385–391
    cascaded, 410–415
    doped-fiber, *see* fiber amplifiers
    gain of, 362
    in-line, 202, 378, 410–415, 426, 488, 498
    LAN, 409
    noise in, 365
    parametric, 453, 513
    power, 366, 408
    properties of, 361–367
    Raman, 379–385
    saturation characteristics of, 364
    semiconductor laser, *see* semiconductor laser amplifiers
    tilted-stripe, 370
amplifier noise, *see* noise
amplitude-phase coupling, 114, 121
amplitude-shift keying, *see* modulation format
angled-facet structure, 370
anti-reflection coating, 302, 368

anticorrelation, 120, 216
antireflection coating, 91, 107
APD, 147–154
    physical mechanism behind, 148
    bandwidth of, 150
    design of, 149
    enhanced shot noise in, 167
    excess noise factor for, 167
    gain of, 150
    optimum gain for, 169, 174
    reach-through, 151
    responsivity of, 150
    SAGCM, 152
    SAGM, 152
    SAM, 151
    staircase, 152
    superlattice, 153
APD gain, 150
apodization technique, 442
argon-ion laser, 392
ASCII code, 7
asynchronous transfer mode, *see* ATM
ATM, 13, 292, 295
attenuation coefficient, 57
Auger coefficient, 84
Auger recombination, 83, 113
autocorrelation, 346
autocorrelation function, 119, 164, 165
avalanche breakdown, 150
avalanche photodetector, *see* APD
average-soliton regime, 486–488, 499, 500, 508, 514, 519

bandgap discontinuity, 81
bandwidth
    amplifier, 362, 363, 415
    APD, 150
    Brillouin-amplifier, 386
    fiber, 55, 211
    filter, 157, 297, 302, 405, 414